FUNDAMENTALS OF WIRELESS COMMUNICATION ENGINEERING TECHNOLOGIES

WILEY SERIES ON INFORMATION AND COMMUNICATION TECHNOLOGY

Series Editors: T. Russell Hsing and Vincent K. N. Lau

The Information and Communication Technology (ICT) book series focuses on creating useful connections between advanced communication theories, practical designs, and end-user applications in various next generation networks and broadband access systems, including fiber, cable, satellite, and wireless. The ICT book series examines the difficulties of applying various advanced communication technologies to practical systems such as WiFi, WiMax, B3G, etc., and considers how technologies are designed in conjunction with standards, theories, and applications.

The ICT book series also addresses application-oriented topics such as service management and creation and end-user devices, as well as the coupling between end devices and infrastructure.

T. Russell Hsing, PhD, is the Executive Director of Emerging Technologies and Services Research at Telcordia Technologies. He manages and leads the applied research and development of information and wireless sensor networking solutions for numerous applications and systems. Email: thsing@telcordia.com

Vincent K.N. Lau, PhD, is Associate Professor in the Department of Electrical Engineering at the Hong Kong University of Science and Technology. His current research interest is on delay-sensitive cross-layer optimization with imperfect system state information. Email: eeknlau@ee.ust.hk

Wireless Internet and Mobile Computing: Interoperability and Performance
Yu-Kwong Ricky Kwok and Vincent K. N. Lau

RF Circuit Design
Richard C. Li

Digital Signal Processing Techniques and Applications in Radar Image Processing
Bu-Chin Wang

The Fabric of Mobile Services: Software Paradigms and Business Demands
Shoshana Loeb, Benjamin Falchuk, and Euthimios Panagos

Fundamentals of Wireless Communication Engineering Technologies
K. Daniel Wong

FUNDAMENTALS OF WIRELESS COMMUNICATION ENGINEERING TECHNOLOGIES

K. Daniel Wong

WILEY

A JOHN WILEY & SONS, INC., PUBLICATION

Published by John Wiley & Sons, Inc., Hoboken, New Jersey.
Published simultaneously in Canada.

Library of Congress Cataloging-in-Publication Data:

Wong, K. Daniel.
Fundamentals of wireless communication engineering technologies / K. Daniel Wong.
p. cm. – (Information and communication technology series ; 98)
Includes bibliographical references.
ISBN 978-0-470-56544-5
1. Wireless communication systems. 2. Wireless communication systems–Examinations–Study guides. I. Title.
TK5103.2.W59 2011
384.5–dc23 2011013591

1 0 9 8 7 6 5 4 3 2

To my parents and Almighty God

CONTENTS

II RADIO FREQUENCY, ANTENNAS, AND PROPAGATION

III WIRELESS ACCESS TECHNOLOGIES

FOREWORD

Wireless communications is one of the most advanced and rapidly advancing technologies of our time. The modern wireless era has produced an array of technologies, such as mobile phones and WiFi networks, of tremendous economic and social value and almost ubiquitous market penetration. These developments have in turn created a substantial demand for engineers who understand the basic principles underlying wireless technologies, and who can help move the field forward to meet the even greater demands for wireless services and capacity expected in the future. Such an understanding requires knowledge of several distinct fields upon which wireless technologies are based: radio frequency physics and devices; communication systems engineering; and communication network architecture.

This book, by a leading advocate of the IEEE Communications Society's Wireless Communication Engineering Technologies certification program, offers an excellent survey of this very broad set of fundamentals. It further provides a review of basic foundational subjects, such as circuits, signals and systems, as well as coverage of several important overlying topics, such network management, security, and regulatory issues. This combination of breadth and depth of coverage allows the book to serve both as a complete course for students and practicing engineers, and as an entrée to the field for those wishing to undertake more advanced study or do research in a particular aspect of the field. Thus, *Fundamentals of Wireless Communication Engineering Technologies* is a very welcome addition to the pedagogical literature in this important field of technology.

H. VINCENT POOR

Princeton, New Jersey

PREFACE

This book presents a broad survey of the fundamentals of wireless communication engineering technologies, spanning the field from radio frequency, antennas, and propagation, to wireless access technologies, to network and service architectures, to other topics, such as network management and security, agreements, standards, policies and regulations, and facilities infrastructure.

Every author has to answer two major questions: (1) What is the scope of coverage of the book, in terms of breadth of topics and depth of discussion of each topic, focus and perspective, and assumptions of prior knowledge of the readers? and (2) Who are the intended readers of the book? I am honored to have been a member of the Practice Analysis Task Force convened by IEEE Communications Society to draft the syllabus and examination specifications of IEEE Communication Society's Wireless Communication Engineering Technologies (WCET) certification program. The scope of coverage of this book has been strongly influenced by the syllabus of the WCET program.

This book is designed to be helpful to three main groups of readers:

- Readers who would like to understand a broad range of topics in practical wireless communications engineering, from fundamentals and theory to practical aspects. For example, wireless engineers with a few years of experience in wireless might find themselves deeply involved with one or two aspects of wireless systems, but not actively keeping up-to-date with other aspects of wireless systems. This book might help such engineers to see how their work fits into the bigger picture, and how the specific parts of the overall system on which they work relate to other parts.
- Electrical engineering or computer science students with an interest in wireless communications, who might be interested to see how the seemingly dry, abstract theory they learn in class is actually applied in real-world wireless systems.
- Readers who are considering taking the WCET exam to become Wireless Certified Professionals. This group could include readers who are not sure if they would take the exam but might decide after reviewing the scope of coverage of the exam.

I hope this book can be a helpful resource for all three groups of readers. For the third group of readers, those with an interest in the WCET exam, several appendices may be useful, including a list of where various formulas from the WCET glossary are discussed in the text (Appendix B), and a few exam tips (Appendix C). However, the rest of the book has been written so that it can be read beneficially by any of the aforementioned groups of readers.

The book is divided into four main sections, three of which cover important areas in wireless systems: (1) radio frequency, antennas, and propagation; (2) wireless access technologies; and (3) network and service architectures. The fourth main section includes the remaining topics. The first three main parts of the book each begins with an introductory chapter that provides essential foundational material, followed by three chapters that go more deeply into specific topics. I have strived to arrange the materials so that the three chapters that go deeper into specific topics build on what is covered in the introductory chapter for that area. This is designed to help students who are new to an area, or not so familiar with it, to be able to go far on their own in self-study, through careful reading first of the introductory chapter, and then of the subsequent chapters. Numerous cross-references are sprinkled throughout the text, for example, so that students who are reading about a topic that relies on some foundational knowledge can see where the foundational knowledge is covered in the relevant introductory chapter. Also, references might be from the relevant introductory chapter to places where specific topics are covered in more detail, which may help motivate students to understand the material in the introductory chapter, as they can see how it is applied later.

The amount of technical knowledge that a wireless engineer "should know" is so broad that it is practically impossible to cover everything in one book, much less to cover everything at the depth that might satisfy every reader. In this book we have tried to select important topics that can be pulled together into coherent and engaging stories and development threads, rather than simply to present a succession of topics. For example, the results of some of the examples are used in later sections or chapters of the book. We also develop various notions related to autocorrelation and orthogonality with an eye to how the concepts might be needed later to help explain the fundamentals of CDMA.

Thanks to Diana Gialo, Simone Taylor, Sanchari Sil, Angioline Loredo, Michael Christian, and George Telecki of Wiley for their editorial help and guidance during the preparation of the manuscript, and to series editors Dr. Vincent Lau and Dr. T. Russell Hsing for their support and helpful comments. Thanks are also due to Dr. Wee Lum Tan, Dr. Toong Khuan Chan, Dr. Choi Look Law, Dr. Yuen Chau, HS Wong, Lian Pin Tee, Ir. Imran Mohd Ibrahim, and Jimson Tseng for their insightful and helpful reviews of some chapters in the book.

There is a web site for this book at http://www.danielwireless.com/wcet, where various supplementary materials, including a list of corrections and updates, will be posted.

K. DANIEL WONG
PH.D. (STANFORD), CCNA, CCNP (CISCO), WCP (IEEE)

Palo Alto, California

I

PRELIMINARIES

1

INTRODUCTION

In this chapter we provide a brief and concise review of foundational topics that are of broad interest and usefulness in wireless communication engineering technologies. The notation used throughout is introduced in Section 1.1, and the basics of electrical circuits and signals are reviewed in Section 1.2, including fundamentals of circuit analysis, voltage or current as signals, alternating current, phasors, impedance, and matched loads. This provides a basis for our review of signals and systems in Section 1.3, which includes properties of linear time-invariant systems, Fourier analysis and frequency-domain concepts, representations of bandpass signals, and modeling of random signals. Then in Section 1.4, we focus on signals and systems concepts specifically for communications systems. The reader is expected to have come across much of the material in this chapter in a typical undergraduate electrical engineering program. Therefore, this chapter is written in review form; it is not meant for a student who is encountering all this material for the first time.

Similarly, reviews of foundational topics are provided in Chapters 2, 6, and 10 for the following areas:

- *Chapter 2:* review of selected topics in electromagnetics, transmission lines, and testing, as a foundation for radio frequency (RF), antennas, and propagation
- *Chapter 6:* review of selected topics in digital signal processing, digital communcations over wireless links, the cellular concept, spread spectrum, and othogonal frequency-division multiflexing (OFDM), as a foundation for wireless access technologies

Fundamentals of Wireless Communication Engineering Technologies, First Edition. K. Daniel Wong.

- *Chapter 10:* review of selected topics in fundamental networking concepts, Internet protocol (IP) networking, and teletraffic analysis, as a foundation for network and service architectures

Compared to the present chapter, the topics in Chapters 2, 6, and 10 are generally more specific to particular areas. Also, we selectively develop some of the topics in those chapters in more detail than we do in this chapter.

1.1 NOTATION

In this section we discuss the conventions we use in this book for mathematical notation. A list of symbols is provided in Appendix D.

$\mathcal{R}$ and $\mathcal{C}$ represent the real and complex numbers, respectively. Membership in a set is represented by $\in$ (e.g., $x \in \mathcal{R}$ means that x is a real number). For $x \in \mathcal{C}$, we write $\Re\{x\}$ and $\Im\{x\}$ for the real and imaginary parts of x, respectively.

log represents base-10 logarithms unless otherwise indicated (e.g., $\log_2$ for base-2 logarithms), or where an expression is valid for all bases.

Scalars, which may be real or even complex valued, are generally represented by italic type (e.g., x, y), whereas vectors and matrices will be represented by bold type (e.g., **G**, **H**). We represent a complex conjugate of a complex number, say an impedance Z, by Z^*. We represent the magnitude of a complex number x by $|x|$. Thus, $|x|^2 = xx^*$.

For $x \in \mathcal{R}$, $\lfloor x \rfloor$ is the largest integer n such that $n < x$. For example, $\lfloor 5.67 \rfloor = 5$ and $\lfloor -1.2 \rfloor = -2$.

If $\mathbf{G}$ is a matrix, $\mathbf{G}^\mathrm{T}$ represents its transpose.

When we refer to a matrix, vector, or polynomial as being *over* something (e.g., *over the integers*), we mean that the components (or coefficients, in the case of polynomials) are numbers or objects of that sort.

If $x(t)$ is a random signal, we use $< x(t) >$ to refer to the time average and $\overline{x(t)}$ to refer to the ensemble average.

1.2 FOUNDATIONS

Interconnections of electrical elements (resistors, capacitors, inductors, switches, voltage and current sources) are often called a *circuit*. Sometimes, the term *network* is used if we want "circuit" to apply only to the more specific case of where there is a closed loop for current flow. In Section 1.2.1 we review briefly this type of electrical network or circuit. Note that this use of "network" should not be confused with the very popular usage in the fields of computer science and telecommunications, where we refer to computer networks and telecommunications networks (see Chapters 9 to 12 for further discussion). In Chapter 2 we will see how transmission lines (Section 2.3.3) can be modeled as circuit elements and so can be part of electrical networks and circuits.

In *electronic* networks and circuits, we also have components with *gain* and/or *directionality*, such as semiconductor devices, which are known as *active* components (as opposed to *passive* components, which have neither gain nor directionality). These are outside the scope of this book, except for our discussion on RF engineering in Chapter 3. Even there, we don't discuss the physics of the devices or compare different device technologies. Instead, we take a "signals and systems" perspective on RF, and consider effects such as noise and the implications of nonlinearities in the active components.

1.2.1 Basic Circuits

Charge, Q, is quantified in coulombs. Current is charge in motion:

$$I = \frac{dQ}{dt} \qquad \text{amperes} \tag{1.1}$$

The direction of current flow can be indicated by an arrow next to a wire. For convenience, I can take a negative value if current is flowing in the direction opposite from that indicated by the arrow.

Voltage is the difference in electric potential:

$$V = RI \qquad \text{volts} \tag{1.2}$$

Like current, there is a direction associated with voltage. It is typically denoted by $+$ and $-$. $+$ is at higher potential than $-$, and voltage drops going from $+$ to $-$. For convenience, V can take a negative value if a voltage drop is in the direction opposite from that indicated by $+$ and $-$

- Power:

$$P = \frac{V^2}{R}, \qquad P = I^2 R \qquad \text{watts} \tag{1.3}$$

- Resistors in series:

$$R = R_1 + R_2 + \cdots + R_n \tag{1.4}$$

- Resistors in parallel:

$$R = \frac{R_1 R_2 \cdots R_n}{R_1 + R_2 + \cdots + R_n} \tag{1.5}$$

1.2.2 Capacitors and Inductors

A capacitor may be conceived of in the form of two parallel plates. For a capacitor with capacitance C farads, a voltage V applied across its plates results in charges $+Q$ and $-Q$ accumulating on the two plates.

$$Q = CV \tag{1.6}$$

$$I = \frac{dQ}{dt} = C\frac{dV}{dt} \tag{1.7}$$

A capacitor acts as an open circuit under direct-current (dc) conditions.

- Capacitors in series:

$$C = \frac{C_1 C_2 \cdots C_n}{C_1 + C_2 + \cdots + C_n} \tag{1.8}$$

- Capacitors in parallel:

$$C = C_1 + C_2 + \cdots + C_n \tag{1.9}$$

An inductor is often in the form of a coil of wire. For an inductor with inductance L henries, a change in current of dI/dt induces a voltage V across the inductor:

$$V = L\frac{dI}{dt} \tag{1.10}$$

An inductor acts as a short circuit under dc conditions.

- Inductors in series:

$$L = L_1 + L_2 + \cdots + L_n \tag{1.11}$$

- Inductors in parallel:

$$L = \frac{L_1 L_2 \cdots L_n}{L_1 + L_2 + \cdots + L_n} \tag{1.12}$$

As hinted at by (1.3), an ideal capacitor or ideal inductor has no resistance and does not dissipate any power as heat. However, a practical model for a real inductor has an ideal resistor in series with an ideal inductor, and they are both in parallel with an ideal capacitor.

1.2.3 Circuit Analysis Fundamentals

A *node* in a circuit is any place where two or more circuit elements are connected. A *complete loop* or *closed path* is a continuous path through a circuit that begins and ends at the same node.

Kirchhoff's Current Law. The sum of all the currents entering is zero. This requires at least one current to have a negative sign if one or more of the others is positive. Alternatively, we say that the sum of all the current entering a node is equal to the sum of all the current leaving a node.

Kirchhoff's Voltage Law. The sum of all the voltage drops around any complete loop (or closed path) is zero. This requires at least one voltage drop to have a negative sign if one or more of the others is positive.

1.2.3.1 Equivalent Circuits Often, a subcircuit is connected to the rest of the circuit through a pair of terminals, and we are interested to know what the voltage and current are across these terminals, not how the subcircuit is actually implemented. Norton and Thévenin equivalent circuits can be used for this purpose, for any circuit comprising linear elements. A *Thévenin equivalent circuit* comprises a single voltage source, V_T, in series with a single resistor, R_T. A *Norton equivalent circuit* comprises a single current source, I_N, in parallel with a single resistor, R_N. A Thévenin equivalent circuit can be converted to a Norton equivalent circuit, or vice versa, by a simple source transformation.

1.2.4 Voltage or Current as Signals

A voltage or current can be interpreted as a signal (e.g., for communications purposes). We usually write t explicitly to emphasize that it is a function of t [e.g., $v(t)$ or $i(t)$ for a voltage signal or current signal, respectively].

If $x(t)$ is a signal, we say that $x(t)$ is

- An *energy signal* if

$$0 < \int_{-\infty}^{\infty} x^2(t)\, dt < \infty \tag{1.13}$$

- A *power signal* if

$$0 < \lim_{T\to\infty} \frac{1}{T} \int_{-\infty}^{\infty} x^2(t)\, dt < \infty \tag{1.14}$$

A *periodic* signal is a signal for which a $T \in \mathcal{R}$ can be found such that

$$x(t) = x(t+T) \qquad \text{for } -\infty < t < \infty \tag{1.15}$$

and the smallest such T is called the *period* of the signal.

The duration of a signal is the time interval from when its begins to be nonnegligible to when its stops being nonnegligible.† Thus, a signal can be of finite duration or of infinite duration.

Sinusoidal Signals. Any sinusoid that is a function of a single variable (say, the time variable, t; later, in Section 2.1.1.4, we see sinusoids that are functions of both

† We say nonnegligible rather than nonzero to exclude trivial blips outside the duration of the signal.

temporal and spatial variables) can be written as

$$A\cos(\omega t+\phi)=A\cos(2\pi ft+\phi)=A\sin(2\pi ft+\phi+\pi/2)=A\angle\phi \tag{1.16}$$

where A is amplitude ($A\in\mathcal{R}$), ω is *angular frequency* (radians/second), f is *frequency* (cycles/second, i.e., hertz or s^{-1}), ϕ is *phase angle*, and where the last equality shows that the shorthand notation $A\angle\phi$ can be used when f and the sinusoidal reference time are known implicitly. The period T is

$$T=\frac{1}{f}=\frac{2\pi}{\omega} \tag{1.17}$$

Continuous-Wave Modulation Signals. A continuous-wave modulation signal is a sinusoidal signal that is *modulated* (changed) in a certain way based on the information being communicated. Most communications signals are based on continuous-wave modulation, and we expand on this important topic in Section 1.4.

Special Signals. A fundamental building block in continuous-time representation of digital signals is the rectangular pulse signal, a rectangular function given by

$$\Pi(t)=\begin{cases}1 & \text{for } |t|\le 1/2\\ 0 & \text{for } |t|>1/2\end{cases} \tag{1.18}$$

The triangle signal is also commonly used, but not as frequently. It is denoted by

$$\Lambda(t)=\begin{cases}1-|t| & \text{for } |t|\le 1\\ 0 & \text{for } |t|>1\end{cases} \tag{1.19}$$

$\Pi(t)$ and $\Lambda(t)$ are shown in Figure 1.1.

The sinc signal is given by

$$\text{sinc}(t)=\begin{cases}(\sin\pi t)/\pi t & \text{for } |t|\ne 0\\ 1 & \text{for } t=0\end{cases} \tag{1.20}$$

Although it may be described informally as $(\sin\pi t)/\pi t$, $(\sin\pi t)/\pi t$ is actually undefined at $t=0$, whereas sinc(t) is 1 at $t=0$. The sinc function is commonly seen in communications because it is the Fourier transform of the rectangular pulse signal. Note that in some fields (e.g., mathematics), sinc(t) may be defined as $(\sin t)/t$, but

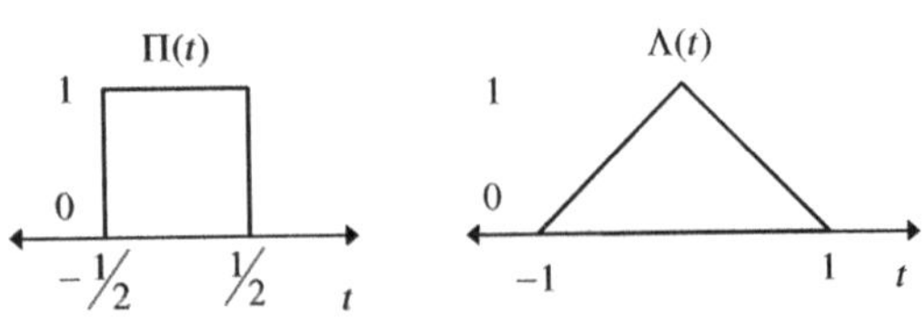

FIGURE 1.1 $\Pi(t)$ and $\Lambda(t)$ functions.

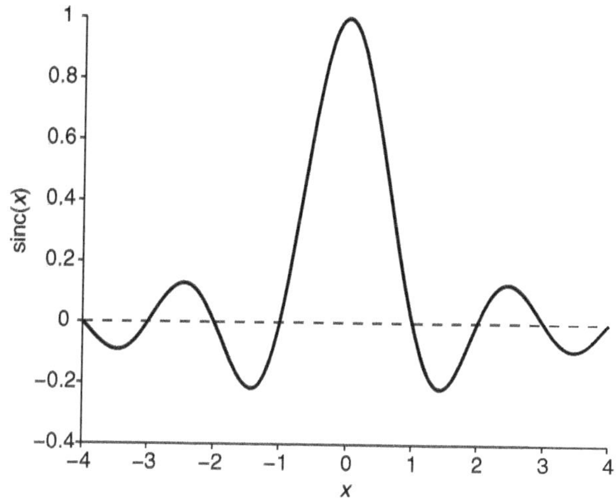

FIGURE 1.2 Sinc function.

here we stick with our definition, which is standard for communications and signal processing. The sinc function is shown in Figure 1.2.

Decibels. It is sometimes convenient to use a log scale when the range of amplitudes can vary by many orders of magnitude, such as in communications systems where the signals have amplitudes and powers that can vary by many orders of magnitude. The standard way to use a log scale in this case is by the use of decibels, defined for any signal voltage or current signal $x(t)$ as

$$10 \log x^2(t) = 20 \log x(t) \tag{1.21}$$

If the signal $s(t)$ is known to be a power rather than a voltage or current, we don't have to convert it to a power, so we just take $10 \log s(t)$. If the power quantity is in watts, it is sometimes written as dBW, whereas if it is in milliwatts, it is written as dBm. This can avoid ambiguity in cases where we just specify a dimensionless quantity A, in decibels, as $10 \log A$.

1.2.5 Alternating Current

With alternating current (ac) the voltage sources or current sources generate time-varying signals. Then (1.3) refers only to the *instantaneous power*, which depends on the instantaneous value of the signal. It is often also helpful, perhaps more so, to consider the *average power*. Let $v(t) = V_0 \cos 2\pi f t$, where V_0 is the maximum voltage (and correspondingly, let I_0 be the maximum current), then the average power P_{av} is

$$P_{\text{av}} = \frac{V_0^2}{2R}, \qquad P_{\text{av}} = \frac{I_0^2 R}{2} \tag{1.22}$$

Equation (1.22) can be obtained either by averaging the instantaneous power directly over one cycle, or through the concept of *rms voltage* and *rms current*. The rms voltage is defined for any periodic signal (not just sinusoidally periodic) as

$$V_{\text{rms}} = \sqrt{\frac{1}{T}\int_0^T v^2(t)\,dt} \tag{1.23}$$

Then we have (again, for any periodic signal, not just sinusoidally periodic)

$$P_{\text{av}} = \frac{V_{\text{rms}}^2}{R}, \qquad P_{\text{av}} = I_{\text{rms}}^2 R \tag{1.24}$$

which looks similar to (1.3). For sinusoidally time-varying signals, we have further,

$$V_{\text{rms}} = \frac{V_0}{\sqrt{2}}, \qquad I_{\text{rms}} = \frac{I_0}{\sqrt{2}} \tag{1.25}$$

1.2.6 Phasors

When working with sinusoidal signals, it is often convenient to work with the *phasor representation* of the signals. Of the three quantities amplitude, phase, and frequency, the phasor representation includes only the amplitude and phase; the frequency is implicit.

Starting from our sinusoid in (1.16) and applying Euler's identity (A.1), we obtain

$$A\cos(2\pi ft + \phi) = A\Re\left\{e^{j(2\pi ft+\phi)}\right\} = \Re\left\{Ae^{j(2\pi ft+\phi)}\right\} \tag{1.26}$$

We just drop the $e^{j2\pi ft}$ and omit mentioning that we need to take the real part, and we have a phasor,

$$Ae^{j\phi} \tag{1.27}$$

Alternatively, we can write the equivalent,

$$A(\cos\phi + j\sin\phi) \tag{1.28}$$

which is also called a phasor. In either case, we see that a phasor is a complex number representation of the original sinusoid, and that it is easy to get back the original sinusoid by multiplying by $e^{j2\pi ft}$ and taking the real part. A hint of the power and convenience of working with phasor representations can be seen by considering differentiation and integration of phasors. Differentiation and integration with respect to t are easily seen to be simple multiplication and division, respectively, by $j2\pi f$.

Rotating Phasors. Sometimes it helps to think of a phasor not just as a static point in the complex plane, but as a rotating entity, where the rotation is at frequency f revolutions (around the complex plane) per second, or w radians per second. This is consistent with the $e^{j2\pi ft}$ term that is implicit in phasors. The direction of rotation is as illustrated in Figure 1.3.

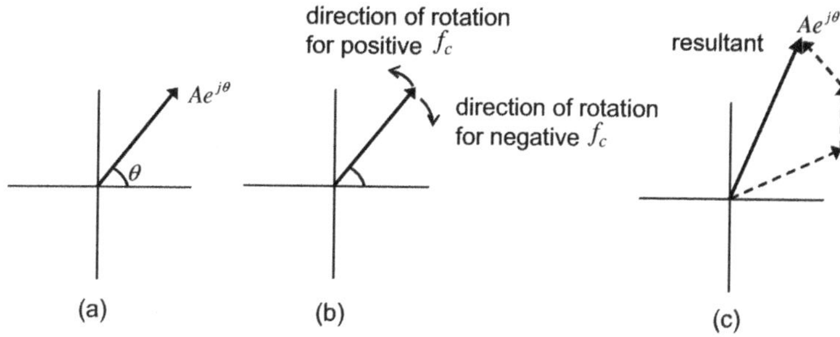

FIGURE 1.3 (a) Phasor in the complex plane; (b) rotating phasors and their direction of rotation; (c) vector addition of phasors.

Expressing Familiar Relationships in Terms of Phasors. Returning to familiar relationships such as (1.2) or (1.3), we find no difference if $v(t)$, $i(t)$ are in phasor representation; however, for capacitors and inductors we have

$$I = j2\pi fCV \quad \text{and} \quad V = j2\pi fLI \tag{1.29}$$

Thus, if we think in terms of rotating phasors, then from (1.29) we see that with a capacitor, I rotates 90° ahead of V, so it *leads* V (and V *lags* I), whereas with an inductor, V leads I (I lags V).

Meanwhile, Kirchhoff's laws take the same form for phasors as they do for non-phasors, so they can continue to be used. Thévenin and Norton equivalent circuits can also be used, generalized to work with impedance, a concept that we discuss next.

1.2.7 Impedance

From (1.29) it can be seen that in phasor representation, resistance, inductance, and capacitance all have the same form:

$$V = ZI \tag{1.30}$$

Thus, the concept of *impedance*, Z, emerges, where Z is R for resistance, $j2\pi fL$ for inductance, and $1/j2\pi fC$ for capacitance, and Z is considered to be in ohms. The complex part of Z is also known as *reactance*.

Impedance is a very useful concept. For example, Thévenin's and Norton's equivalent circuits work in the same way with phasors, except that impedance is substituted for resistance.

1.2.8 Matched Loads

For a linear circuit represented by a Thévenin equivalent voltage V_T and Thévenin equivalent impedance Z_T, the maximum power is delivered to a load Z_L when

$$Z_L = Z_T^* \tag{1.31}$$

(NB: It is the complex conjugate of Z_T, not Z_T itself, in the equation.) This result can be obtained by writing the expression for power in terms of Z_L and Z_T, taking partial derivatives with respect to the load resistance and load reactance, and setting both to 0.

1.3 SIGNALS AND SYSTEMS

Similarly, suppose that we have a system (e.g., a circuit) that takes an input $x(t)$ and produces an output $y(t)$. Let $\longrightarrow$ represent the operation of the system [e.g., $x(t) \longrightarrow y(t)$]. Suppose that we have two different inputs, $x_1(t)$ and $x_2(t)$, such that $x_1(t) \longrightarrow y_1(t)$ and $x_2(t) \longrightarrow y_2(t)$. Let a_1 and a_2 be any two scalars. The system is *linear* if and only if

$$a_1 x_1(t) + a_2 x_2(t) \longrightarrow a_1 y_1(t) + a_2 y_2(t) \tag{1.32}$$

The phenomenon represented by (1.32) can be interpreted as the *superposition* property of linear systems. For example, given knowledge of the response of the system to various sinusoidal inputs, we then know the response of the system to any linear combination of sinusoidal signals. This makes Fourier analysis (Section 1.3.2) very useful.

A system is *time-invariant* if and only if

$$x(t - t_0) \longrightarrow y(t - t_0) \tag{1.33}$$

Systems that are both linear and time invariant are known as *LTI* (linear time-invariant) *systems*.

A system is *stable* if bounded input signals result in bounded output signals.

A system is *causal* if any output does not come before the corresponding input.

1.3.1 Impulse Response, Convolution, and Filtering

An impulse (or unit impulse) signal is defined as

$$\delta(t) = \begin{cases} 1, & t = 0 \\ 0, & t \neq 0 \end{cases} \tag{1.34}$$

and also where

$$\int_{-\infty}^{\infty} \delta(t) = 1 \tag{1.35}$$

Strictly speaking, $\delta(t)$ is not a function, but to be mathematically rigorous requires measure theory or the theory of generalized functions. $\delta(t)$ could also be thought of as

$$\lim_{T \to \infty} T\Pi\,(tT) \tag{1.36}$$

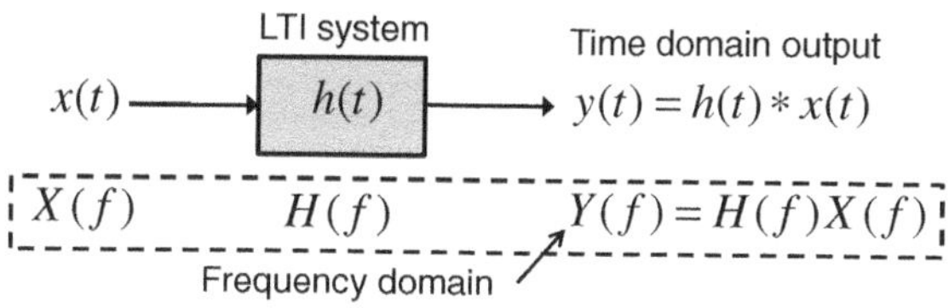

FIGURE 1.4 Mathematical model of an LTI system.

Thus, we often view it as the limiting case of a narrower and narrower pulse whose area is 1.

All LTI systems can be characterized by their *impulse response*. The impulse response, $h(t)$, is the output when the input is an impulse signal; that is,

$$\delta(t) \longrightarrow h(t) \tag{1.37}$$

Convolution: The output of an LTI system with impulse response $h(t)$, given an input $x(t)$, is

$$y(t) = h(t) * x(t) = \int_{\tau=-\infty}^{\tau=\infty} x(\tau)h(t-\tau)\,d\tau = \int_{\tau=-\infty}^{\tau=\infty} h(\tau)x(t-\tau)\,d\tau \tag{1.38}$$

This is shown as the output of the LTI system in Figure 1.4.

With (1.38) in mind, whenever we put a signal $x(t)$ into an LTI system, we can think in terms of the system as *filtering* the input to produce the output $y(t)$, and $h(t)$ may be described as the impulse response of the filter. Although the term *filter* is used in the RF and baseband parts of wireless transmitters and receivers, $h(t)$ can equally well represent the impulse response of a communications channel (e.g., a wire, or wireless link), in which case we may then call it the *channel response* or simply the *channel*.

1.3.1.1 Autocorrelation It is sometimes useful to quantify the similarity of a signal at one point in time with itself at some other point in time. Autocorrelation is a way to do this. If $x(t)$ is a complex-valued energy signal (a real-valued signal is a special case of a complex-valued signal, where the imaginary part is identically zero, and the complex conjugate of the signal is equal to the signal itself), we define the autocorrelation function, $R_{xx}(\tau)$, as

$$R_{xx}(\tau) = \int_{-\infty}^{\infty} x(t)x^*(t+\tau)\,dt \qquad \text{for } -\infty < \tau < \infty \tag{1.39}$$

For a complex-valued periodic power signal with period T_0,

$$R_{xx}(\tau) = \frac{1}{T_0}\int_{-T_0/2}^{T_0/2} x(t)x^*(t+\tau)\,dt \qquad \text{for } -\infty < \tau < \infty \tag{1.40}$$

whereas for a complex-valued power signal, in general,

$$R_{xx}(\tau) = \lim_{T\to\infty} \frac{1}{T} \int_{-T/2}^{T/2} x(t)x^*(t+\tau)\,dt \qquad \text{for } -\infty < \tau < \infty \tag{1.41}$$

1.3.2 Fourier Analysis

Fourier analysis refers to a collection of related techniques where:

- A signal can be broken down into sinusoidal components *(analysis)*
- A signal can be constructed from constituent sinusoidal components *(synthesis)*

This is very useful in the study of linear systems because the effects of such a system on a large class of signals can be studied by considering the effects of the system on sinusoidal inputs using the superposition principle. (NB: The term *analysis* here can be used to refer either to just the breaking down of a signal into sinusoidal components, or in the larger sense to refer to the entire collection of these related techniques.)

Various Fourier *transforms* are used in analysis, and *inverse transforms* are used in synthesis, depending on the types of signals involved. For most practical purposes, there is a one-to-one relationship between a time-domain signal and its Fourier transform, and thus we can think of the Fourier transform of a signal as being a different *representation* of the signal. We usually think of there being two domains, the *time domain* and the *frequency domain*. The (forward) transform typically transforms a *time-domain representation* of a signal into a *frequency-domain representation*, whereas the inverse transform transforms a frequency-domain representation of a signal into a time-domain representation.

1.3.2.1 (Continuous) Fourier Transform The (continuous) Fourier transform of a signal $x(t)$ is given by

$$X(f) = \int_{-\infty}^{\infty} x(t)e^{-j2\pi ft}\,dt \tag{1.42}$$

where $j = \sqrt{-1}$, and the inverse Fourier transform is given by

$$x(t) = \int_{-\infty}^{\infty} X(f)e^{j2\pi ft}\,df \tag{1.43}$$

Table 1.1 gives some basic Fourier transforms.

1.3.2.2 Fourier Series For periodic signals $x(t)$ with period T, the Fourier series (exponential form) coefficients are the set $\{c_n\}$, where n ranges over all the integers,

TABLE 1.1 Fourier Transform Pairs[a]

Time Domain, $x(t)$	Frequency Domain, $X(f)$
$\delta(t)$	1
1	$\delta(f)$
$\delta(t-t_0)$	$e^{-j2\pi f t_0}$
$e^{\pm j2\pi f_0 t}$	$\delta(f \mp f_0)$
$\cos 2\pi f_0 t$	$\frac{1}{2}[\delta(f-f_0)+\delta(f+f_0)]$
$\sin 2\pi f_0 t$	$\frac{1}{2j}[\delta(f-f_0)-\delta(f+f_0)]$
$u(t)=\begin{cases}1 & \text{for } t>0\\ 0 & \text{for } t<0\end{cases}$	$\frac{1}{2}\delta(f)+\frac{1}{j2\pi f}$
$e^{-at}u(t), a>0$	$\frac{1}{a+j2\pi f}$
$te^{-at}u(t), a>0$	$\frac{1}{(a+j2\pi f)^2}$
$e^{-a\lvert t\rvert}, a>0$	$\frac{2a}{a^2+(2\pi f)^2}$
$\Pi\left(\frac{t}{T}\right)$	$T \operatorname{sinc} fT$
$B \operatorname{sinc} Bt$	$\Pi\left(\frac{f}{B}\right)$
$\Lambda\left(\frac{t}{T}\right)$	$T \operatorname{sinc}^2 fT$
$\sum_{k=-\infty}^{\infty} \delta(t-kT)$	$\frac{1}{T}\sum_{n=-\infty}^{\infty} \delta\left(f-\frac{n}{T}\right)$

[a] $\Pi(t)$ and $\Lambda(t)$ are the rectangle and triangle functions defined in Section 1.2.4. $\sum_{k=-\infty}^{\infty} \delta(t-kT)$ is also known as an impulse train.

and c_n is given by

$$c_n = \frac{1}{T}\int_{-T/2}^{T/2} x(t)e^{-j2\pi f_0 n t}\,dt \tag{1.44}$$

where $f_0 = 1/T$, and the Fourier series representation of $x(t)$ is given by

$$x(t) = \sum_{n=-\infty}^{\infty} c_n e^{j2\pi f_0 n t} \tag{1.45}$$

1.3.2.3 Relationships Between the Transforms The (continuous) Fourier transform can be viewed as a limiting case of Fourier series as the period T goes to ∞, and the signal thus becomes aperiodic. Since $f_0 = 1/T$, let $f = nf_0 = n/T$. Using (1.44), then

$$\begin{aligned}
\lim_{T\to\infty} c_n T &= \lim_{T\to\infty} \int_{-T/2}^{T/2} x(t)e^{-j2\pi nt/T}\, dt \\
&= \int_{-\infty}^{\infty} x(t)e^{-j2\pi ft}\, dt \\
&= X(f)
\end{aligned} \tag{1.46}$$

Since $1/T$ goes to zero in the limit, we can write $1/T$ as Δf. $\Delta f \to 0$ as $T \to \infty$. Then (1.45) can be written as

$$\begin{aligned}
x(t) &= \sum_{n=-\infty}^{\infty} T\frac{1}{T}c_n e^{j2\pi f_0 nt} \\
&= \sum_{n=-\infty}^{\infty} (c_n T)e^{j2\pi nf_0 t}\frac{1}{T} \\
&= \sum_{n=-\infty}^{\infty} (c_n T)e^{j2\pi n(\Delta f)t}\Delta f
\end{aligned} \tag{1.47}$$

$$\lim_{\Delta f\to 0} x(t) = \int_{-\infty}^{\infty} X(f)e^{j2\pi ft}\, df \tag{1.48}$$

where we used the substitution from (1.46) in the last step.

1.3.2.4 Properties of the Fourier Transform

Table 1.2 lists some useful properties of Fourier transforms. Combining properties from the table with known Fourier transform pairs from Table 1.1 lets us compute many Fourier transforms and inverse transforms without needing to perform the integrals (1.42) or (1.43).

TABLE 1.2 Properties of the Fourier Transform

Concept	Time Domain, $x(t)$	Frequency Domain, $X(f)$
Scaling	$x(at)$	$\frac{1}{\lvert a\rvert}X\left(\frac{f}{a}\right)$
Time shifting	$x(t-t_0)$	$X(f)e^{-j2\pi ft_0}$
Frequency shifting	$x(t)e^{j2\pi f_0 t}$	$X(f-f_0)$
Modulation	$x(t)\cos(j2\pi f_0 t+\phi)$	$\frac{1}{2}\left(X(f-f_0)e^{j\phi}+X(f+f_0)e^{-j\phi}\right)$
Differentiation	$\frac{d^n x}{dt^n}$	$(j2\pi f)^n X(f)$
Convolution	$x(t) * y(t)$	$X(f)Y(f)$
Multiplication	$x(t)y(t)$	$X(f) * Y(f)$
Conjugation	$x^*(t)$	$X^*(-f)$

1.3.3 Frequency-Domain Concepts

Some frequency-domain concepts are fundamental for understanding communications systems. A miscellany of comments on the frequency domain:

- In the rotating phasor viewpoint, $e^{j2\pi f_0 t}$ is a phasor rotating at f_0 cycles per cycle. But $\mathcal{F}[e^{j2\pi f_0 t}] = \delta(f - f_0)$. Thus, frequency-domain components of the form $\delta(f - f_0)$ for any f_0 can be viewed as rotating phasors.
- Negative frequencies can be viewed as rotating phasors rotating clockwise, whereas positive frequencies rotate counterclockwise.
- For LTI systems, $Y(f) = X(f)H(f)$, where $Y(f)$, $X(f)$, and $H(f)$ are the Fourier transforms of the output signal, input signal, and impulse response, respectively. See Figure 1.4.

1.3.3.1 Power Spectral Density *Power spectral density* (PSD) is a way to see how the signal power is distributed in the frequency domain. We have seen that a periodic signal can be written in terms of Fourier series [as in (1.45)]. Similarly, the PSD $S_x(f)$ of periodic signals can be expressed in terms of Fourier series:

$$S_x(f) = \frac{1}{T} \sum_{n=-\infty}^{\infty} |c_n|^2 \delta\left(t - \frac{n}{T}\right) \tag{1.49}$$

where c_n are the Fourier series coefficients as given by (1.44).

For nonperiodic power signals $x(t)$, let $x_T(t)$ be derived from $x(t)$ by

$$x_T(t) = x(t)\Pi(t/T) \tag{1.50}$$

Then $x_T(t)$ is an energy signal with a Fourier transform $X_T(f)$ and an energy spectral density $|X_T(f)|^2$. Then the power spectral density of $x(t)$ can be defined as

$$S_x(f) = \lim_{T\to\infty} \frac{1}{T} |X_T(f)|^2 \tag{1.51}$$

Alternatively, we can apply the *Wiener–Kinchine theorem*, which states that

$$S_x(f) = \int_{-\infty}^{\infty} R_{xx}(\tau)e^{-j2\pi f\tau}\, d\tau \tag{1.52}$$

In other words, the PSD is simply the Fourier transform of the autocorrelation function. It can be shown that (1.51) and (1.52) and equivalent. Either one can be used to define the PSD and the other can be shown to be equivalent. Whereas (1.51) highlights the connection with the Fourier transform of the signal, (1.52) highlights the connection with its autocorrelation function.

Note that the Wiener–Kinchine theorem applies whether or not $x(t)$ is periodic. Thus, in the case that $x(t)$ is periodic with period T, clearly also $R_{xx}(\tau)$ is periodic with the same period. Let $R'_{xx}(t)$ be equal to $R_{xx}(t)$ within one period, $0 \le t \le T$, and zero elsewhere, and let $S'_x(f)$ be the power spectrum of $R'_{xx}(t)$. Note that

$$\begin{aligned}R_{xx}(t) &= \sum_{k=-\infty}^{\infty} R'_{xx}(t-kT)\\ &= \sum_{k=-\infty}^{\infty} R'_{xx}(t) * \delta(t-kT)\\ &= R'_{xx}(t) * \sum_{k=-\infty}^{\infty} \delta(t-kT)\end{aligned} \tag{1.53}$$

Then

$$\begin{aligned}S_x(f) &= \mathcal{F}(R_{xx}(\tau))\\ &= \mathcal{F}\left(R'_{xx}(t) * \sum_{k=-\infty}^{\infty} \delta(t-kT)\right)\\ &= \mathcal{F}\left(R'_{xx}(t)\right) \mathcal{F}\left(\sum_{k=-\infty}^{\infty} \delta(t-kT)\right)\\ &= S'_x(f)\frac{1}{T}\sum_{n=-\infty}^{\infty} \delta\left(f-\frac{n}{T}\right)\end{aligned} \tag{1.54}$$

One-Sided vs. Two-Sided PSD. The PSD that we have been discussing so far is the *two-sided PSD*, which has both positive and negative frequencies. It reflects the fact the a real sinusoid (e.g., a cosine wave) is the sum of two complex sinusoids rotating in opposite directions at the same frequency (thus, at a positive and a negative frequency). The *one-sided PSD* is a variation that has no negative frequency components and whose positive frequency components are exactly twice those of the two-sided PSD. The one-sided PSD is useful in some cases: for example, for calculations of noise power.

1.3.3.2 Signal Bandwidth Just as in the time domain, we have a notion of duration of a signal (Section 1.2.4), in the frequency domain we have an analogous notion of *bandwidth*. A first-attempt definition of bandwidth might be the interval or range of frequencies from when the signal begins to be nonnegligible to when it stops being nonnegligible (as we sweep from lower to higher frequencies). This is imprecise but can be quantified in various ways, such as:

- *3-dB bandwidth* or *half-power bandwidth*
- *Noise-equivalent bandwidth* (see Section 3.2.3.2)

Often, it is not so much a question of finding the *correct* way of defining bandwidth but of finding a useful way of defining bandwidth for a particular situation.

Bandwidth is fundamentally related to channel capacity in the following celebrated formula:

$$C = B \log\left(1 + \frac{S}{N}\right) \tag{1.55}$$

The base of the logarithm determines the units of capacity. In particular, for capacity in bits/second,

$$C = B \log_2\left(1 + \frac{S}{N}\right) \tag{1.56}$$

To obtain capacity in bits/second, we use (1.56) with B in hertz and S/N on a linear scale (not decibels).

This concept of capacity is known as *Shannon capacity*. Later (e.g., in Section 6.3.2) we will see other concepts of capacity.

1.3.4 Bandpass Signals and Related Notions

Because bandpass signals have most of their spectral content around a carrier frequency, say f_c, they can be written in an envelope-and-phase representation:

$$x_b(t) = A(t)\cos[2\pi f_c t + \phi(t)] \tag{1.57}$$

where $A(t)$ and $\phi(t)$ are a slowly varying envelope and phase, respectively.

Most communications signals while in the communications medium are continuous-wave modulation signals, which tend to be *bandpass* in nature.

1.3.4.1 In-phase/Quadrature Description A bandpass signal $x_b(t)$ can be written in envelope-and-phase form, as we have just seen. We can expand the cosine term using (A.8), and we have

$$\begin{aligned} x_b(t) &= A(t)\left[\cos(2\pi f_c t)\cos\phi(t) - \sin(2\pi f_c t)\sin\phi(t)\right] \\ &= x_i(t)\cos(2\pi f_c t) - x_q(t)\sin(2\pi f_c t) \end{aligned} \tag{1.58}$$

where $x_i(t) = A(t)\cos\phi(t)$ is the *in-phase* component, and $x_q(t) = A(t)\sin\phi(t)$ is the *quadrature* component. Later, in Section 6.1.8.1, we prove that the in-phase and quadrature components are orthogonal, so can be used to transmit independent bits without interfering with each other.

If we let $X_i(f) = \mathcal{F}[x_i(t)]$, $X_q(f) = \mathcal{F}[x_q(t)]$, and $X_b(f) = \mathcal{F}[x_b(t)]$, then

$$X_b(f) = \frac{1}{2}\left[X_i(f + f_c) + X_i(f - f_c)\right] - \frac{j}{2}\left[X_q(f + f_c) - X_q(f - f_c)\right] \tag{1.59}$$

1.3.4.2 Lowpass Equivalents There is another useful representation of bandpass signals, known as the *lowpass equivalent* or *complex envelope* representation. Going from the envelope-and-phase representation to lowpass equivalent is analogous to going from a rotating phasor to a (nonrotating) phasor; thus we have

$$x_{\text{lp}}(t) = A(t)e^{j\phi(t)} \tag{1.60}$$

which is analogous to (1.27). An alternative definition given in some other books is

$$x_{\text{lp}}(t) = \frac{1}{2}A(t)e^{j\phi(t)} \tag{1.61}$$

which differs by a factor of 1/2. [This is just a matter of convention, and we will stick with (1.60).]

The lowpass equivalent signal is related to the in-phase and quadrature representation by

$$x_{\text{lp}}(t) = x_i(t) + jx_q(t) \tag{1.62}$$

and we also have

$$x_b(t) = \Re\left[x_{\text{lp}}(t)e^{j2\pi f_c t}\right] \tag{1.63}$$

In the frequency domain, the lowpass equivalent is the positive-frequency part of the bandpass signal, translated down to dc (zero frequency):

$$\begin{aligned} X_{\text{lp}}(f) &= \left[X_i(f) + jX_q(f)\right] \\ &= 2X_b(f + f_c)u(f + f_c) \end{aligned} \tag{1.64}$$

where $u(f)$ is the step function (0 for $f < 0$, and 1 for $f \geq 0$).

Interestingly, we can represent filters or transfer functions with lowpass equivalents, too, so we have

$$Y_{\text{lp}}(f) = H_{\text{lp}}(f)X_{\text{lp}}(f) \tag{1.65}$$

where

$$H_{\text{lp}}(f) = H_b(f + f_c)u(f + f_c) \tag{1.66}$$

1.3.5 Random Signals

In well-designed communications systems, the signals arriving at a receiver appear random. Thus, it is important to have the tools to analyze random signals. We assume that the reader has knowledge of basic probability theory, including probability distribution or density, cumulative distribution function, and expectations [4].

Then a *random variable* can be defined as mapping from a sample space into a range of possible values. A sample space can be thought of as the set of all outcomes of an experiment. We denote the sample space by Ω and let ω be a variable that can represent each possible outcome in the sample space. For example, we consider a coin-flipping experiment with outcome either heads or tails, and we define a random variable by

$$X(\omega) = \begin{cases} 1 & \text{if } \omega = \text{heads} \\ 2 & \text{if } \omega = \text{tails} \end{cases} \tag{1.67}$$

where the domain of ω is the set {heads, tails}. If $P(\text{heads}) = 2/3$ and $P(\text{tails}) = 1/3$, then $P(X = 1) = 2/3$ and $P(X = 2) = 1/3$. The *average* (also called *mean*, or *expected value*) of X is $(2/3)(1) + (1/3)(2) = 4/3$. Note that when we write just X, we have omitted the ω for notational simplicity.

1.3.5.1 Stochastic Processes Now we consider cases where instead of just mapping each point in the sample space, ω, to a value, we map each ω to a function. To emphasize that the mapping is to a function, and that this is therefore not the same as a normal random variable, it is called a *stochastic process* or *random process*. It could also be called a *random function*, but that could be confused with *random variable*, so it may be best to stick with *random variable* in general and *stochastic process* in cases where the mapping is to a function. Depending on the application, we may think of a stochastic process as a *random signal*.

For example, a stochastic process could be defined by a sinusoid with a random phase (e.g., a phase that is uniformly distributed between 0 and 2π):

$$x(t, \omega) = \cos(2\pi ft + \phi) \tag{1.68}$$

where $\phi(\omega)$ is a random variable distributed uniformly between 0 and 2π (and where we usually omit writing the ω, for convenience). Stochastic processes in wireless communications usually involve a time variable, t, and/or one or more spatial variables (e.g., x, y, z), so we can write $f(x, y, z, t, \omega)$ or just $f(x, y, z, t)$ if it is understood to represent a stochastic process.

The entire set of functions, as ω varies over the entire sample space, is called an *ensemble*. For any particular outcome, $\omega = \omega_i$, $x(t)$ is a specific *realization* (also known as *sample*) of the random process. For any given fixed $t = t_0$, $x(t_0)$ is a random variable, X_0, that represents the ensemble at that point in time (and hence a stochastic process can be viewed as an uncountably infinite set of random variables). Each of these random variables has a density function $f_{X_0}(x_0)$ from which its *first-order statistics* can be obtained. For example, we can obtain the mean $\int x f_{X_0}(x)\,dx$, the variance, and so on. The relationship between random variables associated with two different times t_0 and t_1 is often of interest. For example, let their joint distribution be written as $f_{X_0,X_1}(x_0, x_1)$; then, if

$$f_{X_0,X_1}(x_0, x_1) = f_{X_0}(x_0) f_{X_1}(x_1) \tag{1.69}$$

the two random variables are said to be *independent* or *uncorrelated*. The *second-order statistics* may be obtained from the joint distribution. This can be extended to the joint distribution of three or more points in time, so we have the *n*th*-order statistics*.

As an example of these ideas, assume that at a radio receiver we have a signal $r(t)$ that consists of a deterministic signal $s(t)$ in the presence of additive white Gaussian noise (AWGN), $n(t)$. If we model the AWGN in the usual way, $r(t)$ is a stochastic process:

$$r(t) = s(t) + n(t) \tag{1.70}$$

Because of the nature of AWGN, $n(t_1)$ and $n(t_2)$ are uncorrelated for any $t_1 \neq t_2$. Furthermore, since AWGN is Gaussian distributed, the first-order statistics depend on only two parameters (i.e., the mean and variance). Since $\overline{n(t)} = 0$ for all t, we just need to know the variance, $\sigma^2(t_1)$, $\sigma^2(t_2)$, and so on. Must we have $\sigma^2(t_1) = \sigma^2(t_2)$ for $t_1 \neq t_2$? We discuss this in Section 1.3.5.4. Here, we have just seen that a deterministic communications signal that is corrupted by AWGN can be modeled as a stochastic process.

1.3.5.2 Time Averaging vs. Ensemble Averaging Averages are still useful for many applications, but since in this case we now have multiple variables over which an average may be taken, it often helps to specify to which average we are referring. If we are working with a specific realization of the random signal, we can take the *time average*. For a periodic signal (in time, t) with period T_0,

$$< x(t) >= \frac{1}{T_0} \int_0^{T_0} x(t)\, dt \tag{1.71}$$

If it is not a periodic signal, we may still consider a time average as given by

$$< x(t) >= \lim_{T\to\infty} \int_{-T/2}^{T/2} \frac{x(t)}{T}\, dt \tag{1.72}$$

Besides the time average, we also have the *ensemble average*, over the entire ensemble, resulting in a function (unlike the time average, which results in a value). For a discrete probability distribution this may be written as

$$\overline{x(t)} = \sum p_{x,t} x \tag{1.73}$$

where $p_{x,t}$ is the probability of event $x(t)$ at time t. The ensemble average for a continuous probability distribution can be written as

$$\overline{x(t)} = \int f_{X_t}(x) x\, dx \tag{1.74}$$

In this book we generally use $< \cdot >$ to denote time averaging or spatial averaging, and $\bar{\cdot}$ to denote ensemble averaging.

1.3.5.3 Autocorrelation As we saw in Section 1.3.1.1, for deterministic signals the autocorrelation is a measure of the similarity of a signal with itself.

The autocorrelation function of a stochastic process $x(t)$ is

$$R_{xx}(t_1, t_2) = \overline{x(t_1)x(t_2)} \tag{1.75}$$

Unlike the case of deterministic signals, this is an ensemble average and in general is a function of two variables representing two moments of time rather than just a time difference. In general, it requires knowledge of the joint distribution of $x(t_1)$ and $x(t_2)$. Soon we will see that these differences go away when $x(t)$ is an ergodic process.

1.3.5.4 Stationarity, Ergodicity, and Other Properties Going back to example (1.70), we saw that $n(t)$ was uncorrelated at any two different times. However, do the mean and variance have to be constant for all time? Clearly, they do not. In that radio receiver example, suppose that the temperature is rising. To make things simple, we suppose that the temperature is rising monotonically as t increases. Then, as we will see in Section 3.2, Johnson–Nyquist noise in the receiver is increasing monotonically with time. Thus,

$$\sigma^2(t_1) < \sigma^2(t_2) \qquad \text{for } t_1 < t_2$$

If, instead,

$$\sigma^2(t_1) = \sigma^2(t_2) \qquad \text{for all } t_1 \neq t_2$$

there is a sense in which the stochastic process $n(t)$ is stationary—its variance doesn't depend on time.

The concept of stationarity has to do with questions of how the statistics of the signal change with time. For example, consider a random signal at m time instances, $t_1, t_2, \ldots, t_m$. Suppose that we consider the joint distribution $f_{X_{t_1},X_{t_2},\ldots,X_{tm}}(x_1, x_2, \ldots, x_m)$. Then a stochastic process is considered *strict-sense stationary* (SSS) if it is invariant to time translations for all sets $t_1, t_2, \ldots, t_m$, that is,

$$f_{X_{t_1+\tau},X_{t_2+\tau},\ldots,X_{tm+\tau}}(x_1, x_2, \ldots, x_m) = f_{X_{t_1},X_{t_2},\ldots,X_{tm}}(x_1, x_2, \ldots, x_m) \tag{1.76}$$

A weaker sense of stationarity is often seen in communications applications. A stochastic process is *weak-sense stationary* (WSS) if

1. The mean value is independent of time.
2. The autocorrelation depends only on the time difference $t_2 - t_1$ (i.e., it is a function of $\tau = t_2 - t_1$), so it may be written as $R_{xx}(\tau)$ [or $R_x(\tau)$ or simply $R(\tau)$] to keep this property explicit.

The class of WSS processes is larger than and includes the complete class of SSS processes. Similarly, there is another property, ergodicity, such that the class of SSS processes includes the complete class of ergodic processes. A random process is *ergodic* if it is SSS and if all ensemble averages are equal to the corresponding time averages. In other words, for ergodic processes, time averaging and ensemble averaging are equivalent.

Autocorrelation Revisited. For random processes that are WSS (including SSS and ergodic processes), the autocorrelation becomes $R(\tau)$, where τ is the time difference. Thus, (1.75) becomes

$$R_{xx}(\tau) = \overline{x(t)x(t+\tau)} \tag{1.77}$$

which is similar to (1.39).

Furthermore, for ergodic processes, we can even do a time average, so the autocorrelation then converges to the case of the autocorrelation of a deterministic signal (in the case of the ergodic process, we just pick any sample function and obtain the autocorrelation from it as though it were a deterministic function).

1.3.5.5 Worked Example: Random Binary Signal Consider a random binary wave, $x(t)$, where every symbol lasts for T_s seconds, and independently of all other symbols, it takes the values A or $-A$ with equal probability. Let the first symbol transition after $t = 0$ be at T_{trans}. Clearly, $0 < T_{\text{trans}} < T_s$. We let T_{trans} be distributed uniformly between 0 and T_s.

The mean at any point in time t is

$$E[x(t)] = A(0.5) + (-A)(0.5) = 0 \tag{1.78}$$

The variance at any point in time t is

$$\sigma^2 = E[x^2(t)] - (E[x(t)])^2 = A^2 - 0 = A^2 \tag{1.79}$$

To figure out if it is WSS, we still need to see if the autocorrelation is dependent only on $\tau = t_2 - t_1$. We analyze the two autocorrelation cases:

- If $|t_2 - t_1| > T_s$, then $R_{xx}(t_1, t_2) = 0$ by the independence of each symbol from every other symbol.
- If $|t_2 - t_1| < T_s$, it depends on whether t_1 and t_2 lie in the same symbol (in which case we get σ^2) or in adjacent symbols (in which case we get zero).

What is the probability, P_a, that t_1 and t_2 lie in adjacent symbols? Let $t_1' = t_1 - kT_s$ and $t_2' = t_2 - kT_s$, where k is the unique integer such that we get both $0 \le t_1' < T_s$ and $0 \le t_2' < T_s$. Then, $P_a = P(T_{\text{trans}}$ lies between t_1' and $t_2') = |t_2 - t_1|/T_s$.

$$E[x(t_1)x(t_2)] = A^2(1 - P_a) = A^2\left(1 - \frac{|t_2 - t_1|}{T_s}\right) = A^2\left(1 - \frac{|\tau|}{T_s}\right) \tag{1.80}$$

Hence, it is WSS. And using the triangle function notation, we can write the complete autocorrelation function compactly as

$$R_{xx}(\tau) = A^2\Lambda(\tau/T_s) \tag{1.81}$$

This is shown in Figure 1.5.

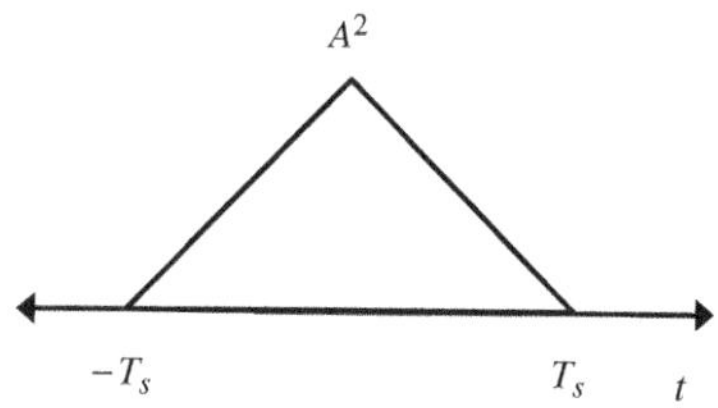

FIGURE 1.5 Autocorrelation function of the random binary signal.

1.3.5.6 Power Spectral Density of Random Signals For a random signal to have a meaningful power spectral density, it should be wide-sense stationary.

Each realization of the random signal would have its own power spectral density, different from other realizations of the same random process. It turns out that the (ensemble) average of the power spectral densities of each of the realizations, loosely speaking, is the most useful analog to the power spectral density of a deterministic signal. To be precise, the following procedure can be used on a random signal, $x(t)$, to estimate its PSD, $S_x(f)$. Let us denote the estimate by $\widetilde{S}_x(f)$.

1. Observe $x(t)$ over a period of time, say, 0 to T; let $x_T(t)$ be the truncated version of $x(t)$, as specified in (1.50), and let $X_T(f)$ be the Fourier transform of $x_T(t)$. Then its energy spectral density may be computed as $|X_T(f)|^2$.
2. Observe many samples $x_T(t)$ repeatedly, and compute their corresponding Fourier transforms $X_T(f)$ and energy spectral densities, $|X_T(f)|^2$.
3. Compute $\widetilde{S}_x(f)$ by computing the ensemble average $\overline{(1/T)\,|X_T(f)|^2}$.

One may wonder how to do step 2 in practice. Assuming that $x(t)$ is ergodic, then $\overline{(1/T)\,|X_T(f)|^2}$ is equivalent to time averaging, so we get a better and better estimate $\widetilde{S}_x(f)$ by obtaining $x_T(t)$ over many intervals of T from the same sample function, and then computing

$$\widetilde{S}_x(f) = \left\langle \frac{1}{T}\,|X_T(f)|^2 \right\rangle \tag{1.82}$$

This procedure is based on the following definition of the PSD for random signals:

$$S_x(f) = \lim_{T\to\infty} \frac{1}{T}\,\overline{|X_T(f)|^2} \tag{1.83}$$

which is analogous to (1.51).

Also, as with deterministic signals, the Wiener–Kinchine theorem applies, so

$$S_x(f) = \int_{-\infty}^{\infty} R_{xx}(\tau) e^{-j2\pi f\tau}\, d\tau \tag{1.84}$$

which can be shown to be equivalent to (1.83).

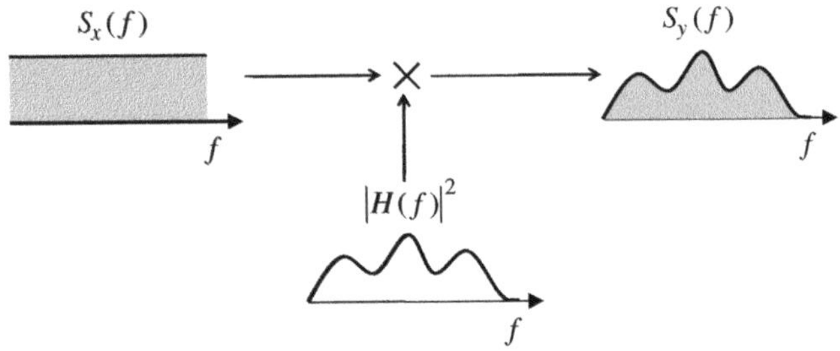

FIGURE 1.6 Filtering and the PSD.

1.3.5.7 Worked Example: PSD of a Random Binary Signal Consider the random binary signal from Section 1.3.5.5. What is the power spectral density of the signal? What happens as T_s approaches zero?

We use the autocorrelation function, as in (1.81), and take the Fourier transform to obtain

$$S_x(f) = A^2 T_s \operatorname{sinc}^2(fT_s) \tag{1.85}$$

As T_s gets smaller and smaller, the autocorrelation function approaches an impulse function. At the same time, the first lobe of the PSD is between $-1/T_s$ and $1/T_s$, so the it becomes very broad and flat, giving it the appearance of the classic "white noise."

1.3.5.8 LTI Filtering of WSS Random Signals Once we can show that a random signal is WSS, the PSD behaves "like" the PSD of a deterministic signal in some ways; for example, when passing through a filter we have (Figure 1.6)

$$S_y(f) = |H(f)|^2 S_x(f) \tag{1.86}$$

where $S_x(f)$ and $S_y(f)$ are the PSDs of the input and output signals, respectively, and $H(f)$ is the LTI system/channel that filters the input signal.

For example, if $S_x(f)$ is flat (as with white noise), $S_y(f)$ takes on the shape of $H(f)$. In communications, a canonical signal might be a "random" signal around a carrier frequency f_c, with additive white Gaussian noise (AWGN) but with interfering signals at other frequencies, so we pass through a filter (e.g., an RF filter in an RF receiver) to reduce the magnitude of the interferers.

1.3.5.9 Gaussian Processes A *Gaussian process* is one where the distribution $f_{X_t}(x)$ is Gaussian and all the distributions $f_{X_{t_1}, X_{t_2}, \ldots, X_{t_m}}(x_1, x_2, \ldots, x_m)$ for all sets $t_1, t_2, \ldots, t_m$ are joint Gaussian distributions.

For a Gaussian process, if it is WSS, it is also SSS.

1.3.5.10 Optimal Detection in Receivers An important example of the use of random signals to model communications signals is the model of the signal received

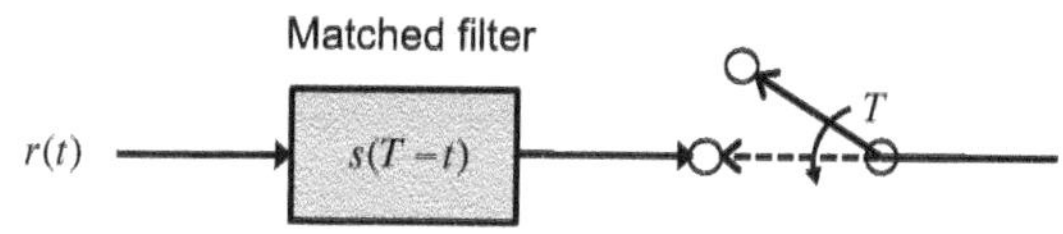

FIGURE 1.7 Matched filter followed by symbol rate filtering.

at a digital communications receiver. We give examples of modulation schemes used in digital (and analog) systems in Section 1.4. But here we review some fundamental results on optimal detection.

Matched Filters. We consider the part of a demodulator after the frequency down-translation, such that the signal is at baseband. Here we have a *receiving filter* followed by a sampler, and we want to optimize the receiving filter. For the case of an AWGN channel, we can use facts about random signals [such as (1.86)] to prove that the optimal filter is the *matched filter*. By *optimal* we are referring to the ability of the filter to provide the largest signal-to-noise ratio at the output of the sampler at time $t = T$, where the signal waveform is from $t = 0$ to T.

If the signal waveform is $s(t)$, the matched filter is $s(T - t)$ [or more generally, a scalar multiple of $s(T - t)$].

The proof is outside the scope of this book but can be found in textbooks on digital communications. The matched filter is shown in Figure 1.7, where $r(t)$ is the received signal, and the sampling after the matched filtering is at the symbol rate, to decide each symbol transmitted.

Correlation Receivers. Also known as *correlators*, correlation receivers provide the same decision statistic that matched filters provide (Exercise 1.5 asks you to show this). If $r(t)$ is the received signal and the transmitted waveform is $s(t)$, the correlation receiver obtains

$$\int_0^T r(t)s(t)\,dt \tag{1.87}$$

1.4 SIGNALING IN COMMUNICATIONS SYSTEMS

Most communications systems use continuous-wave modulation as a fundamental building block. An exception is certain types of ultrawideband systems, discussed in Section 17.4.2. In continuous-wave modulation, a sinusoid is modified in certain ways to convey information. The unmodulated sinusoid is also known as the *carrier*. The earliest communications systems used analog modulation of the carrier.

These days, with source data so often in digital form (e.g., from a computer), it makes sense to communicate digitally also. Besides, digital communication has advantages over analog communication in how it allows error correction, encryption, and other processing to be performed. In dealing with noise and other channel impairments, digital signals can be recovered (with bit error rates on the order of 10^{-3} to 10^{-6}, depending on the channel and system design), whereas analog signals are only degraded.

Generally, we would like digital communications with:

- Low bandwidth signals—so that it takes less "space" in the frequency spectrum, allowing more room for other signals
- Low-complexity devices—to reduce costs, power consumption, and so on.
- Low probability of errors

The trade-offs between these goals is the focus of much continuing research and development.

If we denote the carrier frequency by f_c and the bandwidth of the signal by B, the design constraints of antennas and amplifiers are such that they work best if $B \ll f_c$, so this is usually what we find in communications systems. Furthermore, f_c needs to be within the allocated frequency band(s) (as allocated by regulators such as Federal Communications Commission in the United States; see Section 17.4) for the particular communication system. The signals at these high frequencies are often called *RF* (radio-frequency) *signals* and must be handled with care with special RF circuits; this is called *RF engineering* (more on this in Chapter 3).

1.4.1 Analog Modulation

Amplitude modulation (AM) is given by

$$A_c(1 + \mu x(t)) \cos 2\pi f_c t \tag{1.88}$$

where the information signal $x(t)$ is normalized to $|x(t)| \leq 1$ and μ is the *modulation index*. To avoid signal distortion from *overmodulation*, μ is often set as $\mu < 1$. When $\mu < 1$, a simple *envelope detector* can be used to recover $x(t)$. AM is easy to detect, but has two drawbacks: (1) The unmodulated carrier portion of the signal, A_c, represents wasted power that doesn't convey the signal; and (2) Letting B_b and B_t be the baseband and transmitted bandwidths, respectively, then for AM, $B_t = 2B_b$, so there is wasted bandwidth in a sense. Schemes such as DSB and SSB attempt to reduce wasted power and/or wasted bandwidth.

Double-sideband modulation (DSB), also known as *double-sideband suppressed-carrier modulation* to contrast it with AM, is AM where the unmodulated carrier is not transmitted, so we just have

$$A_c x(t) \cos 2\pi f_c t \tag{1.89}$$

Although DSB is more power efficient than AM, simple envelope detection unfortunately cannot be used with DSB. As in AM, $B_t = 2B_b$.

Single-sideband modulation (SSB) achieves $B_t = B_b$ by removing either the upper or lower sideband of the transmitted signal. Like DSB, it suppresses the carrier to avoid wasting power. Denote the *Hilbert transform* of $x(t)$ by $\tilde{x}(t)$; then

$$\tilde{x}(t) = x(t) * \frac{1}{\pi t}$$

and we can write an SSB signal as

$$A_c\,[x(t)\cos\omega_c t \pm \tilde{x}(t)\sin\omega_c t] \tag{1.90}$$

where the plus or minus sign depends on whether we want the lower sideband or upper sideband.

Frequency modulation (FM), unlike linear modulation schemes such as AM, is a nonlinear modulation scheme in which the frequency of the carrier is modulated by the message.

1.4.2 Digital Modulation

To transmit digital information, the basic modulation schemes transmit blocks of $k = \log_2 M$ bits at a time. Thus, there are $M = 2^k$ different finite-energy waveforms used to represent the M possible combinations of the bits. Generally, we want these waveforms to be as "far apart" from each other as possible within certain energy constraints. The *symbol rate* or *signaling rate* is the rate at which new symbols are transmitted, and it is denoted R. The data rate is often denoted by R_b bits/second (also written bps), and it is also called the *baud rate*. Clearly, $R_b = kR$. The symbol period T_s is the inverse of the symbol rate, and is the time spent transmitting each symbol before it is time for the next symbol.

A bandlimited channel with bandwidth B can support only up to the *Nyquist rate* of signaling, $R_{\text{Nyquist}} = 2B$. Thus, the signaling rate is constrained by

$$R \leq R_{\text{Nyquist}} = 2B \tag{1.91}$$

Digital modulation schemes, especially when the modulation is of the phase or frequency of the carrier, are often referred to as *shift keying* [e.g., amplitude shift keying (ASK), phase shift keying (PSK), and frequency shift keying (FSK)]. Use of the word *keying* in this context may have come from the concept of Morse code keys for telegraph but is useful for distinguishing digital modulation from analog modulation (e.g., FSK refers to a frequency-modulated digital signal, whereas FM refers to the traditional analog modulation signal that goes by that name). Nevertheless, the distinction is not always retained [e.g., a popular family of digital modulation schemes often goes by the name QAM (rather than QASK)].

1.4.2.1 Pulse Shaping A digital modulator takes a simple continuous-time representation of our digital signal and outputs a continuous-time version of our signal, as will be seen in Section 1.4.2.2. How do we prepare our discrete-time digital data to enter a digital modulator? One way of converting from discrete time to continuous time is to let our data be represented by different baseband pulses for different values. For example, using a basic "rectangle" function, a 1 might be represented as

$p(t) = \pi(t/T_s)$ and a 0 by $-p(t) = -\pi(t/T_s)$ going into the digital modulator; this type of signaling, where one pulse is the exact negative of the other, is called *binary antipodal signaling*.

A problem with using a simple rectangle function in this way is that the spectral occupancy of the signals coming out of the digital modulator would be high—the Fourier transform of the rectangle function is the sinc function, which has relatively large spectral sidebands. Thus, it would be inefficient for use in bandwidth-critical systems such as wireless systems. Thus, it is important to use other *pulse-shaping functions*, $p(t)$, that can shape the spectral characteristics to use available spectrum more efficiently. However, not just any $p(t)$ can be used, because it also needs to be chosen to avoid adding *intersymbol interference* unnecessarily between nearby symbols. For example, if we (foolishly) used $p(t) = \pi(t/2T_s)$, every symbol would "spill over" into the preceding and/or subsequent symbol (in time) and interfere with them. There is a *Nyquist criterion* for $p(t)$ to avoid intersymbol interference that can be found in digital communications textbooks. Within the constraints of this criterion, the *raised cosine pulse*, illustrated in Figure 1.8, has emerged as a popular choice for $p(t)$. The frequency and time domains are shown in the subplots at the top and bottom of the figure, respectively. In the frequency domain we see the raised cosine shape from which the function gets its name. The *roll-off factor*, α, is a parameter that determines how sudden or gradual the "roll-off" of the pulse is. In one extreme, $\alpha = 0$, we have a "brick wall" shape in frequency and the familiar sinc function in time (the light solid line on the plots). At the other extreme, $\alpha = 1$, we have the most roll-off, so, the bandwidth expands to twice as much as the $\alpha = 0$ case, as can be seen in the top subplot, with the thick solid line. The case of $\alpha = 0.5$ is also plotted in dashed lines in both subplots, and it can be seen to be between the two extremes. For smaller α, the signal occupies less bandwidth, but the time sidelobes are higher, potentially resulting in more intersymbol interference and errors in practical receivers. For larger α, the signal occupies more bandwidth but has smaller time sidelobes. In practice, to achieve the raised cosine transfer function, a matching pair of *square-root raised cosine filters* are used in the transmitter and receiver, since the receiver would have a matched filter (Section 1.3.5.10). The product of the two square-root raised cosine filters (in the frequency domain) gives the raised cosine shape at the output of the matched filter in the receiver.

1.4.2.2 Digital Modulation Schemes We show examples of common digital modulation schemes. We write examples of these waveforms using lowpass equivalent representation (Section 1.3.4.2) for convenience. In all cases, $p(t)$ is the *pulse-shaping function*.

Pulse amplitude modulation (PAM) uses waveforms of the form

$$A_m p(t) \qquad \text{for } m = 1, 2, \ldots, M \tag{1.92}$$

For optimal spacing, the A_m are arranged in a line with equal spacing between consecutive points.

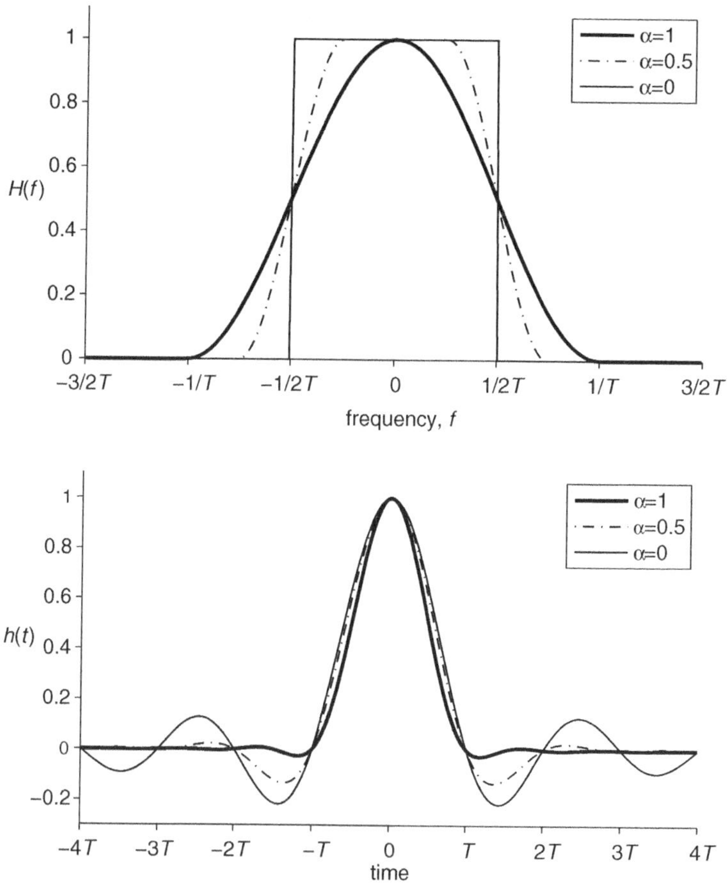

FIGURE 1.8 Family of raised cosine pulses.

To conserve bandwidth, *SSB PAM* may be used:

$$A_m[p(t) + j\tilde{p}(t)] \qquad \text{for } m = 1, 2, \ldots, M \tag{1.93}$$

Quadrature amplitude modulation (QAM), where different bits are put in the in-phase ($A_{i,m}$) and quadrature ($A_{q,m}$) streams, can be written

$$(A_{i,m} + jA_{q,m})p(t) \qquad \text{for } m = 1, 2, \ldots, M/2 \tag{1.94}$$

Normally, wireless systems would use a form of QAM [e.g., 4-QAM (often just called QAM for short), 16-QAM, 32-QAM, 64-QAM] rather than PAM. Between QAM and PAM, QAM is more efficient because PAM does not exploit the quadrature dimension to transmit information. (For a review of the in-phase and quadrature concept, and to see why different bits can be put in in-phase and quadrature, refer to Sections 1.3.4.1 and 6.1.8.1.) The values $A_{i,m}$ and $A_{q,m}$ for $m = 1, 2, \ldots, M/2$ are chosen to be as

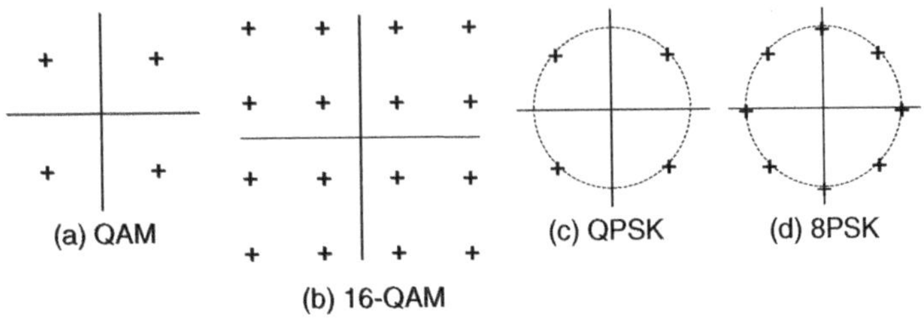

FIGURE 1.9 Signal constellations for various digital modulation schemes.

far apart from one another (in signal space) as they can be, given an average power constraint. This is because the farther apart they are, the lower the bit error rates. Examples of 4-QAM and 16-QAM are shown in Figure 1.9.

Phase shift keying (PSK) uses waveforms of different phases to represent the different bit combinations:

$$e^{j\theta_m} p(t) \qquad \text{for } m = 1, 2, \ldots, M \tag{1.95}$$

Binary PSK (BPSK) is PSK with $m = 1$, *quadrature PSK* (QPSK) is PSK with $m = 2$, and *8-PSK* is PSK with $m = 3$. QPSK is very popular in wireless systems because it is more efficient than BPSK. 8-PSK is seen in EDGE (Section 8.1.3), for example. QPSK and 8-PSK are shown in Figure 1.9.

1.4.2.3 Signal Constellations A good way to visualize the waveforms in a digital modulation scheme is through the *signal constellation* diagram. We have seen that the (lowpass equivalent of the) M possible waveforms in general (except for modulation schemes like PAM) lie in the complex plane. We can therefore plot all the points in the complex plane, and the result is known as the *signal constellation*, some examples of which are shown in Figure 1.9. Notice that the signal constellation of 4-QAM happens to be the same as that of QPSK.

When we discuss wireless access technologies, we elaborate on selected aspects of digital modulation (Section 6.2), especially those having to do with design choices typically encountered in wireless systems.

1.4.3 Synchronization

In a digital receiver, two main types of synchronization are needed at the physical layer (there may also be other types of synchronization at higher layers, e.g., frame synchronization, multimedia synchronization, etc.):

- Carrier phase synchronization
- Symbol timing synchronization and recovery

Carrier phase synchronization is about figuring out, and recovering, a carrier signal frequency and phase. *Symbol timing synchronization and recovery* is about figuring out the locations (in time) of the temporal boundaries between symbols. It is also known as *clock recovery*.

EXERCISES

1.1 The form of the Fourier series given in Section 1.3.2 is the exponential form. Show how this is equivalent to the trigonometric form

$$x(t) = a_0 + \sum_{n=1}^{\infty} a_n \cos 2\pi f_0 nt + b_n \sin 2\pi f_0 nt \tag{1.96}$$

Express c_n in terms of a_n and b_n.

1.2 Instead of the random binary waveform we saw in Section 1.3.5.5, we have a random digital waveform. So it takes not just two values, 1 and -1, but a range of values over a distribution: say, a Gaussian distribution with mean 0 and variance σ^2. Find the autocorrelation function of the random digital waveform. How does it compare with the autocorrelation function of the random binary waveform given by (1.81)?

1.3 Suppose we have a signal $x(t)$ that is multiplied by a sinusoid, resulting in the signal $y(t) = x(t) \cos 2\pi ft$. Assume that $x(t)$ is independent of the sinusoid but could otherwise be a (deterministic or random) signal with autocorrelation function $R_{xx}(\tau)$. Show that the autocorrelation of $y(t)$ is given by

$$R_{yy}(\tau) = R_{xx}(\tau)\left(\frac{1}{2}\cos 2\pi ft\right) \tag{1.97}$$

1.4 Continuing from Exercise 1.3, what is the effect on the power spectral density of multiplication by a sinusoid? In other words, express the power spectral density of $y(t)$ in terms of the power spectral density of $x(t)$. This is a fundamental and useful result, since it means that we can up-convert and down-convert signals to and from carrier frequencies, and the autocorrelation function and power spectral density behave in this predictable way.

1.5 Show that a matched filter followed by sampling at $t = T$ produces the same output as a correlation receiver.

REFERENCES

1. A. B. Carlson. *Communication Systems*, 3rd ed. McGraw-Hill, New York, 1986.
2. L. Couch. *Digital and Analog Communication Systems*, 7th ed. Prentice Hall, Upper Saddle River, NJ, 2007.

3. J. W. Nilsson. *Electric Circuits*, 3rd ed. Addison-Wesley, Reading MA, 1990.
4. A. Papoulis. *Probability, Random Variables, and Stochastic Processes*. McGraw-Hill, New York, 1991.
5. B. Sklar. *Digital Communications: Fundamentals and Applications*, 2nd ed. Prentice Hall, Upper Saddle River, NJ, 2001.

II

RADIO FREQUENCY, ANTENNAS, AND PROPAGATION

2

INTRODUCTION TO RADIO FREQUENCY, ANTENNAS, AND PROPAGATION

In this chapter we review selected topics in electromagnetics that provide foundational support for our coverage of radio frequency (Chapter 3), antennas (Chapter 4), and propagation (Chapter 5). We begin in Section 2.1 with a review of some mathematical tools for computing scalar and vector quantities that are typically used in basic electromagnetics. We then review electrostatics and magnetostatics in Section 2.2. Time-varying situations, wave propagation, and transmission lines are examined in Section 2.3. A brief comparison of different notions of impedance is presented in Section 2.4, followed by an introduction to test and measurement equipment in Section 2.5.

2.1 MATHEMATICAL PRELIMINARIES

Here we review briefly some mathematical tools for working with scalar and vector functions in three dimensions.

2.1.1 Multidimensional/Multivariable Analysis

In Section 1.2.4 the signals concerned were measured at one place in a circuit (e.g., the voltage between two fixed points), where spatial dimensions were not important. Furthermore, the signals were all scalar functions. Now we extend our signal concepts, as well as concepts of sinusoids and phasors, into one or more spatial dimensions. Furthermore, the signals may be vector functions, not just scalar functions. For example,

Fundamentals of Wireless Communication Engineering Technologies, First Edition. K. Daniel Wong.

we might have a scalar function, ρ, of coordinates x, y, and z (and time t) $\rho(x, y, z, t)$, that we may also write as $\rho(\mathbf{A}, t)$, where A is a vector representing a spatial coordinate [e.g., (x, y, z)]. Such a function is also used to represent a *scalar field*. Or we might have a vector function, $\mathbf{H}$, that we may write as $\mathbf{H}(x, y, z, t)$ or $\mathbf{H}(\mathbf{A}, t)$. Such a function is used to represent a *vector field*. It has amplitude and direction at every point in space and time within the domain of the function.

2.1.1.1 Basic Vector Operations Let **A** and **B** be vectors, $A = |\mathbf{A}|$, $B = |\mathbf{B}|$, θ_{AB} the angle between **A** and **B**, and $\mathbf{u_n}$ the unit vector normal (perpendicular) to **A** and **B**.

- Dot product:

$$\mathbf{A} \cdot \mathbf{B} = AB \cos \theta_{AB} \tag{2.1}$$

- Cross product:

$$\mathbf{A} \times \mathbf{B} = \mathbf{u_n} \, |AB \sin \theta_{AB}| \tag{2.2}$$

2.1.1.2 Coordinate Systems The cylindrical coordinate system, (r, ϕ, z), is shown in Figure 2.1 and the spherical coordinate system, (R, θ, ϕ), in Figure 2.2. For conversion between coordinate systems, see Exercise 2.1. In coordinate systems such as the cylindrical and spherical, we often want to convert a differential change in the

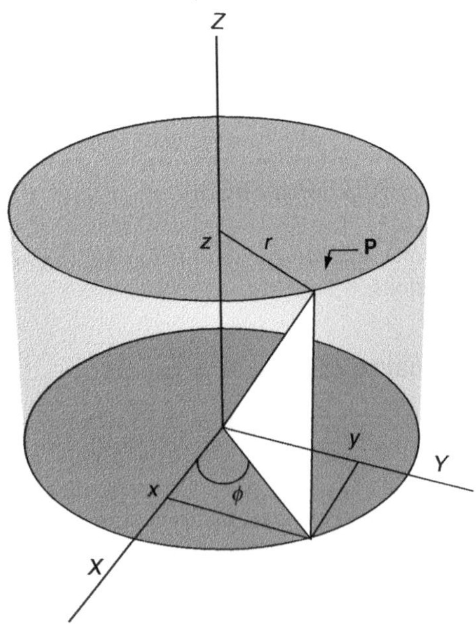

FIGURE 2.1 Cylindrical coordinates.

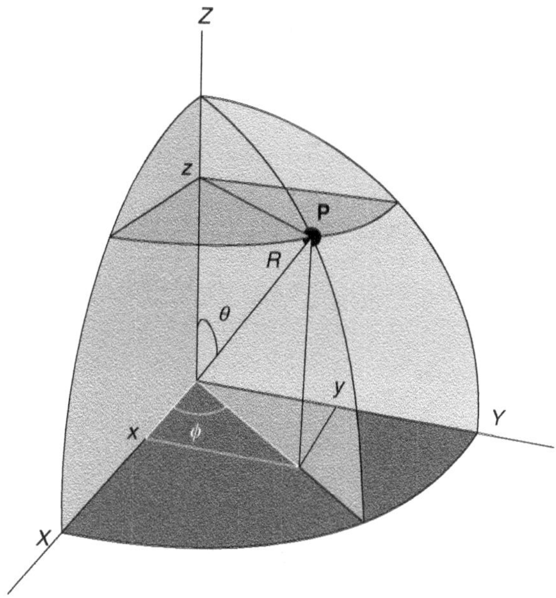

FIGURE 2.2 Spherical coordinates.

coordinates to a differential change in length. Let us denote the unit vectors by **u** with the appropriate subscripts. Metric coefficients (for length conversions) are:

- Cartesian coordinates:

$$dl = \mathbf{u_x}\, dx + \mathbf{u_y}\, dy + \mathbf{u_z}\, dz \tag{2.3}$$

- Cylindrical coordinates:

$$dl = \mathbf{u_r}\, dr + \mathbf{u}_\phi r\, d\phi + \mathbf{u_z}\, dz \tag{2.4}$$

- Spherical coordinates:

$$dl = \mathbf{u_R}\, dR + \mathbf{u}_\theta R\, d\theta + \mathbf{u}_\phi R \sin\theta\, d\phi \tag{2.5}$$

2.1.1.3 Gradient, Divergence, and Curl

- Gradient [for a scalar function of space coordinates, e.g., $V(u_1, u_2, u_3)$]:

$$\nabla \equiv \left(\mathbf{u_1} \frac{\partial}{h_1\, \partial u_1} + \mathbf{u_2} \frac{\partial}{h_2\, \partial u_2} + \mathbf{u_3} \frac{\partial}{h_3\, \partial u_3} \right) \tag{2.6}$$

In Cartesian coordinates it becomes

$$\nabla \equiv \left(\mathbf{u_x} \frac{\partial}{\partial x} + \mathbf{u_y} \frac{\partial}{\partial y} + \mathbf{u_z} \frac{\partial}{\partial z} \right) \tag{2.7}$$

and in cylindrical or spherical coordinates, the appropriate metric coefficients h_1, h_2, and h_3 must be applied.

- Divergence [for a vector field, e.g., $\mathbf{A}(u_1, u_2, u_3)$]:

$$\text{div}\mathbf{A} = \lim_{\Delta v \to 0} \frac{\oint_S \mathbf{A} \cdot d\mathbf{s}}{\Delta v} \tag{2.8}$$

In Cartesian coordinates,

$$\text{div}\mathbf{A} = \frac{\partial A_x}{\partial x} + \frac{\partial A_y}{\partial y} + \frac{\partial A_z}{\partial z} \tag{2.9}$$

Thus, symbolically, div can be written as div $\mathbf{A} = \nabla \cdot \mathbf{A}$. But this is just symbolic, since it makes sense only in Cartesian coordinates. The real definition is given by (2.8).

- Curl (for a vector field):

$$\text{curl}\mathbf{A} = \lim_{\Delta s \to 0} \frac{1}{\Delta s} \left[\mathbf{u_n} \oint_C \mathbf{A} \cdot dl \right] \tag{2.10}$$

Just as div can be written symbolically as $\nabla \cdot \mathbf{A}$, curl can also be written symbolically as curl $\mathbf{A} = \nabla \times \mathbf{A}$, which makes sense in Cartesian coordinates.

2.1.1.4 *Sinusoids, Waves, and Phasors* For a sinusoid that is a function of both position and time, we may expand (1.16) to obtain

$$\xi(x, t) = A\cos(kx - \omega t + \phi) = A\cos(kx - 2\pi f t + \phi) \tag{2.11}$$

where k is the *spatial frequency*, which is to the spatial dimension what the temporal frequency ω is to the temporal dimension. With the introduction of the spatial dimension, this sinusoid could be thought of as a wave, although the concept of waves includes more than just a simple sinusoid like this one. It also encompasses the superposition of multiple sinusoids.

Previously, we introduced $T = 1/f = 2\pi/\omega$ as the period of the sinusoid. We see that this relationship remains if we fix k and x so that kx gets absorbed into the phase with ϕ. Thus, at any particular fixed spatial location (fixed x), the period remains as $T = 1/f = 2\pi/\omega$. However, if we now fix t and ω instead, then ωt gets absorbed into the phase with ϕ, and we have, at any particular fixed moment in time, the "spatial period," much more commonly called the *wavelength*, given by

$$\lambda = \frac{2\pi}{k} \tag{2.12}$$

Analogous to the period in time, the wavelength is the smallest (spatial) distance such that

$$\xi(x) = \xi(x + \lambda) \qquad \text{for } -\infty < x < \infty \tag{2.13}$$

For the case that $\xi(x, t)$ represents a *traveling wave* (also known as a *propagating wave*), we can relate λ and f through the velocity of the wave (also known as the *phase velocity*), v:

$$\lambda = \frac{v}{f} \tag{2.14}$$

When we want the sinusoidal function to represent phenomena in three spatial dimensions (e.g., a propagating electromagnetic wave), we can replace the scalar function in (2.11) with a vector function. For convenience, we often align the axes so that the direction of propagation is along one of the axes (e.g., the z-axis), in which case we could write

$$\xi(z, t) = \mathbf{A}\cos(kz - \omega t + \phi) = \mathbf{A}\cos(kz - 2\pi f t + \phi) \tag{2.15}$$

where $\mathbf{A} = A\mathbf{u_x}$, for example.

In this way, the concept of phasors introduced in Section 1.2.6 could be extended so that:

- Phasors can be both a sinusoidal function of time and a function of space.
- Besides having amplitude and phase, they also have a direction. Thus, we go from the scalar phasors of Section 1.2.6 to vector phasors.

For example, we could represent an electric field as

$$\mathbf{E}(x, y, z, t) = \Re\left[\mathbf{E}(x, y, z)e^{j2\pi f t}\right] \tag{2.16}$$

where $\mathbf{E}(x, y, z, t)$ is a vector phasor.

2.2 ELECTROSTATICS, CURRENT, AND MAGNETOSTATICS

In this section we briefly review electrostatics from Sections 2.2.1 to 2.2.4, electric current in Section 2.2.5, and magnetostatics from Sections 2.2.6 to 2.2.8. Section 2.2.9 provides a summary of the various symbols introduced in the section.

2.2.1 Electrostatics in Free Space

Differential Form

$$\nabla \cdot \mathbf{E} = \frac{\rho}{\epsilon_0} \tag{2.17}$$

$$\nabla \times \mathbf{E} = 0 \tag{2.18}$$

where ρ is the volume charge density of free charges in C/m^3 and ϵ_0 is the permittivity of free space (a.k.a. vacuum permittivity; $\epsilon_0 \approx 1/36\pi \times 10^{-9}$ in F/m).

Integral Form. Gauss's law:

$$\oint_S \mathbf{E} \cdot d\mathbf{s} = \frac{Q}{\epsilon_0} \tag{2.19}$$

$$\oint_C \mathbf{E} \cdot dl = 0 \tag{2.20}$$

Coulomb's Law. Force between q_1 and q_2 is given by

$$\mathbf{F} = \mathbf{u_R} \frac{q_1 q_2}{4\pi\epsilon_0 R^2} \tag{2.21}$$

Alternatively, force on q is given by

$$\mathbf{F} = q\mathbf{E} \tag{2.22}$$

where the unit is newtons.

For a conductor under static conditions, (2.19) and (2.20) can be used to show:

- That the **E** field on the conductor surface is normal to the surface everywhere, and we can write $E_\perp = \rho_s/\epsilon_0$, where $E_\perp$ is the normal component and ρ_s is the surface charge density.
- That the tangential component of the **E** field on the conductor surface is zero.

2.2.2 Voltage

Since E is curl-free, it can be written as the gradient of a scalar field. We define the scalar field to be *electric potential* V and

$$\mathbf{E} = -\nabla V \tag{2.23}$$

Then the units of E are volts per meter. We don't have the space to discuss this further, just to note that electric potential has physical significance, related to the work that needs to be done to move a charge from point to point.

Poisson's equation in free space is

$$\nabla^2 V = -\frac{\rho}{\epsilon_0} \tag{2.24}$$

2.2.2.1 Worked Example: Electric Potential at Distance r from the Spherical Conductor As discussed earlier, the **E** field must be normal and hence pointing radially outward from the spherical conductor. Since the surface area of a sphere is $4\pi r^2$, then $\rho_s = Q/4\pi r^2$ and using (2.19) gives us

$$|\mathbf{E}| = E_\perp = \frac{Q}{4\pi\epsilon_0 r^2} \tag{2.25}$$

Then, taking the point at infinity as the zero-voltage reference point, we have

$$V = -\int_\infty^r E_\perp\, dr = \frac{Q}{4\pi\epsilon_0 r} \tag{2.26}$$

2.2.2.2 *Worked Example: Two Connected Spherical Conductors* Consider two spherical conductors that are electrically connected by a perfectly conducting wire. Let the radii of the spheres be r_1 and r_2, respectively. Assume that the spheres are far enough apart that the charge distribution on each is not influenced by the field caused by the charge distribution on the other; thus, the charge distribution on each is uniform. Let Q coulombs of charge be deposited in the spheres. Find (a) the charges on each sphere; (b) the charge density at the surface of each sphere; (c) the electric field intensity at the surface of each sphere.

Let the charge on the spheres be Q_1 and Q_2, respectively, so that $Q = Q_1 + Q_2$. Since the two spheres are connected by the wire, they are at the same potential, and the potential is given by (2.26), so we have

$$\frac{Q_1}{4\pi\epsilon_0 r_1} = \frac{Q_2}{4\pi\epsilon_0 r_2} \tag{2.27}$$

So

$$Q_1 = \frac{r_1}{r_1 + r_2} Q \quad \text{and} \quad Q_2 = \frac{r_2}{r_1 + r_2} Q \tag{2.28}$$

Then the charge densities are

$$\rho_{s,1} = \frac{Q_1}{4\pi r_1^2} = \frac{Q}{4\pi r_1(r_1 + r_2)} \quad \text{and} \quad \rho_{s,2} = \frac{Q_2}{4\pi r_2^2} = \frac{Q}{4\pi r_2(r_1 + r_2)} \tag{2.29}$$

so

$$E_{\perp,1} = \frac{Q}{4\pi\epsilon_0 r_1(r_1 + r_2)} \quad \text{and} \quad E_{\perp,2} = \frac{Q}{4\pi\epsilon_0 r_2(r_1 + r_2)} \tag{2.30}$$

Thus, if sphere 1 is bigger than sphere 2, it will have proportionately more charge but a smaller surface charge density and electric field intensity at its surface.

2.2.3 Electrostatics in the Case of Dielectrics/Insulators

Dielectrics are also known as *insulators*. When there are dielectrics, there will be polarization charge densities, resulting in a *polarization vector,* **P**. For convenience, we introduce **D**, given by

$$\mathbf{D} = \epsilon_0 \mathbf{E} + \mathbf{P} \tag{2.31}$$

Thus, in considering **D**, we can ignore polarization as it is only affected by free charges (whereas **E** is reduced by the polarization in the dielectric):

$$\nabla \cdot \mathbf{D} = \rho \tag{2.32}$$

If material is linear and isotropic, $\mathbf{P} = \epsilon_0 \chi_e \mathbf{E}$, where χ_e is *electric susceptibility*. Then

$$\mathbf{D} = \epsilon_0(1 + \chi_e)\mathbf{E} = \epsilon_0 \epsilon_r \mathbf{E} = \epsilon \mathbf{E} \tag{2.33}$$

ϵ_r is known as *dielectric constant* or *relative permittivity, and* ϵ is *absolute permittivity* or *permittivity*.

2.2.3.1 *Dielectric Breakdown* Equation (2.31) holds only when the electric field intensity is below a critical amount, the *dielectric strength* of the material. If the electric field exceeds the dielectric strength, *dielectric breakdown* occurs; then it becomes conducting. When air breaks down at 3×10^6 V/m, sparking or corona discharge occurs.

2.2.4 Electrostatics Summary

In summary, for electrostatics we have

$$\nabla \cdot \mathbf{D} = \rho \qquad \text{C/m}^3 \tag{2.34}$$

$$\nabla \times \mathbf{E} = 0 \tag{2.35}$$

Further, if material is linear and isotropic,

$$\mathbf{D} = \epsilon \mathbf{E} \tag{2.36}$$

2.2.5 Currents

There are conduction currents, electrolytic currents, and convection currents. Ohm's law governs only conduction currents. In circuits, it is $V = RI$. Ohm's law in point form is

$$\mathbf{J} = \sigma \mathbf{E} \qquad \text{A/m}^2 \tag{2.37}$$

where σ is *conductivity* in A/V·m or S/m. The reciprocal of σ is resistivity.

The principle of *conservation of charge* leads to the *equation of continuity*:

$$\nabla \cdot \mathbf{J} = -\frac{\partial \rho}{\partial t} \qquad \text{A/m}^3 \tag{2.38}$$

2.2.6 Magnetostatics Introduction

In the case of a moving charge, not only is there electric force, as from (2.22), but there is magnetic force. Hence, we have

$$\mathbf{F} = q(\mathbf{E} + \mathbf{u} \times \mathbf{B}) \qquad \text{N} \tag{2.39}$$

2.2.7 Magnetostatics in Free Space

B is magnetic flux density in Wb/m^2 or teslas (a weber is a volt-second):

$$\nabla \cdot \mathbf{B} = 0 \tag{2.40}$$

$$\nabla \times \mathbf{B} = \mu_0 \mathbf{J} \tag{2.41}$$

where μ_0 is the *permeability* of free space ($\mu_0 = 4\pi \times 10^{-7}$ H/m).

2.2.8 Magnetostatics in the Case of Magnetic Materials

Just as in dielectrics you had a polarization vector, so in magnetic materials you have a magnetization vector. Let the magnetization vector be $\mathbf{M}$. Then

$$\mathbf{B} = \mu_0 \mathbf{H} + \mathbf{M} \qquad \text{A/m} \tag{2.42}$$

Thus, we can just deal with the effects of free current, $\mathbf{J}$, as in

$$\nabla \times \mathbf{H} = \mathbf{J} \tag{2.43}$$

2.2.9 Symbols

We recap some of the symbols introduced in prior sections:

C is capacitance in F/m

D is electric displacement or electric flux density in C/m^2

E is electric field intensity in V/m

J is current density in A/m^2

ϵ_0 is permittivity of free space in F/m

ϵ_r is *dielectric constant* or *relative permittivity* (dimensionless); permittivity relative to free space

ϵ is *absolute permittivity* or *permittivity* in F/m; how much the medium "permits" some charge q to create an electric field

μ_0 is the *permeability* of free space in H/m

μ is *absolute permeability* in H/m

ρ is volume charge density of free charges in C/m^3

σ is conductivity in A/V·m or S/m

2.3 TIME-VARYING SITUATIONS, ELECTROMAGNETIC WAVES, AND TRANSMISSION LINES

We begin in this section with Maxwell's equations (Section 2.3.1) and proceed on to electromagnetic (EM) waves (Section 2.3.2). Then we discuss transmission lines (Section 2.3.3), standing-wave ratios (Section 2.3.4) and S-parameters (Section 2.3.5).

2.3.1 Maxwell's Equations

In differential form, Maxwell's equations are

$$\nabla \times \mathbf{E} = -\frac{\partial \mathbf{B}}{\partial t} \tag{2.44}$$

$$\nabla \times \mathbf{H} = \mathbf{J} + \frac{\partial \mathbf{D}}{\partial t} \tag{2.45}$$

$$\nabla \cdot \mathbf{D} = \rho \tag{2.46}$$

$$\nabla \cdot \mathbf{B} = 0 \tag{2.47}$$

In integral form, Maxwell's equations are

$$\oint_C \mathbf{E} \cdot dl = -\frac{d\Phi}{dt} \tag{2.48}$$

$$\oint_C \mathbf{H} \cdot dl = I + \oint_S \frac{\partial \mathbf{D}}{\partial t} \cdot ds \tag{2.49}$$

$$\oint_S \mathbf{D} \cdot ds = Q \tag{2.50}$$

$$\oint_S \mathbf{D} \cdot ds = 0 \tag{2.51}$$

In linear, isotropic, homogeneous media, Maxwell's equations can be written as (vector) phasors:

$$\nabla \times \mathbf{E} = -j2\pi f\mu\mathbf{H} \tag{2.52}$$

$$\nabla \times \mathbf{H} = \mathbf{J} + j2\pi f\epsilon\mathbf{E} \tag{2.53}$$

$$\nabla \cdot \mathbf{E} = \rho/\epsilon \tag{2.54}$$

$$\nabla \cdot \mathbf{H} = 0 \tag{2.55}$$

where we have written the four equations just in terms of E and H alone for convenience, because in linear and isotropic media, $\mathbf{D} = \epsilon\mathbf{E}$ and $\mathbf{B} = \mu\mathbf{H}$.

2.3.2 Electromagnetic Waves

A *source-free region* is one where $\rho = 0$ and $\mathbf{J} = 0$. Assume a source-free region where the medium is linear, isotropic, homogeneous, and nonconducting. Then using (2.44) and (2.45), we have

$$\nabla \times \nabla \times \mathbf{E} = -\mu \frac{\partial}{\partial t}(\nabla \times \mathbf{H}) = -\mu\epsilon \frac{\partial^2 \mathbf{E}}{\partial t^2} \tag{2.56}$$

But $\nabla \times \nabla \times \mathbf{E} = \nabla(\nabla \cdot \mathbf{E}) - \nabla^2 \mathbf{E} = -\nabla^2 \mathbf{E}$, where $\rho = 0$, so we have

$$\nabla^2 \mathbf{E} - \mu\epsilon \frac{\partial^2 \mathbf{E}}{\partial t^2} = 0 \tag{2.57}$$

Similarly, we can derive

$$\nabla^2 \mathbf{H} - \mu\epsilon \frac{\partial^2 \mathbf{H}}{\partial t^2} = 0 \tag{2.58}$$

These are wave equations, and the speed of the wave is $v = 1/\sqrt{\mu\epsilon}$. In particular, in free space we have

$$\nabla^2 \mathbf{E} - \frac{1}{c^2} \frac{\partial^2 \mathbf{E}}{\partial t^2} = 0 \tag{2.59}$$

where

$$c = \frac{1}{\sqrt{\mu_0 \epsilon_0}} \approx 3 \times 10^8 \text{ m/s} \tag{2.60}$$

NB: Because (2.57) and (2.58) are linear, we can apply the superposition principle and add waves to get the resultant (we do this everywhere, e.g., in adding the transmitted and reflected wave in transmission lines, in adding the contribution of different paths in multipath propagation environments, in analyzing the behavior of an antenna array).

The *intrinsic impedance* of a medium is $\eta = \sqrt{\mu/\epsilon}$. For free space, we have

$$\eta_0 = \sqrt{\frac{\mu_0}{\epsilon_0}} \approx 120\pi \approx 377\ \Omega \tag{2.61}$$

Working in phasor notation, (2.57) and (2.58) become

$$\nabla^2 \mathbf{E} + k^2 \mathbf{E} = 0 \tag{2.62}$$

and

$$\nabla^2 \mathbf{H} + k^2 \mathbf{H} = 0 \tag{2.63}$$

where

$$k = 2\pi f\sqrt{\mu\epsilon} = \frac{2\pi f}{v} = \frac{2\pi}{\lambda} \tag{2.64}$$

since $\lambda = v/f$.

2.3.2.1 *Flow of Electromagnetic Power and the Poynting Vector*

We define the Poynting vector

$$\mathcal{P} = \mathbf{E} \times \mathbf{H} \qquad \text{W/m}^2 \tag{2.65}$$

It is a power flux density vector associated with an electromagnetic field. $\mathcal{P}$ points in the direction of the flow of electromagnetic power and its amplitude is power flux density.

For computing the average power flux density in a propagating wave:

$$\mathcal{P}_{\text{av}} = \frac{1}{2}\Re(\mathbf{E} \times \mathbf{H}^*) \qquad \text{W/m}^2 \tag{2.66}$$

which is analogous to the following from circuit theory:

$$P_{\text{av}} = \frac{1}{2}\Re(VI^*) \qquad \text{W} \tag{2.67}$$

Now consider the case of time-harmonic waves, and in particular, a uniform plane wave propagating in a lossy medium in the $+z$-direction, where (in phasor notation)

$$\mathbf{E}(z) = \mathbf{u}_x E_0 e^{-(\alpha + j\beta)z} \tag{2.68}$$

Then if the intrinsic impedance of the medium is $\eta = |\eta|e^{j\theta_\eta}$, we have

$$\mathbf{H}(z) = \mathbf{u}_y \frac{E_0}{|\eta|} e^{-\alpha z} e^{-j(\beta z + \theta_\eta)} \tag{2.69}$$

Thus, in the lossless case, $\alpha = 0$ and we have

$$\mathcal{P} = \mathbf{u}_z \frac{E_0^2}{|\eta|} \tag{2.70}$$

2.3.3 Transmission-Line Basics

For efficient transmission of electromagnetic waves from one point to another point, the electromagnetic waves must be directed or guided. Transmission lines are one way to do that. They are especially useful when signals are at RF and we cannot just use basic circuits (see Section 3.1.2 for further discussion). In this section we introduce transmission lines and work through just enough of the equations to provide a foundation to introduce the very important concept of standing-wave ratios in Section 2.3.4.

Three of the most common transmission-line structures are:

- *Parallel plate*: two parallel conducting plates that are separated by a dielectric slab of uniform thickness (e.g., microstrips in printed circuit board technology).
- *Two-wire transmission line*: a pair of parallel conducting wires that are separated by a uniform distance (e.g., flat lead-in lines connecting TVs and antennas).
- *Coaxial transmission line*: inner conducting wire and coaxial outer conducting sheath that are separated by a dielectric medium.

Microstrips are commonly found in microwave integrated circuits. They consist of a conducting strip over a ground plane, with a dielectric material between the strip and the plane. They are small, cheap, and easily produced, but suffer from higher losses and can handle less power than can other transmission lines, such as coaxial cables. Microstrip transmission lines are sometimes also known as striplines, but sometimes they are considered different from striplines. When considered different, the term *striplines* is used to refer specifically to a variant with two ground planes, one on each side of the conducting strip. The ground planes then sandwich the dielectric material, and the conducting strip is embedded in the dielectric material [2].

Shown in Figure 2.3, microstrip transmission lines are closely related to microstrip patch antennas (Section 4.2.7.1). With one set of parameters, the microwave energy is better contained within the structure, and it is used as a transmission line, whereas with another set of parameters, the structure radiates and it is used as an antenna. W, ϵ_r, and h (width of the microstrip, dielectric constant of the dielectric, and height of the dielectric) are important parameters, whereas other parameters, such as thickness t and conductivity σ of the strip, are not as important.

Next, we derive a very useful and convenient model of transmission lines in Section 2.3.3.1. The model is quite accurate for coaxial and two-wire transmission lines. It also works well for parallel-plate transmission lines where the two plates are of equal width with negligible fringing effects. However, it should be used with caution for modeling microstrip transmission lines, since the metal strip might not be very wide. The model provides a reasonable approximation when $h \ll W$ and for

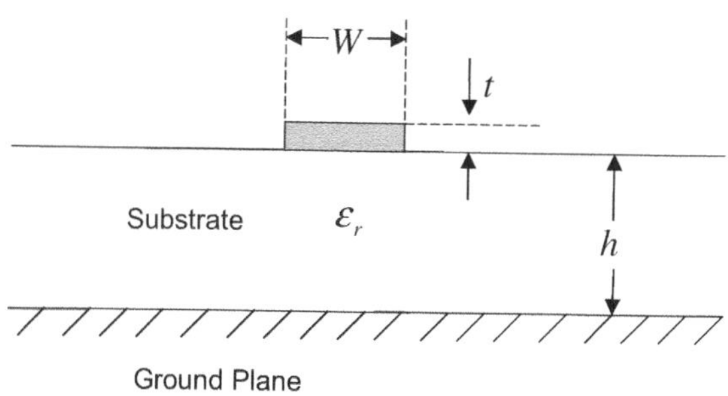

FIGURE 2.3 Model of a microstrip transmission line.

lower microwave frequencies. As for higher frequencies such as millimeter-wave, more complicated full-wave analysis might be recommended [1].

2.3.3.1 *Modeling the Behavior of a Transmission Line*

The transmission line can be modeled as broken up into short segments each of length Δx, as shown in Figure 2.4. Consider the segment between x and $x + \Delta x$. Let the voltage across the first side (at x) be $v(x, t)$ and the current in be $i(x, t)$, and the voltage across the other side (at $x + \Delta x$) be $v(x + \Delta x, t)$ and the current out be $i(x + \Delta x, t)$. Let L, R, C, and G be the inductance, resistance, capacitance, and conductance per unit length. Then the inductance, resistance, capacitance, and conductance in our small segment of circuit can be represented as $L\,\Delta x$ and $R\,\Delta x$ in series, and $C\,\Delta x$ and $G\,\Delta x$ in parallel.

Applying Kirchhoff's current and voltage laws, then dividing by Δx and taking the limit as $\Delta x \to 0$, we have a couple of partial differential equations in x and t. The steady-state sinusoidally time-varying solutions can be written in phasor notation,

$$v(x, t) = \mathrm{Re}\left[V(x)e^{j2\pi ft}\right] \tag{2.71}$$

$$i(x, t) = \mathrm{Re}\left[I(x)e^{j2\pi ft}\right] \tag{2.72}$$

and it turns out that the solution is a "wave equation,"

$$\frac{d^2V(x)}{dx^2} - \gamma^2 V(x) = 0 \tag{2.73}$$

$$\frac{d^2I(x)}{dx^2} - \gamma^2 I(x) = 0 \tag{2.74}$$

and the wave propagation constant γ is

$$\gamma = \sqrt{(R + j2\pi fL)(G + j2\pi fC)} = \alpha + j\beta \tag{2.75}$$

where α is an attenuation constant in nepers per unit length and β is a phase constant in radians per unit length.

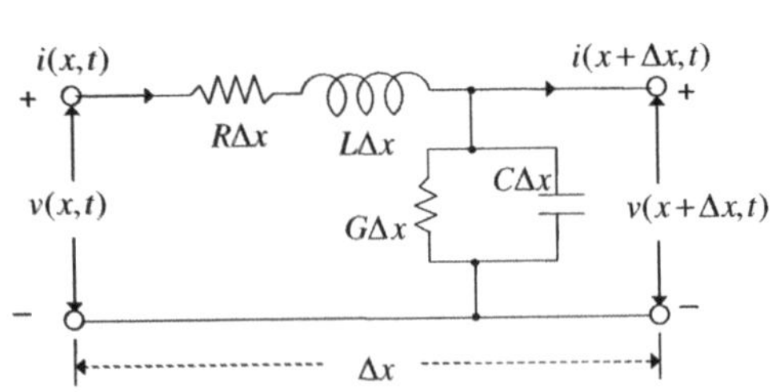

FIGURE 2.4 Model of a transmission line.

Usually, R and G are small, and for the *lossless* case, R and G are zero. Equations (2.73) and (2.74) can be solved by the following functions:

$$V(x) = V^+(x) + V^-(x) \tag{2.76}$$

$$= V_0^+ e^{-\gamma x} + V_0^- e^{\gamma x} \tag{2.77}$$

$$I(x) = I^+(x) + I^-(x) \tag{2.78}$$

$$= I_0^+ e^{-\gamma x} + I_0^- e^{\gamma x} \tag{2.79}$$

where we see the V and I for a forward (V^+,I^+) wave traveling in the $+x$ direction, and a backward wave (V^-,I^-) traveling in the $-x$ direction. The ratio V_0^+/I_0^+ is very important, and we call it the *characteristic impedance* of the transmission line, Z_0. It can easily be shown that

$$Z_0 = \frac{V_0^+}{I_0^+} = \frac{R + j2\pi L}{\gamma} = \frac{\gamma}{G + j2\pi C} = \sqrt{\frac{R + j2\pi L}{G + j2\pi C}} \tag{2.80}$$

2.3.4 Standing-Wave Ratios

When transmitting, say, to an antenna, along a cable, you have an *incident wave*, also known as a *forward wave*, and a *reflected wave* [as we saw in (2.77) and (2.79)]. A *standing wave* results from the superposition of the incident and reflected waves. The *standing-wave ratio* (SWR) is the ratio of the peak to the trough of the standing wave. It could be a voltage ratio or a current ratio (the numerical ratio should be the same whether we consider voltage or current). Since it could be a voltage ratio, the SWR is also often called the *voltage standing-wave ratio* (VSWR). Calling it VSWR also helps remove ambiguity, as sometimes the *power standing-wave ratio* (PSWR) is also seen, where PSWR is the square of the VSWR. Figure 2.7 shows an example of standing waves in a transmission line.

Although we have seen that the voltage and current, as functions of position, are given by (2.77) and (2.79), we need a bit more theory to understand the ratios of the incident and reflected waves so that we can get a grip on the SWR. We begin by examining the special case where impedances are matched (Section 2.3.4.1) so there is no reflected wave. This will be followed in Section 2.3.4.2 by examining the more general case where the reflected wave exists.

2.3.4.1 Impedance Matching and the Transmission Line

A transmission line will have no reflections only if it is infinitely long, *or* if it is connected to a matched load (Figure 2.5). This brings us to the subject of impedance matching in the context of transmission lines. In matching source and loads with transmission lines, it is important to distinguish between:

- The connection from a source to the transmission line on one side.
- The connection from the transmission line to a load on the other side.

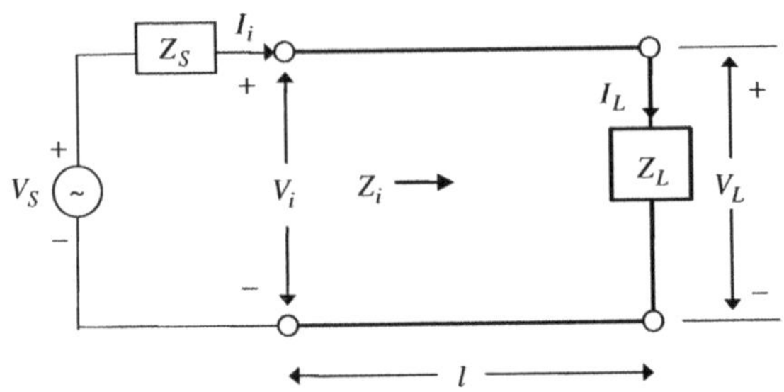

FIGURE 2.5 Characterizing a transmission line.

In the former case (i.e., connecting a source to a transmission line), for maximum power transfer the input impedance looking into the transmission line, Z_i [see (2.81), which we will get to shortly], should equal the complex conjugate of the output impedance of the source (i.e., $Z_i = Z_s^*$, where Z_s is the source impedance). This is what we might expect from basic circuit theory, and coincides with the concept of impedance matching from basic circuit theory. It is also indicative of one way in which we can treat transmission lines as circuit elements in basic circuits. In the latter case, the load connected to a transmission line should have input impedance **equal to the characteristic impedance** of the transmission line for matched load and best efficiency (i.e., $Z_0 = Z_L$). This is *different* from what we might expect from basic circuit theory, so we have to be careful. It is not the complex conjugate of the characteristic impedance but the characteristic impedance itself. This is because this matching is based on a different principle from normal circuit conjugate impedance matching. This matching is based on eliminating the reflected wave, which can result in serious power losses. In fact, for transmission lines, $Z_0 = Z_L$ is more important (to reduce power loss from reflected waves) than conjugate matching on the source side. Having said this, we now proceed to discuss Z_i.

For a transmission line of length l, with characteristic γ and Z_0, if it is connected to a load with input impedance Z_L, the input impedance Z_i of the transmission line and load combination is given by [5]

$$Z_i = Z_0 \frac{Z_L + Z_0 \tanh \gamma l}{Z_0 + Z_L \tanh \gamma l} \quad \Omega \tag{2.81}$$

In the lossless case, $\gamma = j\beta$ and $\tanh j\beta l = j \tan \beta l$, so [5]

$$Z_i = Z_0 \frac{Z_L + Z_0 j \tan \beta l}{Z_0 + Z_L j \tan \beta l} \quad \Omega \tag{2.82}$$

Notice that when we have a load matched to the transmission line (i.e., $Z_L = Z_0$), the input impedance from (2.81) is $Z_i = Z_0$. Thus, the combination of the transmission line and the load in this special case looks exactly the same (same input impedance, same voltage and current distribution over the line), as if the transmission line is infinitely long, with no termination; and there is no reflected wave.

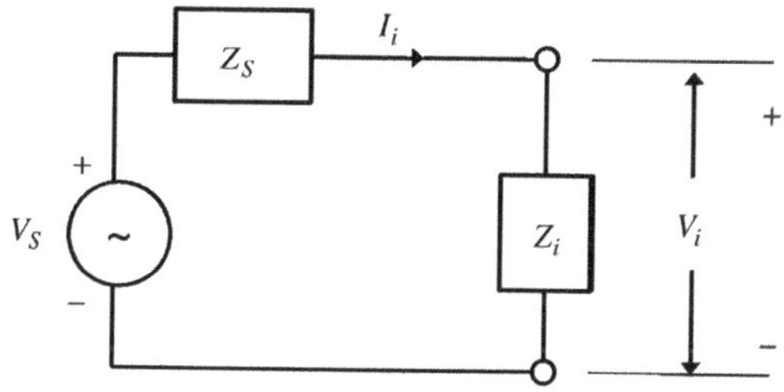

FIGURE 2.6 Replacing the transmission line with its input impedance.

From the perspective of the source, the transmission line and load could be replaced with a load of impedance Z_i [given by (2.81)], as shown in Figure 2.6. This equivalent circuit gives the same input current I_i and input voltage V_i as if we had the transmission line and load there.

2.3.4.2 Characterizing Transmission Lines with Reflected Waves

Define *voltage reflection coefficient* Γ of the load impedance Z_L (it changes depending on the load attached) as the ratio of the complex amplitudes of the reflected and incident voltage waves at the load. It can be shown that

$$\Gamma = |\Gamma| e^{j\theta_\Gamma} = \frac{Z_L - Z_0}{Z_L + Z_0} \tag{2.83}$$

Notice that $\Gamma = 0$ when $Z_L = Z_0$. In general, Γ is a complex number with magnitude 1 or less. Meanwhile, the current reflection coefficient is the negative of the voltage reflection coefficient.

Then we can define the SWR (or VSWR) as

$$S = \frac{|V_{\max}|}{|V_{\min}|} = \frac{1 + |\Gamma|}{1 - |\Gamma|} = \frac{|I_{\max}|}{|I_{\min}|} \tag{2.84}$$

Meanwhile, the inverse relationship is

$$|\Gamma| = \frac{S - 1}{S + 1} \tag{2.85}$$

To help with the visualization of standing waves, we plot the standing waves (in a specific case of resistive termination and lossless line that we describe next) in Figure 2.7 as a function of distance from the load. Both the voltage and current are plotted. As can be seen, the voltage SWR and current SWR are the same, and this is also true in general. VSWR is often expressed as a ratio (e.g., $S = 1$ is expressed as $1:1$, $S = 1.5$ as $1.5:1$, etc.).

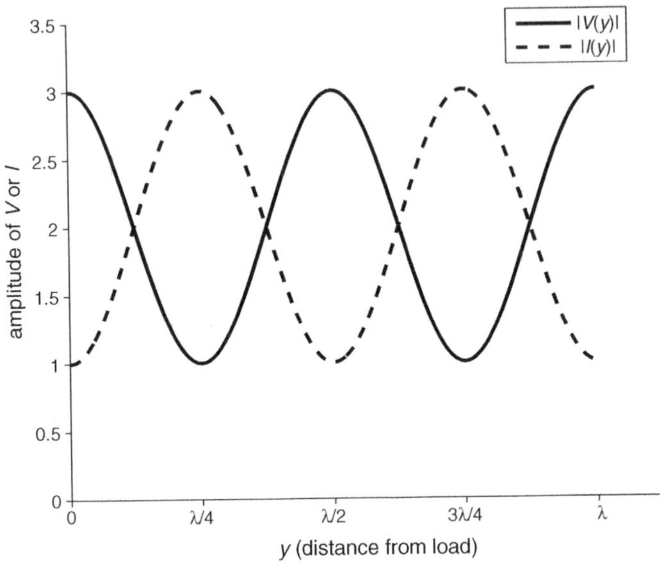

FIGURE 2.7 Standing waves in a transmission line.

Resistive Termination. For the case of resistive termination and lossless line, we have Γ real, given by

$$\Gamma = \frac{Z_L - Z_0}{Z_L + Z_0} \tag{2.86}$$

where both Z_L and Z_0 are real. Important special cases include:

- $Z_L = Z_0$ (matched load), $\Gamma = 0$, $S = 1$.
- $Z_L = 0$ (short circuit), $\Gamma = -1$, $S \rightarrow \infty$.
- $Z_L \rightarrow \infty$ (open circuit), $\Gamma = 1$, $S \rightarrow \infty$.

In general, we try to achieve matched load conditions, or close to matched load conditions, to keep the VSWR low, since the higher the VSWR, the more power is lost. For example, in connecting an RF transmitter/receiver to an antenna with an RF cable, we might try to achieve a VSWR $<2:1$.

Figure 2.7 shows an example with resistive termination and lossless line, where $Z_L > Z_0$.

2.3.4.3 *SWR Summary* The SWR is a very important value for practical purposes. A low SWR is desirable to minimize loss of signal power in cables, such as those between RF equipment and antenna. The SWR depends on:

- The cable length
- Impedance matching
- Losses (from the resistance and conductance of the transmission line)

As a practical matter, because all lines have losses, it is best to measure SWR near the receiving side (e.g., near the antenna). Loss factors will attenuate the reflected wave so that if SWR is measured near the transmitting side, the reflected wave would be most attenuated and the transmitted wave least attenuated there. Thus, the SWR measured there may be artificially low.

2.3.5 S-Parameters

We have seen a number of concepts that can be considered part of a more general, abstract concept. A transmission line, for example, was assumed implicitly to have two interfaces: the input interface and the output interface. These can be called *ports*, and hence the transmission line is an example of a *two-port network*. In general, a port can be said to be a point where current enters or exits an electronic network. Figure 2.8 shows a two-port network.

A *scattering parameter* (or S-parameter, for short), S_{mn}, is the ratio of voltage out of port m to voltage into port n, when all unused ports are attached to matched loads (matched to the system impedance of the network). For a two-port network, m and n take values 1 or 2. The S-parameters for a two-port network are illustrated in Figure 2.9. Thus, S_{11} is the reflection coefficient under matched load conditions (with port 2 being attached to a matched load). If the network represents an amplifier, port 1 is the input, port 2 is the output, and the ports are attached to the appropriate loads, then S_{21} is the gain. Note that the S-parameters may depend on frequency, temperature, control voltage, and so on, so these should be specified as necessary. It

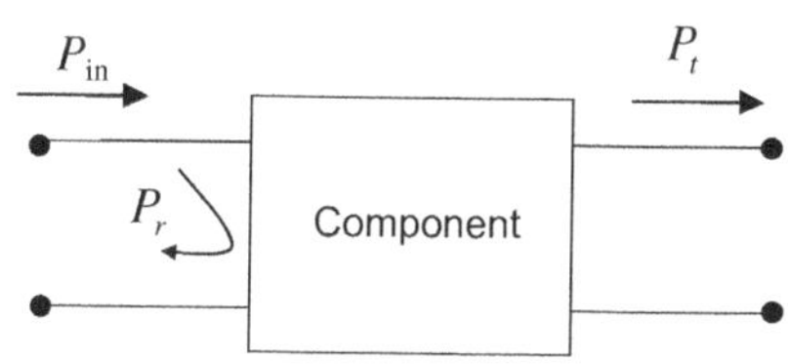

FIGURE 2.8 Two-port network.

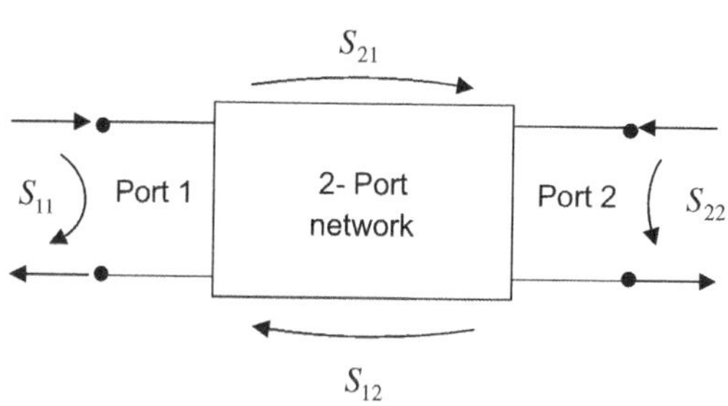

FIGURE 2.9 S parameters.

is possible to generalize from two-port networks to four-port networks and networks with other numbers of ports. In those cases, *m* and *n* would range over the number of ports.

2.4 IMPEDANCE

In this chapter and Chapter 1, we have seen a variety of notions of impedance:

- (Electrical) impedance
- (Intrinsic) impedance
- (Wave) impedance
- (Characteristic) impedance
- Input and output impedance

They are all measured in ohms, and unfortunately all are called impedance, but they refer to different concepts. Nevertheless, they can all take complex values and are ratios of phasors.

Electrical impedance applies in electrical circuits, where we have a voltage across two points and current flow across the same two points. Then electrical impedance is the complex-valued ratio of the phasors V and I, $Z = V/I$.

Intrinsic impedance is a property of a medium (e.g., air) being simply related to ϵ and μ of a medium, as $\eta = \sqrt{\mu/\epsilon}$. It gives us the ratio of the electric and magnetic fields when there is a single wave traveling in one direction in that medium. It is represented most often by η, or η_0 for a specific medium. The intrinsic impedance of air, for example, is about 377 Ω, whereas the electrical impedance between two points separated by air may be on the order of thousands of ohms (assuming that the voltage is large enough to exceed the dielectric breakdown voltage).

Wave impedance is the ratio of the transverse component of the electric and magnetic fields, **E** and **H**, respectively. For a single wave traveling in one direction, wave impedance equals the intrinsic impedance everywhere (with the possible exception of a sign; in some variations of the definition, wave impedance may equal η for a single wave traveling in the $+z$ direction and $-\eta$ for a single wave traveling in the $-z$ direction). It can be different from the intrinsic impedance (e.g., when we have two waves traveling in opposite directions). In that case, wave impedance would also be a function of position. Cheng [3] calls it *wave impedance of the total field*, which helps distinguish it from intrinsic impedance.

Characteristic impedance, usually written as Z_0, is a convenient single-parameter characterization of a transmission line. Characteristic impedance should be used only to refer to a transmission line. It is the ratio of V to I measured at the input of the transmission line if there is no reflected wave. Thus, it differs from intrinsic impedance and wave impedance, which have to do with field strengths. Also, it differs from electrical impedance in that characteristic impedance equals electrical impedance (of the transmission line and load combined) only when the load is matched to the transmission line. Otherwise, the relationship between Z_0 and electrical impedance, Z_i, is given by (2.81). The electrical impedance Z_i in this type of context is also called the input impedance.

Input impedance of a circuit or device has to do with the electrical impedance seen looking into it (i.e., at its input). More precisely, for a circuit or device, it is the Thévenin equivalent impedance at the input. For a transmission line, it is the ratio of the resultant V (from forward and reflected waves) and I (from forward and reflected waves), which can be expressed in terms of Z_0 and Z_L as given by (2.81).

Unfortunately, there are some cases where these careful distinctions are not made; for example, "characteristic impedance" is used to refer to intrinsic impedance or wave impedance outside transmission lines. Hence, the reader should read everything with caution and use context to help clarify meaning.

2.5 TESTS AND MEASUREMENTS

For testing and validation purposes in RF, antennas, and propagation, various devices and tools are available.

An *oscilloscope* allows electrical signals to be viewed in the time domain (e.g., as a plot of voltage against time). Whereas an oscilloscope show a time-domain representation of signals, a *spectrum analyzer* shows a corresponding frequency-domain representation of signals. Spectrum analyzers can be used for many types of measurements, making them especially useful for RF testing. Spectrum analyzers are used in RF engineering for harmonic distortion measurement, intermodulation distortion measurement, measurement of modulation sidebands, and so on.

A *network analyzer* measures the characteristics of a device, system, or network. This is in contrast to a spectrum analyzer or an oscilloscope, both of which measure and analyze a *signal*. (Of course, such measurements are not disjoint from what a network analyzer does in the sense that related or similar information about a device, system, or network could be obtained; for example, the frequency response of a device, system, or network could be deduced after measuring an input spectrum and its corresponding output spectrum using a spectrum analyzer.)

A *time-domain reflectometer* (TDR), as the name suggests, sends a short (time-domain) pulse into a cable, device, or system and measures the reflections, if any. The amplitude, duration, and shape of the reflected wave gives information about the length of the cable, its characteristic impedance, and so on.

Further details on oscilloscopes, spectrum analyzers, network analyzers, and TDRs are provided in Section 2.5.2.

2.5.1 Function Generators

It is often useful in testing to be able to generate various functions (e.g., sine waves, square waves) to be used as system inputs. Basic sine-wave oscillators may generate only sine waves (and perhaps square waves, too), whereas a function generator might have additional features (e.g., also generating other waveforms, such as triangular waves and modulated waves, frequency sweep, and dc offset adjustment). With frequency sweep, the instantaneous frequency of the wave will change with time, sweeping through a range of frequencies (e.g., linearly or logarithmically with time).

More sophisticated function generators, sometimes called *arbitrary waveform generators*, can generate arbitrary waveforms. Usually, this means that waveforms are generated using direct digital synthesis. Thus, any arbitrary waveform can be stored digitally and the waveform is synthesized using digital-to-analog conversion followed by lowpass filtering and amplification.

While a function generator is general-purpose and has an upper limit in the tens of MHz, other, more specialized generators are also used in testing labs, including pulse generators and RF signal generators. A *pulse generator* specializes in producing pulses and square waves of high precision and high quality. Since such pulses would have a very wide bandwidth, they would not be produced as cleanly by a regular function generator as by a pulse generator. It may range up to 1 GHz.

The signals produced by a general-purpose function generator are too low in frequency for RF testing, so RF signal generators are used that can produce signals from several kilohertz to several gigahertz. Besides producing sinusoids in the RF range, they may also provide modulated signals, including various digital modulation formats. Usually, these generators need very accurate and stable frequency generation for obvious reasons (e.g., for testing adjacent channel rejection of receivers, phase noise and inaccurate frequencies from the signal generator will corrupt the results). Moreover, the RF signal generators need to be capable of producing very low sidebands. To test receiver sensitivity, the amplitude of the signal produced also needs to be very accurate.

2.5.2 Measurement Instruments

Measurement instruments always affect the circuit they are measuring, however slightly. To minimize the impact on the circuit they are measuring, many measurement instruments are designed with a very high input impedance. This draws as little current as possible (less loading). Furthermore, this maximizes the voltage transfer to the measurement instrument, making readings of open-circuit voltage more accurate.

At RF, components are usually matched. Input and output impedances are both 50 Ω (or 75 Ω in some cases). For such systems, the measurement instruments would also use a matching input impedance of 50 Ω. Some signals can be interpreted as mixed ac and dc signals, with an ac signal being offset by a dc value. An *ac coupled* device only gives the ac portion of a signal, removing the dc offset through the use of a coupling capacitor at the input.

Each instrument has only finite resolution and finite accuracy. Resolution has to do with the granularity of changes in the measured value that can be detected, whereas accuracy has to do with how close to the correct value the measurement is. The resolution and accuracy of the instrument should be considered to see if it is

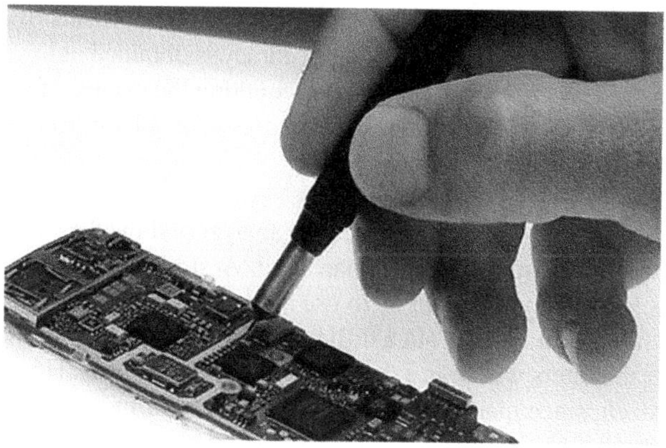

FIGURE 2.10 RF probe. (Courtesy of Aeroflex Inc.)

appropriate for its intended use. The bandwidth and rise time of the measurement instruments should be considered relative to the particular signal to be measured.

2.5.2.1 Voltmeters, Multimeters, and RF Probes Voltmeters can be either dc or ac. Although ac voltmeters normally give readings as root mean square (rms), there are a couple of ways these measurements could be made, so in some ac voltmeters, the readings are calibrated correctly only for sinusoids. So-called "true rms" ac voltmeters give correct rms readings for nonsinusoidal waves as well, but the range of frequencies over which the readings are accurate is always finite. As an alternative to the use of ac voltmeters, high-frequency ac measurements can be made with an *RF probe* together with a dc voltmeter. Usually, the RF probe is a peak detector, and so may be calibrated only for sine waves. Figure 2.10 shows an RF probe. *Multimeters* are popular because they combine voltmeter, ammeter, and ohmmeter. They may also include a few other useful features, such as a frequency counter (often limited in bandwidth).

2.5.2.2 Oscilloscopes Oscilloscopes are very versatile. They can be used to display *eye diagrams* of digitally modulated signals. An eye diagram is a visual tool used to detect the impact of intersymbol interference, noise, and so on, on the quality of a digital signal.

2.5.2.3 Frequency Counters The frequency of a periodic signal is how often it repeats itself, so it is natural to expect that measurement of frequency could be accomplished by something as simple as counting. In a frequency counter, a time base controls the opening of a gate, during which time cycles of the signal are counted. Alternatively, especially for low frequencies, the time base and the signal being measured could be swapped internally by the frequency counter so that the signal being measured controls the gate, and then the counting will be counting the number of cycles of the timebase that occur during one period of the signal. Hence, the period of the signal is measured. While period and frequency are reciprocals of each other, so that obtaining either one is sufficient, more accuracy is obtained at lower frequencies counting the period and at higher frequencies counting the cycles, of the signal being measured. Usually, the time base comes from a crystal oscillator, so it is essential that it be very stable and accurate.

2.5.2.4 *Spectrum Analyzers* Spectrum analyzers show the amplitude of a signal as a function of frequency rather than time. A straightforward first attempt at building a spectrum analyzer might be to use a filter bank of relatively narrow filters. Each of these filters would filter out a different narrowband from all the other filters, where together the filters will span a desired range of frequencies. However, it is not a practical approach, since many filters would be needed for most cases. For example, if the desired range of frequencies is 0 to 2 MHz and each filter is 1 kHz wide, 2000 of these filters would be needed.

Practical spectrum analyzers use methods that do not require thousands of filters. One method is to use the fast Fourier transform (FFT) to obtain the spectrum. To obtain the spectrum of a continuous-time signal, the signal would first have to be sampled. Unlike with the filter-bank approach, this approach is susceptible to aliasing, so the FFT should be preceded by an antialiasing lowpass filter. Another method is to design the spectrum analyzer like a heterodyne radio receiver, with a high-quality fixed IF filter taking input from a mixer that mixes the signal being analyzed with the output of a tunable local oscillator. When such an approach is taken and the tunable local oscillator is swept automatically across a range of desirable frequencies, the spectrum analyzer may be known as a *swept spectrum analyzer*.

A fundamental property of a spectrum analyzer is the resolution bandwidth. If there are multiple distinct spectral components within the resolution bandwidth at a particular band, the spectrum analyzer will not be able to resolve them into distinct components. Making the resolution bandwidth narrower allows the spectrum analyzer to resolve components to a finer level and also reduces the noise in the measurements (since the noise equivalent bandwidth would be smaller). However, narrower resolution bandwidth comes with a longer settling time.

Spectrum analyzers come in different shapes and sizes, and there are even handheld spectrum analyzers that can be brought out into the field, for example, to make measurements at wireless base stations. An example of a handheld spectrum analyzer is shown in Figure 2.11.

2.5.2.5 *Network Analyzers and TDRs* Network analyzers can be used to measure the S-parameters of a two-port network (Section 2.3.5). A *scalar network analyzer* measures amplitude only and not phase, whereas a *vector network analyzer* also measures phase. A network analyzer can generate its own signal to input to the system, or take an external signal for that purpose.

Both TDRs and network analyzers can be used to obtain frequency response. In the case of network analyzers, the inputs are often narrowband (but the frequency may be swept over a range). In the case of the TDR, it sends a very narrow pulse, and the frequency response is obtained by computing the Fourier transform of the reflected

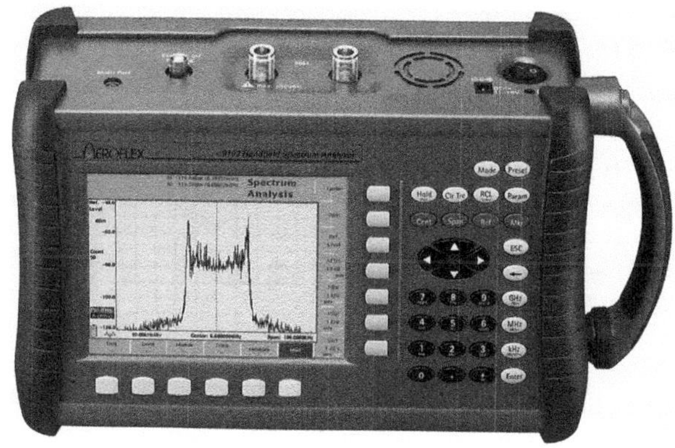

FIGURE 2.11 Handheld spectrum analyzer. (Courtesy of Aeroflex Inc.)

signal. The narrow pulse will have a finite width, say, 10 ps, so in the frequency domain the impulse response is multiplied by the Fourier transform of the narrow pulse, which would be on the order of 100 GHz.

2.5.2.6 Antenna Couplers It can be difficult in some cases to measure the RF signal directly from a transmitter to its antenna. For example, the antenna might be integrated into a mobile phone together with the transmitter and receiver. An alternative approach, then, is to use an *antenna coupler* to measure the signal indirectly as it is being radiated from the antenna. The antenna coupler may be a broadband antenna that is placed very close to the device under test.

2.5.3 Mobile Phone Test Equipment

So far, we have discussed oscilloscopes, spectrum analyzers, network analyzers, time-domain reflectometers, and so on, that have wide applicability to many different application scenarios in electrical engineering and electronics. There are also devices such as SWR meters and antenna couplers that might have a narrower range of applications. However, even these are not specific to any wireless system standard. In contrast, there is also an entire range of test and measurement devices that incorporate specifics about various wireless systems (e.g., GSM, CDMA, LTE, etc.) to enable more specific tests and measurements to be performed on devices for those particular wireless systems.

These test and measurement devices can be very useful in determining if a mobile phone, for example, is in conformance with specific system specifications on spectrum mask, sensitivity, selectivity, and so on, at the RF level. Some of these devices can also test and measure at other layers and are related to other aspects of the systems, such as BER performance, radio link protocols, and network protocols. Thus, these test and measurement devices can be useful for certifying phones and equipment in the wireless infrastructure and also for repair purposes. An example of one such mobile phone test equipment is shown in Figure 2.12.

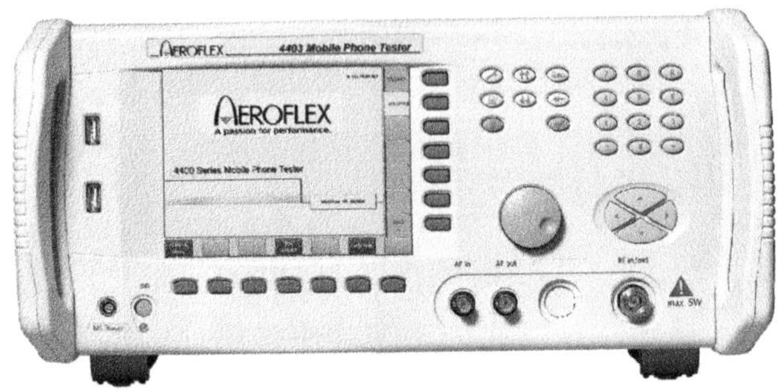

FIGURE 2.12 Example of mobile phone test equipment. (Courtesy of Aeroflex Inc.)

EXERCISES

2.1 Observe the geometry in Figures 2.1 and 2.2. Given a point (x, y, z) in Cartesian coordinates, what are the cylindrical coordinates of the same point? And the spherical coordinates?

2.2 Now work the other way around and convert from a point in cylindrical coordinates to Cartesian coordinates, and from spherical coordinates to Cartesian coordinates.

2.3 Consider an electromagnetic wave propagating in a lossless medium. Suppose that at a point P, the **E** field is given by $\mathbf{E} = \mathbf{u_x} E_0$ and the **H** field by $\mathbf{H} = \mathbf{u_y} H_0$. In what direction is the wave propagating? If $E_0 = 377$ mV/m and the medium is air, what is H_0? What is the Poynting vector? What is the average power flow per unit area at P?

2.4 In general, what is the range of possible values for SWR, S? What would be the corresponding range of values for $|\Gamma|$? Referring to Figure 2.7, what is the SWR? What is Γ?

REFERENCES

1. K. Chang. *RF and Microwave Wireless Systems*. Wiley, Hoboken, NJ, 2000.
2. K. Chang, I. Bahl, and V. Nair. *RF and Microwave Circuit and Component Design for Wireless Systems*. Wiley, Hoboken, NJ, 2002.
3. D. K. Cheng. *Fields and Wave Electromagnetics*. Addison-Wesley, Reading, MA, 1990.
4. S. Ramo, J. Whinnery, and T. Van Duzer. *Fields and Waves in Communication Electronics*. Wiley, New York, 1984.
5. M. Sadiku. *Elements of Electromagnetics*. Oxford University Press, New York, 2006.

3

RADIO-FREQUENCY ENGINEERING

Radio-frequency (RF) *engineering* is about systems that operate at radio frequencies such as microwave frequencies. Our approach to RF will be from a "signals and systems" perspective. The RF portion of radio transmitters and receivers will be viewed as a subsystem of wireless systems. Thus, the relationships of the RF portion to other parts of the overall wireless system design will be pertinent. For example, radio receiver sensitivity depends on the RF design, among other factors, and it has a direct influence on link budgets. Also, when we design wireless access technologies, we need to be aware of the RF subsystem capabilities; for example, a high peak-to-average power ratio (PAPR, Section 6.5.2) in signals causes distortion and/or inefficiencies in the RF subsystem because of the nature of RF amplifiers.

RF generally includes other aspects, such as the device technologies and RF circuits (including active circuits and passive circuits). These aspects are, however, outside our scope in this book, although the interested reader may consult some of the reference listed at the end of the chapter for details on those aspects. What is of interest in our "signals and systems" approach are aspects such as consideration of the noise contributions of the subsystems and indicators of dynamic range, sensitivity, selectivity, and so on. We are interested in these aspects because of their relationship to other aspects of overall wireless system design, and because there are various engineering trade-offs in the RF devices. Real electronic components introduce noise and have other imperfections, such as nonlinearities. While the nature of noise, nonlinearities, and so on, is intimately related to the devices themselves, the results on the system can be studied and quantified at the systems level based on models of these effects without a detailed look at device physics, and this is the path we take.

Fundamentals of Wireless Communication Engineering Technologies, First Edition. K. Daniel Wong.

For those who are new to RF, it should be noted that RF signals are difficult to handle compared to dc and low-frequency signals, which one might be more used to. Unlike in the case of a dc current, whenever you have time-varying currents and electromagnetic fields, various phenomena such as radiation and coupling can occur. It can be difficult to control these effects, so we have to handle the RF signals with care. Furthermore, practical realities of our devices and subsystems, such as nonlinearities, can result in severe performance degradation if we are not careful.

We begin with a big-picture look at RF and introduce relevant analytical assumptions and techniques in Section 3.1. We then tackle the problem of noise in Section 3.2. Besides noise, another characteristic of RF engineering is dealing with nonlinearities, which we consider in Section 3.3. Important parts of RF systems, such as mixers, oscillators, and amplifiers, are discussed in Sections 3.4, 3.5, and 3.6, respectively, before we wrap up with a brief look at some other RF components in Section 3.7.

3.1 INTRODUCTION AND PRELIMINARIES

We begin with a look at the RF subsystem of a typical radio (Section 3.1.1). This will give us a good view of how amplifiers, filters, mixers, and so on, may be organized in a typical RF subsystem. We then elaborate on the "handle with care" aspects of RF and how it differs from low-frequency circuits (Section 3.1.2). Finally, we introduce some mathematical preliminaries, such as how we model RF subsystems (Section 3.1.3) and how we perform analysis on nonlinear effects (Section 3.1.4).

3.1.1 Superheterodyne Receiver

The most popular wireless receiver architecture is known as the *superheterodyne receiver*. Figure 3.1 shows a block diagram of a superheterodyne radio receiver. In the figure we see amplifiers, mixers, frequency synthesizers, and filters. These are all fundamental building blocks of the RF part of radios. Broadly speaking, an amplifier amplifies the power of a signal; a mixer is used to up-convert or down-convert a signal, by multiplying (also described as mixing) it with a periodic signal, such as would be produced by a frequency synthesizer. A frequency synthesizer may be as simple as an oscillator, or it may include an oscillator together with additional circuitry. A filter selects a band of frequencies to pass through, and attenuates signal components at other frequencies.

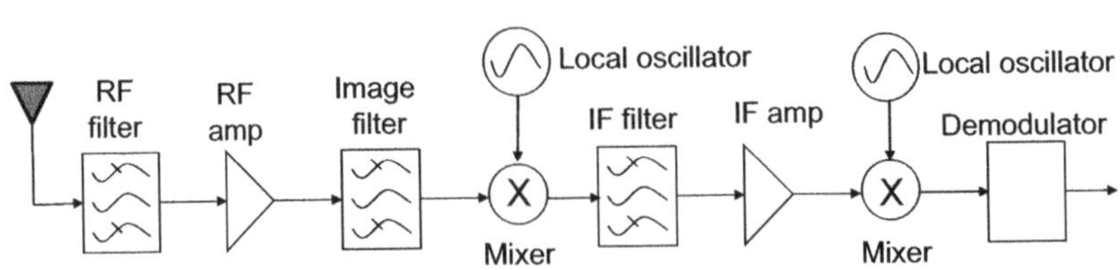

FIGURE 3.1 Superheterodyne receiver.

As mentioned in Section 1.4, the communications signals transmitted over wireless are at very high frequencies and so are often referred to as being "at RF." To demodulate the signals and detect what was transmitted, the RF section of the receiver often needs to bring the signal down to around baseband. For this discussion we use the term *channel* to refer to a particular frequency channel (e.g., the 200-kHz channel in GSM) that a single transmitter transmits, whereas we use the term *band* to refer to the range of frequencies over which other transmitters of the same network or system may be transmitting (e.g., 890 to 915 MHz on the uplink in GSM-900). The superheterodyne receiver brings the signal from RF down to around baseband in two stages; first, it down-converts from RF to an *intermediate frequency* (IF), and second, it down-converts from IF to around baseband. Having two stages of down-conversion introduces some challenges, but in many cases the difficulties are considered to be more than offset by the advantages, which include:

- It is very difficult to design good filters at RF to suppress interference from outside the channel, because:
  - The carrier frequency at RF, f_c, is typically very high, so $f_c \gg B$, where B is the signal bandwidth, making it a challenge to design highly selective filters at RF; on the other hand, the IF could be chosen such that the *fractional bandwidth* B/f_{IF} allows for highly selective filtering at IF.
  - Another impediment to highly selective filtering at RF is that without the IF, the RF filter would need to be tunable, as the receiver would normally need to receive RF signals from within the entire band of the network or system, but the RF filter would need to pick out the channel from within the band. However, when the highly selective filtering for the channel is done at IF, the RF filter can be more relaxed, to pass through the entire band of the network or system. In this case the RF filter is sometimes called a *preselector filter*.
- The tuning to receive RF signals from within the entire band needs only to be done in the front end of the receiver, so the rest of the circuitry can be fixed regardless of f_c; that is, it does not need to adjust to different f_c values (e.g., we can have a highly selective IF filter at a fixed frequency and efficient amplification at IF for the same reason).
- The separation of RF and IF provides some insulation to the receiver circuitry (at IF) from high-power RF signals coming out of the transmitter (e.g., if the radio is both transmitting and receiving signals) and from stray feedback from the output of the receiver IF circuitry itself (which would be amplified as compared to the signals coming in to the receiver) to the input of the receiver.

A major disadvantage of the superheterodyne receiver is that signals at other frequencies (besides f_c) could end up at f_{IF}, which would be impossible to filter out no matter how selective the IF filter is, because these *spurious* signals are at the same frequency as the desired signal. These spurious signals include the *image signal* and the *1/2-IF signal*. In Section 3.4 we see how these spurious signals can arise.

There are alternatives to the superheterodyne architecture (see the references at the end of the chapter for more details), but the superheterodyne architecture remains the most popular and important architecture at the moment.

3.1.2 RF—Handle with Care!

A first, introductory class in electrical engineering (e.g., such as one that introduces some of the topics reviewed in Chapter 1) would usually cover circuits with dc and very low-frequency ac, such as 60 Hz. The wavelength of a 60-Hz electromagnetic wave is about 5000 km, which is many orders of magnitude larger than any circuit the student will see in electrical engineering lab. When the physical dimensions of the circuits are so much smaller than the wavelength (i.e., at high frequencies such as RF), the basic static circuit analysis is sufficient to model the circuit accurately. In particular, the typical *lumped circuit* model used in such cases, with lumped elements (resistors, capacitors, inductors, etc.) connected by round wires of negligible resistance, is inadequate at RF. Instead, we see distributed elements such as transmission lines (Section 2.3.3), and various other RF components as in Section 3.7, plus the filters, amplifiers, mixers, oscillators, and so on, designed to work over a range of frequencies.

One way of explaining why the ratio of circuit size to wavelength matters is from the perspective of time retardation effects [6]. For circuits that are small compared to wavelength, the electric and magnetic fields can be considered *quasi-static*, distributed like static fields as a function of position (even if there is time variation). Mutual coupling, as in mutual inductance or mutual capacitance, can be planned and isolated in specific lumped elements. Consider a loop of wire. Current in any part of the wire produces a magnetic field that travels to any other part of the loop where there is negligible difference in phase (since the loop is so much smaller than the wavelength). So the loop acts as an inductor. For circuits that are not small compared to wavelength, however, considering the same loop of wire, the magnetic field that arrives at any part of the loop from other parts of the loop might have significant phase differences due to the finite propagation time and the relatively small wavelength. So the loop radiates and acts as an antenna. In RF, therefore, one reason we have to handle our signals with care is that otherwise various unintended radiation, coupling, and other effects can severely degrade the performance of the circuit or render it useless.

Another way of explaining why lumped circuit analysis doesn't work at RF is that at high frequencies, the *skin effect* means that most of the electromagnetic energy does not penetrate a regular round wire made of a conducting material (such as copper), but stays mostly on the surface. The impedance increases with frequency, and the round wire becomes very inefficient for carrying RF signals. Traveling mostly at the surface of the wire, the signal power is also easily lost through radiation.

Nonlinearities in devices and subsystems is another aspect of RF that we have to be careful about, and we discuss that in Section 3.1.4. Also, impedance matching is very important (as we have already seen in Section 2.3.4.2, we try to match transmission lines to the devices they connect to, to minimize VSWR and power loss).

3.1.3 RF Devices and Systems: Assumptions and Limitations

For purposes of modeling and analysis, a system can be:

- Linear or nonlinear.
- Memoryless or dynamic; in dynamic systems, the output can depend on the input at times other than the present (i.e., it can have memory). Convolution is the classic example of an operation related to dynamic LTI systems.
- Time invariant or time varying.

In many cases in signal processing and communications signal processing, the systems are modeled and analyzed as linear, time invariant, and dynamic.

In studying RF, however, typically the effect of nonlinearities is of interest. Nonlinear, time-varying, and dynamic systems are too difficult to analyze at a first attempt, so usually people look at nonlinear, time-varying, and memoryless systems. Unless stated otherwise, this is what we assume here.

3.1.3.1 Two-Port Networks In talking about modeling our systems, subsystems, and devices as linear or nonlinear, we made the implicit assumption that there is an input and there is an output (a system is linear if the output behaves in certain ways when the input changes in some ways; this would not make sense without an input or output!). Stated explicitly, then, many of the systems, subsystems, and devices considered in this chapter are modeled as *two-port networks*, which we introduced briefly in Section 2.3.5. The word *network* here is used in the sense of electrical or electronic network (as introduced at the beginning of Section 1.2). We carry this assumption to our considerations of noise figure, for example, where we always consider noise figure in the context of there being some input signal and some output signal. This is the most useful way to consider noise contributions of the RF components, since we want to know the noise contributed when there is a signal flowing into the component and a signal flowing out.

The concept of modeling subsystems as two-port networks is very useful in allowing us to abstract away from the details of the subsystem and just consider its impact on the system and on signals passing through it as a "black box." There will also be some devices and subsystems that have more than two ports, such as directional couplers (Section 3.7.1) and circulators (Section 3.7.2). We will point out the number of ports in these cases when we discuss them.

3.1.4 Effect of Nonlinearities

Nonlinearities in real devices lead to harmonic distortion and intermodulation distortion.

3.1.4.1 Harmonics Given the assumptions just stated (nonlinear systems), the phenomena of harmonic terms and intermodulation terms naturally emerge once we work through the analysis. First, we consider harmonics. Relative to a sinusoid at a given frequency, say, $\cos \omega t$, the term $\cos n\omega t$, for each n, is known as the nth *harmonic*. The term $\cos \omega t$ is also known as the *fundamental*.

For a nonlinear and memoryless system, the output $y(t)$ can be expressed in terms of the input $x(t)$ as

$$y(t) = a_1 x(t) + a_2 x^2(t) + a_3 x^3(t) + \cdots \tag{3.1}$$

If the system is time varying, the coefficients a_1, a_2, and so on, can be time varying.

If $x(t) = V \cos \omega t$, then (for negligible fourth- and higher-order terms) we use a Taylor series approximation to obtain

$$y(t) \approx \frac{a_2 V^2}{2} + \left(a_1 V + \frac{3a_3 V^3}{4} \right) \cos \omega t + \frac{a_2 V^2}{2} \cos 2\omega t + \frac{a_3 V^3}{4} \cos 3\omega t + \cdots \tag{3.2}$$

We see that the output is a sum of terms of the form $\cos n\omega t$ (i.e., a sum of harmonic terms). The nth harmonic is roughly proportional to V^n, but the coefficients a_n become very small as n increases, so usually only the low-order harmonics, such as the second and third, are of consequence.

3.1.4.2 *Intermodulation Effects* So, we have seen harmonics, but to see intermodulation terms, we need to input at least two sinusoids. Basically, the intermodulation terms (also known as *intermodulation products*) arise when there are two or more sinusoids and they end up "modulating" each other, due to multiplication and mixing of sinusoidal terms when we expand terms in (3.2). They may also be known informally as "intermods." In the case of the two-sinusoid signal, after it passes through a memoryless nonlinear device, there will be terms (although some may be very small) of the form

$$\cos(m\omega_1 + n\omega_2)t \tag{3.3}$$

where m and n can be any integers, even negative integers or zero. These are all intermodulation terms (and thus, harmonics are a special case of intermodulation terms where $m = 0$ or $n = 0$). The *order* of the intermodulation term is $|m| + |n|$. People most often talk about the second- and third-order intermodulation terms.

We examine the very important case where we have a sum of two sinusoids (i.e., $x(t) = A \cos \omega_1 t + B \cos \omega_2 t$, where ω_1 and ω_2 are the two different frequencies of the two sinusoids). In general, they need not be in phase alignment and there could be a phase offset, but that is not relevant as far as intermodulation distortion is concerned, so we write them this way for convenience.

We assume that the system is nonlinear and memoryless, so it is modeled by (3.1). We look first at the term $a_2 x^2(t)$ of (3.2) (from which the second-order intermodulation

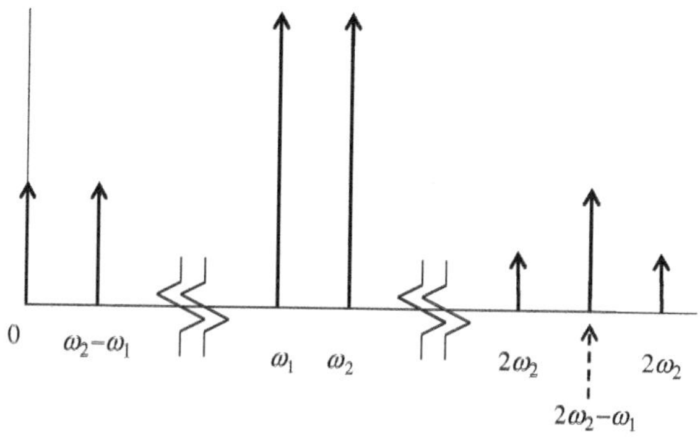

FIGURE 3.2 Second-order products of two equal-power sinusoids.

terms will arise), and we have

$$a_2x^2(t) = a_2\,(A\cos\omega_1 t + B\cos\omega_2 t)^2 \tag{3.4}$$

$$= a_2\left\{\frac{A^2+B^2}{2} + \frac{A^2}{2}\cos 2\omega_1 t + \frac{B^2}{2}\cos 2\omega_2 t\right.$$
$$\left. +AB\,[\cos(\omega_2-\omega_1)\,t + \cos(\omega_1+\omega_2)\,t]\right\} \tag{3.5}$$

The first term is a dc term; the second and third terms are intermodulation terms that are second harmonic terms, at $2\omega_1$ and $2\omega_2$ (Figure 3.2). The last two terms are also intermodulation terms, and they are at the sum and difference frequencies (i.e., $\omega_1+\omega_2$ and $\omega_2-\omega_1$), respectively.

Next, we look at the term $a_3x^3(t)$ of (3.2) (from which the third-order intermodulation terms will arise). We can compute it by partially reusing (3.5) and multiplying by $A\cos\omega_1 t + B\cos\omega_2 t$. Thus, we have

$$a_3x^3(t) = a_3\left\{\frac{A^2+B^2}{2} + \frac{A^2}{2}\cos 2\omega_1 t + \frac{B^2}{2}\cos 2\omega_2 t\right.$$
$$\left. +AB\,[\cos(\omega_2-\omega_1)\,t + \cos(\omega_1+\omega_2)\,t]\right\}(A\cos\omega_1 t + B\cos\omega_2 t)$$
$$= \frac{a_3}{4}\left\{(3A^3+6AB^2)\cos\omega_1 t + (3B^3+6A^2B)\cos\omega_2 t\right.$$
$$+\,3\left[A^2B\cos(2\omega_1-\omega_2)t + AB^2\cos(2\omega_2-\omega_1)t\right]$$
$$+\,3\left[A^2B\cos(2\omega_1+\omega_2)t + AB^2\cos(\omega_1+2\omega_2)t\right]$$
$$\left.+\,A^3\cos 3\omega_1 t + B^3\cos 3\omega_2 t\right\} \tag{3.6}$$

We see that there are six third-order intermodulation terms: $2\omega_1-\omega_2$, $2\omega_2-\omega_1$, $2\omega_1+\omega_2$, $\omega_1+2\omega_2$, $3\omega_1$, and $3\omega_2$ (Figure 3.3).

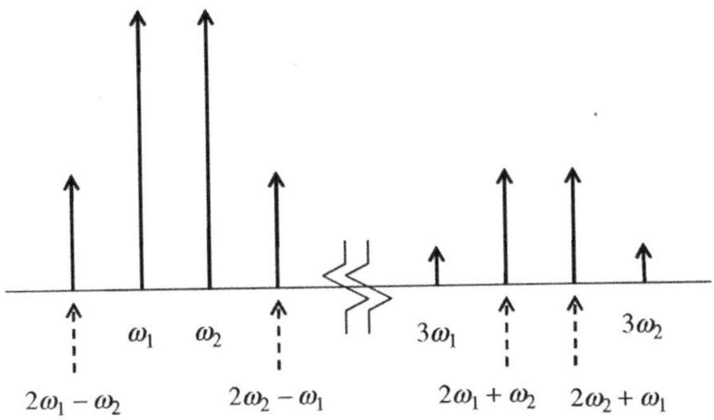

FIGURE 3.3 Third-order products of two equal-power sinusoids.

System Implications. How do such effects, arising from nonlinearities in active devices, affect system performance? What are the implications on RF system design? Suppose that ω_1 and ω_2 are close together in frequency. When we examine the second- and third-order intermodulation terms together, we see that most of the frequencies are relatively far from ω_1 and ω_2, except for $2\omega_1 - \omega_2$ and $\omega_1 - 2\omega_2$, which may be very close to ω_1 and ω_2. Thus, it may be very hard or impractical to filter out these terms. Moreover, the intermodulation terms can grow at a faster rate than the fundamentals, as input power increases, effectively resulting in an upper limit to the input signal power that can be used with the nonlinear system. We discuss such issues in more detail in Section 3.3. Another problem that can occur is that intermodulation terms that are far from the desired signal can end up near the desired signal after passing through a mixer, as we will see in Section 3.4.

3.2 NOISE

Radio signals arriving at a receiver like a mobile phone may be very weak (e.g., on the order of −100 dBm). At such signal levels, any noise added to the signal in an RF subsystem can be a very serious problem. In the receiver, the RF subsystem precedes the digital demodulation and detection. *After* demodulation and detection, we can make the signal power high enough that we don't have to worry much about noise in the rest of the receiver, provided that we do sensible things with our circuits. Thus, the RF subsystem is where the potential problems with noise are most critical. It can mean the difference between the receiver being able to recover the transmitted signal and the receiver struggling with a noisy signal that may not be usable.

Between the receiver and the transmitter, where is noise a greater concern? If the same amount of noise power is added in both the transmitter and the receiver, the added noise would be a much higher percentage of the signal power in the receiver than in the transmitter, since the signal power is highest in the transmitter. In other words, the effect on SNR of noise added in the transmitter is much less than the effect on SNR of the same amount of noise added in the receiver.

Thus, we focus especially on the RF subsystem of receivers. We cannot eliminate noise completely, but we can limit the amount of noise added, by careful design. So we need to characterize the noise and to be able to compute the noise generated internally in the RF subsystem and transferred out of the RF subsystem to the baseband demodulator. A closely related perspective is that we want to characterize the output SNR (output of the RF subsystem) with reference to the input SNR (of the RF subsystem), since for a given minimum SNR input to the baseband demodulator, the ratio of the output SNR to input SNR will determine the *receiver sensitivity*. Receiver sensitivity is the minimum SNR at input to the receiver that the system can handle, and we quantify these ideas in Section 3.2.5.3.

3.2.1 Types of Noise

There are different types of noise, including:

- *Johnson–Nyquist noise*. Whenever there is an electrical conductor that is not at absolute zero (0 K) the charge carriers (usually, the electrons) will experience a "random" motion reflecting the energy associated with the nonzero temperature. This motion, which may be described as thermal agitation, results in Johnson–Nyquist noise. The random nature of the thermal agitation results in the statistical characterization of the noise as additive white Gaussian. Because of the reasons for Johnson–Nyquist noise, it is also known as *thermal noise*.
- *Shot noise*. When there is a current carried by discrete charge carriers (usually, the electrons), there will be random fluctuations of the current. Shot noise may be modeled as an additive white Gaussian process.
- *Flicker noise*. From random trapping of charges, this noise has a probability density function that varies inversely with frequency and hence is known as *$1/f$ noise*. It is an issue primarily at very low frequencies because of its $1/f$ characteristic.

In some areas of technology, shot noise is a dominant mechanism, but for radio electronics, the main concern is with thermal noise, although flicker noise might be a problem in some cases where very low frequency signals get mixed with RF or IF signals.

3.2.2 Modeling Thermal Noise

Thermal noise generated in a resistor can be modeled as a noise voltage source followed by an ideal, noiseless resistor with the same resistance, as shown on the left of Figure 3.4. Note that the average noise voltage is zero, but the rms noise voltage is nonzero. Quantum mechanics leads us to the following rms value for the noise voltage in this model:

$$\overline{V_n} = \sqrt{4kTBR} \tag{3.7}$$

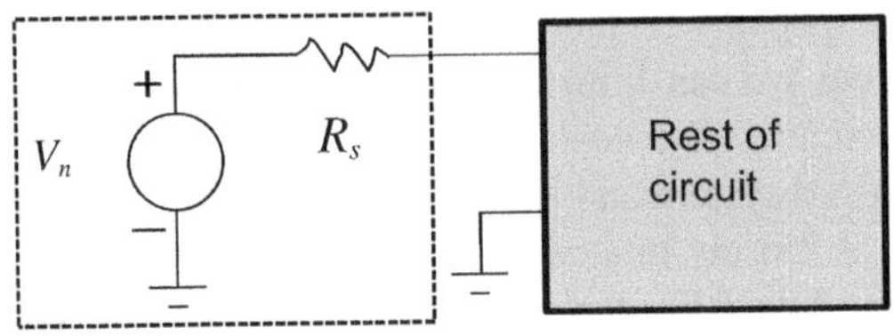

FIGURE 3.4 Noise source connected to the rest of the circuit.

When T is in kelvin and R is in ohms, Boltzmann's constant k is 3.8×10^{-23} J/K. B is the system bandwidth or measurement bandwidth. See Section 3.2.3.3 for a worked example.

Why is the rms noise voltage dependent on B? It could also be said the the rms noise voltage *per hertz* is $\sqrt{4ktR}$, but this seems to imply infinite noise power if we let B go to infinity! In reality, the total noise power of a resistor is not infinite because (3.7) applies only in the region below 100 GHz. Equation (3.7) is an approximation that breaks down at extremely high frequencies or extremely low temperatures [5]. We emphasize that to obtain an actual noise power, we must have some finite measurement bandwidth (see Section 3.2.3.2). Thus, as long as we make it a habit to ask ourselves what the system bandwidth or measurement bandwidth is, and that bandwidth is a reasonable value and we apply it correctly, we should be OK in using (3.7).

3.2.3 Transferred Thermal Noise Power

So we have the RMS noise voltage of the resistor. What about the actual power that a noisy RF subsystem contributes to the RF circuit? (NB: We are talking about average power here, since the noise voltage is a random time-varying signal.)

We have to be careful to differentiate the two concepts of power:

1. How much noise power is generated in the device; that is, in the case of thermal noise, what is the power of the thermal agitation of the charge carriers? We have already discussed how blind application of (3.7) would result in an infinite value for this quantity, and why that is incorrect. In any case, the next concept of noise power is more pertinent to RF engineering.
2. How much noise power is transferred to the "remaining circuit"? All loads and measuring devices would have some measurement bandwidth B, and for reasonable values of B this is an important parameter related to how much noise power is transferred to the remaining circuit. Furthermore, the concept of noise power transferred to the remaining circuit can be divided into:
   - Available power (usually, the maximum that could be transferred).
   - Delivered power (actual power transferred). Delivered power is generally less than or equal to available power; the two are equal under matched conditions.

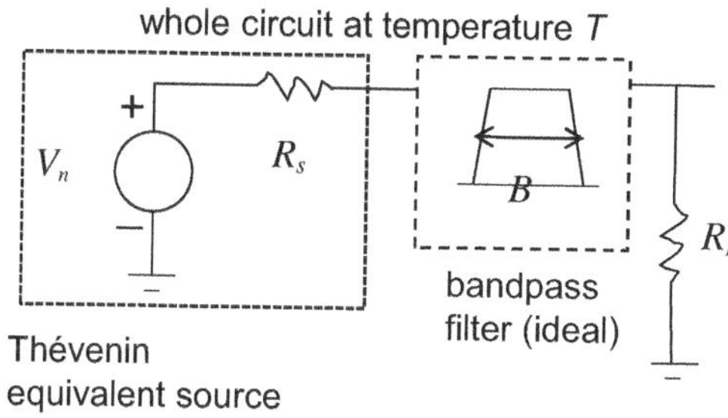

FIGURE 3.5 Noise transferred from a noisy resistor.

Usually, it is the delivered noise power transferred to the remaining circuit that is of primary interest. Continuing the example of a noisy resistor, how much power would be transferred to a load? Suppose that we connect a load with (Thévenin equivalent) resistance R_L and bandwidth B in series with our noisy resistor, as shown in Figure 3.5. Then the power transferred to the load (the delivered power) is

$$\frac{R_L}{R+R_L}\frac{\overline{V_n^2}(t)}{R+R_L} = \frac{4kTRR_LB}{(R+R_L)^2} \tag{3.8}$$

which is easily shown to be largest when $R = R_L$ (or we can just apply the principle of matched loads). In that case we have

$$\frac{4kTR^2B}{4R^2} = kTB \tag{3.9}$$

Thus, for a load with arbitrary resistance R_L, the power delivered is $4kTRR_LB/(R+R_L)^2$, whereas under matched load conditions, the delivered power is equal to the available power, which is kTB. (NB: A matching kTB is also dissipated in our noisy resistor). It is interesting to note that the available power, kTB, is independent of R.

3.2.3.1 *Noise Power Spectral Density* There are two concepts of noise power spectral density of thermal noise in resistors that are in use in RF engineering, and we must be careful not to confuse the two. We mention these two concepts here as a service to readers who might otherwise be potentially confused when reading about noise power from multiple sources.

The first is based on the concept of available noise power. Given that the available power is kTB, as we have just seen, we can define the *available noise power spectral density* as kT (which, when multiplied by a measurement bandwidth B, gives kTB), in watts/hertz.

The second concept defines the PSD of thermal noise as the square of the rms noise voltage per hertz [7], so, squaring (3.7) and dividing by B, we have

$$\overline{V_n^2}(t) = 4kTR \tag{3.10}$$

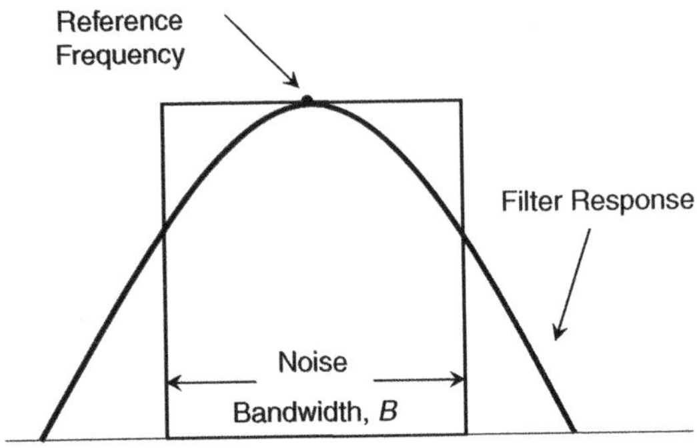

FIGURE 3.6 Noise equivalent bandwidth.

Note that $4kTR$ is actually, dimensionally, a voltage squared per hertz! However, for voltage signals, one convention uses V^2/Hz for the PSD (which would be numerically equal to the actual PSD in watts/hertz if the voltage were hypothetically to be applied across a 1-Ω resistor).

We observe the equivalence of the two concepts by recalling how we started with (3.7) and ended up with (3.9) in Section 3.2.3.

3.2.3.2 *Noise Equivalent Bandwidth* When we talk about the bandwidth B in noise calculations, it appears as if we have an ideal rectangular filter of width B. If we view a measurement device or a load as a filter, it could be considered to have a *noise equivalent bandwidth* (or *simply noise bandwidth*), B, as seen in Figure 3.6. The idea is that we obtain the area, A, under the curve representing the frequency response of the filter. Then, we observe the maximum value, x, of the frequency response of the filter. Next, we construct a rectangle with the same area A, so the sides of the rectangle are x and A/x. Then $B = A/x$, representing the bandwidth of an ideal filter with the same area A.

3.2.3.3 *Worked Example* Find the noise voltage in a circuit containing only a 5-Ω resistor at room temperature. Take the bandwidth as 1 kHz:

$$\overline{V_n} = 2\sqrt{(1.38 \times 10^{-23}\ \text{J/K})(290\ \text{K})(1000\ \text{Hz})(5\ \Omega)} = 2.8\ \text{nV} \tag{3.11}$$

3.2.4 Equivalent Noise Source Models

In this subsection we assume that all blocks are matched. This is a reasonable assumption for real subsystems when maximum signal power transfer is desired. We have seen how a noisy resistor can transfer power kTB into a matched load, as given by (3.9). We now construct equivalent noise source models for other noisy devices and systems. The main requirement to use this type of model is that these noise sources can be modeled as "white," that is, with a relatively flat noise spectral density. The noise contribution of these devices or subsystems,[†] even something active like an amplifier (not just passive loads like resistors) can be modeled as coming from an equivalent noisy resistor that contributes the same amount of noise. The model allows us to set the noise power from the equivalent resistor by increasing or decreasing T regardless of the actual operating temperature. Since the noise power from the equivalent resistor, kTB, is proportional to T, we can think of T as an adjustable parameter that allows us to set the appropriate noise power to equal that of the noisy subsystem. Since the temperature parameter, T, wouldn't be the real operating temperature, it is called an *equivalent temperature*, T_e. T_e can be expressed as a function of the available noise power at the output, $P_{\text{noise,out}}$:

$$T_e = \frac{P_{\text{noise,out}}}{kB} \tag{3.12}$$

Then, as far as its noise contributions are concerned, the subsystem can then be thought of as equivalent to a resistor at temperature T_e, since both would transfer noise power $P_{\text{noise,out}}$ to a matched load.

3.2.4.1 Input Referencing Our equivalent noise source model is not yet complete, because the subsystem is actually generating the noise internally, but we need to put the equivalent noise source somewhere when we model it. The usual way we handle this is by *input referencing*; that is, the noise generated inside the subsystem is referenced back (or referred back) to the input, as though it were generated before the subsystem, and then as though the subsystem itself was noiseless but the noise generated passes through it (and gets amplified, etc., depending on what the subsystem does). In other words, the subsystem is decomposed into two parts:

- An equivalent noise source that has kT_eB of available noise power to transfer to a matched load.
- A noiseless subsystem that is exactly the same as the original subsystem except that it generates no noise.

We assume that the subsystem is connected to a matched load, so the noise power delivered equals the noise power available.

What matters is that the model results in the correct noise power being generated at the output of the subsystem. Thus, we must be careful to adjust the input noise power of the equivalent noise source to produce the correct output noise power. In particular, if the subsystem is an amplifier with gain G, we should replace (3.12) with

$$T_e = \frac{P_{\text{noise,out}}}{kBG} \tag{3.13}$$

[†] For convenience, rather than saying "device or subsystem" many times over the next few pages, we will just say "subsystem," where it should be understood that the subsystem might be a single device.

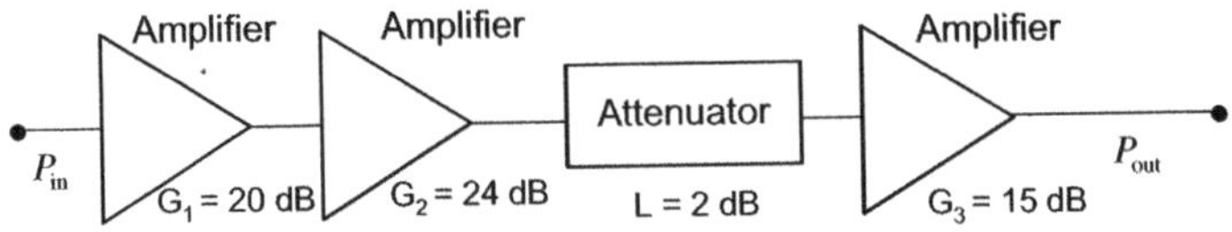

FIGURE 3.7 Example of a cascade of systems.

3.2.4.2 Cascades We often have cascades of subsystems (e.g., as shown in Figure 3.7), and it is important to be able to figure out the overall noise contribution of a cascade of subsystems, not just a single subsystem. If we have two subsystems in cascade, we can consider the two as one noisy subsystem, with one overall equivalent temperature to represent the total noise generated in the subsystem. Let the gains of the two subsystems be G_1 and G_2, respectively, and their noise temperatures be T_1 and T_2, respectively. Let $P_{\text{noise1,out}}$ and $P_{\text{noise2,out}}$ be the noise power at the output of the cascade that comes from subsystems 1 and 2, respectively. Then, clearly,

$$P_{\text{noise1,out}} = (T_1 kBG_1)G_2 \tag{3.14}$$

(because it passes through the second subsystem after it comes out of the first),

$$P_{\text{noise2,out}} = T_2 kBG_2 \tag{3.15}$$

and

$$P_{\text{noise,out}} = T_1 kBG_1 G_2 + T_2 kBG_2 \tag{3.16}$$

Thus, referring the sum of these two noise contributions back to the input, we need to divide by the overall gain of the cascade, G_1G_2, and to get the equivalent noise temperature, we need to express it in the appropriate form:

$$P_{\text{noise1+noise2,in}} = \frac{P_{\text{noise,out}}}{G_1 G_2} = T_1 kB + \frac{T_2 kB}{G_1} = kB\left(T_1 + \frac{T_2}{G_1}\right) \tag{3.17}$$

Thus, the noise temperature of the cascade is

$$T_{\text{cascade2}} = T_1 + \frac{T_2}{G_1} \tag{3.18}$$

This analysis is extended straightforwardly to three or more subsystems in cascade. The formula is given in (3.28). From such formulas, we can see that we would generally prefer to have the larger amplifications early in the cascade so that the big gains appear in more denominators in the formula rather than later. Another way to put it is that we get less noise added overall if the big gains are earlier in the cascade because they do not amplify the noise contributed by all subsequent subsystems, whereas an amplifier toward the end of the cascade would be amplifying the noise contributions from all earlier subsystems.

3.2.5 Noise Figure

So the noise contributions of a subsystem, or cascade of subsystems, can be modeled by equivalant noise sources with the appropriate equivalent temperatures. There is another related way to quantify the noise contributions of a subsystem, and that is by the *noise figure* (also known as the *noise factor*).

The context for discussions on noise figure assumes that the subsystem in question (also described as the *device under test*) is connected to both a signal source and a load. Unless otherwise stated, matched impedances are assumed on both sides. It is essential to note that the source resistor will contribute noise, kTB. Since the noise contribution from the source resistor depends on T, it is often assumed that the measurements are done at *room temperature*, taken as $T_0 = 290$ K or sometimes $T_0 = 300$ K.

Thus, under matched impedances on both sides and room-temperature conditions, the noise figure, F, may be defined as

$$F = \frac{\text{SNR}_{\text{input}}}{\text{SNR}_{\text{output}}} \tag{3.19}$$

(often given in decibels, although it is in absolute values for many calculations). Another definition for F is

$$F = \frac{\text{measured noise power out of the subsystem at room temperature}}{\text{power out of the subsystem if the subsystem was ideal}} \tag{3.20}$$

If G is the gain of the subsystem, the equivalence of the two definitions can be seen in

$$F = \frac{\text{SNR}_{\text{input}}}{\text{SNR}_{\text{output}}} = \frac{S_{\text{in}}}{kT_0B} \bigg/ \frac{GS_{\text{in}}}{GkT_0B + GkT_eB} = \frac{GkT_0B + GkT_eB}{GkT_0B} \tag{3.21}$$

where T_0 is room temperature and the equivalent noise of the subsystem is kT_eB, referred to the input, and the noise measured at the output is therefore $GkT_0B + GkT_eB$.

As a fringe benefit of (3.21), we can also use it to write F in terms of T_e and T_0, since

$$F = \frac{GkT_0B + GkT_eB}{GkT_0B} = \frac{T_0 + T_e}{T_0} = 1 + \frac{T_e}{T_0} \tag{3.22}$$

Alternatively,

$$T_e = (F - 1)T_0 \tag{3.23}$$

Again going back to (3.21), we can see another reason why input referencing is convenient when we work with noise figures. If we let

$$N_{\text{in}} = kT_0B \tag{3.24}$$

be the noise power at the input and

$$N_{\text{out}} = GkT_0B + GkT_eB \tag{3.25}$$

be the noise power at the output, then substituting into (3.21), we have

$$F = \frac{N_{\text{out}}}{GN_{\text{in}}} = \frac{GN_{\text{out,input-ref}}}{GN_{\text{in}}} = \frac{N_{\text{out,input-ref}}}{N_{\text{in}}} \tag{3.26}$$

where $N_{\text{out,input-ref}}$ is the output noise, referenced to the input. Here we see that F is independent of G, so a benefit of referring the noise back to the input is that, then, F doesn't depend on gains in the system.

3.2.5.1 Noise Figure for Different Types of Devices

We have already seen how the noise figure is computed for subsystems with a gain G (e.g., amplifiers). For passive devices such as transmission lines or attenuators operating at room temperature, $F = L$, where L is the loss. This is because the signal gets attenuated by L dB, whereas noise measured at the output equals noise measured at the input. Thus SNR decreases by exactly L dB going through the passive device. Equivalently, we have $F = -G$.

For antennas, it depends on what an antenna "sees." For antennas that are terrestrial, room temperature is the norm (i.e., they bring in atmospheric noise with a T_e of room temperature). For antennas pointing to space (e.g., for satellite systems), it is typically said that they "see" a noise temperature of 50 K.

3.2.5.2 Cascade

When there are two or more elements in cascade, we can use the Friis formulas

$$F = F_1 + \frac{F_2 - 1}{G_1} + \frac{F_3 - 1}{G_1G_2} + \cdots \tag{3.27}$$

or, equivalently,

$$T = T_1 + \frac{T_2}{G_1} + \frac{T_3}{G_1G_2} + \cdots \tag{3.28}$$

where we are using equivalent temperatures.

Derivation: Starting with $FGkT_0B$ as the noise out of a device with gain G and noise figure F, if we have F_1 and G_1 in cascade with F_2 and G_2, the noise out of the first device is $F_1G_1kT_0B$. This noise then becomes $F_1G_1G_2kT_0B$ coming out of the second device. Meanwhile, second device adds $G_2kT_eB = F_2G_2kT_0B - G_2kT_0B$ (we think of it as that only the first device in the cascade gets the "real" input noise kT_0B and all the others only have their added noise). Let $G = G_1G_2$. Then

$$FGkT_0B = F_1G_1G_2kT_0B + F_2G_2kT_0B - G_2kT_0B. \tag{3.29}$$

$$FG = F_1G_1G_2 + F_2G_2 - G_2. \tag{3.30}$$

$$F = F_1 + (F_2 - 1)/G_1. \tag{3.31}$$

This can be extended for three or more subsystems in cascade.

3.2.5.3 Receiver Sensitivity Suppose that the signal detector (after RF) needs a minimum SNR of SNR_{min} in order to work (e.g., to achieve some BER target). Then the smallest SNR that will provide the signal detector with at least SNR_{min} must be larger than SNR_{min}, because we expect components in the RF circuitry to add noise and hence to reduce the SNR. Receiver sensitivity, then, is the smallest SNR entering the RF stage in the receiver that would provide the signal detector with at least SNR_{min}. Thus, in RF design, low (better) receiver sensitivity is an important goal, so the receiver can work with signals with smaller SNR.

As could be expected, receiver sensitivity relates directly to the noise figure, where a larger noise figure means a higher (worse) receiver sensitivity. We relate noise figure, F, bandwidth, B, and minimum needed SNR, SNR_{min}, as follows. From the definition of the noise figure;

$$F = \frac{\text{SNR}_{\text{in}}}{\text{SNR}_{\text{out}}} = \frac{S/N_{\text{in}}}{\text{SNR}_{\text{out}}} \tag{3.32}$$

where S and N_{in} are the signal power and input noise power, respectively. Expressing N_{in} in terms of measurement bandwidth and input noise spectral density $N_{\text{in/Hz}}$ and rearranging terms, we have

$$S = N_{\text{in/Hz}} \times B \times F \times \text{SNR}_{\text{out}} \tag{3.33}$$

Now, for the sensitivity, we simply substitute SNR_{out} with SNR_{min} and S becomes the sensitivity. Furthermore, we write it all in decibels, to get

$$S_{\text{in,min|dBm}} = N_{\text{in/Hz|dBm/Hz}} + F|_{\text{dB}} + \text{SNR}_{\text{min|dB}} + 10\log B \tag{3.34}$$

Usually, we assume matched conditions and room temperature at the input, so then we have

$$N_{\text{in/Hz|dBm/Hz}} = kT = -174 \text{ dBm/Hz} \tag{3.35}$$

So we have the following useful relationship:

$$S_{\text{in,min|dBm}} = -174 \text{ dBm/Hz} + F|_{\text{dB}} + \text{SNR}_{\text{min|dB}} + 10\log B \tag{3.36}$$

NB: Sensitivity is one of those quantities that we come across from time to time where the numerical value may be opposite that of an informal description of it; when we say that receiver A is more sensitive than receiver B, the numerical value of sensitivity of A is *less* than that of receiver B, and vice versa.

3.2.5.4 Noise Floor The concept of noise floor is closely related to the concept of sensitivity. Denoting noise floor by N_{floor}, we define *noise floor* as

$$N_{\text{floor}} = -174 \text{ dBm/Hz} + F|_{\text{dB}} + 10\log B \tag{3.37}$$

which can be thought of as the minimum noise that will be contributed by the system, measured at the output at room temperature under matched load conditions and for a measurement bandwidth B. Thus, sensitivity and noise floor are related by

$$S_{\text{in,min}|\text{dBm}} = N_{\text{floor}} + \text{SNR}_{\text{min}|\text{dB}} \tag{3.38}$$

3.2.5.5 *Assumptions, Gotchas, and so on* Here are some things to remember about noise calculations, especially for those who might be new to this topic.

- The noise figure is always measured at room temperature, 290 K.
- Always assume matching load conditions at the input and output.

These are the typical assumptions, but you can also talk about a more general concept of noise figure that doesn't necessarily have to be at room temperature or under matched load conditions. Razavi uses this more general concept and shows how in this case, the noise figure becomes dependent on the source resistance, for example [7]. The traditional cases of matched loads removes ambiguity. So, in practice, we make the traditional assumptions.

3.3 SYSTEM ISSUES RELATED TO NONLINEARITY

Nonlinear systems such as amplifiers work ideally (i.e., output power increases linearly with input power, in dB) only over a certain range of input powers. Quantitatively, the 1-dB compression point (Section 3.3.1.1) is one way to characterize an upper limit to the input power. Another issue is that intermodulation products grow faster than the fundamentals as input power increases, resulting in another upper limit to the input power above which the contributions from intermodulation products would be considered excessive. In this section we show how these phenomena result from nonlinearity and how they can be quantified.

3.3.1 Gain Compression

Nonlinearity can be viewed as a variation from the small signal gain of a sub-system. In most cases, the gain is "compressive" (it saturates). Using our standard model, (3.2), the nonlinearity is seen as the $3a_3A^3/4$ term in $a_1 + 3a_3A^3/4$, where in the usual compressive case, $a_3 < 0$.

3.3.1.1 *1-dB Compression Point* Gain compression is typically quantified by the concept of a 1-dB compression point, the point at which the output drops by 1 dB from the extrapolation of the linear gain region. The linear region, where $P_{\text{out}} = G + P_{\text{in}}$ (in decibels), is shown in Figure 3.8 together with its extrapolation. However, since gain is compressive, it moves away from linear when the input signal becomes large enough, and the 1-dB compression point is also shown in the figure. In general, points such as the 1-dB compression point may be referred to the input or output power (as it is associated with both an input and an output power), and both are shown in the diagram.

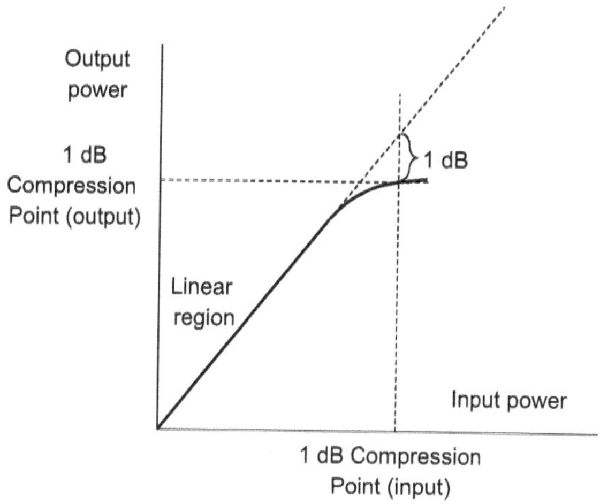

FIGURE 3.8 One-decibel compression point.

3.3.2 Size of Intermodulation Products

To find the 1-dB compression point, we could just start with a small input power, increase it gradually, and observe the output power increase linearly until it bends and reaches the compression point. To observe amplification of intermodulation products, we can use the *two-tone test*. As we saw in Section 3.1.4.2, we need at least two sinusoids input to a nonlinear system to see intermodulation products at the output other than harmonics. In particular, we are looking for the two third-order intermodulation products that are closest to the fundamentals ($2\omega_1 - \omega_2$ and $2\omega_2 - \omega_1$, as discussed earlier). Thus, we can start by putting two sinusoids (a pure sinusoid can, alternatively, be called a *tone*) into the input at small power, in a setup such as the one shown in Figure 3.9.

At small powers, the intermodulation products are "buried in the noise." As we increase the input power of the fundamentals, we expect the power of the third-order intermodulation products to increase three times as fast as the power of the fundamentals. At a particular power level, the third-order intermodulation products will "emerge out of the noise floor" (Figure 3.10) and are clearly visible as distinct

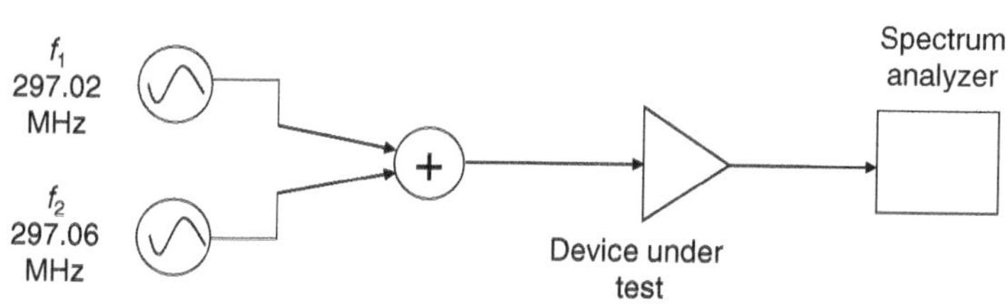

FIGURE 3.9 Two-tone test.

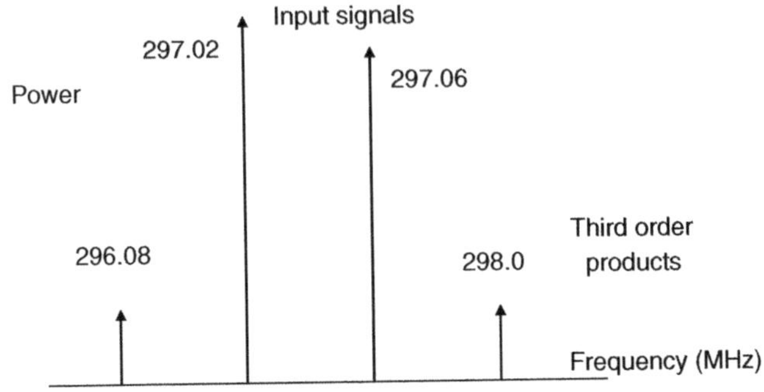

FIGURE 3.10 Third-order products close to the two tones.

spectral components on a spectrum analyzer. The point where this happens is important as part of SFDR, which is discussed in Section 3.3.3. However, we consider first what happens as we continue to increase the input power. As can be seen in Figure 3.11, the two curves will theoretically intersect at some point of higher input power, called the *third-order intercept point*, often abbreviated as IP3. In reality, gain compression occurs before that (often, 12 to 15 dB before getting to IP3 [5]; Egan [2] provides a theoretical basis for estimating the difference between the 1-dB compression point and IP3 at 10.6 dB, but cautions that it is only a simple attempt that depends on a number of assumptions, so in practice there will be deviations). Nevertheless, the slope of both lines should be computable from a couple of measurements, and the IP3 can therefore be obtained by extrapolation. In fact, if we trust our model, just one measurement is needed, of the output power of the fundamental and of the third-order intermodulation product, for one value of the input power in the linear region. Then, taking the slopes of the curves as 1 and 3, we can extrapolate to the intersection point, IP3.

The IP3 is sometimes presented as a figure of merit, because the larger the IP3, the higher the input power before the third-order intermodulation products in the output

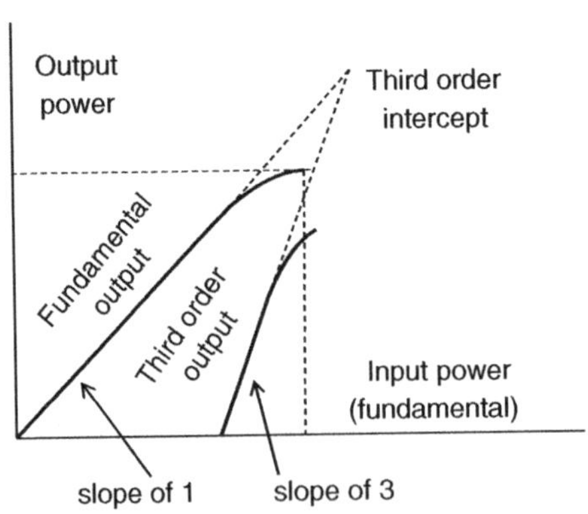

FIGURE 3.11 Third-order intercept.

become too large. However, there can be some ambiguities in the meaning of IP3, such as whether it is output referenced or input referenced, and whether the x-axis represents the power of one or both of the fundamentals (making a 3-dB difference in the IP3) [2].

3.3.3 Spur Free Dynamic Range

Intuitively, the dynamic range of a subsystem such as an amplifier, or of a radio receiver as a whole, is from some minimum usable input power to some maximum usable input power. The question is what we consider to be "usable." Two possibilities for the lower end are:

- It could be taken as the sensitivity [as defined in (3.38)].
- Since sensitivity depends on the SNR_{min} of the detector stage, to define a minimum usable input signal that is independent of an SNR_{min} specification, the minimum usable input is sometimes taken as the noise floor.

For the upper end, a popular concept of maximum usable input power is that input power in which the third-order intermodulation products just begin to emerge from the noise floor (more precisely, the third-order intermodulation products equal the noise floor). This concept of usable input power range is called the *spur free dynamic range* or *spurious free dynamic range* (SFDR).

Let P_{IIP3} and P_{OIP3} be the third-order intercept point, referenced at the input and output of the amplifier, respectively. Let SNR_{min} be the minimum acceptable SNR. Hence, the minimum input power to meet the minimum SNR requirement (i.e., the receiver sensitivity) is

$$P_{\text{in,min}} = N_{\text{floor}} + \text{SNR}_{\text{min}} \tag{3.39}$$

Now, given any small input power (i.e., so we are in the linear operating region) of the fundamental, P_{in}, denote the corresponding output power of the fundamental as P_{out}, and denote the output power of the corresponding third-order products as P_{OIM3}. Then, clearly, $P_{\text{out}} = G + P_{\text{in}}$ and $P_{\text{OIM3}} = G + P_{\text{IIM3}}$, where P_{IIM3} represents the input-referenced third-order product. Even though P_{IIM3} can be shown as a point on the x-axis, we must be careful to remember that to interpret both curves in Figure 3.12 correctly, we have to take the x-axis as the input power of the *fundamental*, and then the two curves represent the output power of the fundamental and the third-order intermodulation (IM) product, respectively.

Since we know that the slope of the third-order IM and fundamental curves are 3 and 1, respectively,

$$\frac{P_{\text{OIP3}} - P_{\text{OIM3}}}{P_{\text{OIP3}} - P_{\text{out}}} = 3 \tag{3.40}$$

$$3\left(P_{\text{OIP3}} - P_{\text{out}}\right) = P_{\text{OIP3}} - P_{\text{OIM3}} \tag{3.41}$$

$$2\left(G + P_{\text{IIP3}}\right) - 3\left(G + P_{\text{in}}\right) = -P_{\text{OIM3}} \tag{3.42}$$

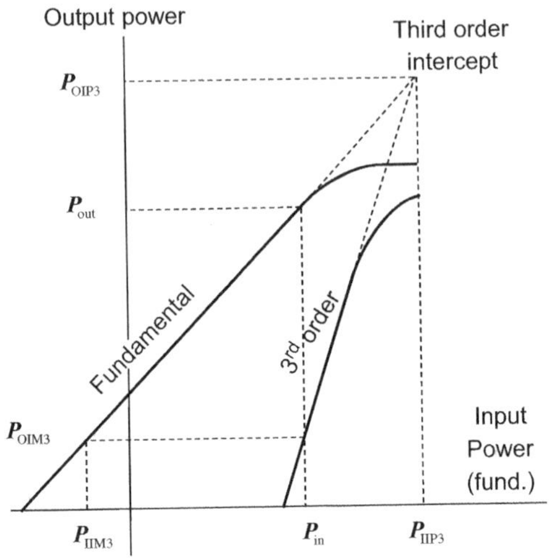

FIGURE 3.12 Computing the spur free dynamic range.

$$2P_{IIP3} - 3P_{in} = -(P_{OIM3} - G) \tag{3.43}$$

$$P_{in} = \frac{1}{3}(2P_{IIP3} + P_{IIM3}) \tag{3.44}$$

Thus, if we now set $P_{IIM3} = N_{floor}$, then P_{in} would be the maximum input signal for purposes of the SFDR definition, which we call $P_{in,max}$. Thus,

$$P_{in,max} = (1/3)(N_{floor} + 2P_{IIP3}) \tag{3.45}$$

Therefore, we now have

$$\begin{aligned} \text{SFDR} &= P_{in,max} - P_{in,min} && (3.46)\\ &= (1/3)(N_{floor} + 2P_{IIP3}) - N_{floor} - \text{SNR}_{min} && (3.47)\\ &= (2/3)(P_{IIP3} - N_{floor}) - \text{SNR}_{min} && (3.48) \end{aligned}$$

An alternative expression for SFDR, when $P_{in,min} = N_{floor}$, is therefore

$$\text{SFDR} = (2/3)(P_{IIP3} - N_{floor}) \tag{3.49}$$

3.3.3.1 IP3 of a Cascade When there is a cascade of subsystems, we can find the OIP3 of the system if we have gain G and P_{OIP3} of each subsystem. (NB: If we are given P_{IIP3}, we can convert to P_{OIP3} before using the formula.) Let the subsystems be numbered from 0 to N, and let i be an index. Let $P'_{OIP3,i}$ be the OIP3 of the cascade of subsystems 0 to i, where $0 < i < N$ (and $P_{OIP3,0}$ is the OIP3 of subsystem 0). Then we can compute the system OIP3 recursively using the formula

$$P'_{\mathrm{OIP3},i} = \left(\frac{1}{P'_{\mathrm{OIP3},i-1} G_i} + \frac{1}{P_{\mathrm{OIP3},i}} \right)^{-1} \tag{3.50}$$

NB: These values are all linear (i.e., not decibel values).

3.4 MIXING AND RELATED ISSUES

Mixers are typically used to "multiply" or "mix" a signal with a single-frequency signal (such as the output of an oscillator). As such, a mixer is a three-port subsystem (two input ports and one output port). As we have seen in Section 3.1.4, nonlinearities in the RF circuit can result in intermodulation products, including harmonics. Some of the intermodulation products, like some third-order intermodulation products, can be very close to the frequencies of the desired signals, making them difficult to filter out. Additional challenges arise when we look at the mixers in the RF circuit.

- Mixers are themselves nonlinear devices, and that is why they can provide an output that has, among other terms, the product of the two inputs.
- Mixers perform frequency translation, so undesired signals that were far from the desired signal (in frequency) might end up close to the desired signal at the output of the mixer.

Suppose that we are using a mixer to down-convert a cellular signal from an RF frequency f_{RF} to an IF frequency f_{IF}. So we are interested only in the *difference frequency* out of the mixer (the sum frequency is always greater than either of the input frequencies, so it can be used for up-conversion, but never for down-conversion). To take the difference frequency out of the mixer as f_{IF}, we can either set the local oscillator frequency to $f_{\mathrm{LO}} = f_{\mathrm{RF}} - f_{\mathrm{IF}}$ or to $f_{\mathrm{LO}} = f_{\mathrm{RF}} + f_{\mathrm{IF}}$. In the following, we denote the two cases by $f_{\mathrm{LO}} < f_{\mathrm{RF}}$ and $f_{\mathrm{LO}} > f_{\mathrm{RF}}$, respectively. At the mixer output we have

$$f_{\mathrm{IF}} = |f_{\mathrm{RF}} - f_{\mathrm{LO}}| \tag{3.51}$$

(where we take absolute values to allow for the case that $f_{\mathrm{LO}} > f_{\mathrm{RF}}$). Now if there are other signal components at some other frequencies (not RF), they will appear at

$$f_{\mathrm{out}} = |f_{\mathrm{other}} - f_{\mathrm{LO}}| \tag{3.52}$$

where f_{other} is some other frequency, and $f_{\mathrm{other}} \neq f_{\mathrm{RF}}$. Now if $f_{\mathrm{IF}} = f_{\mathrm{out}}$ (or even if they are close), the undesired signal interferes with the desired signal. To be more precise, when $f_{\mathrm{IF}} = f_{\mathrm{out}}$, we call f_{other} the *image frequency*, which we may denote by f_{image}. These constraints define a unique image frequency. f_{image} and f_{RF} are the same distance from f_{LO} (the distance being equal to f_{IF}) and on opposite sides of f_{LO}.

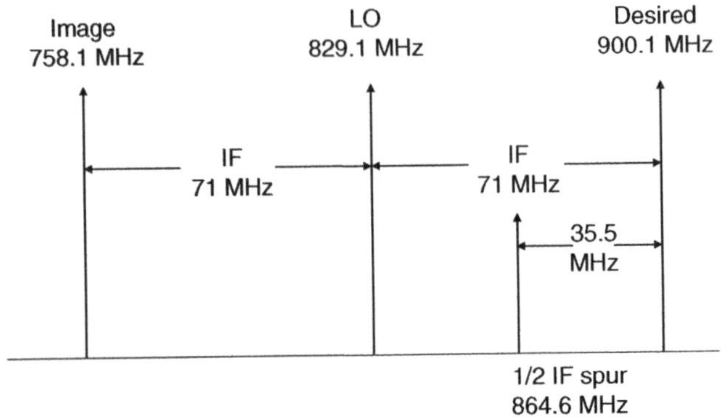

FIGURE 3.13 Image frequency and half-IF spur.

An example is provided in Figure 3.13, where $f_{\mathrm{RF}} = 900.1$ MHz (labeled "Desired" in the figure), $f_{\mathrm{LO}} = 829.1$ MHz and $f_{\mathrm{image}} = 758.1$ MHz. The other possibility is where f_{RF} and f_{image} are interchanged. The figure also illustrates another possible interfering frequency, the *1/2-IF spur*. The 1/2-IF spur arises because of the presence of the second harmonic of f_{LO}, which we denote by $f_{\mathrm{LO},2} = 2f_{\mathrm{LO}}$. There are two frequencies that $f_{\mathrm{LO},2}$ translates to f_{IF}. These are $2f_{\mathrm{LO}} \pm f_{\mathrm{IF}}$, both of which are easily removed by filtering. However, we may also have second harmonics of undesired signals. In particular, $2f_{\mathrm{LO}} \pm f_{\mathrm{IF}}$ are the second harmonics of

$$f_{\mathrm{other},1} = f_{\mathrm{LO}} + \frac{1}{2} f_{\mathrm{IF}} \tag{3.53}$$

and

$$f_{\mathrm{other},2} = f_{\mathrm{LO}} - \frac{1}{2} f_{\mathrm{IF}} \tag{3.54}$$

In the event that $f_{\mathrm{LO}} < f_{\mathrm{RF}}$, then (3.53) becomes

$$f_{1/2-\mathrm{IFspur}} = f_{\mathrm{RF}} - \frac{1}{2} f_{\mathrm{IF}} \tag{3.55}$$

where we have renamed $f_{\mathrm{other},1}$ as $f_{1/2-\mathrm{IFspur}}$. This is the case shown in Figure 3.13. The 1/2-IF spur could present a very serious problem because it is so close (only 1/2 the IF, i.e., 22.5 MHz in our example) to the desired signal that it could be hard to filter off well. As for the other case, that $f_{\mathrm{LO}} > f_{\mathrm{RF}}$, Exercise 3.5 works out where the 1/2-IF spur is.

In general, the mixing not only of the signal desired and the local oscillator signal, but also second and higher harmonics (of both the local oscillator and desired signals) lead to a range of frequencies from which other signals could be translated to interfere with the desired signal at the mixer output, and these should be studied in any careful radio design. However, the image frequency and the 1/2-IF spur are the best known of these, given the strength of the fundamentals, and even second harmonics in many cases, and given the closeness of the 1/2-IF spur to the signal desired. These problems can be made more challenging in the presence of phase noise (Section 3.5.1).

3.5 OSCILLATORS AND RELATED ISSUES

Oscillators are used to generate the continuous-wave signal to be modulated in radio transmitters and for up-coverting the frequency to RF, and they are also used in receivers for down-converting received signals (for up-conversion or down-conversion, the oscillator output would be one input to the mixer).

3.5.1 Phase Noise

Oscillators are not perfect. Thus, the output of an ideal oscillator is a delta function at the desired frequency, f_0, but the output of a real oscillator would be smeared out around f_0. Figure 3.14 shows the spectral distribution of an oscillator with phase noise. We see that some of the power of the oscillator signal is not exactly at f_0.

The effects of phase noise include:

- It degrades SNR (and correspondingly, BER); that is, it *desensitizes* the receiver, so it is sometimes said that the sensitivity is reduced (the numerical value of sensitivity goes up).
- It causes jitter in the received signal that can cause problems for timing recovery.
- It degrades receiver selectivity.

Phase noise also degrades receiver selectivity, because of a phenomenon called *reciprocal mixing*, which occurs when there is a strong interferer in an adjacent frequency channel, and the LO signal has enough phase noise that the down-converted interferer signal spills into the carrier frequency of the down-converted desired signal. This effect is illustrated in Figure 3.15.

In the case of a nearby transmitter (e.g., a transmitter in the same device as the receiver), phase noise *in the transmitter signal* could seriously interfere with a desired signal which may also be many orders of magnitude smaller than the transmitter

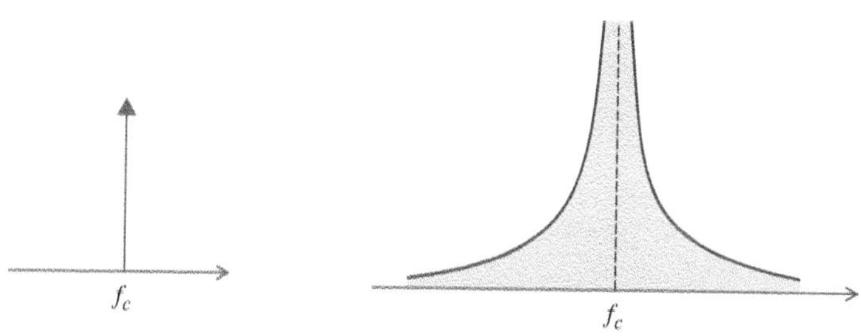

FIGURE 3.14 Phase noise.

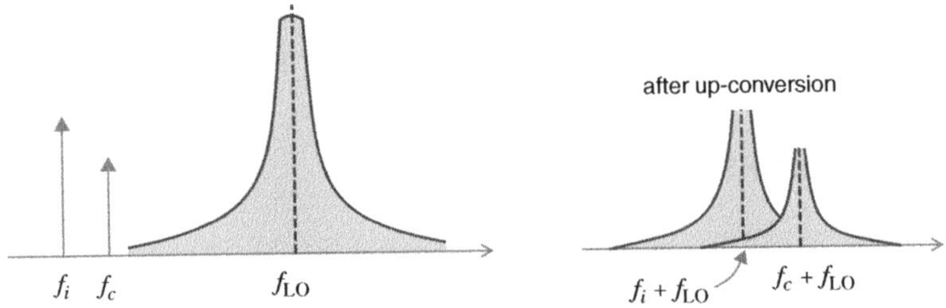

FIGURE 3.15 Reciprocal mixing.

signal (Figure 3.16). This is different from reciprocal mixing in that the problem is in the transmitter, so even if the receiver local oscillator is ideal (no phase noise), the transmitter signal may still interfere significantly with the receiver signal.

Phase noise in a receiver causes the SNR to decrease, leading to a more challenging symbol detection problem for the detector, resulting in higher bit error rates. One way that this can be viewed is by observing the effect of phase noise on the signal constellation at the receiver, as shown in Figure 3.17. The distance between constellation points is effectively reduced because of the phase noise. AWGN also causes constellation points to be smeared, but in a more circular pattern. The figure actually shows three different effects (in reality, a given system would have one of the effects only, on all constellation points; however, rather than draw three separate diagrams, we more compactly show all three in one diagram). The constellation point on the bottom left is the case where there is no AWGN and no phase noise. Where there is phase noise, the constellation point would move along the edge of the circle, as shown in the top left constellation point. The top right constellation point exhibits the effects of AWGN alone, without phase noise. Here, not only the phase, but also the amplitude, of the signal is affected. The bottom right constellation point exhibits both AWGN and phase noise.

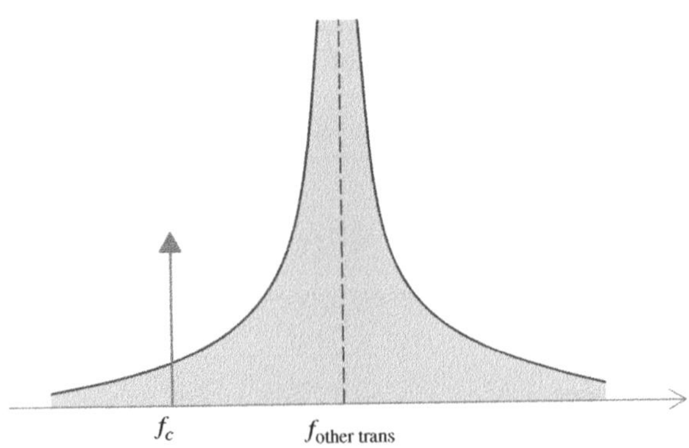

FIGURE 3.16 Phase noise in nearby transmitter.

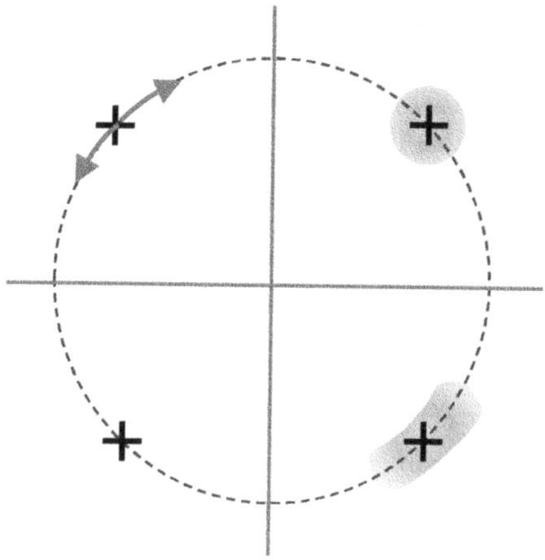

FIGURE 3.17 Effects of phase noise and AWGN on the signal constellation of a QPSK signal.

Furthermore, if timing recovery in a received frame is affected because of the jitter from phase noise, so that it introduces an offset of ϕ_Δ, then every constellation point received in that frame would be offset by ϕ_Δ.

3.6 AMPLIFIERS AND RELATED ISSUES

There are two main types of amplifiers in common use in RF engineering: *low-noise amplifiers* (LNAs) and *power amplifiers* (PAs). Although both attempt to amplify the input signal, the design space (range of parameters and design requirements) are quite different. Sometimes, RF engineers also talk about *low-level linear amplifiers*, which are in between LNAs and PAs in their requirements, and which may precede PAs.

3.6.1 Low-Noise Amplifiers

Low-noise amplifiers are typically found toward the front stages of an RF receiver. As such, they need to amplify a very weak signal while at the same time adding as little noise as possible—hence the name LNA. Typically, the noise figure is 2 dB, although it may be as low as 1 dB. However, the gain may be limited (15 dB is a typical value), because of the constraint from the low-noise figure.

Reverse isolation might be important, to suppress stray signals from the local oscillator that may leak to the antenna. A typical value for reverse isolation is 20 dB.

3.6.2 Power Amplifiers

Power amplifiers are used on the transmitter side to transmit at relatively high powers. Thus, they usually consume the most power of any subsystem of an RF transceiver. The gain may be 20 to 30 dB, and the output power may be in the range 20 to 30 dBm.

There are a variety of classes of PAs, each with different trade-offs. On one extreme, *class A amplifiers* are the most linear, but least efficient, following the input signal most closely. *Class B amplifiers* follow the input signal only during half of the input signal cycle, achieving more efficiency with less linearity. *Class C amplifiers* can achieve more efficiency than class B amplifiers, but due to the nonlinearity, are useful only if the modulation is of the constant envelope variety.

3.7 OTHER COMPONENTS

In low-frequency and dc circuits, it is straightforward to combine two or more signals at a point in the circuit, or to split a signal into two or more paths. We just arrange for electrical contact by touching the relevant wires together. At RF, we cannot do the same thing. However, there are common components that can be used for such purposes and more. For example, a power divider (e.g., a three-way power divider as shown in Figure 3.18) can be used to split an incoming signal into three output paths.

3.7.1 Directional Couplers

A directional coupler is a popular four-port device. As a four-port device, it may be represented as in Figure 3.19. Typically, a signal is input to port 1, and the following happens at the other three ports:

- *Port 2* (*straight-through*). Most of the signal comes out of this port, with a small amount of coupling loss.

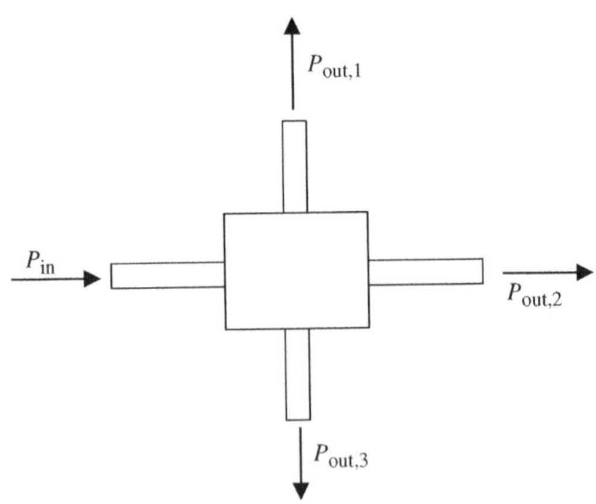

FIGURE 3.18 Three-way power divider.

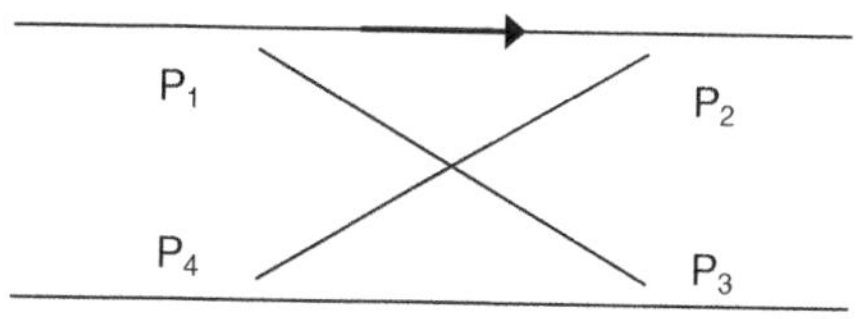

FIGURE 3.19 Four-port reference diagram for discussion of directional couplers and circulators.

- *Port 3* (*coupling*). Some of the signal comes out of this port, due to coupling within the device.
- *Port 4* (*isolation*). Very little of the signal comes out of this port (generally, the less, the better).

Thus, a directional coupler is useful for various applications, such as where we want to "tap" the signal, or "siphon" off part of it (e.g., to connect to a test and measurement device). Then we would input the signal into port 1, with most of it coming out of port 2 and the "siphoning" off being at port 3.

3.7.2 Circulators

A circulator is a three-port device, with the following behavior:

- When a signal goes into port 1, most of it comes out of port 2, with very little out of port 3.
- When a signal goes into port 2, most of it comes out of port 3, with very little out of port 1.
- When a signal goes into port 3, most of it comes out of port 1, with very little out of port 2.

3.7.2.1 Duplexers A duplexer is used to connect both a transmitter and a receiver to the same antenna or antenna system. When signal is coming from the transmitter, it should ideally go only to the antenna and not to the receiver. When signal is coming from the antenna, it should ideally go only to the receiver and not to the transmitter. Hence, a circulator can be used to provide this behavior.

EXERCISES

3.1 In computing the noise figure for a chain of subsystems, there is a small shortcut that can be used when passive lossy devices such as transmission lines are in the chain. Suppose that we have such a device as the ith subsystem, and another subsystem as the $(i+1)$th, with gains given by $G_i = -L$ and G_{i+1}, respectively, and if the noise figure of the $(i+1)$th subsystem is F_{i+1} (all in decibels).

Then the ith and $(i+1)$th subsystems can be reduced to one subsystem with the noise figure given by

$$F|_{\text{dB}} = L + F_{i+1}|_{\text{dB}}$$

This simple addition of the noise figures in dB obviates the usual need to convert from dB to linear to use the Friis formula. One way to show the validity of this shortcut is to use the Friis formula on the the subsystems. Do it!

3.2 Consider part of an RF system that has a bandpass filter (1.5 dB loss, 150 MHz centered at 2.4 GHz), followed by two amplifiers, one after the other (10 dB gain, $F = 2$ dB, and 15 dB gain, $F = 1.5$ dB), at room temperature (assume 290 K). What is the noise figure of the system?

3.3 Let us think about the noise floor and what it really means, especially since we use it as part of our concept of SFDR. We interpret (3.38): Is it (a) noise power entering the subsystem input from earlier stages, (b) noise power contributed by the subsystem, input referenced, (c) noise power contributed by the subsystem, output referenced, or (d) something else? In the definition of SFDR, what does it mean for the IM3 product to be equal to the noise floor? In particular, is it an input-referred IM3 product or an output-referred IM3 product that is equal to the noise floor? Why?

3.4 A receiver has a 10-dB noise figure, a 1-MHz bandwidth, a 5-dBm third-order intercept point, and a 0-dB SNR_{min}. Compute its sensitivity and SFDR. Next, add a preamplifier with 24 dB gain and 5 dB NF. What is the sensitivity now?

3.5 In the text we have worked out the location of the 1/2-IF spur for the case $f_{\text{LO}} < f_{\text{RF}}$. Now we consider the other case, $f_{\text{LO}} > f_{\text{RF}}$. Do either of (3.53) and (3.54) give the location of a 1/2-IF spur that is close to the desired signal? If so, and if $f_{\text{LO}} = 829.1$ MHz and $f_{\text{IF}} = 71$ MHz, as in Figure 3.13, what is $f_{1/2-\text{IFspur}}$?

REFERENCES

1. K. Chang. *RF and Microwave Wireless Systems*. Wiley, Hoboken, NJ, 2000.
2. W. F. Egan. *Practical RF System Design*. IEEE-Wiley, Hoboken, NJ, 2003.
3. T. S. Laverghetta. *Microwaves and Wireless Simplified*, 2nd ed. Artech House, Norwood, MA, 2005.
4. D. K. Misra. *Radio-Frequency and Microwave Communication Circuits: Analysis and Design*. Wiley, Hoboken, NJ, 2001.
5. D. M. Pozar. *Microwave Engineering*, 3rd ed. Wiley, Hoboken, NJ, 2005.
6. S. Ramo, J. Whinnery, and T. Van Duzer. *Fields and Waves in Communication Electronics*. Wiley, New York, 1984.
7. B. Razavi. *RF Microelectronics*. Prentice Hall, Upper Saddle River, NJ, 1998.

4

ANTENNAS

An antenna is a device that is used in both wireless transmitters and receivers. In a wireless transmitter, an antenna converts guided electromagnetic signals (usually, from a transmission line) into propagating electromagnetic wave signals. In a wireless receiver, an antenna converts propagating wireless electromagnetic wave signals (arriving at the receiver) into guided electromagnetic signals.

When used for transmitting wireless signals, instead of letting the same amount of power be radiated in every direction, antennas often direct the signals, for more efficient communications. The directivity, antenna gain, and so on, are ways of quantifying this phenomenon, which is also known as the antenna pattern. In Section 4.1.8 we take a further and more quantitative look at these directional characteristics. When used for receiving wireless signals, the directional characteristics are the *same* as for transmitting. This reciprocity between the patterns for transmitting and receiving is one of the reciprocity characteristics of antennas.

By reciprocity principles, the following are the same:

1. The impedance for transmitting and receiving
2. The directional characteristics/patterns when used for transmitting and for receiving

To improve our understanding of antennas and their use, we need to examine different ways to characterize them, only part of which are their directional characteristics. Thus, we examine various characterizations of antennas in Section 4.1, which will provide us with different ways to talk about antennas. Having provided this foundation, in Section 4.2 we examine some of the many types of antennas in existence.

Fundamentals of Wireless Communication Engineering Technologies, First Edition. K. Daniel Wong.

Another dimension of flexibility and control is available to us if we use arrays of antennas, not just single antennas, so we consider antenna arrays in Section 4.3. We end the chapter with a brief look at some practical issues in using and connecting antennas, and in feeding them.

4.1 CHARACTERIZATION

We begin with some aspects of three-dimensional geometry useful for antenna work, then compare the near field to the far field, and discuss polarization. Then we consider a range of topics related to the antenna pattern. We also consider concepts of aperture.

4.1.1 Basic 3D Geometry

Although most people are probably more familiar with two-dimensional (2D) geometry than three-dimensional (3D) geometry, we live in a 3D world. In many areas of engineering, 2D suffices, but sometimes 3D geometry is necessary and helpful. The study of antennas is one area where some basic 3D concepts are helpful.

Often, we try to put 3D phenomena such as radiation patterns of antennas on paper as 2D diagrams. A common way to do this is to take cross sections of the phenomena in question (Figure 4.1). In studying antennas we often see references to these cross sections as representing various planes (we can imagine these as the intersection of a plane of infinite extent with the 3D shape). The *azimuth plane* is the horizontal plane parallel to the ground. The *elevation plane* is the vertical plane perpendicular

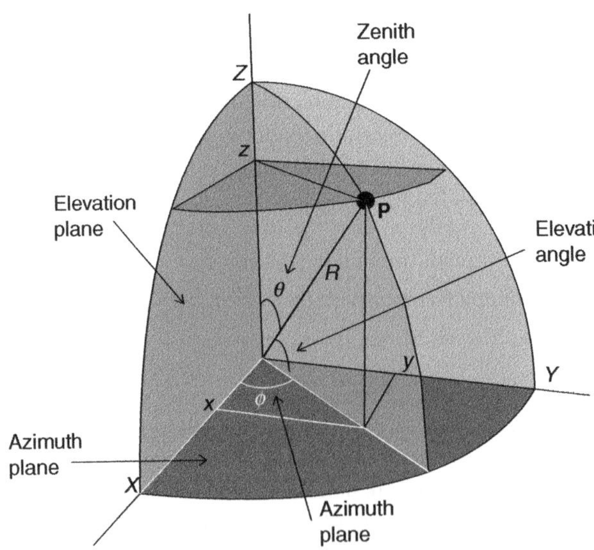

FIGURE 4.1 Geometrical terms related to a discussion of antenna patterns.

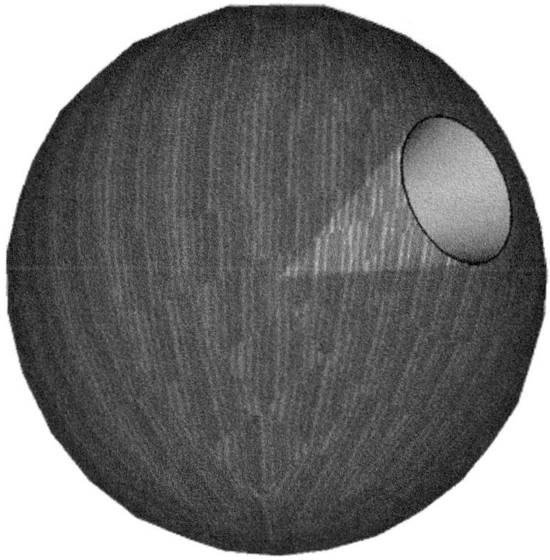

FIGURE 4.2 Small solid angle.

to the ground. The angle from a reference direction in the azimuth plane to a point in the azimuth plane is called the *azimuth angle* and denoted by ϕ; the angle in the elevation plane, from its intersection with the azimuth plane to a point in the elevation plane, is called the *elevation angle*. It is more common to use *zenith angle* instead of *elevation angle*. The zenith angle is the angle in the elevation plane from the perpendicular to the azimuth plane to a point in the elevation plane. It is denoted by θ. Any point around the antenna can be described in spherical coordinates as (R, θ, ϕ), where R is the distance from the antenna.

Besides the use of planes in 3D to understand antenna-related phenomena, the use of the concept of *solid angle* (i.e., 3D angle) is also helpful. A *steradian* is a common way to quantify a solid angle. It is defined in an analogous way to its 2D counterpart, the radian. In the 2D case, working with regular (2D) angles (also known as *planar angles*), 1 radian is the angle subtended by an arc whose length is exactly equal to the radius of the circle. In the 3D case, circles and length of arcs generalize to spheres and areas of regions on the surface of the sphere, respectively. One steradian is the solid angle subtended by a region whose area is equal to the square of the radius of the sphere. The concepts of solid angle and steradian are illustrated in Figure 4.2. The figure shows a sphere from which a cone has been cut. The tip of the cone is at the center of the sphere, and the other end of the cone intersects with the surface of the sphere. The walls of the cone and the tip of the cone make a solid angle, analogous to how two lines intersecting at a point make a (2D) planar angle.

4.1.2 Near Field and Far Field

Differences can be observed in the behavior of electromagnetic waves at different distances from an antenna, no matter what antenna we use. It is convenient to think of

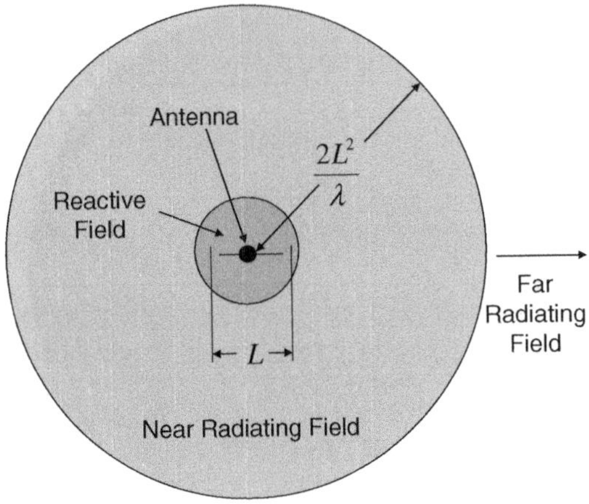

FIGURE 4.3 Far field, near radiating field, and reactive field.

there being two regions, the *Fresnel region* (also known as the *near-field region*) and the *Fraunhofer region* (also known as the *far-field region*). In the near field, we are close enough to the antenna that coupling effects significantly affect field patterns, whereas in the far field, we are far away enough from the antenna that the waves are just propagating radially outward (from the location of the antenna), no matter what antenna is used. In the far field, the shape of the field pattern is independent of distance. In the near field, it may depend on distance. In the near field, you have *coupling*, also known as *reciprocating* (or *oscillating*) energy flow, whereas energy flows outward in the far field. Reciprocating energy is reactive energy that is trapped near the antenna, as in a resonator. Thus, we need to be in the far field to measure how much energy is radiated far from the antenna.

The near field is sometimes further divided into two regions: the *reactive field* and the *near radiating field*. In the reactive field, coupling effects dominate, whereas there are both coupling and radiating effects in the near radiating field. All three regions are shown in Figure 4.3.

The boundary between the far field and near radiating field is often taken as

$$d_{\text{boundary}} = \frac{2L^2}{\lambda} \tag{4.1}$$

where L is the maximum dimension of the antenna. (NB: This rule of thumb works best if $L > \lambda$ [1].) As for the boundary between the near radiating field and reactive field, it may be taken [1] as $0.62\sqrt{L^3/\lambda}$. These are just popular approximations and we shouldn't expect to find abrupt transitions between regions as we make observations around these distances. Table 4.1 summarizes the three regions.

TABLE 4.1 Near and Far Fields

	Alternative Term	Characteristics
Reactive field	Fresnel region	Couping effects dominate
Near radiating field	Fresnel region	Some radiation, but energy flow not entirely radial
Far field	Fraunhofer region	Radiation, energy flow directed radially outward

4.1.3 Polarization

Linear polarization (of an electromagnetic wave) refers to cases where the **E** field is pointing in one direction. Horizontal polarization and vertical polarization refer to linear polarization that is parallel to the horizon or ground or is perpendicular to it, respectively. Why is the polarization described in terms of the direction of the electric field rather than of the magnetic field? This is just a matter of convention.

In the broad sense, circular polarization is sometimes used to refer to all cases where the direction of **E** is changing with time, which strictly speaking, includes both *elliptical polarization* and *circular polarization*. For circular or elliptical polarization to occur, the wave needs to be a superposition of waves that have **E** pointing in different directions and that are out of time phase (otherwise, if they are in phase, it can easily be seen to be linear polarization, e.g., by rotating the axes appropriately so that we have **E** pointing in just one direction). For example, we might have

$$\mathbf{E}(z) = \mathbf{a}_x E_x e^{-jkz} + \mathbf{a}_y E_y e^{j\theta_0} e^{-jkz} \tag{4.2}$$

where θ_0 is not an integer multiple of 2π. Circular polarization is, strictly speaking, a special case of elliptical polarization when the ellipse becomes a circle (e.g., $E_x = E_y$, $\theta_0 = 90°$). Sometimes, this is described as a question of looking at the *axial ratio* of the ellipse, where the axial ratio is the ratio of the major to minor axes. If it is almost 1, we could regard the polarization as circular.

4.1.3.1 Antenna Polarization

Closely related to the concept of wave polarization is the concept of *antenna polarization* (Figure 4.4). The word *polarization* originally refers to a property of waves, whereas an antenna is not a wave, so how can an antenna have a polarization? The idea of antenna polarization is that the

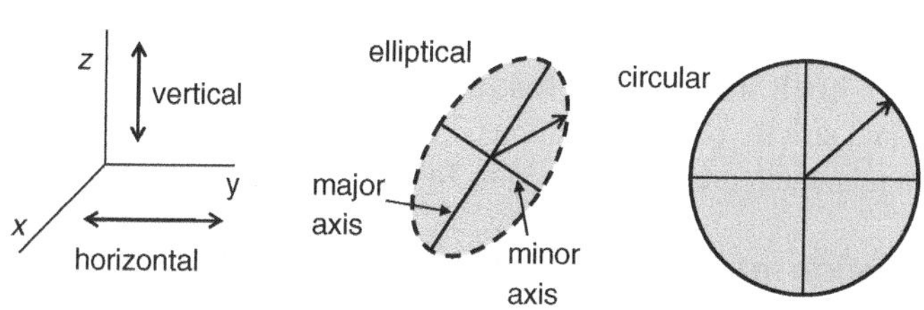

FIGURE 4.4 Antenna polarizations.

antenna, when used for transmission, will be transmitting waves of a certain polarization. Equivalently, and reciprocally, the antenna would best receive waves of a certain polarization (see Section 4.1.3.2 for more discussion on how it can receive waves with some polarizations better than others). [NB: Antenna polarization is not an absolute property of an antenna; the *same* antenna, in a different orientation, could be used to send and receive waves of a different polarization (see Section 4.2.4 for an example).]

Examples of antenna polarization include:

- For a dipole antenna, the polarization is the same as the orientation of the antenna. Thus, if the dipole is horizontal, its polarization is horizontal, and if it is vertical, its polarization is vertical.
- A turnstile antenna (see Section 4.2.4) can be used as a circularly polarized antenna.

For better reception of radio waves, antennas in receivers should, if possible, be oriented according to the polarization of the transmitted signal. TV antennas often have a horizontal orientation, to receive broadcast TV signals that are often horizontally polarized [5]. Car antennas often have a vertical orientation, to best receive vertically polarized AM broadcast signals. FM radio signals are transmitted circularly polarized, so the orientation doesn't matter as much. Multiple reflections in the propagation path change the polarization angles, so even if a signal is transmitted in a certain linear polarization, it may arrive linearly polarized at a different angle, or circularly polarized.

4.1.3.2 Polarization Loss and Mismatch Suppose that an antenna is linearly polarized. It will best receive waves that are linearly polarized in the same direction. Otherwise, the signal received will be reduced by $\cos\theta_p$, where θ_p is the angle between the antenna polarization and wave polarization. Thus, as we go from $\theta_p = 0$ to $\theta_p = 45^\circ$, we go from no loss to a 3-dB loss, and when we get to 90°, the polarization loss is infinite (in theory; in practice, polarization loss could be 20 to 30 dB for such severe mismatch cases).

If the wave is linearly polarized and the antenna is circularly polarized, or vice versa the polarization loss is 3 dB (the circularly polarized wave can be decomposed into two linearly polarized components, and a linearly polarized antenna picks up only one of the two). If both are circularly polarized, there is no loss if they are circularly polarized in the same sense. However, if they are in opposite senses (one right hand, one left hand), the loss is theoretically infinite. In practice, polarization loss on the order of 20 to 30 dB may be observed (because the signal and antenna may not be completely and strictly one polarization). Polarization loss due to different types of polarization mismatch are summarized in Table 4.2.

4.1.4 Radiation Intensity, Patterns, and Directivity

Antennas do not radiate the same amount of power in every direction. The *antenna pattern* describes how the amount of power radiated differs in different directions

TABLE 4.2 Polarization Loss

Wave Polarization	Antenna Polarization[a]	Polarization Loss
Linear	Linear, $\theta_p = 0$	0 dB
	Linear, $\theta_p = 45°$	3 dB
	Linear, $\theta_p = 90°$	∞ in theory (20–30 dB in practice)
	Circular	3 dB
Circular	Linear	3 dB
	Circular, same sense	0 dB
	Circular, opposite sense	∞ in theory (20–30 dB in practice)

[a] θ_p is difference in polarization (linear case).

around an antenna. The Poynting vector, introduced in Section 2.3.2.1, points radially outward from an antenna in the far field and gives power per unit area. Therefore, patterns may be expressed in (time-average) power per unit area (i.e., the Poynting vector). Patterns may also be expressed in (time-average) power per steradian (unit solid angle) (i.e., *radiation intensity*). This is written $U(\theta, \phi)$. If R is the distance from the antenna and $S(\theta, \phi)$ is the magnitude of the Poynting vector, then

$$U(\theta, \phi) = R^2 S(\theta, \phi) \qquad \text{W/sr} \tag{4.3}$$

Radiation intensity is a very convenient concept. Even as $S(\theta, \phi)$ falls off with R^2, radiation intensity is independent of distance. Thus, we do not need to worry about specifying a distance associated with an antenna pattern. As long as we are in the far field, radiation intensity just depends on the angles, not the distance from the antenna. Common variations of the pattern (besides being just the radiation intensity) are the radiation intensity in decibels and the normalized radiation intensity. Plotting normalized radiation intensity gives us the *normalized power pattern* [3]. We denote it by $P_n(\theta, \phi)$, and it is the ratio of U to the *maximum* radiation intensity:

$$P_n(\theta, \phi) = \frac{U(\theta, \phi)}{U(\theta, \phi)_{\max}} \tag{4.4}$$

Even though antenna patterns are often plotted in two dimensions, these are two dimensional cross sections of 3D phenomena. So it helps to be able to visualize the 3D picture, at least mentally. For our basic 3D geometry, we refer back to Figure 4.1. For an illustration of how a 3D pattern is typically plotted, we refer to Figure 4.5. Imagine that we have an antenna at the origin; then the figure shows a (3D) antenna pattern that might be obtained. It also shows how the 3D pattern can be projected to the azimuth and elevation planes to obtain the (2D) azimuth pattern and elevation pattern. In particular, the azimuth plane is shown intersecting the 3D pattern, and that cross section becomes the azimuth pattern.

Notice in Figure 4.5 that the azimuth plane pattern is perfectly symmetrical, being a circle (i.e., a constant function of azimuth angle ϕ, i.e., independent of ϕ). This means that power radiates equally for all ϕ. Thus, the antenna is called *omnidirectional*. Notice also that in the elevation plane, the pattern is a nonconstant function of the

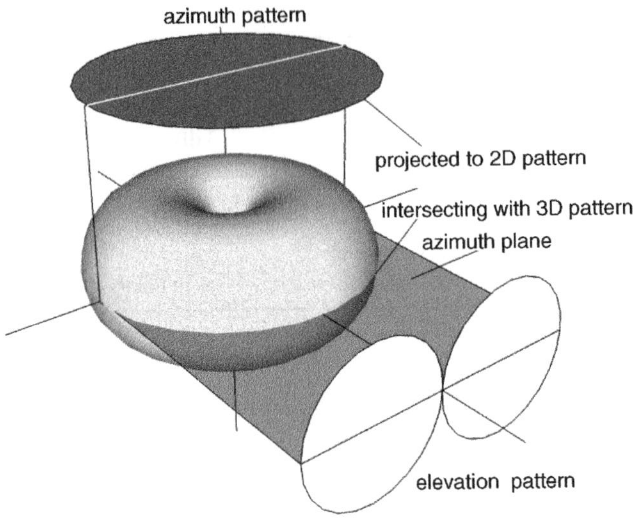

FIGURE 4.5 Azimuth plane and elevation plane of antenna patterns.

zenith angle, θ, and thus it is directional in that plane. Nevertheless, it is still common to call an antenna with such a pattern omnidirectional, since at the important angle of maximum radiation ($\theta = 90°$), it is a constant function in the azimuth plane.

The *directive gain* of an antenna, $D(\theta, \phi)$, is the ratio of U to the *average* radiation intensity:

$$D(\theta, \phi) = \frac{U(\theta, \phi)}{U(\theta, \phi)_{\mathrm{av}}} = \frac{U(\theta, \phi)}{P_r/4\pi} = \frac{4\pi U(\theta, \phi)}{P_r} \tag{4.5}$$

where P_r is the total power radiated. Then the *directivity* of the antenna is the maximum directive gain:

$$D = \frac{U(\theta, \phi)_{\mathrm{max}}}{U(\theta, \phi)_{\mathrm{av}}} = \frac{4\pi U(\theta, \phi)_{\mathrm{max}}}{P_r} = \frac{S_{\mathrm{max}}}{S_{\mathrm{av}}} \tag{4.6}$$

The direction of maximum directive gain is sometimes called the *boresight direction*.

NB: There are multiple definitions of "directive gain" versus "directivity" that people use. We have just followed one convention, which is to use the two terms to distinguish the two concepts (a) $4\pi U/P_r$ and (b) max $4\pi U/P_r$, calling the former *directive gain* and the latter *directivity*. Readers should be aware of another convention, which argues that "directivity" is a newer term that has replaced "directive gain," so "directivity" is used for both concepts, where (b) is implied when no direction is given, else (a) would be implied, if a particular direction is stated or implied. In terms of notation, we represent $4\pi U/P_r$ explicitly as a function of θ and ϕ [i.e., $D(\theta, \phi)$] and max $4\pi U/P_r$ as simply D, independent of θ and ϕ.

D is often expressed in decibels, referring to unity. Examples of D are shown in Table 4.3 (more on dBi and dBd in Section 4.1.9).

TABLE 4.3 Examples of Antenna Directivities

	D	D (dBi)	D (dBd)
Isotropic	1	0	−2.15
Half-wave dipole	1.64	2.15	0
Short dipole	1.5	1.76	−0.39
Small loop	1.5	1.76	−0.39

4.1.4.1 Lobes Taking a step back from looking at just one direction, the direction of maximum gain, we consider the radiation pattern of an antenna as a whole. By visual inspection one can often spot regions of relatively higher intensity, separated by regions of relatively lower intensity (of course, this would not apply to isotropic antennas; for omnidirectional antennas, there are no lobes in the plane where it is omnidirectional, whereas one can talk about lobes in the other plane). The highest radiation intensity is found in the *main lobe* (also known as the *major lobe*), while other lobes are called *side lobes* (also known as *minor lobes*).

If there is a small lobe 180° from the main lobe, it may be called a *back lobe*. The *front-to-back ratio* is the ratio of the maximum signal from the front of the antenna (main lobe peak) to the maximum signal from the back.

4.1.5 Beam Area

The *beam area* (or beam solid angle), Ω_A, is given by

$$\Omega_A = \int_0^{2\pi} \int_0^{\pi} P_n(\theta, \phi)\, d\Omega \qquad \text{sr} \tag{4.7}$$

where $d\Omega = \sin\theta\, d\theta\, d\phi$.

Beam area relates to directivity as

$$D = \frac{4\pi}{\Omega_A} \tag{4.8}$$

4.1.6 Antenna Gain

Antenna gain, referred to a (lossless) isotropic source, is given by

$$G = E_{\text{ant}} D \tag{4.9}$$

where E_{ant}, known as the *efficiency factor of antenna* or *radiation efficiency*, is dimensionless, and $0 \le E_{\text{ant}} \le 1$. E_{ant} quantifies the ohmic loss, so $E_{\text{ant}} = 1$ only if the antenna is lossless.

The difference between G and D: Gain, G, includes the efficiency (and ohmic losses), whereas directivity, D, doesn't include it; so $G < D$. For link budget calculations, in the Friis formula, and so on, which of the two do we use? G.

If we let P_{in} be total input power, of which P_{rad} is radiated and P_{loss} is ohmic power loss, then

$$E_{\text{ant}} = \frac{G}{D} = \frac{P_{\text{rad}}}{P_{\text{in}}} \tag{4.10}$$

Let R_{rad} be the *radiation resistance*, the equivalent resistance that would dissipate P_{rad} when the current in the resistance is equal to the maximum current along the antenna. (NB: The radiation resistance is a hypothetical resistance because the P_{rad} is actually radiated by the antenna, rather than dissipated as heat.) We have

$$\frac{P_{\text{rad}}}{P_{\text{in}}} = \frac{P_{\text{rad}}}{P_{\text{rad}} + P_{\text{loss}}} = \frac{I^2 R_{\text{rad}}}{I^2(R_{\text{rad}} + R_{\text{loss}})} \tag{4.11}$$

Therefore,

$$E_{\text{ant}} = \frac{R_{\text{rad}}}{R_{\text{rad}} + R_{\text{loss}}} \tag{4.12}$$

Normally, R_L is small, and E_{ant} is close to 1 [2].

4.1.7 Aperture

In a receiving antenna, think of how much of the propagating wave's energy can be captured. Let the magnitude of the Poynting vector be $S = |\mathcal{P}|$, and letting P be the power in the terminating impedance in the receiver, we have

$$A = \frac{P}{S} \tag{4.13}$$

Let V be the induced voltage when the antenna is oriented for maximum response and the incident wave has the same polarization as the antenna (i.e., to avoid polarization mismatch). We write the terminating or load impedance as $Z_T = R_T + jX_T$ and the antenna impedance as $Z_A = R_A + jX_A$. We assume a matched load scenario, so $R_T = R_A = R_{\text{rad}} + R_{\text{loss}}$ and $X_T = -X_A$.

Then, the *effective aperture* A_e is

$$A_e = \frac{V^2}{4S(R_{\text{rad}} + R_{\text{loss}})} \qquad \text{m}^2 \text{ or } \lambda^2 \tag{4.14}$$

For lossless antennas, $R_{\text{loss}} = 0$, we have the *maximum effective aperture*,

$$A_{\text{em}} = \frac{V^2}{4SR_{\text{rad}}} \qquad \text{m}^2 \text{ or } \lambda^2 \tag{4.15}$$

4.1.8 Antenna Gain, Directivity, and Aperture

It can be shown that

$$\lambda^2 = A_{\text{em}}\Omega_A \tag{4.16}$$

so A_{em} depends entirely on Ω_A and the wavelength.

From (4.16) and (4.8), we get immediately

$$D = \frac{4\pi}{\lambda^2} A_{\text{em}} \tag{4.17}$$

$$G = E_{\text{ant}} D = \frac{4\pi}{\lambda^2} A_e \tag{4.18}$$

4.1.9 Isotropic Radiators and EIRP

An *isotropic radiator* is an ideal antenna which radiates power with unit gain uniformly in all directions. An isotropic radiator should not be confused with an omnidirectional antenna. An omnidirectional antenna transmits uniformly in all directions *in one plane* only. Thus, a dipole antenna is omnidirectional (in the plane perpendicular to the antenna), but it is not isotropic.

Any antenna can be compared with an isotropic antenna. The isotropic antenna has unit gain. Suppose that our antenna has gain G. Thus, if both the isotropic antenna and ours were to be operating with the same transmitter power P_t, our antenna would be transmitting G times more power in the direction of maximum gain. For the isotropic antenna to transmit the same power in the direction of maximum gain (of our antenna), it would have to transmit G times more total power (i.e., to use transmitter power $P_t G$ instead of just P_t). This concept is useful enough to merit a name, EIRP. The *effective isotropic radiated power* (EIRP), often given as

$$\text{EIRP} = P_t G_t \tag{4.19}$$

indicates simply how much power an isotropic antenna would have to radiate to have effectively the same power in the direction of maximum gain.

So if you have a regulatory limit on EIRP, using directional antennas will not help because G_t goes up with them. But if the limit is EIRP/Hz, you can use very wide bandwidths (think of spread spectrum transmissions) to remain under the limit.

Alternatively, *effective radiated power* (ERP) compares the radiated power with a half-wave dipole antenna rather than an isotropic antenna. Since an antenna may be compared with multiple reference antennas (isotropic and dipole) the terms *dBi* and *dBd* are used to distinguish the cases. Antenna gains in dBi are with respect to an isotropic antenna, whereas gains in dBd are with respect to a half-wave dipole. See Table 4.3 for an example.

4.1.10 Friis Formula for Received Signal Strength

We have seen the Poynting vector, $\mathcal{P}$, in Section 2.3.2.1. In free space and far-field conditions, with an isotropic antenna, the power flux density is equal everywhere on the surface of an imaginary sphere of radius d, pointing radially outward. Letting P_d be the power flux density at a distance d from a transmitting antenna, and noting that the surface area of a sphere of radius d is $4\pi d^2$, then

$$P_d 4\pi d^2 = P_t \tag{4.20}$$

Then, for the same assumptions, but using a nonisotropic antenna with gain G_t, in the direction of maximum directive gain,

$$P_d = \frac{\text{EIRP}}{4\pi d^2} = \frac{P_t G_t}{4\pi d^2} \tag{4.21}$$

For a receiver a distance d from the transmitter and in the far field, the received power is related to the effective aperture by

$$P_r = P_d A_e \tag{4.22}$$

and the effective aperture is related to the receiving antenna gain by (4.18). Substituting for P_d and A_e from (4.21) and (4.18), we therefore have

$$P_r = \frac{P_t G_t}{4\pi d^2} \frac{G_r \lambda^2}{4\pi} = \frac{P_t G_t G_r \lambda^2}{(4\pi d)^2} \tag{4.23}$$

4.1.10.1 *Self-Impedance, Mutual Impedance, and Resonance* Two types of impedance can be associated with any antenna. *Self-impedance* is the impedance measured at the feed-point terminals of an antenna located in complete isolation from other conductors. *Mutual impedance*, on the other hand, is the contribution to the impedance measured at the feed-point terminals of an antenna, from interactions of the field, parasitic effects, and coupling, with conductors in the near field of the antenna. Even the ground could be one of those other conductors (even though it is a lossy conductor). Mutual impedance can distort the antenna pattern and change the impedance seen at the feed point. Understanding mutual impedance is essential for understanding a Yagi-Uda antenna, for instance.

Self-impedance (assuming no mutual impedance) is voltage applied to the feed point divided by the current flowing into the feed point. When in phase, the impedance is purely resistive. Then the antenna is termed *resonant*. But an antenna need not be resonant to be effective. It is more important to have good impedance matching and low VSWRs than for the antenna to be resonant.

4.1.11 Bandwidth

You may have heard of terms such as *broadband antenna* and *narrowband antenna*. These refer to the *bandwidth* of the antennas (relatively wide in the case of broadband antennas and relatively narrow in the case of narrowband antennas). What would we consider the bandwidth of an antenna to be? Antennas typically operate best over a certain range of frequencies. Usually, that means that certain characteristics of the antenna are within some acceptable range. These characteristics might include the pattern, the beamwidth, and the input impedance, among others. Since different characteristics are related to frequency in different ways, the usable bandwidth depends on what is important for a particular application. Thus, there is no single definition of bandwidth for antennas.

However, there are some definitions of bandwidth for antennas that have been found useful in many cases. Such useful definitions of bandwidth for antennas include:

- Pattern bandwidth, based on criteria related to gain, beamwidth, and so on.
- Impedance bandwidth, based on criteria related to input impedance and radiation efficiency.
- The range of frequencies over which VSWR ≤ 2 or VSWR ≤ 1.5, a popular concept of antenna bandwidth. This is sometimes also called the impedance bandwidth.

In practice, measurements should be made to examine how such parameters as antenna pattern, efficiency, gain, and input impedance vary over the intended range of frequencies. Then it becomes more a subjective judgment as to what the operational bandwidth is than a precise specification.

4.2 EXAMPLES

Antennas come in all shapes and sizes. Here we select a few important examples to survey briefly.

4.2.1 Dipole Antennas

Dipole antennas are also known as *Hertz antennas*, since Hertz used such an antenna to prove Maxwell's equations [5]. A dipole antenna is two wires or hollow tubes at 180° from each other, thus creating two "poles" and hence giving it its name. Figure 4.6 shows the dipole on the left in horizontal orientation (radiating or receiving vertically). The same dipole is shown in the center oriented vertically, radiating or receiving horizontally.

To find the antenna pattern, we need to know the current distribution. Often, the current distribution is approximated by simple functions, thus simplifying the analysis. For example, a sinusoidal current distribution is a good approximation of the actual current distribution if we can make the following assumptions:

- The antenna is fed symmetrically (at the center) by a balanced two-wire transmission line.
- The antenna is "thin", i.e., the wire diameter is much smaller than the wavelength λ.

4.2.1.1 Half-Wavelength Dipole Antennas The half-wavelength dipole is one of the most commonly used antennas. As its name suggests, it is nominally half a wavelength long (with reference to the optimal frequency of operation); that is, as a dipole it has two halfs, each $\lambda/4$ long. Often, a half-wavelength dipole is cut about 5% shorter "than the theoretical value to account for (capacitive) fringing effects at the ends" [5]. An uncut $\lambda/2$ dipole has a terminal impedance of $Z = 73 + j42.5\ \Omega$. The feed point is at the center, where current is highest (zero current at ends) and where the antenna impedance is about 72 Ω (for a cut $\lambda/2$ dipole). Most of the antenna impedance is radiation resistance.

4.2.1.2 *Very Short Dipoles* Dipoles much shorter than $\lambda/2$ are possible, and they are sometimes called *Hertzian dipoles*. Such a dipole can either be:

- An infinitesimal current element $I\,dl$, which does not exist in real life, or
- A short linear antenna which when radiating is assumed to carry constant current along its length.

Sometimes, a very short dipole might also be called a *short dipole*.

4.2.1.3 *In Between the Half-Wave and Hertzian Dipoles* So far, we have the very short Hertzian dipole, where $L \ll \lambda$ and with uniform current distribution, and the half-wave dipole, with sinusoidal current distribution. An in-between length, say $L < \lambda/4$, sometimes called a *short dipole*, can be modeled as having a roughly triangular current distribution.

4.2.2 Grounded Vertical Antennas

The *grounded vertical antenna* also known as the *Marconi antenna* or *quarter-wave vertical antenna*, was invented by Guglielmo Marconi. It is a $\lambda/4$ antenna over a ground plane. The antenna is fed at the bottom, not at the center, so it can be described as a *monopole*, in contrast to dipoles, which are center-fed. The ground plane reflects the behavior of the monopole like a mirror, so the antenna looks like one-half of a dipole. Thus, the currents and voltage pattern of the monopole are the same as one-half a $\lambda/2$ dipole, but the voltage at the input is only half that of a $\lambda/2$ dipole. Therefore, the input impedance is half that of the $\lambda/2$ dipole: 37 Ω.

The ground plane should be a good conductor for ideal behavior, in which case it reflects energy radiated from the vertical pole in mirrorlike fashion. However, in many applications, the ground plane conductivity is quite poor, so the pattern of the quarter-wave vertical antenna departs from the ideal pattern. This antenna, sometimes called a *whip*, is illustrated in Figure 4.6 as the rightmost antenna. The dashed line beneath it shows the fictitious "other half" of the dipole that the ground plane "creates" through its reflections. It can be compared to the dipole just to its left.

4.2.3 Folded Dipoles

Many TV receiving antennas use a folded dipole as the active element. The folded dipole is shown in Figure 4.6, as the second antenna from the left. The fold can help make the folded dipole more sturdy than a regular dipole.

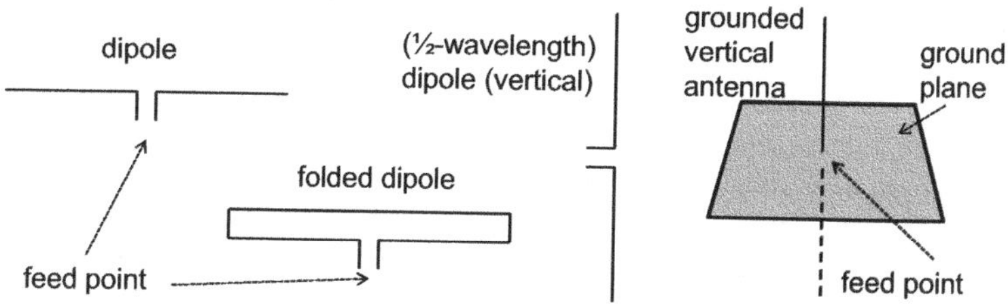

FIGURE 4.6 Various antennas illustrated.

A 1/2-wavelength folded dipole input current has an input impedance fourfold higher than that of a 1/2-wavelength dipole. Since $72 \times 4 = 288\ \Omega$, folded dipoles are often used with 300-Ω balanced twin-lead transmission lines.

4.2.4 Turnstiles

A circularly polarized antenna can be made by placing two dipoles $\lambda/4$ apart along the axis of the direction of wave propagation. For example, one dipole can be parallel to the x axis in the x–z plane, the other can be parallel to the y axis in the y–z plane, and they are spaced $\lambda/4$ apart in the direction of z. If used for transmission, a circular polarized wave is transmitted in the z direction. Note that the "antenna" could be thought of as an array (see Section 4.3), and it is the combination of the two constituent dipoles. The concept is illustrated on the left side in Figure 4.7.

Alternatively, instead of spacing the two dipoles $\lambda/4$ apart, they can be placed at the same z coordinate. Indeed, there is a name for antennas with this combination. A *turnstile antenna* is two dipoles placed at right angles to each other and fed 90° out of phase. If mounted horizontally, the structure looks like a turnstile. Waves propagating in the direction perpendicular to the plane of the antenna would be circularly polarized. The circular polarization would be in either the right- or left-hand sense, depending on which of the dipoles leads the other by 90°. Variations of this type of antenna are sometimes used for satellite communications.

Another use of turnstile antennas is based on their horizontal characteristics (assuming that the antenna is oriented horizontally). The pattern in the azimuth plane would be almost omnidirectional. When observed horizontally, the polarization is

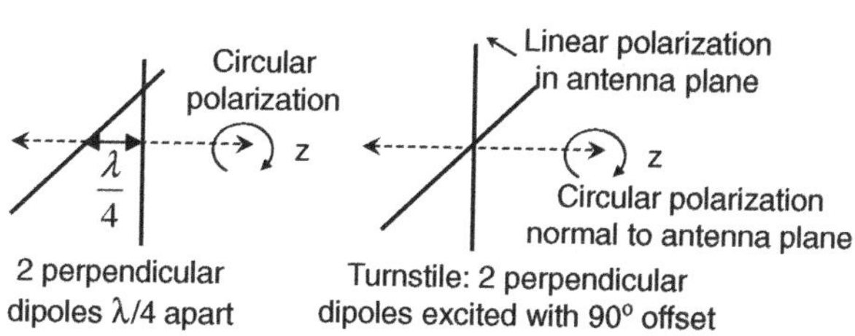

FIGURE 4.7 Turnstile antenna and circular polarization.

linear and horizontal. Variations of turnstile antennas have been used in FM broadcasting, where the nearly omnidirectional pattern with horizontal polarization is attractive.

NB: The characteristics are not so much a question of horizontal or vertical mounting as of the plane of observation. If observed in the plane of the crossed dipoles, it is omnidirectional with linear polarization in that plane. If observed in the direction perpendicular to the crossed dipoles, we have circular polarization. Thus, we cannot say whether the antenna itself is linearly or circularly polarized. It depends on where the observation is made.

4.2.5 Loop Antennas

Loop antennas are generally small in size and wide in bandwidth. The radiation pattern looks like a toroid. Some loop antennas consist of just one loop, but more turns can be added to increase the sensitivity of the antenna. Voltage induced in the antenna is directly proportional to the number of turns.

Considered fundamental, loop antennas invite comparison to a dipole. They can be thought of in a number of ways; for example, a small square loop can be thought of as consisting of four short linear dipoles. Or they can be thought of as a short magnetic dipole.

4.2.6 Parabolic Dish Antennas

In some applications, such as TV broadcasting or base station antennas, omnidirectional antennas, or antennas with wide beams, are needed, since the signal needs to be transmitted over a large area or be received from a large area. In other applications, however, very high directivity and gain are needed. One example is point-to-point microwave radio links, and another is satellite communications. Many other antennas, even those with good directivity such as Yagi-Udas, still have a beamwidth that causes a transmitted signal to spread out, especially when received far away. An excellent solution for applications such as point-to-point microwave links or satellite communucations, then, are parabolic dish antennas (Figure 4.8). In practice, a parabolic dish would need to be fed from a source, and there are many ways to feed the dish. As such, parabolic dish antennas are also called *parabolic reflectors*.

Parabolic dishes can provide extremely high gain and directivity because of their *ray collimating* property; that is, the radio waves are sent out from a transmitting parabolic dish parallel to each other rather than spreading out. Of course, the propagation environment can cause dispersion, reflections, refraction, and so on, but in a line-of-sight environment, the transmissions could potentially reach very far, including to and from a satellite orbiting high over the Earth.

The directivity is

$$D = \epsilon_{\text{ap}} \left(\frac{2\pi r}{\lambda} \right)^2 \tag{4.24}$$

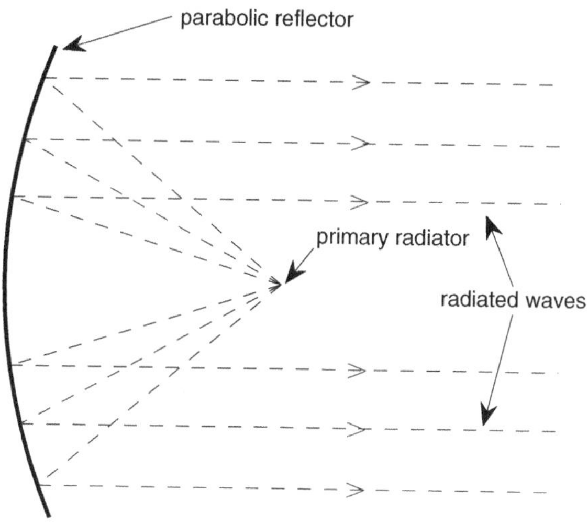

FIGURE 4.8 Parabolic reflector.

where r is the radius of the dish and ϵ_{ap} is the *aperture efficiency* or *illumination efficiency*. (NB: The aperture efficiency is not to be confused with the antenna efficiency. Recall that directivity does not depend on antenna efficiency, but gain does). Aperture efficiency is a product of several factors, including the fraction of total power that is radiated by the feed (there is often some *spillover* from the feed past the reflector), blockage (the feed and its supporting structures may be partially blocking the signal from the reflector, as when the feed is at the focal point), and nonuniformity of the feed pattern over the reflector surface.

4.2.7 Mobile Device Antennas

Mobile phone antennas have to be chosen under tight design constraints. Unlike base station antennas, which are mounted on cell towers, mobile phone antennas are packaged with mobile phones that consumers carry around. Thus, there are tight constraints on *size*, *weight*, *visibility*, and *cost*. In particular, the antennas need to be small, lightweight, low in profile, and cheap. Furthermore, the antennas need to be relatively efficient (to help with battery energy consumption), and in some cases, broadband (especially for multimode phones that operate in multiple bands, where often the same antenna is used for all the bands). Additionally, given the proximity to human users, the antenna pattern can be affected significantly by the presence of a person and can cause significant amounts of radiation to penetrate the person's body (e.g., head and brain), so these effects need to be studied as well. The penetration into the human is often quantified by the *specific absorption rate* (SAR).

Historically, monopole antennas were used, and it was considered acceptable to see the antenna sticking out of the phone structure. Nowadays, the most common mobile device antenna family is that of the patch antenna, also known as a planar antenna. Specifically, the *planar inverted F antenna* (PIFA) and its variants are popular in mobile devices. These are more complicated than a basic patch antenna, so we discuss rectangular patch antennas just briefly in Section 4.2.7.1.

For larger devices such as laptops and tablets, the design challenges are relaxed in comparison to mobile phones. Sometimes, part of the frame can be used as part of the antenna.

4.2.7.1 Patch Antennas *Patch antennas* are also known as *microstrip antennas*. A patch antenna consists of a conducting patch bonded to a dielectric substrate over a ground plane. Like microstrip transmission lines, one reason for the popularity of microstrip antennas is that they can be printed on a printed circuit board. Furthermore, they are cheap to produce, small, lightweight, and have a low profile. Figure 4.9 shows a rectangular patch antenna. On the left is a view from the side, with the patch over the substrate, which in turn lies over the ground plane. It also shows the width W of the patch, the height h, and the permittivity ϵ_r. When we look at the patch antenna from the top, we can also see the length L. In this example it is fed by a microstrip transmission line. However, patch antennas need not just be fed by microstrip transmission lines. For example, they could also be fed by coaxial transmission lines. Typically, L is chosen to be about half of the wavelength of the desired carrier frequency within the dielectric medium (since the wavelength in free space is different from the wavelength in the dielectric medium). Just like the half wavelength dipole, though, L may need to be slightly shorter in practice, because fringing fields may make the patch appear longer. The height is usually chosen to be a small fraction of the wavelength in the dielectric.

We have already seen microstrip *transmission lines* in Section 2.3.3, and they sound much like microstrip antennas. What allows one to be used as a transmission line and another as an antenna? One difference is the choice of L and W. The patch antenna looks like a very wide (relatively large W) transmission line that is open-circuited and radiating out of both ends. When used as a transmission line, on the other hand, both ends would typically be connected to matched loads. Also, the radiation in the patch antenna happens at the edges, where the fringing fields are located

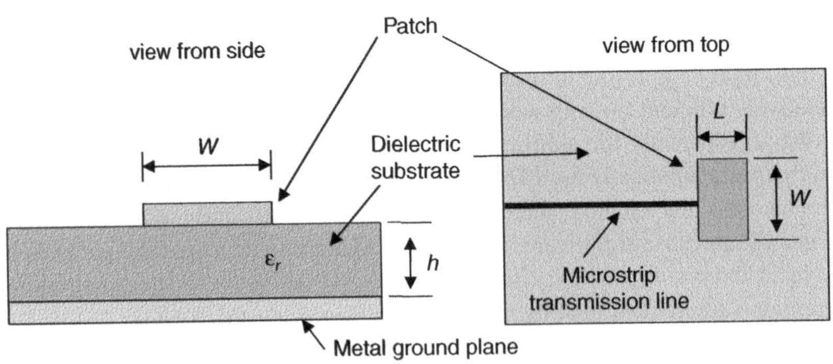

FIGURE 4.9 Rectangular patch antenna.

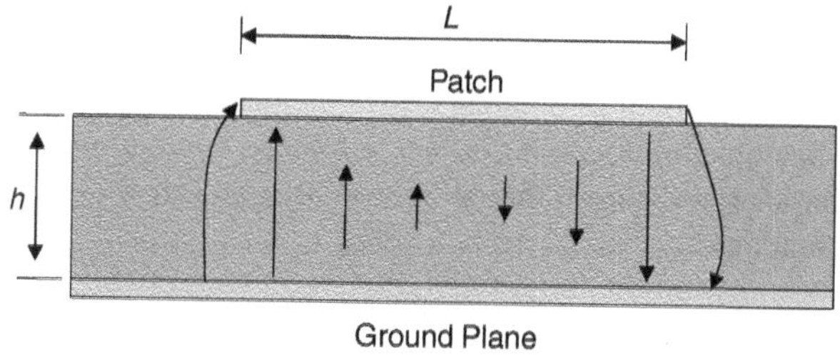

FIGURE 4.10 How a patch antenna radiates.

(Figure 4.10). These field lines are bowed, bending outward, so radiation escapes from there. The choice of ϵ_r makes a difference here, in that smaller ϵ_r results in more bending of the fringe fields, which results in more radiation. Hence, smaller ϵ_r is good for radiation, whereas microstrip transmission lines would be better off with larger ϵ_r, keeping more of the energy inside.

4.3 ANTENNA ARRAYS

Given some specified desired antenna pattern, input impedance, and so on, if it does not match the parameters of some known antenna (e.g., a half-wavelength dipole), it is a challenge to create a new antenna that matches the parameters. In general, it is difficult to design a new antenna for each new application. It is often more convenient to work with *antenna arrays*. Antenna arrays consist of multiple antennas close to each other and treated as a system. The component antennas may be called *array elements*. Almost arbitrarily, complex antenna patterns can be generated by arrays, by suitable choice of spatial distribution and phase–amplitude relationships. One way of thinking about it is that in order to achieve a particular antenna pattern, we would like to control the spatial current distribution in our antenna structure. Having antenna arrays allows us to control spatial current distribution much more accurately and easily than if we were to design a monolithic (single) antenna to try to achieve the same spatial current distribution.

So, antenna arrays are collections of simple antennas (often, dipoles) with a particular spatial distribution and that are excited by voltages and currents with particular phase and amplitude relationships. The array elements need not all be the same, although they often are dipoles of the same type and length. It is helpful when the array elements are all the same, so we can apply powerful concepts such as *array factor* and *pattern multiplication* for rapid and efficient analysis and prediction of the behavior of various antenna arrays. Common and useful arrangements of antenna arrays include:

- *Linear*. The antennas are all arranged in one line in a plane.
- *Planar*. The antennas are distributed over a plane.
- *Circular*. The antennas are arranged in a circle.

In this book, we only have room to introduce linear arrays, which we discuss in Section 4.3.1.

A more general definition of antenna arrays includes configurations where there may also be reflectors and directors. In fact, a famous configuration, the Yagi-Uda, is often considered an array even though it has only one active element; the rest are reflectors and directors.

4.3.1 Linear Arrays

We consider an N-element array consisting of identical array elements arranged in a line and want to consider the antenna pattern of the antenna array. For convenience, we assume that the array elements are arranged along the z axis, and for simplicity assume that there is a constant distance, d, between adjacent elements. More specifically, we can assume that the elements are located at $z = 0$, $z = d$, up to $z = (N - 1)d$, as shown in Figure 4.11. We assume no coupling between the array elements (which would complicate the analysis). Thus, we can simply apply the superposition principle and obtain the **E** field at a location in the far field as the sum of the contributions from each individual array element (as though the individual element were just by itself). We make our observations of **E** field at a location P in the far field that is far enough away that we can assume that:

- The (zenith) angle θ_i between the ith element and P is a constant, θ.
- The distance R_i from the ith element to P is a constant, R, as far as considering the *amplitude* of the **E** field at P.
- The distance R_i from the ith element to P is $R_0 + id\cos\theta$, as far as considering the *phase* of the **E** field at P. Thus, in radians, the phase difference of the wave from adjacent elements is $(2\pi/\lambda)d\cos\theta$.

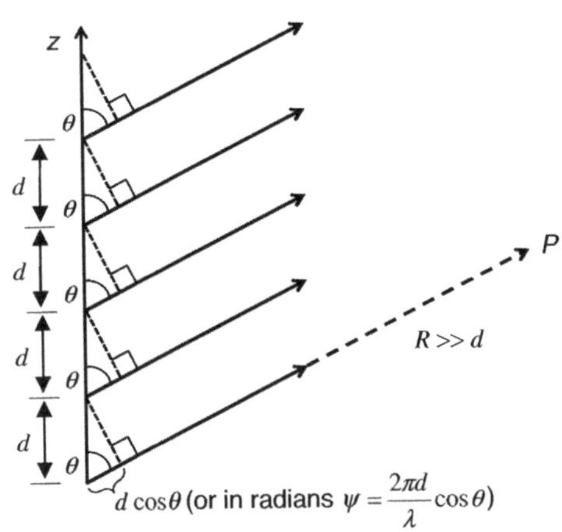

FIGURE 4.11 Linear array.

These assumptions are reasonable for $d \ll R$. In particular, we note that the discrepancy between our treatment of the distances R_i for amplitude and phase determination is reasonable because very small changes in distance make very little difference to the amplitude, but can make significant differences to the phase.

The elements are all fed the same signal, with the same amplitude but with a phase offset between adjacent elements. In particular, as we move along the array from one end to the other, each element leads the element before it by β and lags the element after it by β, where β is a small phase difference. Then the **E** field at P can be written as a sum of phasors:

$$\mathbf{E} = \mathbf{E_0}\left(1 + e^{j\psi} + e^{j2\psi} + \cdots + e^{j(N-1)\psi}\right) \tag{4.25}$$

where $\mathbf{E_0}$ is the **E**-field contribution from the array element located at the origin, and where

$$\psi = \frac{2\pi}{\lambda} d \cos\theta + \beta \tag{4.26}$$

Recognizing that the right side of (4.25) is a geometric series, we can simplify (4.25) to

$$\mathbf{E} = \mathbf{E_0}\frac{1 - e^{jN\psi}}{1 - e^{j\psi}} \tag{4.27}$$

Notice that the total **E** field is the product of the **E** field from one element, with a factor that is a function of the geometry of the array and the relative excitation times of the various array elements. We call that factor the *array factor*. A useful principle that can be applied to arrays where all the elements are the same (e.g., all dipoles) is *pattern multiplication*:

$$\text{array pattern} = \text{single element pattern} \times \text{array factor} \tag{4.28}$$

Thus, in our example, the array factor was

$$\frac{1 - e^{jN\psi}}{1 - e^{j\psi}}$$

4.3.1.1 Broadside Arrays When we have a linear array, we often would like the direction of maximum radiation to be perpendicular to the axis of the array. In this case the array would be called a *broadside array*. To obtain a broadside array, then, we observe that (4.25) is maximum when each term $e^{ji\psi} = 1$, so we need

$$ji\psi = 2\pi m \tag{4.29}$$

for any integer m. For a broadside array, $\theta = \pi/2$, so the first term of (4.26) is zero. Thus, we set $\beta = 0$, giving $\psi = 0$ for maximum broadside radiation.

However, we have to be careful, because for broadside arrays we don't want the maximum radiation to be found at other angles as well. But (4.29) is satisfied by other angles besides $\pi/2$. For example, when $d = \lambda$, $\beta = 0$, maximum radiation occurs at

$\theta = 0$ and $\theta = \pi$ as well! Such phenomena are known as *grating lobes*. To avoid grating lobes, we can choose $d < \lambda$.

For example, a row of dipoles arranged end to end in a line, with the orientation of all the dipoles parallel to the line and all fed in phase ($\beta = 0$), can result in a significant increase in gain over a single dipole. Each dipole already radiates maximally at $\theta = \pi/2$. The pattern gets multiplied by the array factor, making it much more directive, especially for large N. See Figure 4.14 for a graphical illustration of this.

4.3.1.2 Endfire Arrays

In contrast to a broadside array, the direction of maximum radiation in *endfire arrays* is parallel to the axis of the array. Thus either $\theta = 0$ or $\theta = \pi$. Solving (4.26) for $\psi = 0$ yields $\beta = -2\pi d/\lambda$ and $\beta = 2\pi d/\lambda$, respectively. Notice that for $d = \lambda/2$, the maximum radiation is in both the $\theta = 0$ and $\theta = \pi$ directions. Selecting $d < \lambda/2$ would avoid this and other cases of grating lobes.

4.3.1.3 Directing a Beam at a Particular Angle

Instead of $\theta = 0, \pi/2$, or π as we have seen for broadside and endfire arrays, suppose that we wish to direct the main beam at angle $\theta = \theta_0$. Again, we can solve (4.26) for $\psi = 0$ and obtain $\beta = -2\pi d \cos\theta/\lambda$.

4.3.2 Yagi-Uda Antennas

The Yagi-Uda is an antenna commonly used for TV reception. It is shown in Figure 4.12. The Yagi-Uda has three different types of elements:

- The *active* or *driven element* is a dipole or folded dipole that has an electrical connection with the RF circuit for transmission or reception.
- The *reflector* is behind the dipole, usually a straight rod cut about 5% longer than the dipole. It has no electrical connection with the active element or the RF circuits.
- The *directors* are in front of the dipole and cut about 5% shorter than the dipole. There may be multiple directors, and they are also termed *parasitic elements*.

There is no electrical connection between the active element, reflector, and directors.

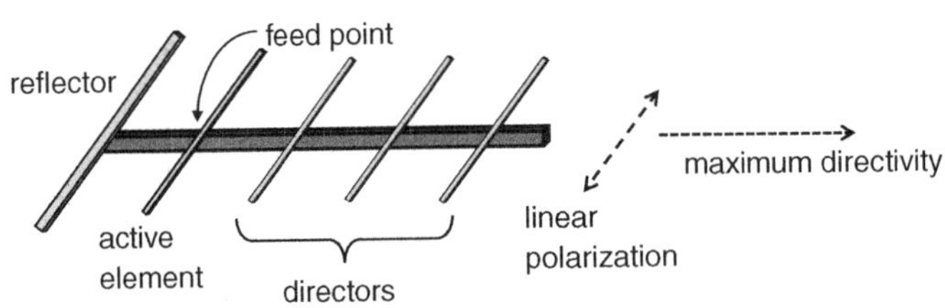

FIGURE 4.12 Yagi-Uda antenna.

4.3.3 Log-Periodic Dipole Arrays

The log-periodic array comprises a horizontal array of half-wavelength dipoles. The longest is cut for the lowest frequency (e.g., channel 2 for VHF TV), and then subsequent dipoles are each cut shorter and positioned closer to the one before it, where the separations and lengths have a constant ratio:

$$\frac{l_2}{l_1} = \frac{l_3}{l_2} = \cdots = \frac{D_2}{D_1} = \frac{D_3}{D_2} = \cdots \tag{4.30}$$

where $l_1 = \lambda/4$, λ is for the lowest frequency desired, and D_1 is given by

$$\alpha = \tan^{-1}\frac{l_1}{D_1} \tag{4.31}$$

where α is called the spread angle.

The array has a broad bandwidth because it has many dipoles resonating at different frequencies. Directivity and gain follow reasoning similar to that of Yagi (i.e., reflectors and directors). If dipole 2 is resonant, dipole 1 acts as a reflector, and dipoles 3 and 4 act as directors. Unlike a Yagi, though, the elements of a log-periodic array are electrically connected to each other. It is a broadband antenna because as the frequency changes, different elements become active, and the others can act as directors. Unlike some other arrays, such as the N-element uniform array we discussed earlier, the log-periodic array is one whose elements are not all the same (even if they are all dipoles, they are of different lengths).

4.3.4 Base Station Antennas

A common omnidirectional base station antenna is shown in Figure 4.13. It is basically a linear array of dipoles, each of which is oriented parallel to the axis of the array.

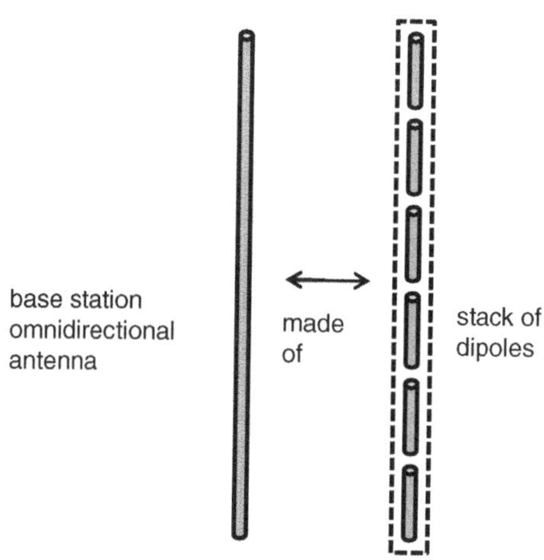

FIGURE 4.13 Typical omnidirectional base station antenna.

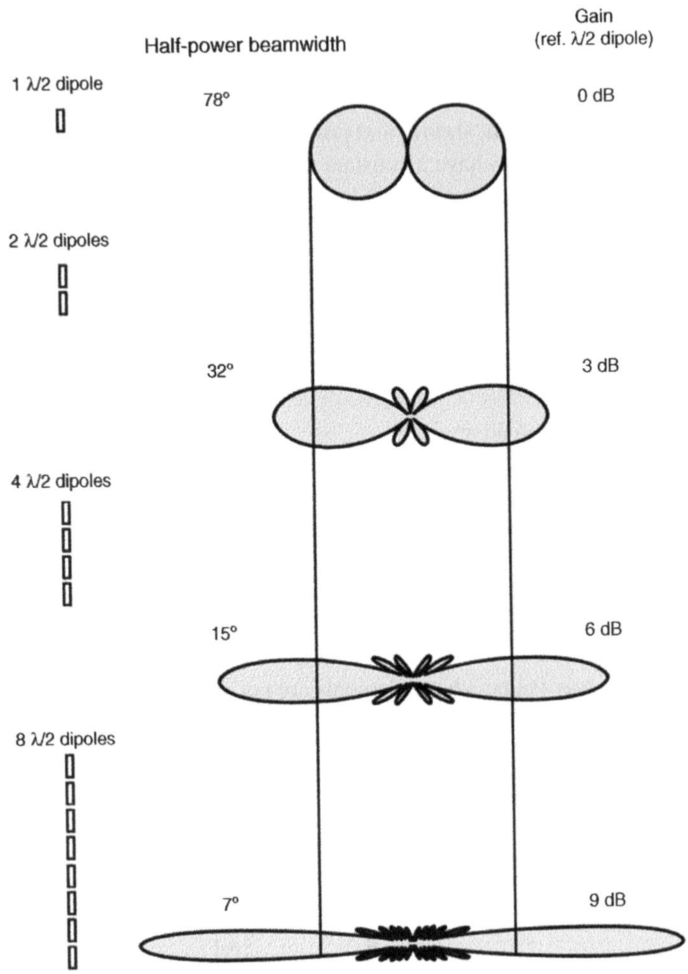

FIGURE 4.14 Increasing directivity of a dipole array as more dipoles are added. (Courtesy of Kathrein.)

The elements are often excited in-phase, so a broadside pattern emerges. Figure 4.14 shows how the directivity of such a dipole array increases as more dipoles are added.

A "better" base station antenna that is also commonly seen is the panel antenna shown in Figure 4.15. This is useful especially in cells where sectors are used, so we don't want the antenna to be omnidirectional. It consists of segments each of which has two parallel dipoles in front of a flat reflector. A single dipole in front of such a reflector would experience a 3-dB gain (3 dBd), and the beam width would change from 360° to about 180°. Having a pair of dipoles in front of the reflector results in a roughly 6 dBd gain, plus a narrowing of the beam to about 90° (the actual numbers depend on the parameters of the specific deployment, e.g., 65° 3-dB beamwidth and

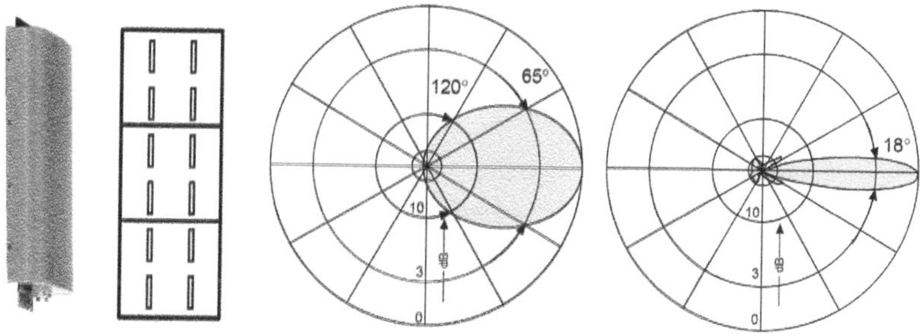

FIGURE 4.15 Typical base station panel antenna. (Courtesy of Kathrein.)

120° 10-dB beamwidth, in the example shown in Figure 4.15). The panel antenna, then, consists of an array of such dipole pairs, each in front of a reflector. By pattern multiplication, we can expect good directivity.

In Figures 4.16 and 4.17, we see both panel antennas and omnidirectional antennas mounted on a base station. Notice that the antennas are arranged in a lower and a higher triangle. We discuss the common triangle arrangements in Section 4.3.4.1.

FIGURE 4.16 Both panel and omnidirectional antennas found on a base station.

FIGURE 4.17 Close-up of some of the antennas on the same base station as in Figure 4.16.

4.3.4.1 Arrangement of Multiple Base Station Antennas For base station panel antennas as in Figure 4.15, the 10-dB beamwidth may be about 120°, making it possible to divide the base station coverage area into three sectors, each spanning about 120°. Such sectorization is commonly done, and the antennas are often arranged in a triangle, as shown in Figure 4.18. The antennas on each edge of the triangle are directed to the sector facing that edge. Often, three antennas can be seen on each edge of the triangle. As shown in the figure, this is because there is one transmitting antenna in the center and two receiving antennas at the sides, near the vertices. The

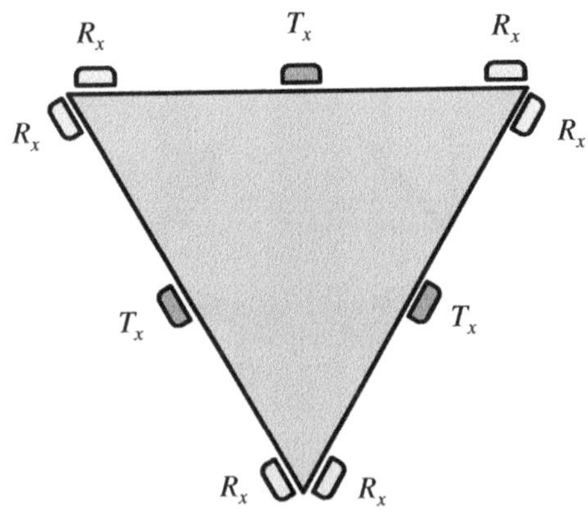

FIGURE 4.18 Typical arrangement of antennas in a base stations.

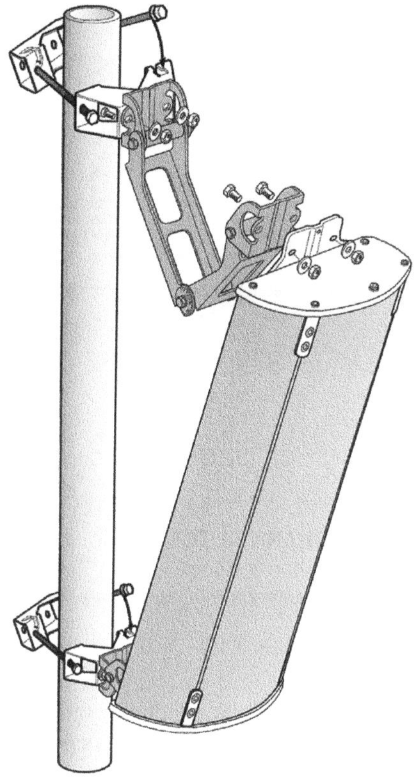

FIGURE 4.19 Panel antenna with mechanical down tilt. (Courtesy of Kathrein.)

two receiving antennas are spread about 12 wavelengths apart for antenna diversity purposes (see Section 5.3.5).

4.3.4.2 Tilting The direction of maximum directive gain should not be completely out from the antenna horizontally (i.e., at zero elevation), but should be pointing slightly downward, because the mobile stations will usually be slightly below the height of the base station antennas.

The tilting can be done *mechanically* or *electrically*. In the case of mechanical tilting, the antenna can be moved physically so that it lies at the desired angle. A panel antenna with mechanical down tilt is shown in Figure 4.19. The corresponding pattern is shown in Figure 4.20. Electrical down tilt is accomplished by using nonzero β (phase shift) for the excitation of the various elements in the array. In particular, a constant, increasing phase shift is applied, moving from top to bottom, to achieve down tilt.

4.3.4.3 Stealth Antennas Cell towers are sometimes considered "ugly" and not as aesthetically pleasing as some other structures. In some locations, especially

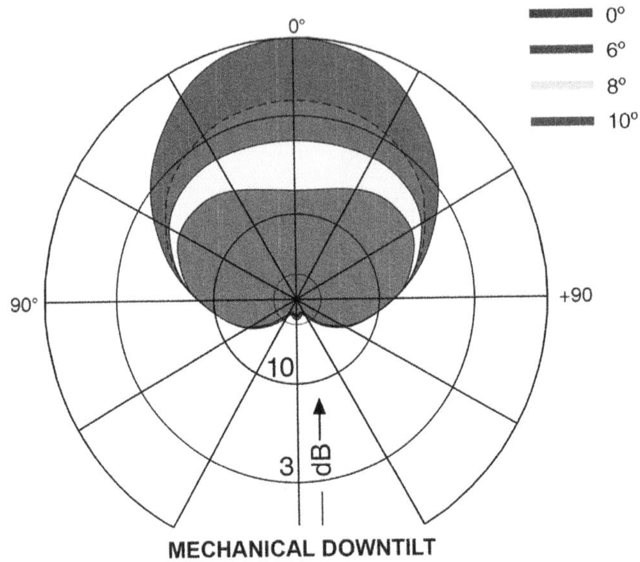

FIGURE 4.20 Pattern of panel antenna with mechanical down tilt. (Courtesy of Kathrein.)

where population density is high, cellular system operators may be requested, or even be required by local government, to hide their towers and antennas, or at least make them less obvious. Thus, various stealth antennas and stealth towers can be found where the towers and/or antennas are disguised as trees, for example. In Figure 4.21 we see a stealth antenna that is made to appear like a rooftop decoration, and in Figure 4.22 we see one that is hidden by painting it with a brick pattern.

FIGURE 4.21 Stealth antenna disguised as a rooftop decoration. (Courtesy of Kathrein.)

FIGURE 4.22 Stealth antenna camouflaged to blend into the surrounding brick pattern. (Courtesy of Kathrein.)

4.3.5 Newer Ideas for Using Multiple Antennas

Traditionally, antenna arrays have been used as a unit, that is, as a single antenna with a possibly complex pattern, a pattern that could be adjusted if phase–amplitude relationships between the array elements are changed. The pattern could be such that there are beams in some directions and nulls in others (this could be called *beamforming*). Differences in the transmissions between array elements would be in the form of phase differences, but they would be transmitting the same signal. On the receiving side, an array might also be used for beamforming, or it might be used to do diversity combining, where the same signal is received at all the antennas (the same signal, but possibly at different phases and amplitudes, due to differences in the channel to each antenna). Since the late 1990s, though, new ideas have been introduced, where *different* signals (consisting of very different bits) may be transmitted on different antennas.

- *More sophisticated and adaptive beamforming*. Beamforming can get very sophisticated when we allow the antenna arrays to adapt to various system conditions, for such purposes as reducing co-channel interference or changing the direction of a beam to track mobile movements; such *adaptive antenna arrays*, also known as *smart antennas*, are discussed briefly in Section 9.2.2.3.

- *Antenna diversity*. One aspect of antenna diversity is a receiver technique to try to obtain low correlation between the signal received at different antennas, and then to combine them advantageously. (Such techniques are discussed in Section 5.3.5.) However, these ideas can be generalized into the concept of *spatial diversity*, which involves such things as *space-time coding*; this is discussed briefly in Section 9.2.2.2.
- *Spatial multiplexing, or multi-input multioutput* (MIMO). Multiple antennas are used to create multiple parallel channels between transmitter and receiver, to allow for higher data rate communications. We discuss MIMO briefly in Section 9.2.2.1.

All these ideas can be considered *multiple antenna techniques*. We discuss them further in Section 9.2.2.

4.4 PRACTICAL ISSUES: CONNECTING TO ANTENNAS, TUNING, AND SO ON

In this section we are not talking about antennas per se or RF per se, but about the connections to antennas.

4.4.1 Baluns

Some transmission lines are inherently *balanced*, and some are inherently *unbalanced*. In balanced transmission lines, the signal is symmetric, whereas in unbalanced transmission lines, the signal is asymmetric, with a ground connection. For example, a twin-lead cable is balanced, whereas coaxial cable is unbalanced (the outer conductor is grounded). Meanwhile, antennas such as dipoles are symmetric. Thus, when connecting an unbalanced cable to a balanced cable or balanced antenna (e.g., a dipole), a *balun* (derived from "balanced/unbalanced") should be used.

4.4.2 Feeder Loss

We have the basic expression for EIRP given by (4.19). What if you have cable losses between the transmitter and the antenna? A more precise expression for EIRP is

$$\text{EIRP} = \frac{P_t G_t}{L_B L_f} \tag{4.32}$$

where L_B and L_f are *branching loss* and *feeder loss*, respectively. L_B is because the transmitter power P_t may be coupled through circulators that couple several transmitters (not just the transmitter of interest) to the antenna feed system. The circulators try to minimize VSWR and blasts of signal power from other transmitters. Nevertheless, a power loss L_B is experienced by this "coupling" and "branching." Feeder loss occurs because the transmitter output is typically some distance from the antenna, so there must be a feed system (cables, waveguides, etc., with their associated VWSRs). If the feeder loss is not known exactly, a "reasonable estimate" is 10 dB per 100 m of feeder [5]. Sometimes, a distinction is made between feeder cables and jumper cables, where a *feeder cable* is a less flexible (and less lossy) cable and a *jumper cable* is a more flexible cable (but more lossy). For example, one rule of thumb is 6.1 dB per 100 m of feeder cable and 21 dB per 100 m of jumper cable [4]. We say a bit more about cables in Section 16.3.1.

In addition to losses due to lossy cables, there may be losses due to reflections, as characterized by high VSWR (Section 2.3.4). In connecting cables between antennas and RF subsystems, therefore, we need to be careful to create matched load conditions to minimize the VSWR so that we do not suffer unnecessary losses from mismatched load conditions that lead to high VSWR and losses from reflections. Various tuning techniques have been developed to help in achieving this goal.

EXERCISES

4.1 Consider a half-wavelength dipole. What is d_{boundary}, the distance from the antenna that we take as the demarcation point between the near and far fields? Express your answer in terms of λ, the wavelength of the signal, and also in terms of L, the antenna length. How about for a quarter-wavelength antenna? Assume that the antennas are exactly half-wavelength and quarter-wavelength, not slightly shorter.

4.2 If we have a parabolic dish antenna of diameter 5 m, transmitting a signal at 5 GHz, what is the directivity of the antenna, assuming that $\epsilon_{\text{ap}} = 0.7$?

4.3 We want to build a half-wavelength dipole to receive a 200-MHz broadcast. What is the optimal length of the dipole, assuming a 95% correction factor?

4.4 Show that (4.27) can be expressed as

$$\mathbf{E} = \mathbf{E}_0 e^{j(N-1)\psi/2} \frac{(\sin N\psi)/2}{(\sin \psi)/2} \tag{4.33}$$

What if the array elements are not located at $z = 0$ to $z = (N-1)d$, but symmetrically about the origin? Where would the array elements be located, and what would be the corresponding array factor if we keep the reference point (for the phases) at the origin?

4.5 Why might a Yagi-Uda antenna sometimes not be considered an antenna array?

REFERENCES

1. C. Balanis. *Antenna Theory: Analysis and Design*, 3rd ed. Wiley, Hoboken, NJ, 2005.
2. D. K. Cheng. *Field and Wave Electromagnetics*. Addison-Wesley, Reading, MA, 1990.
3. J. D. Kraus. *Antennas*. McGraw-Hill, New York, 1988.
4. M. Nawrocki, M. Dohler, and A. H. Aghvami. *Understanding UMTS Radio Network Modelling, Planning and Automated Optimisation: Theory and Practice*. Wiley, Hoboken, NJ, 2006.
5. P. Young. *Electronic Communication Techniques*, 5th ed. Prentice Hall, Upper Saddle River, NJ, 2004.

5

PROPAGATION

All electromagnetic phenomena can be described by Maxwell's equations, so it might be argued that Maxwell's equations are all we need to study these phenomena. Nevertheless, it has been found that certain phenomena occur frequently, and it helps to study and classify them appropriately. Then, physical models and/or mathematical/geometrical/statistical models can be created. Such models are useful for:

- Studying and analyzing the phenomena.
- Predicting electromagnetic wave behavior.
- Creating analytical and simulation models for performance evaluation of wireless systems.
- Design techniques and technologies for communications over wireless channels.

In this chapter, therefore, we examine such models that describe common effects frequently observed. We begin in Section 5.1 with effects such as reflection, refraction, and diffraction that occur by virtue of the wave nature of electromagnetic waves. These effects are not specific to radio waves. We then consider the more specialized case of the cellular system propagation environment. We divide that discussion into two parts: The large-scale effects and related models are examined in Section 5.2, and the small-scale effects and related models are examined in Section 5.3. What we mean by large scale and small scale will become clear when we get to those sections. Finally, we examine briefly how propagation effects may be incorporated into the link budget for radio link design.

Fundamentals of Wireless Communication Engineering Technologies, First Edition. K. Daniel Wong.

5.1 ELECTROMAGNETIC WAVE PROPAGATION: COMMON EFFECTS

We begin this section with a general model for "path loss" in free space (Section 5.1.1), where there are no objects in the environment in free space. Next, we look at four major phenomena that occur when propagating electromagnetic waves interact with objects in the environment: reflection, diffraction, refraction, and scattering (Sections 5.1.2 to 5.1.4). These can be viewed as wave phenomena not specific to radio waves or even electromagnetic waves alone. "Objects in the environment" need not be solid objects—even a change of density of the air can result in some of these effects occurring.

5.1.1 Path Loss

In *free space*, a signal from an isotropic antenna can be expected to decrease in strength for purely geometric reasons and considerations of conservation of energy. In other words, as the surface area of an expanding sphere around the antenna increases with d^2, where d is distance from the antenna, so also the electromagnetic energy must be distributed more and more "thinly" over the surface of the sphere as the wave propagates away from the isotropic antenna, by $1/d^2$. The Friis formula for received signal power in free space was derived as (4.23):

$$P_r(d) = \frac{P_t G_t G_r \lambda^2}{(4\pi)^2 d^2 \Lambda_0} \tag{5.1}$$

where $P_r(d)$ is signal power received at distance d from the transmitter, P_t is transmitted power, G_t and G_r are transmitter and receiver antenna gains (relative to isotropic antennas) respectively, λ is the wavelength (same units as d), and Λ_0 is the system loss factor not related to propagation (i.e., losses in filters, antennas, etc.). As expected, the Friis formula shows the $1/d^2$ characteristic.

The $1/d^2$ drop-off is an ideal case for free space, where there are no losses as the signal propagates. Do real-life measurements confirm such a trend? The empirically observed overall trend is that the received signal power level tends to follow a $d^{-\nabla}$ curve, where d is the distance between the base station and user antennas, and ∇ is a number typically between 2 to 6, referred to as the *path loss exponent*. Since the path loss exponent in (4.23) is 2, one might expect ∇ to be close to 2 in wireless systems, such as the cellular environment. However, the cellular propagation environment is *not* free space and ∇ tends to be closer to 4 in most locations around a base station. This can be explained by the ground reflection model discussed in Section 5.2.1 and by losses through absorption in reflectors (Section 5.1.2) and scatterers (Section 5.1.4).

5.1.2 Reflection and Refraction

When a wave is propagating in a medium and then comes upon an object or a region of space with another medium that is much larger than the wavelength and relatively smooth [this may not always be so; for example, there may be scattering

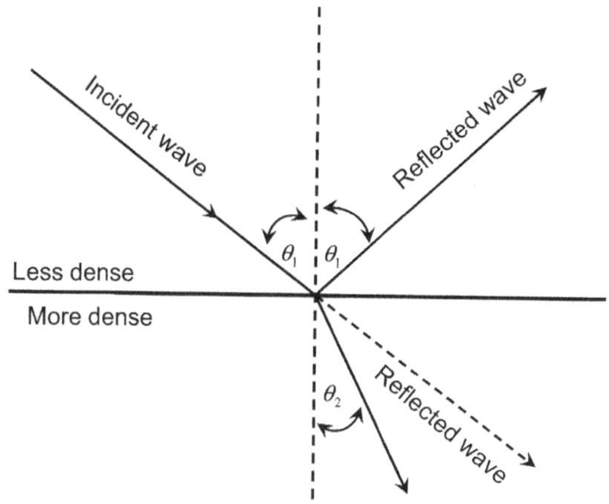

FIGURE 5.1 Reflection and refraction.

(Section 5.1.4)], reflection and/or refraction may occur. *Reflection* is the "bouncing off" (from the object) of the wave, at an angle equal to the angle of incidence. In the case of electromagnetic waves, when a wave is incident upon a perfect conductor, 100% of the wave is reflected. When a wave is incident upon a dielectric, a portion of it is reflected and a portion of it is refracted (it is common for both reflection and refraction to happen at the same time). With *refraction,* a portion of the wave enters the object, or medium. Unlike with reflection, the angle of the refracted wave is in general not equal to the angle of incidence, but depends on a number of factors, as seen in (5.2).

Reflection and refraction can be seen in Figure 5.1. In this example, refraction is toward the perpendicular line, because the wave is going from a less dense to a denser medium. It is also possible to have refraction when the wave is going from a denser to a less dense medium. In that case, it refracts away from the perpendicular line.

Snell's law relates the index of refraction in each medium to the speed of the wave in each medium, and also to the angle of incidence in each medium:

$$\frac{N_1}{N_2} = \frac{v_2}{v_1} = \frac{\sin\theta_2}{\sin\theta_1} \tag{5.2}$$

where v_2 and v_1 are the speed of the wave in medium 2 and medium 1, respectively, and N_1 and N_2 are the index of refraction of medium 1 and medium 2, respectively, and where θ_1 and θ_2 are the incident angles (from the normal, as shown in the diagram).

5.1.2.1 Distance to the Horizon Given that the Earth is roughly spherical in shape, we may naturally ask: How far can we transmit a radio signal from an antenna mounted at a height h above the surface of the Earth?

First, we consider a related question. If we draw a straight line with one end at a height h above the surface of the Earth, what is the longest line that can be drawn, provided that the other end must reach the surface of the Earth at some point, and assuming that the Earth is a perfect sphere? (This is sometimes called the "optical

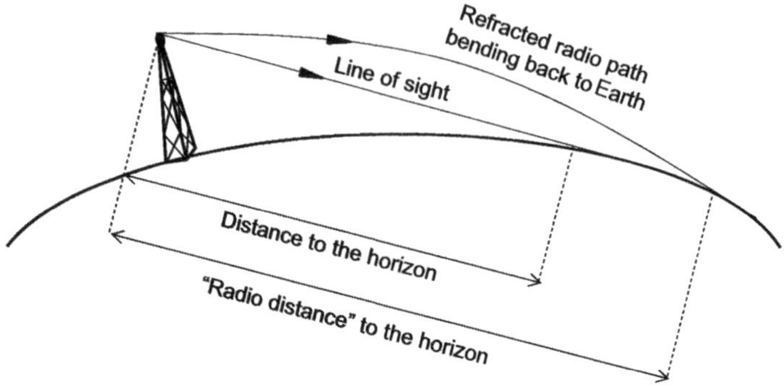

FIGURE 5.2 The optical horizon and radio horizon are not the same.

horizon" because it is the straight-line distance to the horizon.) Clearly, the line must touch the Earth at a tangent, and thus we can use the Pythagorean theorem to find

$$d_{\text{optical horizon}} = \sqrt{2Rh + h^2} \approx \sqrt{2Rh} \tag{5.3}$$

where R is the radius of the Earth, in the same units as h.

In reality, a radio wave can go even farther than the optical horizon before it hits the Earth, because of refraction downward as it travels in the atmosphere (the index of refraction usually decreases with height over the Earth). We can say that this distance, which is farther than the optical horizon, is the true distance to the horizon for a radio signal. See Figure 5.2.

Typically, the effects of refraction are taken into account through the use of a correction factor, K, which is typically taken as 4/3 [7]. Thus, the distance to the horizon is

$$\sqrt{2KRh} \tag{5.4}$$

Suppose that we take R in kilometers and h in meters, and the radius of the Earth as 6378 km; then we have

$$\sqrt{2KRh} = \sqrt{\frac{8}{3}6378\frac{h}{1000}} \approx \sqrt{17h} \qquad \text{km} \tag{5.5}$$

whereas if we take R in miles and h in feet, we have

$$\sqrt{2h} \qquad \text{miles} \tag{5.6}$$

5.1.3 Diffraction

Another way that a radio wave can travel beyond the optical horizon is by bending around an obstacle. Bending around an obstacle, termed *diffraction*, also allows radio waves to be received in shadowed regions not in the line of sight of the radio transmitter. This can happen even without a change in the medium (which can result in

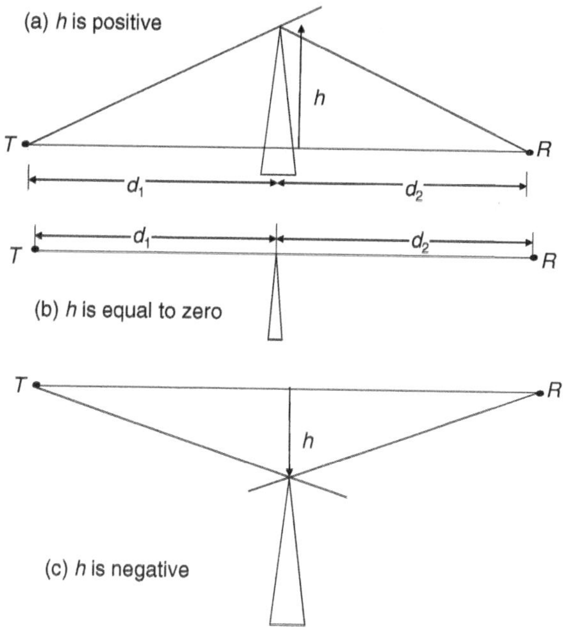

FIGURE 5.3 Knife-edge diffraction.

bending of waves through refraction) or reflection occurring. Diffraction can also refer to waves spreading out after passing through a narrow opening. In general, diffraction is about how waves behave around obstacles.

One way to think of diffraction is based on *Huygens' principle*, which is, in fact, a powerful principle that explains all wave phenomena, including diffraction, interference, refraction, and reflection. In Huygens' model, all points on a wavefront can be viewed as point sources producing secondary wavelets. Th effect of sharp obstructions (e.g., "knife-edge" obstructions, as shown in Figure 5.3) on waves can be explained by Huygens' principle, whereas the behavior appears mysterious if we try to explain it using simply reflection and refraction. For example, if there is a knife edge (as in Figure 5.3; the comments on h are for the derivation in Section 5.1.3.1) surrounded by a uniform medium with no other objects around, waves passing above the knife edge will not refract or reflect (since the medium is uniform and there is nothing off which to reflect). Yet waves bend around the knife-edge obstruction in the phenomenon called diffraction.

5.1.3.1 Fresnel Zones and LOS Clearance Consider a radio transmitter and radio receiver that are in line of sight (LOS) of each other. Even if a straight line can be drawn from the transmitter to the receiver (equivalently, one can see the receiver from the transmitter, or vice versa), the radio signal might still experience a significant loss of power (compared to a truly unobstructed LOS link) if there is an obstruction such as a building, tree, or tower that is close enough to the LOS path. The signal loss

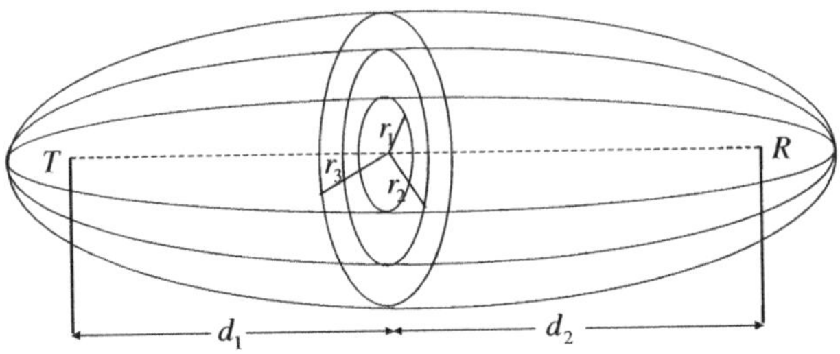

FIGURE 5.4 Fresnel zones.

in such cases is due primarily to diffraction. The concept of *LOS clearance* is about how much space should be left around the LOS path to avoid this.

LOS clearance is often quantified using *Fresnel zones* (also known in the optics literature as *half-period zones*). The Fresnel zones are concentric ellipsoids around the LOS path that represent zones with the following property: Consider a signal from the transmitter to the receiver that diffracts at one point along the way and then arrives at the receiver; it will travel a longer distance than the LOS path, and we call the extra distance the *excess path length*. The first Fresnel zone represents the union of such points where the excess path length is at most $\lambda/2$; the nth Fresnel zone represents the union of such points where the excess path length is between $(n-1)\lambda/2$ and $n\lambda/2$.

The radius of the nth Fresnel zone is

$$r_n = \sqrt{\frac{n\lambda d_1 d_2}{d_1 + d_2}} \tag{5.7}$$

where d_1 and d_2 are the distance along the LOS path between the transmitter and obstruction, and between the obstruction and receiver, respectively. Notice that r_n is a function of d_1 and d_2, not just λ and n. The first three Fresnel zones are shown in Figure 5.4. Next we verify that this radius is consistent with the excess pathlengths required in each Fresnel zone.

Consider an obstruction that comes within h of the LOS path or that blocks the LOS path up to a distance h away from it (Figure 5.5). In the first case we take h to be

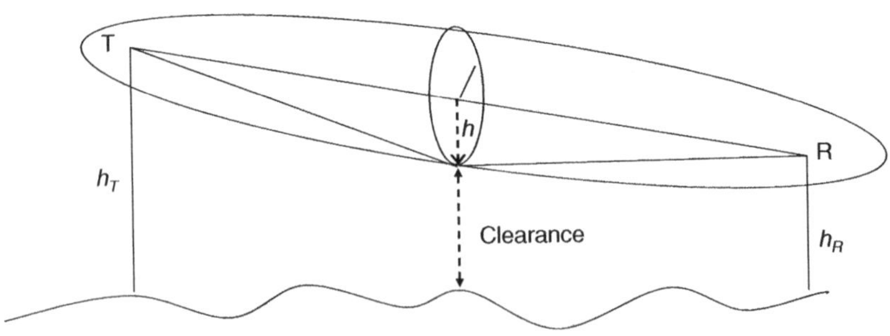

FIGURE 5.5 LOS clearance and the first Fresnel zone.

negative, and in the second case we take it to be positive (see Figure 5.3). In any case, for most of the derivation, we consider h^2, so the sign of h could be plus or minus. A wave that travels from the transmitter to the tip of the obstruction to the receiver would travel a distance that exceeds the LOS pathlength by

$$\Delta \approx \frac{h^2}{2} \frac{d_1 + d_2}{d_1 d_2} \tag{5.8}$$

This excess pathlength, Δ, can be derived by computing the distance from the transmitter to the tip of the obstruction:

$$\sqrt{h^2 + d_1^2} = d_1 \sqrt{1 + \frac{h^2}{d_1^2}} \approx d_1 \left(1 + \frac{h^2}{2d_1^2}\right) \tag{5.9}$$

where the approximation is good for $h \ll d_1$. Doing the same thing with h and d_2, the excess path line then becomes

$$\Delta \approx \frac{h^2}{2d_1} + \frac{h^2}{2d_2} \tag{5.10}$$

from which (5.8) follows. In radians, the excess pathlength would be

$$\phi = \frac{2\pi\Delta}{\lambda} \approx \frac{\pi h^2}{\lambda} \frac{d_1 + d_2}{d_1 d_2} \tag{5.11}$$

Thus, if we let $h = \pm r_n$, we have

$$\phi = \frac{\pi}{\lambda} \frac{n\lambda d_1 d_2}{d_1 + d_2} \frac{d_1 + d_2}{d_1 d_2} = \pi n \tag{5.12}$$

In wavelengths, then, we have $n\lambda/2$, confirming our characterization of Fresnel zones in terms of excess path length.

For LOS clearance purposes (Figure 5.5), the first Fresnel zone is the most important. One rule of thumb is that the entire first Fresnel zone should be clear of obstructions, so we need $h < -r_1$. Another rule of thumb is that $0.6r_1$ should be clear of obstructions; that is, the closest obstruction must be at least $0.6r_1$ away from the LOS path. In terms of h, we need $h < -0.6r_1$. It is sometimes said that "radio LOS" is different from plain LOS, because it needs to include LOS clearance.

5.1.4 Scattering

Radio waves can be scattered by "rough" objects in the environment, such as, trees and rough surfaces. The concept of scattering is perhaps best understood by comparing and contrasting it with reflection off a smooth surface. Consider a large, smooth surface (where the concept of largeness is relative to the wavelength, λ). Supposing that a plane wave hits the surface at a particular angle of incidence. Reflections off such a surface would follow Snell's law of reflection and all be at one angle. We call this *specular reflection*, in contrast to a case where the surface is rough, so the actual

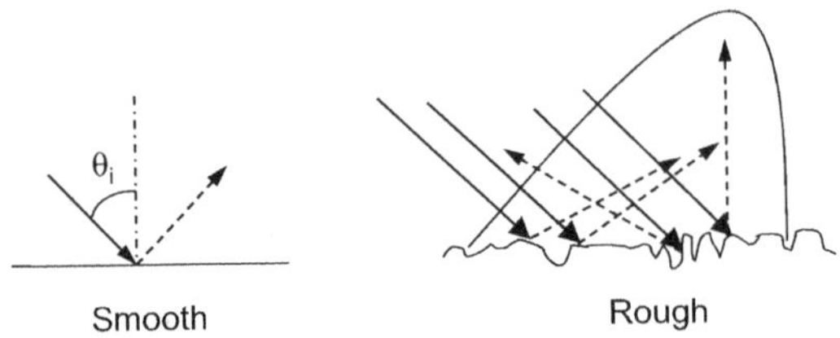

FIGURE 5.6 Specular reflection vs. scattering.

angles of reflection would vary over a range of angles, depending on where on the surface each part of the plane wave arrived. In this case, we have scattering, and the wave is diffused over a range of angles (Figure 5.6).

How do we quantify roughness or smoothness? A standard criterion is the *Rayleigh criterion*. Suppose that the plane wave arrives at the surface at an angle of incidence θ_i, and the largest protrubances are a height/depth h_c; the surface is considered smooth if

$$h_c < \frac{\lambda}{8 \cos \theta_i} \tag{5.13}$$

and otherwise it is considered rough (an alternative formulation considers h_c as the RMS value of the height/depth of the protrubances and replaces the number 8 with the number 32 [12]). The criterion is tightest when θ_i is small, and the limit becomes as small as $\lambda/8$ for $\theta_i = 0$.

5.2 LARGE-SCALE EFFECTS IN CELLULAR ENVIRONMENTS

The signal level arriving from a base station at a mobile device is typically influenced by many factors. Traditionally, a useful way to divide the effects into smaller, more manageable subsets is to consider them as either large- or small-scale effects. Large-scale effects are those associated with distances on the order of many wavelengths, whereas small-scale effects are those associated with distances on the order of a wavelength. As a mobile device moves away from a base station, the signal strength tends to go down, but there will be many fluctuations. The large-scale effects can be thought of as the reflections, refractions, path loss, and so on, that account for the average signal strength around a particular region, whereas the small-scale effects (such as constructive and destructive interference of waves arriving at different times from different angles) account for the fluctuations around the average signal strength.

In this section we consider large-scale effects, we wait until Section 5.3 to consider small-scale effects. To understand large-scale effects, we can choose from analytical models and empirical models. Analytical models attempt to model analytically one or more phenomena (e.g., reflection, refraction, diffraction, scattering) to come up with a useful characterization of the path loss. Empirical models are based on extensive measurements. The ground reflection model is an analytical model (Section 5.2.1), whereas the Okumura model is an empirical model (Section 5.2.2). The Hata model is a quasianalytical model derived from the Okumura model (Section 5.2.3). Analytical models such as the lognormal fading model (Section 5.2.4) can be used with some of the other analytical or empirical models to model *variation* of the actual large-scale fading from what might be obtained from some of the other models. It must be noted, though, that this is still about variation of *large-scale* fading (as distinct from small-scale fading).

5.2.1 Ground Reflection Model

The ground reflection model provides a possible explanation for why $\nabla \approx 2$ closer to the base station, and $\nabla \approx 4$ farther away, assuming perfect, lossless reflection from the ground. Closer to the base station, propagation follows the free-space case. Any ground-reflected paths are significantly longer than the direct path and are therefore significantly weaker, making the propagation conditions similar to those of the free-space case. Farther away, though, there is significant interference of the direct path from the ground-reflected path, because both are of comparable strength, causing a faster drop-off in signal strength. The interference effects can be analyzed as follows: Suppose that we have two antennas a distance d_0 meters apart, as in Figure 5.7. The user antenna is h_1 meters high and the base station antenna is h_2 meters high. There are two main paths that the signal will take from the base station to the user, or vice versa: the direct path and the ground-reflected path. The ground-reflected path makes an angle θ with the ground. In general, θ will be small when d_0 is much bigger than h_1 and h_2, which is the case we are interested in here. However, for clarity in illustration, the size of θ has been exaggerated in Figure 5.7.

The length of the direct path is given by

$$d_{\text{direct}} = d_1 = \sqrt{d_0^2 + (h_2 - h_1)^2} \tag{5.14}$$

The length of the ground-reflected path is given by

$$d_{\text{reflect}} = d_2 + d_3 = \sqrt{d_0^2 + (h_1 + h_2)^2} \tag{5.15}$$

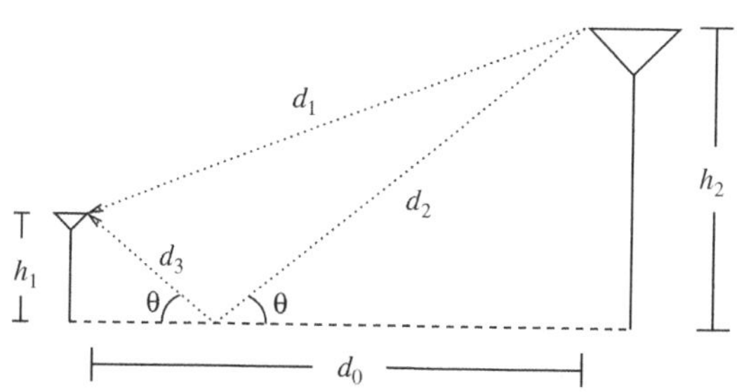

FIGURE 5.7 Basis for the ground reflection model.

The difference in pathlengths is

$$d_{\text{diff}} = d_{\text{reflect}} - d_{\text{direct}} = \sqrt{d_0^2 + (h_1 + h_2)^2} - \sqrt{d_0^2 + (h_2 - h_1)^2} \quad (5.16)$$

$$= d_0 \left[\sqrt{1 + \left(\frac{h_1 + h_2}{d_0}\right)^2} - \sqrt{1 + \left(\frac{h_2 - h_1}{d_0}\right)^2} \right] \quad (5.17)$$

$$\approx d_0 \left[\left\{ 1 + \frac{1}{2}\left(\frac{h_1 + h_2}{d_0}\right)^2 \right\} - \left\{ 1 + \frac{1}{2}\left(\frac{h_2 - h_1}{d_0}\right)^2 \right\} \right] \quad (5.18)$$

$$= \frac{2h_1 h_2}{d_0} \quad (5.19)$$

where the approximation in (5.18) is very good for $d_0 \gg h_1, h_2$.

Assuming that the ground reflection causes a phase shift of π radians [13], the phase difference between the direct and ground-reflected paths is given by

$$\Delta\phi = 2\pi \left\lfloor \frac{1}{\lambda} d_{\text{diff}} + \frac{1}{2} \right\rfloor = 2\pi \left\lfloor \left(\frac{2h_1 h_2}{d_0 \lambda} + \frac{1}{2} \right) \right\rfloor \quad (5.20)$$

where $\lfloor x \rfloor$ is the largest integer k such that $k < x$, where $x \in \mathcal{R}$. For $d_0 \gg h_1, h_2$, (5.20) becomes

$$\Delta\phi = 2\pi \left(\frac{2h_1 h_2}{d_0 \lambda} + \frac{1}{2} \right) \approx \pi \quad (5.21)$$

For $d_0 \gg h_1, h_2$, the direct and ground-reflected paths are both approximately of length d_0. Therefore, we assume that both paths have the same amplitude, resulting separately (if each path were the only path) in power given by (4.23) [i.e., $P_t G_t G_r \lambda^2/(4\pi d)^2 \Lambda_0$, where $d \approx d_0$ for $d_0 \gg h_1, h_2$]. Then the power received is given by

$$P_r(d) = \frac{P_t G_t G_r \lambda^2}{(4\pi)^2 d_0^2 \Lambda_0} \left\{ 2 \sin \left[\frac{1}{2}(\pi - \Delta\phi) \right] \right\}^2$$

$$\approx \frac{P_t G_t G_r \lambda^2}{(4\pi)^2 d_0^2 \Lambda_0} \left(\frac{4\pi h_1 h_2}{d_0 \lambda} \right)^2 \quad (5.22)$$

$$= \frac{P_t G_t G_r h_1^2 h_2^2}{d_0^4 \Lambda_0} \quad (5.23)$$

where the approximation to obtain (5.22) comes from the fact that $\sin\theta \approx \theta$ for $\theta \approx 0$. It should be noted that λ cancels out and is not found in (5.23). However, the main observation is that the path loss exponent here is $\nabla = 4$ instead of $\nabla = 2$ as for free space.

5.2.2 Okumura Model

The Okumura model is an empirical model based on extensive measurements conducted in Tokyo. It is regarded as simple and fairly accurate for cellular systems in cities. The model is meant to be used for systems operating between 150 and 1920 MHz, for distances between 1 and 100 km, and for base station antenna heights between 30 and 1000 m. The model comes in the form of graphs and curves for different parameter ranges. Correction factors for other terrain types are also available in the form of more curves.

5.2.3 Hata Model

The Hata model [5] is modified from the Okumura model, mainly in taking the curves and quantifying them as a set of equations. It is valid from 150 to 1500 MHz for distances up to 20 km. The main formula is meant for urban areas, but there are correction factors to allow it to be used in small and medium-sized city environments. It is considered to be better for systems with large cells (larger than 1 km) than for systems with smaller cells.

The formula is (in dB)

$$\begin{aligned} L = {} & 69.55 + 26.16\log(f_c) - 13.82\log(h_{\mathrm{BS}}) \\ & -\alpha(h_{\mathrm{MS}}) + \left[44.9 - 6.55\log(h_{\mathrm{BS}})\right]\log d - K \end{aligned} \qquad (5.24)$$

where L is the median path loss at distance d, h_{BS} and h_{MS} are the height of base station and mobile station antennas, respectively, and α and K are correction factors, given in Table 5.1.

An example of how Hata's formulas compare with values from Okumura's curves is shown in Figure 5.8, where the solid lines are calculated using Hata's formulas, and the individual points are obtained from Okumura's curves. It shows that Hata did a decent job in providing simple, yet reasonably accurate formulas to approximate Okumura's results.

TABLE 5.1 Hata Model: Correction Factors

Type of Area	$\alpha(h_{\mathrm{MS}})$	K
Open	—	$4.78[\log(f_c)]^2 - 18.33\log(f_c) + 40.94$
Suburban	—	$2[\log(f_c/28)]^2 + 5.4$
Medium-sized to small city	$[1.1\log(f_c) - 0.7]h_{\mathrm{MS}} - [1.56\log(f_c) - 0.8]$	
Large city		
$f_c > 300$ MHz	$3.2[\log(11.75h_{\mathrm{MS}}]^2 - 4.97$	
$f_c < 300$ MHz	$8.29[\log(1.5h_{\mathrm{MS}}]^2 - 1.10$	

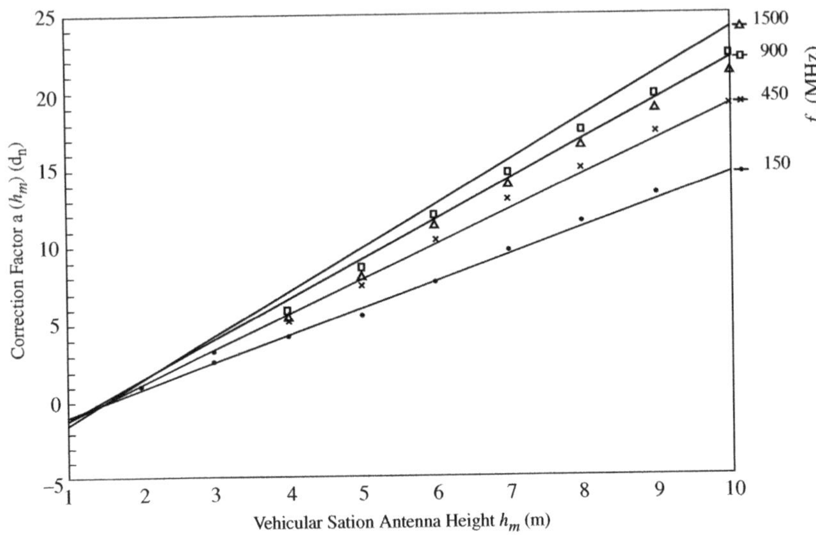

FIGURE 5.8 Comparison of one aspect of Hata's model with Okumura's results. (From [5]; copyright © 1980 by IEEE, reprinted with permission.)

5.2.4 Lognormal Fading

On top of the path loss is a perturbation whose first-order statistics have been found empirically to be lognormally distributed [3], that is, the logarithm of the random variable is normally distributed, with its mean being the path loss. The reason for this perturbation from the basic trend in signal power level of the path loss component is that there exist trees and buildings and other attenuators, reflectors, scatterers, and diffractors in a real cellular environment. A heuristic to explain the lognormal distribution based on physical reasons is as follows: If there are several reflectors or attenuators along a given path, each reflection or attenuation contributes a multiplying factor to the amplitude of the signal, and these multiplying factors can be considered as random variables. In decibels, this would be equivalent to the addition of several random variables, giving an approximately normal distribution by the central limit theorem [10].

The second-order statistics of large-scale lognormal fading are not well understood, but an exponentially shaped correlation function fits some empirical data reasonably well [4]. Large-scale fading† is called that because the correlation distances are on

†Large-scale fading is also known as *slow fading* [1] because the correlation distances are several orders of magnitude larger than those of small-scale fading. Therefore, the rate of change in the fading is "slow" compared to the rate of change of small-scale fading. Analogously, small-scale fading is also known as *fast fading*, but we follow [13] in calling it small-scale fading because the terms "fast" and "slow" with respect to fading are also used to refer to the rate of time variation of time-varying wireless channels (i.e., a fast-fading channel would refer to one with a large Doppler spread). The use of "large-scale" and "small-scale" avoids possible confusion.

the order of hundreds of meters in large cells. (They might be smaller, however, on the order of tens of meters, in microcells. A heuristic justification is that the larger the cells, the more likely it is that users are farther away from the base stations, so the effect of the larger reflectors and attenuators tends to dominate, and these are spaced farther apart than smaller reflectors and attenuators.) This is in contrast to small-scale fading (Section 5.3), with correlation distances on the order of tens of centimeters for typical operating frequencies.

5.3 SMALL-SCALE EFFECTS IN CELLULAR ENVIRONMENTS

The fundamental reason for small-scale effects is that the wireless signal takes multiple paths from the transmitter to the receiver, as shown in Figure 5.9. This leads to the phenomenon of multipath delay spread (Section 5.3.1). A useful, single-number quantification of multipath delay spread is the rms delay spread, σ. Depending on the value of σ compared to the symbol period T_s, we may have flat fading (Section 5.3.2) or frequency selective fading (Section 5.3.3). The mobile wireless channel is time varying. Ways to characterize the time variation of the channel are discussed in Section 5.3.4. A useful technique for mitigating the effects of small-scale fading is diversity combining, and we examine some diversity combining methods in Section 5.3.5.

5.3.1 Multipath Delay Spread

Suppose that $s(t)$ is the signal transmitted. The signal received can be expressed as a sum of the signals arriving on multiple paths:

$$r(t) = \sum_{n=1}^{N} \alpha_n s(t - \tau_n) e^{-j2\pi f_c \tau_n} \tag{5.25}$$

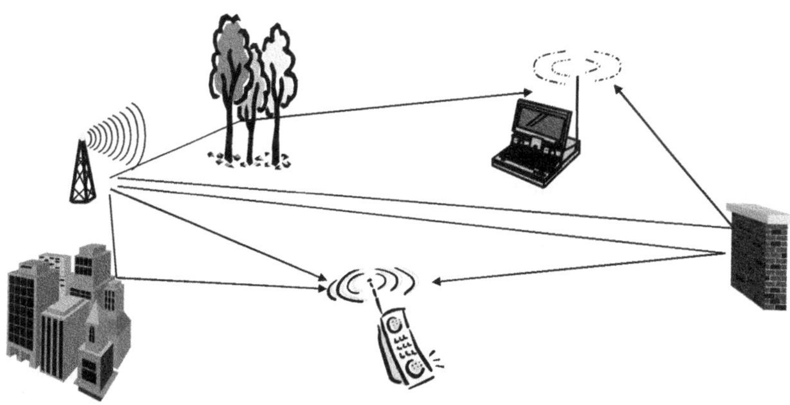

FIGURE 5.9 Multipath phenomenon in wireless transmission.

where the α_n are real values representing the amplitude of the signal on the various paths, and the τ_n are real values representing delays of the various paths, and (5.25) includes up to N different paths. Then the *mean excess delay* is

$$\bar{\tau} = \frac{\sum_n \alpha_n^2 \tau_n}{\sum_n \alpha_n^2} \tag{5.26}$$

Then let

$$\overline{\tau^2} = \frac{\sum_n \alpha_n^2 \tau_n^2}{\sum_n \alpha_n^2} \tag{5.27}$$

then the *rms delay spread* is

$$\sigma = \sqrt{\overline{\tau^2} - \bar{\tau}^2} \tag{5.28}$$

NB: In the computation of both $\bar{\tau}$ and $\overline{\tau^2}$, only the relative values of the α_n^2 are important. Thus, if normalized values are provided (as is often the case where the channel measurements are normalized to the amplitude of the strongest arriving path), they can be used directly in the formulas. However, care must be taken to see if the given values are amplitude or power values. For example, if a *power delay profile* is provided, the values should be used without squaring.

The rms delay spread is a useful characterization of multipath delay spread by a single number proportional to how bad the channel is. Comparing urban, suburban, and indoor environments, the rms delay spread is typically largest in urban environments and smallest in indoor environments. In urban environments, the rms delay spread may be on the order of microseconds (up to 25 μs in cities such as San Francisco, where there are many hills and valleys). In suburban environments, rms delay spread may range from 0.2 to 2 μs. In indoor (in-building) environments, the rms delay spread tends to be on the order of tens to hundreds of nanoseconds.

Figures 5.10 and 5.11 show the two "typical urban" delay spread channels given in the GSM specifications [8]. They can be used, for example, in computer simulations of wireless systems.

5.3.2 Flat Fading

Suppose that the signal transmitted is a narrowband signal (with respect to the environment). In other words, the "frequency response"† of the wireless channel is relatively

†We put quotation marks around the term "frequency response" because the wireless channel is actually a time-varying channel due to the motion of the user. So it does not have a frequency response, which makes sense only for linear, time-invariant (LTI) channels. However, because the wireless channel varies relatively slowly, it is often described as quasi-stationary and it becomes possible to talk of a frequency response, albeit a slowly varying frequency response. For example, in a time-division-multiple-access system, in which users share frequency channels in round-robin fashion, each user transmits and receives data in short bursts in the time slot allocated. For most practical purposes, the frequency response could be considered constant over each burst, although it may be different from burst to burst.

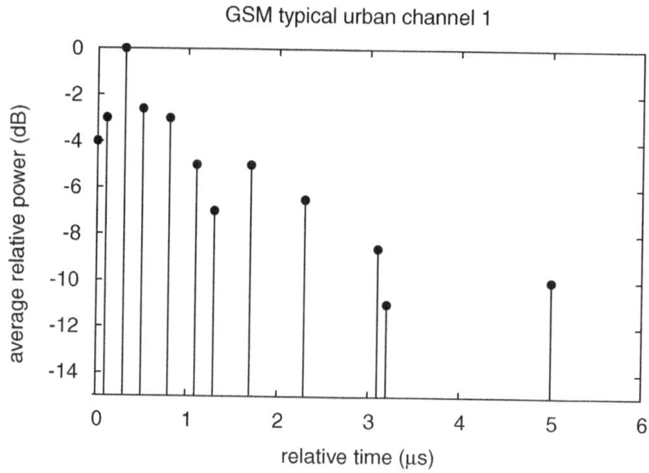

FIGURE 5.10 Typical urban channel in GSM specifications.

flat in the range of frequencies over which the signal has significant power. If the signal transmitted is a digitally modulated waveform, *narrowband* would mean that the symbol period is much greater than the average greatest difference in pathlengths of the significant paths between the transmitter and receiver, where a significant path is one through which a significant fraction of the received power arrives. Then we have *flat fading*, since the wireless channel is relatively flat in the range of frequencies over which the signal has significant power. If this is not the case, we may have frequency-selective fading, which we examine in Section 5.3.3.

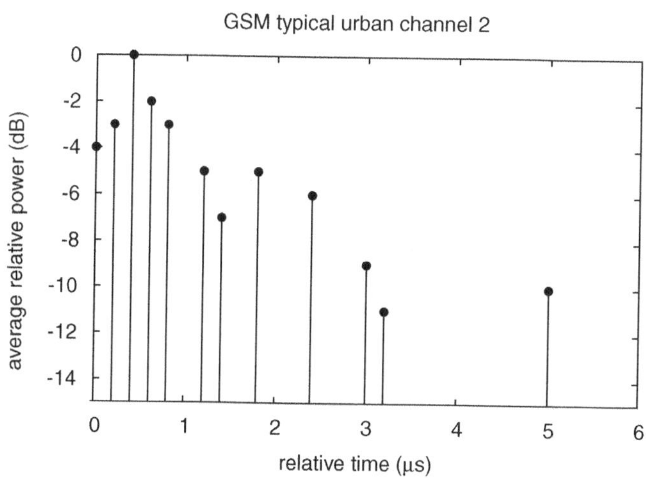

FIGURE 5.11 Typical urban channel in GSM specifications.

Flat fading occurs when

$$\sigma \ll T_s \tag{5.29}$$

For a frequency-domain perspective, define the *coherence bandwidth* to be

$$B_c = \frac{1}{2\pi\sigma} \tag{5.30}$$

and then flat fading occurs when

$$1/T_s \ll B_c \tag{5.31}$$

A reasonable assumption to make is that each path has an independent random phase offset with respect to a reference path and that the random phase offset is distributed uniformly between 0 and 2π. Suppose that the reference path is dominant (i.e., it conveys an average power significantly larger than that conveyed by any other path by itself, although not necessarily larger than the sum total of that conveyed by all the other paths), as when there is a direct path between the antennas. If all the other paths convey approximately the same average power, it can be shown that the distribution of the amplitude of the received signal is Ricean, as the position of the receiving antenna varies. In other words, the magnitude of the signal envelope exhibits Ricean-distributed fluctuations around the local mean signal power level as the receiving antenna moves around. This result was derived by Rice in his seminal paper on the mathematical analysis of random noise [14], and the distribution is given by

$$f_{\text{Ricean}}(x) = \frac{x}{p} \mathrm{I}_0\left(\frac{x p_d}{p}\right) e^{-(x^2+p_d^2)/2p} \qquad x \geq 0 \tag{5.32}$$

where $\mathrm{I}_0(\cdot)$ is the zeroth-order modified Bessel function of the first kind. The critical factor, $\mathcal{K} = p_d^2/2p$, is of fundamental importance in Ricean distributions, because it is the ratio of the dominant component power to the fading component power. $\mathcal{K}$ is also known as the *specular-to-diffuse ratio* [9] or *Rice factor* [6].

In most cases of interest for hand-off algorithms, though, the user is relatively far from the base station, and $p_d \to 0$ and $\mathcal{K} \to 0$ because there is no direct path or dominant component. This case is the important case of Rayleigh fading, where in the absence of a dominant path, the resultant of the contributions from many different paths forms a complex Gaussian process (heuristically justified by the central limit theorem). The amplitude of the signal envelope is well known to be Rayleigh distributed in such a case [2]. It can also be seen that the amplitude is Rayleigh distributed by noting that when $p_d \to 0$, (5.32) reduces to a Rayleigh distribution, given by

$$f_{\text{Rayleigh}}(x) = (x/p)e^{-x^2/2p} \qquad x \geq 0 \tag{5.33}$$

where x is the signal envelope amplitude and p is the mean power of the signal.

Exactly what p is and exactly what it is that is Rayleigh distributed sometimes cause confusion to newcomers. Linking back with our representations from Section 1.3.4.1, we can write a passband signal in in-phase and quadrature form as

$$x_b(t) = x_i(t)\cos(2\pi f_c t) - x_q(t)\sin(2\pi f_c t) \tag{5.34}$$

Now we can define p as the time-average power of $x_b(t)$,

$$p = \overline{|x_b(t)|^2} \tag{5.35}$$

So what is Rayleigh distributed? Not $x_b(t)$, but the *envelope*, $A(t)$, given by

$$A(t) = \sqrt{x_i^2(t) + x_q^2(t)} \tag{5.36}$$

[from the definitions of $x_i(t)$ and $x_q(t)$ in Section 1.3.4.1]. In fact, the average power of $A(t)$ is $\overline{A^2(t)} = 2p$, and its mean is given by $\sqrt{p\pi/2}$. Thus, p is sometimes described as the "average power of the received signal before envelope detection" to emphasize that it is the average power of $x_b(t)$ rather than of $A(t)$, which is $2p$.

The mean of the Rayleigh distribution is $\sqrt{\pi p/2}$, but the Rayleigh fading process can dip 20, 30, or even 40 dB below this mean. This kind of drop in signal strength is known as a *Rayleigh fade*. Even in a system with generous radio link margins, it is almost impossible to communicate reliably if the user is in a Rayleigh fade. However, it is rare that a user will remain in a Rayleigh fade long enough for it to become a serious or insurmountable problem, because users are normally moving. Little can be done about users who are stationary in a Rayleigh fade. Even children who use mobile phones know from experience, though, that just by moving around a little, a bad signal might sometimes improve significantly. What does "moving around a little" mean? In Section 5.3.4 we quantify this.

5.3.3 Frequency-Selective Fading

Suppose we relax the assumption that the signal is narrowband with respect to the environment. If the system is a digital system, relaxing this assumption means that the symbol period may now be comparable to or smaller than some measure of *delay spread*, the spread in arrival times of different paths due to differences in pathlength. This is the case of *for frequency-selective fading*, whereas the narrowband case is known as *flat fading* because the fading affects all frequency components of the signal by approximately the same amount. With frequency-selective fading, intersymbol interference (ISI) is introduced. Typically, we call it a frequency-selective fading situation when

$$\sigma \gg T_s \quad \text{or} \quad 1/T_s \gg B_c \tag{5.37}$$

Figure 5.12 illustrates the very important point that the *same* channel [e.g., represented in time by the impulse response $h(t)$ on the left, and in frequency by $H(f)$ on the right] can be a flat-fading channel or a frequency-selective fading channel. It depends on the signal bandwidth relative to the channel. As we see in (5.31) and (5.37),

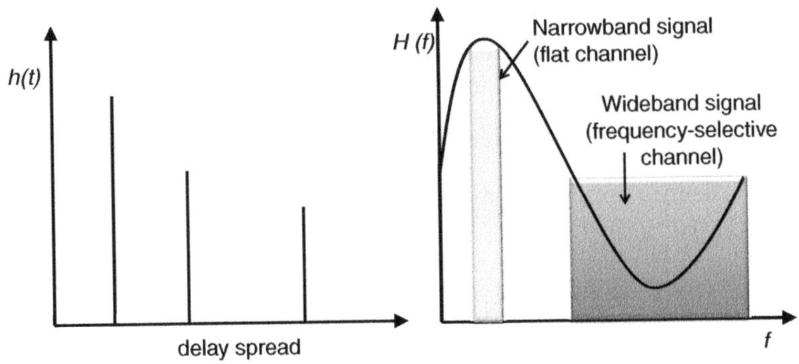

FIGURE 5.12 The same channel may be flat or frequency selective.

we could have the same channel with the same B_c, but depending on $1/T_s$, we could be in either a flat or frequency-selective fading situation with respect to the particular signal we are transmitting.

To build a good system it is necessary to incorporate a way to mitigate the effects of delay spread. Methods include the use of multicarrier techniques, the use of spread-spectrum techniques, and the use of more traditional equalizers. Equalizers filter the received signal to mitigate the effects of the channel [11]. One way of looking at the other two methods is that they break the broadband channel into smaller narrowband channels, use the narrowband channels, and then intelligently recombine the outputs in the receiver. Often the model of a Gaussian wide-sense stationary uncorrelated scattering channel† is applicable. Then the different narrowband subchannels fade independently, each having the behavior described earlier in this section for the narrowband case.

5.3.4 Time Variation: The Doppler Shift

To estimate the second-order statistics of small-scale fading it is commonly assumed that the different paths arrive at the receiving antenna independently and distributed uniformly in angle. As the mobile moves, each angle is associated with a different Doppler shift because of the different rate of change of the pathlength, depending on the angle of the particular path. Let θ be the angle between the direction of motion of the mobile and the path of the incoming signal in the plane parallel to the ground plane, as shown in Figure 5.13(a). Let v_0 be the speed of the mobile and v_1 be the speed of the mobile in the direction directly toward the base station. Then v_1 is also the rate of change of the pathlength toward the base station and is given by

$$v_1 = v_0 \cos\theta \qquad \text{m/s} \tag{5.38}$$

† Gaussian because of the complex Gaussian process from the sum of the contributions from different paths randomly distributed in phase; wide-sense stationary because the changes in the statistics are on a much larger scale; uncorrelated because the fading at different delays is uncorrelated.

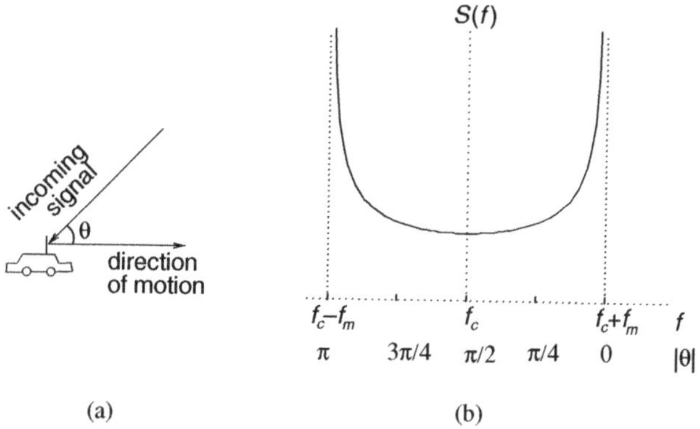

FIGURE 5.13 Doppler spread phenomenon in cellular systems. Part (a) shows an incoming signal at an angle θ to the direction of motion. Part (b) shows the power spectrum of the receiver signal. The angles indicated under the plot show the angles associated with the various Doppler shifts.

or (in cycles/second)

$$v_1 = (v_0/\lambda)\cos\theta \qquad \text{Hz} \tag{5.39}$$

This change in pathlength causes an apparent shift in carrier frequency of the same frequency [i.e., $(v_0/\lambda)\cos\theta$ Hz]. The shift in carrier frequency is known as the *Doppler shift*, and the general phenomenon is known as the *Doppler effect.*

It can be shown that under the assumptions made, the power spectrum of the signal received when a sinusoid is transmitted is given by

$$S(f) = \begin{cases} \frac{C_1}{\sqrt{f_m^2-(f-f_c)^2}} & \text{for } |f-f_c| < f_m \\ 0 & \text{for } |f-f_c| > f_m \end{cases} \tag{5.40}$$

where C_1 is a proportionality constant, f_c is the carrier frequency, and f_m is the maximum Doppler spread, given by

$$f_m = \max v_1 = v_0/\lambda \tag{5.41}$$

The power spectrum is plotted in Figure 5.13(b).

Taking the inverse Fourier transform of the lowpass equivalent of (5.40) gives the autocorrelation of the small-scale fading. The autocorrelation, with means removed [similar to (6.27), but in the continuous case], can then be shown [7] to be equal to

$$J_0^2(2\pi f_m \tau) = J_0^2\left[2\pi\left(\frac{v_1}{\lambda}\tau\right)\right] = J_0^2\left[2\pi\left(\frac{\Xi}{\lambda}\right)\right] \tag{5.42}$$

where $J_0(\cdot)$ is the zeroth-order Bessel function of the first kind, τ is a time interval, Ξ is a spatial interval, and $v_1 = \Xi/\tau$. We can specify the autocorrelation as a function of either τ or Ξ because a temporal difference τ is the same as a spatial difference Ξ,

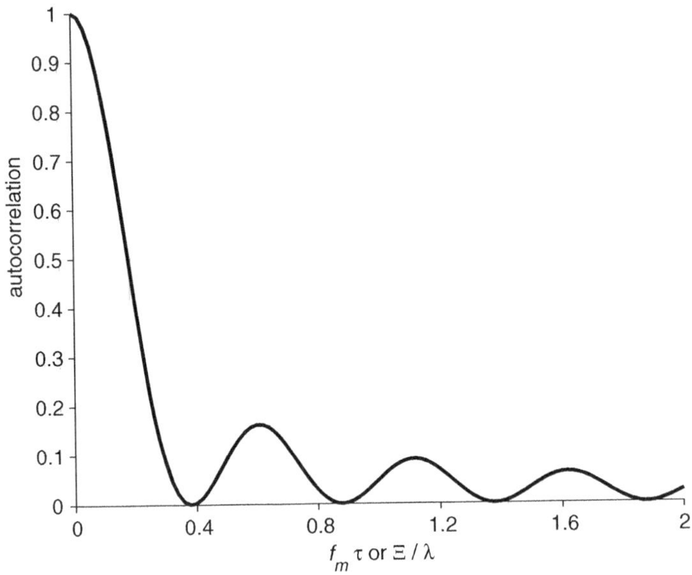

FIGURE 5.14 Autocorrelation of the signal received.

under the assumption that the mobile is moving at constant speed v_1 toward the base station.

Equation (5.42) is plotted in Figure 5.14, from which it can be seen that points separated by more than 0.4λ have a small correlation. Therefore, a useful rule of thumb in the wireless industry is that consecutive samples are assumed to be independent if spaced more than half a wavelength apart.† At a typical carrier frequency of around 900 MHz, half a wavelength is about 15 cm. Such figures are useful in designing diversity combining schemes (see Section 5.3.5), for example.

Another way of quantifying the rate of change of the wireless channel is by the concept of *coherence time*. Coherence time has been variously defined [13] as

$$T_c = \frac{1}{f_m} \quad \text{or} \quad T_c = \frac{9}{16\pi f_m} \quad \text{or} \quad T_c = \sqrt{\frac{9}{16\pi f_m}} = \frac{0.423}{f_m} \tag{5.43}$$

A channel can be said to be *fast fading* if

$$T_c \ll T_s \tag{5.44}$$

A channel can be said to be *slow fading* if

$$T_c \gg T_s \tag{5.45}$$

† Of course, this assumes that the uniform-angle-of-arrival assumption, resulting in (5.40), is valid; in Section 5.3.5.1 we discuss how this may not be applicable at base stations, for example.

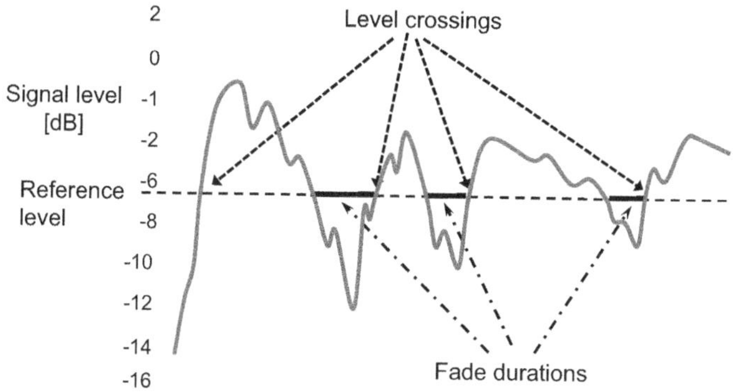

FIGURE 5.15 Level crossings and fade durations.

In a way, coherence time is to Doppler shift what coherence bandwidth is to delay spread. In both cases, a phenomenon is quantified in one domain (e.g., σ in the time domain and f_m in the frequency domain), and coherence time or coherence bandwidth is defined to quantify it in the dual domain. In both cases, the exact quantification is somewhat arbitrary, as these quantities are useful primarily to give wireless system designers an idea of the design space they are working in. If T_s is close to T_c, it is neither a clearly fast fading nor a clearly slow fading channel, and if T_s is close to σ, it is neither a clearly frequency-selective channel nor a clearly flat fading channel.

5.3.4.1 Level Crossing and Fade Duration Statistics

The *level crossing rate* is the rate at which the amplitude of the fading envelope crosses a given threshold level moving upward (Figure 5.15). Normally, the threshold level would be normalized to the root-mean-squared value of the amplitude of the fading envelope. Given a threshold level, R, we let $\rho = R/R_{\text{rms}}$, where R_{rms} is the root-mean-squared value of the amplitude of the fading envelope. Thus, ρ is normalized, and if the fading is Rayleigh distributed, the *level crossing rate* is given by

$$N_R = \sqrt{2\pi} f_m \rho e^{-\rho^2/2} \tag{5.46}$$

The *average fade duration* is the average length of a fade (i.e., the average time that it is below a threshold value R; Figure 5.15). The larger it is, the worse the channel. Again we let $\rho = R/R_{\text{rms}}$. Then the average fade duration is given by

$$\bar{\tau} = \frac{e^{\rho^2} - 1}{\rho f_m \sqrt{2\pi}} \tag{5.47}$$

5.3.5 Diversity Combining

Diversity combining is one way to mitigate the effects of small-scale fading. It is based on the idea that when multiple samples of a random variable are taken, they will take different values: some larger, some smaller. Thus, instead of sampling the fading random variable once, and sometimes being stuck with a bad channel, we obtain multiple independent (or slightly correlated) samples of the same random variable (multiple independent "looks" at the same channel). Then we combine them some way. For example, with selection diversity, the best channel (best instantaneous SNR) is selected.

5.3.5.1 Obtaining Multiple Independent Samples

To obtain multiple independent samples, we may use multiple antennas at the receiver spaced sufficiently far apart enough to reduce correlation of the fading statistics (space diversity or antenna diversity). How far apart should the antennas be? A rule of thumb for base stations is about 10λ or more, whereas for mobiles it is about $\lambda/2$. One way of understanding this asymmetry is to think about how most reflections would occur near to the ground, and hence near the mobiles. The paths between the reflectors and the mobile would span a wide angle as viewed from the mobile. As viewed from the base station, however, since the base station is farther away, the paths would span a much narrower angle. Thus, a small change in mobile position can make a big difference, whereas a larger change in base station antenna position would be needed for similar changes.

Another way to obtain multiple independent samples is to use different antenna polarizations (polarization diversity). For example, at a receiver, two antennas can be placed in the same location, but with one of them optimized to receive horizontally polarized signals and the other optimized to receive vertically polarized signals. If we just had a horizontally polarized antenna, much of the signal that arrives vertically polarized would be lost through polarization loss (Section 4.1.3.1). Similarly, if we just had a vertically polarized antenna, much of the signal that arrives horizontally polarized would be lost through polarization loss. For purposes of diversity, then, we expect that there may be differences in fading in the signals that arrive at one antenna from the signals that arrive at another antenna with different polarization. Informally, we may think of it as meaning that different portions of the total signal are received at each antenna, and these different portions may have undergone different reflections, refractions, and so on. So polarization loss is exploited for diversity purposes.

Other ways to obtain multiple independent samples include transmitting and receiving the same signal at different frequencies spaced more than the coherence bandwidth apart (frequency diversity), and transmitting and receiving the same signal at different times spaced more than the coherence time apart (time diversity), and so on. From our concepts of coherence bandwidth and coherence time, we would expect low correlation in the fading at two points separated by more than the coherence bandwidth or coherence time.

5.3.5.2 Selection Diversity

In selection diversity (Figure 5.16), the instantaneous SNR on each of the diversity branches is monitored continually, and the receiver switches continually to the diversity branch that currently has the highest instantaneous SNR.

If all N branches are assumed to be identical, independently distributed (i.i.d.) Rayleigh fading branches with instantaneous SNR of γ_j, and average SNR, $\Gamma = \bar{\gamma}$,

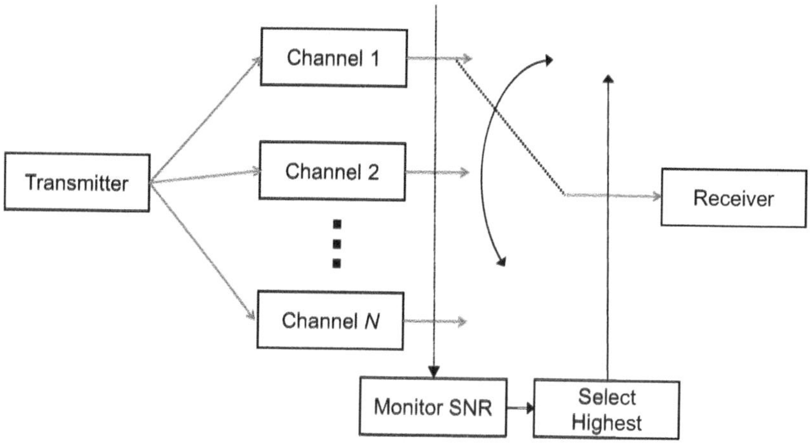

FIGURE 5.16 Selection diversity.

then†

$$p(\gamma_j) = \frac{1}{\Gamma} e^{\gamma_j/\Gamma} \tag{5.48}$$

and for a given value γ_0,

$$P[\gamma_j \leq \gamma_0] = 1 - e^{-\gamma_0/\Gamma} \tag{5.49}$$

so the probability that all branches are below γ_0 at the same time is

$$P[\gamma_1 \leq \gamma_0, \gamma_2 \leq \gamma_0, \ldots, \gamma_N \leq \gamma_0] = (1 - e^{-\gamma_0/\Gamma})^N \tag{5.50}$$

Whereas Γ is the average SNR on each branch, let $\Gamma_{\text{selection}}$ be the average SNR on the instantaneously selected branch (for every moment in time, there is one branch selected, and the instantaneous SNR from that branch is included in the averaging). Then the overall improvement in average SNR is given by

$$\Gamma_{\text{selection}} = \Gamma \sum_{n}^{N} \frac{1}{n} \tag{5.51}$$

5.3.5.3 Equal Gain Combining Rather than simply discarding the other diversity branches and using only the one with the highest instantaneous SNR, we can exploit the energy in the other diversity branches as well, to yield better overall SNR than with selection diversity, provided that we combine the energy in the diversity branches cleverly.

† See Exercise 5.5 for the derivations.

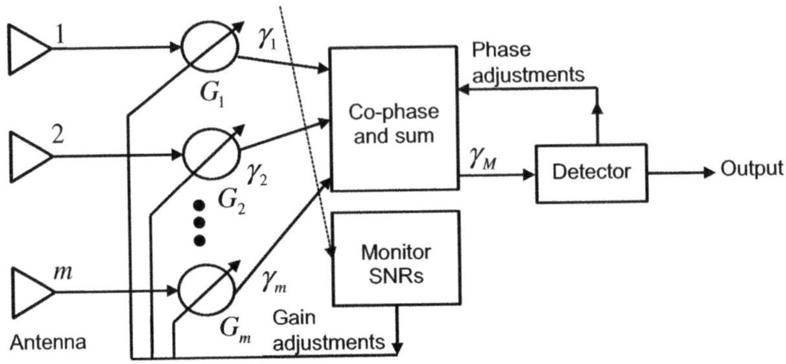

FIGURE 5.17 Maximal ratio diversity.

Then the overall improvement in average SNR, in the special case that all branches have the same average SNR, Γ, is given by

$$\Gamma_{\text{equal gain}} = \Gamma\left[1 + (N-1)\frac{\pi}{4}\right] \tag{5.52}$$

5.3.5.4 Maximal Ratio Combining The best diversity combining scheme is the maximal ratio combining scheme (Figure 5.17). In maximal ratio combining, the diversity branches are co-phased, and then a weighted sum is taken. The weight for each diversity branch is the instantaneous SNR on that diversity branch. Thus, diversity branches with better SNR get a higher weightage, and therefore the result performs better than equal gain combining.

It can be shown that the average output SNR is the sum of the input SNRs (each averaged over time):

$$\Gamma_{\text{max ratio,general}} = \sum_{j=1}^{N} \Gamma_j \tag{5.53}$$

and in the special case that all branches have the same average SNR, Γ, we have

$$\Gamma_{\text{max ratio}} = \sum_{j=1}^{N} \Gamma = N\Gamma \tag{5.54}$$

5.4 INCORPORATING FADING EFFECTS IN THE LINK BUDGET

Would a certain transmitter power be sufficient such that the received signal can be detected with better than a specified BER? It depends on many factors, including the gains of the transmitter and receiver antennas, feeder cable losses, and so on. The link budget refers to a common way to quantify these factors as a list of numbers representing gains and losses in decibels. For example, if we work backward from the receiver to find the minimum transmitter power required on the transmitter side, analysis of the wireless access technologies (the combination of modulation schemes, error coding, etc.) might lead to a value for SNR_{min} required for satisfactory BER. Analysis of the RF design might lead to a noise floor that we add to SNR_{min} to obtain the receiver sensitivity [as given by (3.38)]. Taking into account such factors as cable losses from the receiver antenna, this leads to a minimum value for the signal power received at the antenna, $P_{\text{r,dB}}$, from which we can use (4.23) to obtain $P_{\text{t,dB}}$.

There are two ways that we could improve on simply using (4.23) to obtain $P_{\text{t,dB}}$. First, we could replace the simple $10\nabla \log d_0$ term with path loss obtained from other models: the Hata model or any other model we would like to use. Second, we could add some margins to account for lognormal fading and Rayleigh fading effects. We now consider how we might obtain suitable margins to account for lognormal fading and Rayleigh fading effects in the link budget.

Signal-level measurements are normally discrete-time real-valued samples in decibels. The samples are in decibels because the output of the amplifiers often has a logarithmic characteristic, and because this allows for a wide dynamic range of power levels. Suppose that Y were Rayleigh distributed; then the distribution of $X = 20 \log Y$ would be what we will call anti-log-Rayleigh (ALR). The probability density function of the ALR distribution [15] is given by

$$f_X(x) = (\ln 10)\frac{10^{x/10}}{20p} \exp\left(-\frac{10^{x/10}}{2p}\right) \tag{5.55}$$

where p is a power parameter given by $2p = \text{E}[Y^2]$. $\bar{X} = 10[\log(2p) - \gamma/(\ln 10)]$, where $\gamma \approx 0.577216$ is Euler's gamma constant. It can be shown that

$$X = \bar{X} + \frac{10\gamma}{\ln 10} + 10 \log[-\ln U] \tag{5.56}$$

is ALR-distributed with mean $\bar{X}$, where U is a random variable distributed uniformly between 0 and 1. Hence, (5.56) illustrates the point that Rayleigh fading can be seen as additive noise around a mean $\bar{X}$ in decibels (multiplicative noise in the absolute domain).

Hence, (4.23) can be modified so that the received signal power can be written as

$$\begin{aligned} P_{\text{r,dB}}(\varrho) = {} & P_{\text{t,dB}} + G_{\text{t,dB}} + G_{\text{r,dB}} + 20 \log \lambda - 20 \log(4\pi) \\ & -10\nabla \log d_0 - 10 \log \Lambda_0 + \Upsilon(\varrho) + \Phi(\varrho) \end{aligned} \tag{5.57}$$

where $P_{\text{r,dB}}$, $P_{\text{t,dB}}$, $G_{\text{t,dB}}$, and $G_{\text{r,dB}}$ are P_r, P_t, G_t, and G_r in decibels, respectively, and ϱ is a coordinate vector representing a location a distance d_0 from the base station. $\Upsilon(\varrho)$ is a zero-mean Gaussian random variable with a standard deviation of σ_{lf}, representing the lognormal fading with mean removed, and $\Phi(\varrho)$ is an ALR random variable with mean removed, representing small-scale fading.

Typically, about 10 dB of margin is given to account for σ_{lf}, whereas if both lognormal and small-scale fading are to be accounted for, the combined margin might increase to 20 dB. If diversity combining techniques are used in the system, though, the margin could be reduced.

EXERCISES

5.1 Suppose that we need $h < -r_1$ LOS clearance, where r_1 is the radius of the first Fresnel zone, so we need $|h| > \sqrt{\lambda d_1 d_2/(d_1 + d_2)}$. Often when dealing with point-to-point microwave links, the frequencies are on the order of gigahertz and the distances are on the order of kilometers, but the LOS clearance is on the order of meters. Show that we can use

$$h > 17.3\sqrt{\frac{d_1 d_2}{F(d_1 + d_2)}}$$

where F is frequency in gigahertz and d_1 and d_2 are distances in kilometers.

5.2 Calculate the mean excess delay and rms delay spread for the multipath profile given in Figure 5.18. Estimate the coherence bandwidth of the channel. If a GSM system is using this channel, would it be encountering flat fading or frequency-selective fading? Would an equalizer be needed?

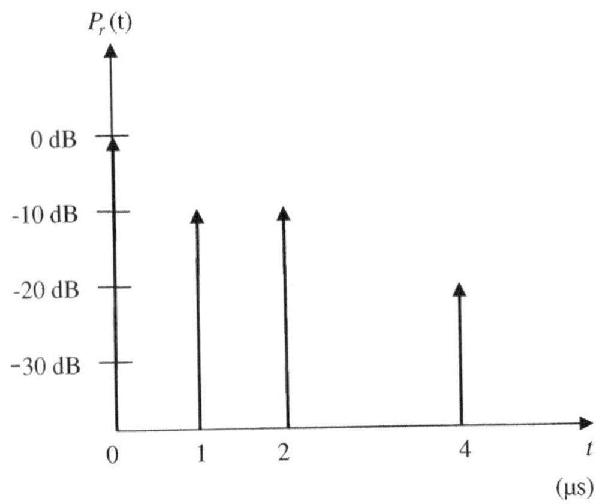

FIGURE 5.18 Multipath profile.

5.3 Check if the Hata model is compatible with the simple path-loss-exponent model, where there is a d^{-n} drop-off of power with distance (e.g., d^{-4}). By "compatible" we mean that they produce results that are within reasonable range of each other. Let's focus on the term in the Hata model that contains d. First, why is it positive whereas the power of d in d^{-n} is negative? Second, let's check some numerical values. In the case that $h_{BS} = 1$, what is the drop-off of power with distance? What value of h_{BS} would result in a path loss exponent that is exactly equal to 4?

5.4 A mobile is moving at 10 m/s away from a base station, and the carrier frequency is 900 MHz. What is the maximum Doppler frequency? What are the level crossing rate and fade duration for $\rho = 0.5$?

5.5 Suppose that x is Rayleigh distributed. Show that

$$P(x \leq X) = 1 - e^{-X^2/2p}$$

Now derive (5.49) by recognizing that γ_j is the square of a Rayleigh-distributed random variable. Finally, derive (5.48) from (5.49).

5.6 Suppose that we have a receiver with three independent diversity branches, with average SNRs of 5, 7, and 10 dB. If maximal ratio combining is used, what is the average SNR of the output of the diversity combiner?

APPENDIX: RICEAN FADING DERIVATION

The magnitude of the envelope of the received signal in a multipath environment is often modeled as a Rayleigh-distributed random process. Assuming that we have a sufficient number of paths with uniformly distributed phase, this is a good model. However, there are exceptions. When there is dominant signal component (such as is provided by a line-of-sight path from transmitter to receiver), it has been argued, both theoretically and from experimental data, that the Ricean distribution will better describe the signal variations.

It is assumed that the dominant component is dominant because it is not as attenuated as the other components. For example, the line-of-sight path is the shortest path between transmitter and receiver. Under small-scale conditions, we can assume that the magnitude of the envelope of the dominant component is constant. Writing the signal received as the sum of the dominant component and everything else (this "everything else" is Rayleigh distributed, as usual), we have

$$re^{j\theta} = ve^{j\beta} + ue^{j\alpha} \tag{5.58}$$

where r and θ are the envelope and phase of the signal received, v and β are the envelope and phase of the dominant component, and u and α are the envelope and phase of everything else.

We can rewrite (5.58) as

$$\begin{aligned} r\cos\theta\cos wt &- r\sin\theta\sin wt \\ &= (v\cos\beta + u\cos\alpha)\cos wt - (u\sin\alpha + v\sin\beta)\sin wt \end{aligned}$$

To simplify the derivation without loss of generality, we realign the time origin so that the dominant component is completely in phase. This means that we set β to zero. We then have

$$r\cos\theta\cos wt - r\sin\theta\sin wt = (v + u\cos\alpha)\cos wt - u\sin\alpha\sin wt$$

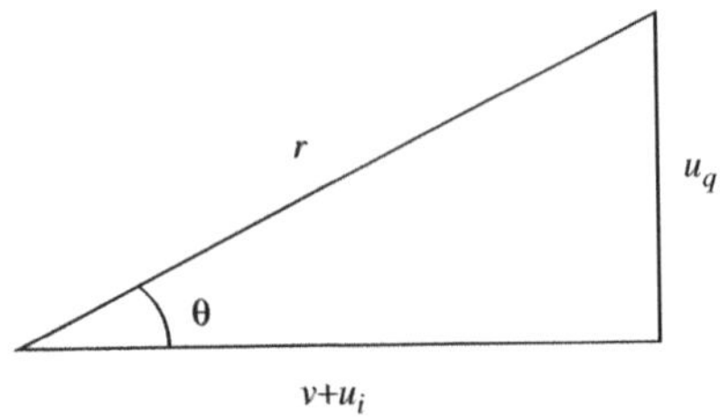

FIGURE 5.19 Relationship between r, θ, u_q, and $u_i + v$.

If we write the in-phase and quadrature components of u as $u_i = u\cos\alpha$ and $u_q = u\sin\alpha$, we have

$$r = \sqrt{(v+u_i)^2 + u_q^2}$$

and

$$\theta = \tan^{-1}\frac{u_q}{v+u_i}$$

Clearly, $u_q = r\sin\theta$ and $u_i = r\cos\theta - v$. The diagram that goes with this is Figure 5.19.

The Jacobian of the transformation is

$$\begin{vmatrix} \dfrac{\partial u_i}{\partial r} & \dfrac{\partial u_i}{\partial \theta} \\ \dfrac{\partial u_q}{\partial r} & \dfrac{\partial u_q}{\partial \theta} \end{vmatrix} = (\cos\theta)(r\cos\theta) - (\sin\theta)(-r\sin\theta)$$
$$= r$$

If we have sufficient paths, we can apply the central limit theorem to argue that u_i and u_q are approximately zero-mean Gaussian distributed. If, further, we make the reasonable assumption that they are uncorrelated, then they must be independent and hence jointly Gaussian distributed:

$$f_{i,q}(u_i, u_q) = \frac{1}{2\pi p}e^{-(u_i^2+u_q^2)/2p} \tag{5.59}$$

and we have

$$\begin{aligned} f(r,\theta) &= rf_{i,q}(r\cos\theta - v, r\sin\theta) \\ &= \frac{r}{2\pi p}e^{-[(r\cos\theta-v)^2+(r\sin\theta)^2]2p} \\ &= \frac{r}{2\pi p}e^{-(r^2+v^2-2rv\cos\theta)/2p} \end{aligned}$$

Therefore,

$$f(r) = \int_0^{2\pi} \frac{r}{2\pi p} e^{-(r^2+v^2-2rv\cos\theta)/2p} d\theta \tag{5.60}$$

$$= \frac{r}{p} \mathrm{I}_0\left(\frac{rv}{p}\right) e^{-(r^2+v^2)/2p} \qquad r \geq 0 \tag{5.61}$$

where

$$\mathrm{I}_0(x) = \frac{1}{2\pi} \int_0^{2\pi} e^{x\cos\theta} d\theta$$

is the zeroth-order modified Bessel function of the first kind. Equation (5.61) is known as the *Ricean distribution*. Incidentally, it was first derived as an attempt to model the distribution of the envelope of a signal v in the presence of random noise u.

It is interesting to compare the Rayleigh and Ricean distributions, (5.33) and (5.32). We see that the Ricean distribution is the Rayleigh distribution weighted by a factor of

$$\mathrm{I}_0\left(\frac{rv}{p}\right) e^{-v^2/2p}$$

Note that the distribution depends on both v and p, the envelope of the dominant component and the power of the underlying Gaussian process of the random component. From (5.59),

$$E[u_i^2] = E[u_q^2] = p \quad \text{and} \quad E[u_i] = E[u_q] = 0$$

The random component (from everything but the dominant component) is Rayleigh distributed, with

$$E[u^2] = 2p \quad \text{and} \quad E[u] = \sqrt{\frac{\pi p}{2}}$$

It turns out (Rice) that for the Ricean-distributed sum, we have

$$E[r^n] = (2p)^{n/2}\Gamma\left(\frac{n+2}{2}\right) \mathrm{F}_1\left(-\frac{n}{2}; 1; -\frac{v^2}{2p}\right)$$

where Γ is the gamma function and F_1 is a hypergeometric function given by

$$\mathrm{F}_1(a; c; z) = 1 + \frac{az}{c1!} + \frac{a(a+1)}{c(c+1)}\frac{z^2}{2!} + \cdots$$

so, in particular,

$$E[r^2] = 2p\left(1 + \frac{v^2}{2p}\right) = 2p + v^2 = E[u^2] + v^2$$

and

$$E[r] = \sqrt{2p}\,\Gamma\left(\frac{3}{2}\right)\mathrm{F}_1\left(-\frac{1}{2}; 1; -\frac{v^2}{2p}\right)$$

When v/p goes to zero, we are reduced to a Rayleigh distribution, as expected, since we essentially don't have a dominant component in this case. When rv/p becomes large (e.g., when $v/p >> 1$ or when r approaches ∞), it can be shown that

$$p(r) \sim \left(1 + \frac{p}{8rv}\right)\sqrt{\frac{r}{2\pi v}}e^{-(r-v)^2/2p}$$

When v/p is very large or we are at the "tail" end of the distribution curve (r is large), the distribution begins to look sort of Gaussian, with mean v and variance p. For v/p large, we can think of this as the case when the dominant component so overwhelms everything else that we have essentially the dominant component with Gaussian-like fluctuations around it.

REFERENCES

1. J.-E. Berg, R. Bownds, and F. Lotse. Path loss and fading models for microcells at 900 MHz. In *IEEE Vehicular Technology Conference*, pp. 666–671, Denver, CO, May 1992.
2. A. B. Carlson. *Communication Systems*. McGraw-Hill, New York, 1986.
3. D. C. Cox, R. R. Murray, and A. W. Norris. 800 MHz attenuation measured in and around suburban houses. *AT&T Bell Laboratory Technical Journal*, 63(6):921–954, July 1984.
4. M. Gudmundson. Correlation model for shadow fading in mobile radio systems. *Electronics Letters*, 27(23):2145–2146, Nov. 1991.
5. M. Hata. Empirical formula for propagation loss in land mobile radio services. *IEEE Transactions on Vehicular Technology*, VT-29(3):317–325, Aug. 1980.
6. M.-J. Ho and G. L. Stüber. Co-channel interference of microcellular systems on shadowed Nakagami fading channels. In *IEEE Vehicular Technology Conference*, pp. 568–571, Secaucus, NJ, May 1993.
7. W. C. Jakes, editor. *Microwave Mobile Communications*. Wiley, New York, 1974. Republished by IEEE Press, Piscatouray, NJ, 1994.
8. 3GPP Technical Specification Group GSM/EDGE Radio Access Network. Radio transmission and reception (release 1999). 3GPP TS 05.05 V8.20.0, Nov. 2005.
9. K. Pahlavan and A. Levesque. *Wireless Information Networks*. Wiley, New York, 1995.
10. A. Papoulis. *Probability, Random Variables, and Stochastic Processes*. McGraw-Hill, New York, 1991.
11. J. Proakis. *Digital Communications*. McGraw-Hill, New York, 1995.

12. A. V. Raisanen and A. Lehto. *Radio Engineering for Wireless Communication and Sensor Applications*. Artech House, Norwood, MA, 2003.
13. T. Rappaport. *Wireless Communications: Principles and Practice*. Prentice Hall, Upper Saddle River, NJ, 1996.
14. S. Rice. Mathematical analysis of random noise: 2. *Bell System Technical Journal*, 23: 46–156, 1945.
15. D. Wong and D. Cox. Estimating local mean signal power level in a Rayleigh fading environment. *IEEE Transactions on Vehicular Technology*, 48(3):956–959, May 1999.

III

WIRELESS ACCESS TECHNOLOGIES

6

INTRODUCTION TO WIRELESS ACCESS TECHNOLOGIES

Our primary emphasis here is on wireless personal communications systems, where there are many mobile devices (also known as mobile stations or mobile phones) that connect to the network through base stations (also known as access points). There may also be many base stations in the system. Such systems would typically support two-way communications (unlike broadcast radio or broadcast television), and the devices would typically be associated with particular individuals (also known as users or subscribers). In an environment where there are multiple mobile devices trying to access the network, a *channelization* scheme is a scheme to separate the communications of the different devices from one another.

In the next few chapters we focus on wireless access technologies. In particular, in Chapter 7 we examine component technologies that are found in many wireless access technologies. Then we see how these component technologies, and other techniques, are incorporated into various standards, in our survey of selected standards (GSM, IS-95 CDMA, and 802.11 "WiFi") in Chapter 8. More recent trends in wireless access technologies may be found in Chapter 9.

In this chapter we build on the foundations from Chapter 1, to provide further foundations more specifically relevant to wireless access technologies. We review digital signal processing in Section 6.1, then explore digital communications over wireless links in Section 6.2. The revolutionary *cellular concept* is covered in Section 6.3. We finish the chapter by discussing spread spectrum (Section 6.4) and OFDM (Section 6.5). Our discussion of spread spectrum (including CDMA) and OFDM here is from the technology perspective; in Chapter 8 we will see how these ideas are actually used in real systems and standards.

Fundamentals of Wireless Communication Engineering Technologies, First Edition. K. Daniel Wong.

6.1 REVIEW OF DIGITAL SIGNAL PROCESSING

The amplitude of signals can be *discrete* or *continuous*. A discrete-amplitude signal takes on only amplitude values from a finite set, whereas a continuous-amplitude signal can assume values from an infinite and continuous set (e.g., all real numbers with magnitude less than a certain limit).

Signals can be *discrete-time* or *continuous-time*. For a continuous-time signal, the time variable is continuous. For a discrete-time signal, the time variable is discrete and usually has uniform spacing. A discrete-time signal can be thought of as obtained from a continuous-time signal through processing called *sampling*, and then the values of the signal at different times are known as *samples*. If it is not important to continually relate the discrete-time variable to a continuous time, for convenience the discrete-time variable is often written as a sequence of integers. Often, the actual time interval between samples is needed only for conversion from discrete time to continuous time, and vice versa, and otherwise in discrete time, the samples can be viewed as a sequence. In this book we represent discrete-time signals with brackets around the time variable (e.g., $x[n]$, where n is an integer), and we represent continuous-time signals with parentheses around the time variable [e.g., $x(t)$, where t is a real number].

An *analog* signal is continuous-amplitude and continuous-time. A *digital* signal is discrete-amplitude and discrete-time. Suppose we have a system that takes an input $x[n]$, and produces an output/response $y[n]$. Let $\longrightarrow$ represent the operation of the system (e.g., $x[n] \longrightarrow y[n]$). Suppose that we have two different inputs, $x_1[n]$ and $x_2[n]$, such that $x_1[n] \longrightarrow y_1[n]$ and $x_2[n] \longrightarrow y_2[n]$. Let a_1 and a_2 be any two scalars. The system is *linear* if and only if

$$a_1x_1[n] + a_2x_2[n] \longrightarrow a_1y_1[n] + a_2y_2[n] \tag{6.1}$$

A system is *time invariant* if and only if (discrete-time case)

$$x[n - n_0] \longrightarrow y[n - n_0] \tag{6.2}$$

6.1.1 Impulse Response and Convolution

An impulse (or unit impulse) signal is defined as

$$\delta[n] = \begin{cases} 1, & n = 0 \\ 0, & n \neq 0 \end{cases} \tag{6.3}$$

All linear time-invariant (LTI) systems can be characterized by their *impulse response*. The impulse response, $h[n]$, is the output when the input is an impulse signal:

$$\delta[n] \longrightarrow h[n] \tag{6.4}$$

If the impulse response goes to zero after a finite time, the system has a *finite impulse response*; otherwise, it has an *infinite impulse response*.

Convolution (discrete-time): The output of an LTI system with impulse response $h[n]$ is

$$y[n] = h[n] * x[n] = \sum_{l=-\infty}^{l=\infty} x[l]h[n-l] = \sum_{l=-\infty}^{l=\infty} h[l]x[n-l] \tag{6.5}$$

6.1.2 Frequency Response

Suppose that the input to a stable LTI system is

$$x[n] = A\cos(\theta n + \phi) \tag{6.6}$$

Then

$$A\cos(\theta n + \phi) \rightarrow A|H(e^{j\theta})|\cos[\theta n + \phi + \angle H(e^{j\theta})] \tag{6.7}$$

where

$$H(e^{j\theta}) = \sum_{l=-\infty}^{\infty} h[l]e^{-j\theta l} = |H(e^{j\theta})|e^{j\angle H(\exp(j\theta))} \tag{6.8}$$

The digital frequency θ is related to the analog frequency f by $\theta = 2\pi fT$, where T is the sampling interval. Substituting θ by $2\pi fT$ and replacing nT (sampling times) by t, we have

$$A\cos(2\pi ft + \phi) \rightarrow A|H(f)|\cos[2\pi ft + \phi + \angle H(f)] \tag{6.9}$$

where instead of writing $H(e^{j2\pi ft})$, the sinusoidal aspect is implicit and we just write $H(f)$:

$$H(f) = \int_{-\infty}^{\infty} h(t)e^{-j2\pi ft}\,dt = |H(f)|e^{j\angle H(f)} \tag{6.10}$$

6.1.2.1 Filters

A *filter* is a transformation of a signal, usually for a specific purpose. A *lowpass filter* preserves the low-frequency components while reducing the high-frequency components, whereas a *highpass filter* preserves the high-frequency components while reducing the low-frequency components. A *bandpass filter* preserves frequency components within a band while reducing frequency components outside the band. Bandpass filters are essential in wireless communications.

Frequency ranges in the frequency response of the filter can be divided into *passband* and *stopband*, where the passband frequencies are the ones being passed through by the filter. Just as we live in a world where objects are not just black or white, in reality, a filter would not only have passband and stopband, with transition regions in between, but the frequency response would not be pure and flat in the passband or stopband (there might be ripples or, at least, nonconstant amplitude).

6.1.3 Sampling: A Connection Between Discrete and Continuous Time

In many cases, discrete-time signals are obtained from continuous-time signals through a process called *sampling*. This is done by obtaining the value of the continuous-time signal at regular moments in time, spaced T' apart, where T' is called the sampling interval and $F' = 1/T'$ is the sampling rate. These values are called *samples*. It is useful to understand the relationships between a continuous-time signal and a discrete-time signal obtained through sampling of that signal, for the various insights and applications it yields. Let $x(t)$ be the continuous-time signal, $x[n]$ be the discrete-time version (where n is the sample index), and $x_s(t)$ be the discrete-time version written as a continuous function of time, explicitly showing the sampling operation. It sometimes helps to think of the discrete-time signal as $x[n]$ and sometimes as $x_s(t)$. We have

$$x[n] = x(nT') \tag{6.11}$$

and

$$\begin{aligned} x_s(t) &= x(t) \sum_{n=-\infty}^{\infty} \delta(t - nT') \\ &= \sum_{n=-\infty}^{\infty} x(nT')\delta(t - nT') \qquad (6.12) \\ &= \sum_{n=-\infty}^{\infty} x[n]\delta(t - nT') \qquad (6.13) \end{aligned}$$

Using the Fourier transform of an impulse train (Table 1.1) and the "multiplication" property (Table 1.2), we obtain the Fourier transform of $x_s(t)$ as

$$X_s(f) = X(f) * \left[\frac{1}{T'} \sum_{n=-\infty}^{\infty} \delta\left(f - \frac{n}{T'}\right) \right] \tag{6.14}$$

$$= \frac{1}{T'} \sum_{n=-\infty}^{\infty} X(f - nF') \tag{6.15}$$

6.1.3.1 Nyquist Sampling Theorem From (6.15) we see that if $X(f)$ is bandlimited to $-F'/2 < f < F'/2$, we just have distinct, nonoverlapping repetitions of $X(f)$ spaced F' apart, over all frequencies. However, if $X(f)$ is not bandlimited to $-F'/2 < f < F/2$, there will be overlapping, known as *aliasing*. When aliasing occurs, the signal is corrupted and cannot be recovered perfectly. On the other hand, when there is no aliasing, the continuous-time signal can be recovered perfectly. In other words, in cases of no aliasing, the samples $x[n]$ uniquely determine the continuous-time signal. This is the essence of the *Nyquist sampling theorem*. It says that if we start with a bandlimited signal where $X(f) = 0$ for $|f| > f_N$, then $x(t)$ is uniquely determined by its samples if the sampling rate, F', satisfies

$$F' > 2f_N \tag{6.16}$$

where f_N is called the *Nyquist frequency* and the sampling rate $2f_N$ is known as the *Nyquist sampling* rate.

6.1.3.2 Reconstructing the Continuous-Time Signal

Given a set of samples of a continuous-time signal, the Nyquist sampling theorem tells us that we can reconstruct the original signal if the sampling rate is $F' > 2f_N$.

6.1.4 Fourier Analysis

As we have seen in Section 1.3.2, Fourier analysis lets us decompose a signal into a sum of sinusoidal components and is a very useful tool for studying linear systems. The Fourier transforms in Section 1.3.2 were for continuous-time signals; however, Fourier analysis can also be useful for discrete-time signals, and we also have Fourier transforms for discrete-time signals.

6.1.4.1 Discrete-Time Fourier Transform

For a discrete-time signal $x[n]$, the discrete-time Fourier transform is given by

$$X\left(e^{j2\pi F}\right) = \sum_{n=-\infty}^{\infty} x[n]e^{-j2\pi Fn} \tag{6.17}$$

and the inverse discrete-time Fourier transform is given by

$$x[n] = \int_{-1/2}^{1/2} X\left(e^{j2\pi F}\right) e^{j2\pi Fn}\, dF \tag{6.18}$$

NB: Different notations can be used, but the one we use in this book, $X\left(e^{j2\pi F}\right)$, emphasizes its distinction from the (continuous-time) Fourier transform, denoted $X(f)$, in that $X\left(e^{j2\pi F}\right)$ is a periodic function of F, with period 1. Alternatively, the domain of the DTFT can be thought of as the unit circle, whereas the domain of the (continuous-time) Fourier transform is the real numbers. We use F to represent the frequency variable for the DTFT, rather than f, since $f \neq F$ when we compare the (continuous-time) Fourier transform and the DTFT.

The DTFT and (continuous-time) Fourier transform can be thought of as related through a simple time/frequency scaling. Whereas the spacing between samples is T' when we compute the Fourier transform, the spacing between samples is in a sense normalized to 1 when we compute the DTFT. Thus, we might expect the transforms to have a scaling relationship (in the frequency domain) as well. Indeed, computing the Fourier transform of (6.13) by direct integration leads to

$$X(f) = \sum_{-\infty}^{\infty} x[n]e^{-j2\pi fnT'}\, dt \tag{6.19}$$

and now, comparing with the expression for the DTFT, (6.17), we see that the relationship is

$$F = fT' \tag{6.20}$$

so

$$X\left(e^{j2\pi F}\right) = \frac{1}{T'} \sum_{n=-\infty}^{\infty} X\left(\frac{F-n}{T'}\right) \tag{6.21}$$

where the replicas of $X(f)$ are scaled by T' and the train of $X(f)$ replicas are spaced at $F = 1$ apart.

6.1.4.2 Discrete Fourier Transform The DTFT is discrete in time but continuous in frequency. The DFT, on the other hand, is discrete in both time and frequency. In particular, the N-point DFT is

$$X_k = \sum_{n=0}^{N-1} x_n e^{-j2\pi kn/N}, \qquad k = 0, \ldots, N-1 \tag{6.22}$$

and the IDFT (inverse DFT) is

$$x_n = \frac{1}{N} \sum_{k=0}^{N-1} X_k e^{j2\pi kn/N}, \qquad n = 0, \ldots, N-1 \tag{6.23}$$

A family of algorithms, collectively known as the *fast Fourier transform* (FFT), are typically used to compute the DFT more efficiently than is possible by direct computation of (6.22).

6.1.5 Autocorrelation and Power Spectrum

Analogous to our development of autocorrelation and power spectra for the continuous-time case, we have similar expressions for discrete-time signal processing. We do not do so here, for lack of space, but point the interested reader to texts such as Oppenheim and Schafer's [4]. Instead, what we do here is provide the expressions analogous to (1.39) and (1.40): namely, (6.24) and (6.25).

For a finite energy sequence, $x[n]$, we have the *aperiodic autocorrelation sequence*,

$$r_{xx}[n] = \sum_{k=-\infty}^{\infty} x[k]x^*[n+k] \tag{6.24}$$

For an infinite energy sequence that is a periodic power sequence with period N, we define the autocorrelation sequence as

$$r_{xx}[n] = \frac{1}{N} \sum_{k=0}^{N-1} x[k]x^*[n+k] \tag{6.25}$$

where the sum can be taken over any period of the sequence. Clearly, $r_{xx}[n]$ is also periodic.

As a special case of (6.25), for a periodic sequence of period N that takes only the values A and $-A$, (6.25) becomes

$$r_{xx}[n] = \frac{A^2}{N}(N_{\text{agree}} - N_{\text{disagree}}) \tag{6.26}$$

where N_{agree} and N_{disagree} are the number of times $x[k]$ and $x[k+n]$ agree or disagree, respectively, as k goes from 0 to $N-1$. By agreement, we mean either $x[k] = x[k+n] = A$ or $x[k] = x[k+n] = -A$, and by disagreement, we mean where $x[k] \neq x[k+n]$. The reason for (6.26) is that each agreement adds A^2 to the sum, and each disagreement adds $-A^2$ to the sum. To normalize (6.26), we may just divide by A^2.

6.1.5.1 Binary Sequences So far, when we have been discussing our discrete-time sequences $x[n]$, we have let $x[n]$ take on arbitrary real, or even complex, values. We have seen how the sequences could be viewed as samples of continuous-time real- or complex-valued signals. However, there is another important class of sequences that we work with a lot in digital communications, and these are sequences whose values are drawn from a *finite* set of values. As an example, we consider binary sequences, for example, sequences of two values, often written as 0's and 1's, such as 00111100.

Now the binary notation of 0's and 1's is very useful for thinking of encoding data that we wish to transmit over a communications system. However, when it comes to the electrical domain, it is in a sense inherently unbalanced, violating some of the implicit assumptions of our formulas, so we cannot use those formulas "as is." For example, consider $y_1[n]$ and $y_2[n]$ as two binary sequences where $y_1[n]$ is all 0's and $y_2[n]$ is all 1's. What would we expect the autocorrelation of each sequence to be? Intuitively, a sequence of all 0's is completely correlated with itself, just as a sequence of all 1's is completely correlated with itself. Yet, when we apply (6.25) (with period 1) to both sequences, we get surprising results: $r_{y_1y_1}[n] = 0$ for $y_1[n]$ and $r_{y_2y_2}[n] = 1$ for $y_2[n]$. Unlike many cases, the mean of our sequence is not 0, but $1/2$ (if 0's and 1's are equally likely), which leads to the problem here.

One solution is to use an alternative expression for autocorrelation, sometimes called *autocorrelation with means removed* (because the mean values $\overline{x[k]}$ are removed):

$$r_{xx}[n] = \frac{1}{N}\sum_{k=0}^{N-1}(x[k] - \overline{x[k]})(x^*[n+k] - \overline{x^*[n+k]}) \tag{6.27}$$

where we subtract the mean, $1/2$, before multiplying. In our example with $y_1[n]$ and $y_2[n]$, we would then have $r[n] = 1/4$ for both of them, and autocorrelation would range from 0 to $1/4$ as a sequence went from being completely uncorrelated to being completely correlated. If we don't like $1/4$, we could always normalize by multiplying by 4.

Another possible solution is to define a binary operator, say $\cdot$, to replace the multiplication in formulas for autocorrelation (like (6.25)) and orthogonality [like (6.32)], where we want $1 \cdot 1 = 1$, $0 \cdot 0 = 1$, $1 \cdot 0 = -1$, and $0 \cdot 1 = -1$, because we want autocorrelation to be about similarity. So by doing this, we have in effect defined a "score" of 1 when the two inputs are the same at any given time, and a "score" of -1 when they are different.

A third solution is to notice that 0's and 1's are often encoded as opposite-valued pulses, so they could be represented as 1's and -1's, and *then* we have 0 mean, so we can happily go back to using expressions such as (6.25) and (6.32). In particular, this kind of representation (as 1's and -1's) is sometimes called *binary antipodal signaling*, as we have seen in Section 1.4.2.1. In particular, with binary antipodal signaling, a waveform $x(t)$ represents 0 and the negative waveform $-x(t)$ represents 1. We use this solution when we talk about autocorrelation properties of PN sequences, orthogonality of Walsh codes, and so no, in CDMA systems.

6.1.6 Designing Digital Filters

Finite impulse response (FIR) filters and *infinite impulse response* (IIR) are the two main categories of filters. As the name implies, FIR filters have a finite impulse response, and IIR filters have an infinite impulse response.

IIR filters may be designed to behave like one of three types of continuous-time filters:

- *Butterworth filters*. These are designed to be maximally flat in the passband. The magnitude response is monotonic in both the passband and the stopband.
- *Chebyshev filters*. The magnitude response is *equiripple* (not monotonic, but varying within some specified approximation error) in the passband while still monotonic in the stopband (this is the *type 1* Chebyshev filter) or equiripple in the stopband while still monotonic in the passband (the *type 2* Chebyshev filter). Often, a Chebyshev filter can be of lower order than a Butterworth filter.
- *Elliptic filters*. The magnitude response is equiripple in both the passband and stopband.

6.1.7 Statistical Signal Processing

6.1.7.1 Autocorrelation The autocorrelation of a random sequence, $x[n]$, is

$$r_{xx}[n, m] = \overline{x[n]x[m]} \tag{6.28}$$

If we are given a random binary sequence whose two values are 0 and 1, we may wish to map these values to 1 and -1 before computing the autocorrelation, as discussed in Section 6.1.5.1.

Wide-Sense Stationarity (WSS). Just as in the continuous-time case, the mean value is independent of time and the autocorrelation depends only on the time difference $m - n$ (i.e., it is a function of $k = m - n$), so it may be written as $r_{xx}[k]$ to keep this property explicit.

6.1.7.2 Worked Example: Random Binary Sequence Consider a random binary sequence $x[n]$, where every symbol, independent of all other symbols, takes the values 1 or -1 with equal probability. In other words, the values in the sequence $x[n]$ are i.i.d. with probability 1/2 of being 1 and probability 1/2 of being -1.

Then $x[n]$ is WSS, and the autocorrelation function is

$$r_{xx}[k] = \begin{cases} 1 & \text{for } k = 0 \\ 0 & \text{for } k \neq 0 \end{cases} \tag{6.29}$$

It is interesting to compare the autocorrelation of this random binary sequence to the autocorrelation of the random binary wave seen in (1.81).

6.1.8 Orthogonality

Two signals, $x(t)$ and $y(t)$, are *orthogonal* (over a period of time, say 0 to T) if

$$\int_0^T x(t)y^*(t)\,dt = 0 \tag{6.30}$$

where in general $x(t)$ and $y(t)$ may be complex-valued signals, and we denote the complex conjugate of $y(t)$ by $y^*(t)$. For example, if $x(t) = \cos(2\pi nt/T)$ and $y(t) = R$, where n is an integer [so $x(t)$ contains exactly n cycles without time T] and R is an arbitrary real number, we can verify that $x(t)$ and $y(t)$ are orthogonal.

$$\int_0^T \cos(2\pi nt/T)R\,dt = \frac{RT}{2\pi n}\sin(2\pi n) = 0 \tag{6.31}$$

An interesting property of sines and cosines (coming from their symmetry about the horizontal axis) is that whenever we integrate them over an integer number of cycles, the integral is zero. This property may appear almost trivial, but it is very useful in working out whether various waveforms are orthogonal, as we see now in some worked examples.

More generally, we say that two functions are orthogonal if the *inner product* of the two functions is zero. Thus, we can also apply orthogonality to discrete-time sequences according to their inner product; for example, for two sequences of length N they are orthogonal if and only if

$$\sum_{n=0}^{N-1} x[n]y^*[n] = 0 \tag{6.32}$$

6.1.8.1 Worked Example Consider two sinusoids of the same frequency, both with the same period T_0. Let $T = nT_0$ be an integer multiple of T_0. Can they be orthogonal if there is a phase offset? If so, what phase offset(s) result in their orthogonality?

Let $x(t) = \cos(2\pi t/T_0)$ and $y(t) = \cos(2\pi t/T_0 + \phi)$; then we have

$$\begin{aligned}\int_0^T x(t)y(t)\,dt &= \int_0^T \frac{1}{2}\left[\cos(4\pi t/T_0 + \phi) + \cos(\phi)\right] dt \\ &= T\cos\phi \end{aligned} \tag{6.33}$$

where we have used (A.4) for the product of cosines and where the left term vanishes because the integration is over an integer number of cycles of a sinusoid. Clearly, (6.33) is zero if and only if $\phi = \pi/2 \pm n\pi$, where n is any integer. Thus, sine and cosine are orthogonal, so the in-phase signaling and quadrature signaling (Section 1.3.4.1) can be treated as two independent channels. At a receiver, we can extract just the in-phase or just the quadrature signal on the basis of the orthogonality of the sine and cosine.

6.1.8.2 *Worked Example* Consider two sinusoids of different frequencies, $f_c \pm f_\Delta/2$:

$$\begin{aligned}&\int_0^T \cos\left[2\pi(f_c - f_\Delta/2)t\right]\cos\left[2\pi(f_c + f_\Delta/2)t\right] dt \\ &\quad = \int_0^T \frac{1}{2}\left\{\cos\left[2\pi(2f_c)t\right] + \cos(2\pi f_\Delta t)\right\} \\ &\quad = \frac{1}{2\pi f_\Delta}\sin(2\pi f_\Delta T) \end{aligned} \tag{6.34}$$

where the first integral drops due to (6.31), so the two sinusoids are orthogonal when $2\pi f_\Delta T$ is an integer multiple of π. Thus, the smallest nonzero Δ for which two such sinusoids are orthogonal is

$$f_\Delta = 1/2T \tag{6.35}$$

6.1.8.3 *Worked Example* Consider two *complex sinusoids* of different frequencies, $f_c \pm f_\Delta/2$. In this example we are dealing with complex-valued signals, so we must take the complex conjugate of one of them when we evaluate (6.1.8). Since the complex conjugate of $e^{j\theta}$ is $e^{-j\theta}$, we have

$$\begin{aligned}\int_0^T e^{j2\pi(f_c - f_\Delta/2)t} e^{-j2\pi(f_c + f_\Delta/2)t}\,dt &= \int_0^T e^{-j2\pi f_\Delta t}\,dt \\ &= \frac{1}{j2\pi f_\Delta}\left(e^{j2\pi f_\Delta T} - 1\right) \\ &= \frac{1}{j2\pi f_\Delta}\left\{\left[\cos(2\pi f_\Delta T) - 1\right] + j\sin(2\pi f_\Delta T)\right\}\end{aligned}$$

For orthogonality, then, both the real and imaginary parts must be zero, so $2\pi f_\Delta T$ must be an integer multiple of 2π. Thus, the smallest nonzero f_Δ for which two such sinusoids are orthogonal is

$$f_\Delta = 1/T \tag{6.36}$$

NB: Compare this with the preceding example, where two cosines can be just $1/2T$ apart.

6.2 DIGITAL COMMUNICATIONS FOR WIRELESS ACCESS SYSTEMS

Designers of wireless systems face challenging requirements and severe constraints that often lead to difficult trade-offs. One aspect of wireless systems where these challenges are encountered is in the choice of modulation scheme and how related issues such as channel estimation and timing recovery are handled.

6.2.1 Coherent vs. Noncoherent

Constant envelope modulation schemes such as QPSK are popular in wireless systems for various reasons. Since the information is encoded in the phase of the signal, accurate carrier phase synchronization at the receiver is crucial. Many wireless systems therefore typically arrange for certain *pilot* symbols, which are fixed and known, to be transmitted along with the other symbols. The pilot symbols can be used for symbol timing recovery and, more generally, for channel estimation. Thus, *coherent demodulation* can be performed, where coherent demodulation means demodulation with channel knowledge that allows carrier phase synchronization.

An alternative to coherent demodulation is *noncoherent demodulation*, where channel knowledge is not required, and in particular, without a carrier phase reference. Noncoherent demodulation has the advantage that complicated channel estimation is not needed (and less power and bandwidth needs to be allocated for the corresponding pilot symbols). However, there is a performance loss of a few decibels compared to coherent demodulation. For example, a common ballpark figure for the performance loss is 3 dB, but a closer examination of the specific case of noncoherent demodulation of *differentially encoded QPSK* (DQPSK) vs. coherent demodulation of regular QPSK yields a figure closer to 2.3 dB. In fact, coherent demodulation could be performed on a DQPSK signal (if channel knowledge is available, e.g., because pilot symbols are available), and the performance of coherently demodulated DQPSK is reported to be between that of QPSK and of noncoherently demodulated DQPSK.

6.2.2 QPSK and Its Variations

Due to the popularity of QPSK and its variants in wireless systems, we discuss briefly QPSK and some variants here.

DQPSK (differentially encoded QPSK) allows *differential decoding*. Each symbol is encoded as a phase difference between the previous symbol and the new symbol, rather than as a fixed constellation point. Since differential decoding can be done,

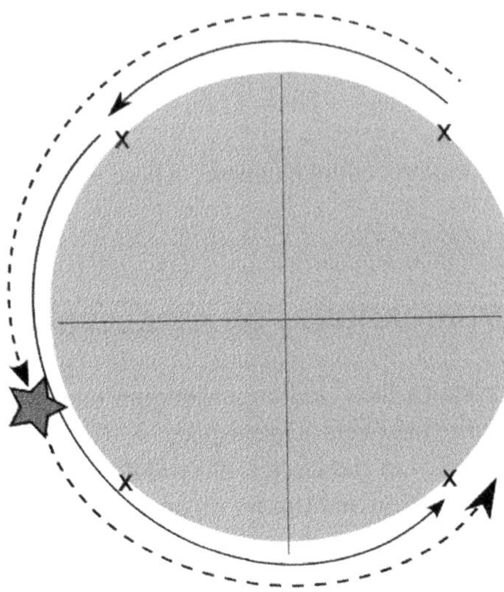

FIGURE 6.1 One error tends to become two errors in DQPSK.

DQPSK can be used with simple hardware. It doesn't need coherent demodulation, and often it is demodulated noncoherently, since an exact phase reference is not necessary. However, there is a roughly 2.3 dB loss in performance compared to QPSK (i.e., it may need E_b/N_0 to be 2.3 dB higher than for QPSK to obtain the same BER). Intuitively, this is because each decoding error tends to result in an additional error, so we would expect pairs of errors to occur frequently. This can be seen from Figure 6.1. In this figure we start at $\pi/4$ and move to $3\pi/4$ (phase shift of $\pi/2$) and then to $7\pi/4$ (phase shift of π). Assume for simplicity that the SNR is reasonably high, so we are just looking at occasional single errors in phase. An example of correct decoding is seen in the solid arcs, which go close to $3\pi/4$ and then close to $7\pi/4$, so the difference in phase would be decoded correctly as $\pi/2$ and π. An example of incorrect decoding is seen in the dashed-line arcs. Suppose that due to noise or other problems, the second point lands closer to $5\pi/4$ (indicated by the solid star) and the third point (correctly) lands close to $7\pi/4$. If QPSK modulation had been used, this would result in one decoding error ($5\pi/4$). Since DQPSK modulation is used, however, it results in two errors: phase shifts of π and $\pi/2$ (instead of $\pi/2$ and π).

A problem with QPSK is that there are often 180° phase transitions (Figure 6.2). When such phase transitions occur abruptly (e.g., at the transition from one symbol to the next), the normally constant envelope signal is going through the zero point of the phasor diagram. Thus, there will be more envelope variation in the signal than if phase transitions were smoother. One of the benefits of QPSK was that it can be used without a linear power amplifier in the transmitter, since it is constant envelope and the information is encoded in the phase. With the envelope variations from abrupt phase 180° transitions, however, we have a choice: Either go back to using linear power amplifiers, or suffer from *spectral regrowth*, where there is an increase in sidebands,

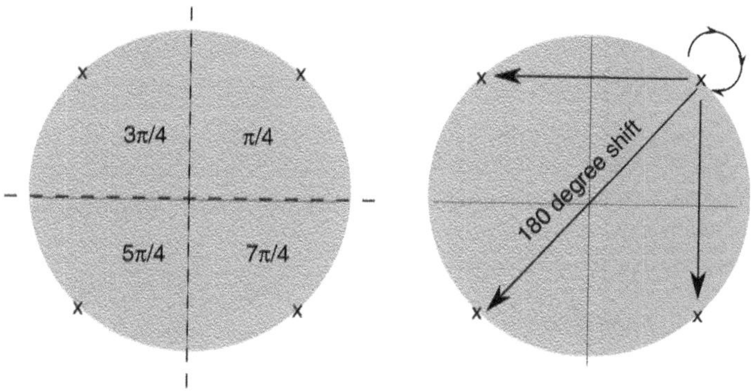

FIGURE 6.2 QPSK with its four decision regions, and 180° phase transition in QPSK.

causing more interference to adjacent channels. Since neither of these two choices is desirable, designers have, instead, found ways to avoid abrupt phase 180° transitions.

OQPSK (offset QPSK) is a form of QPSK where the I and Q transitions are offset by half a symbol, so at any transition, only one of them changes, and therefore the maximum transition is 90°. Figure 6.3 shows how OQPSK avoids going through zero. An OQPSK-encoded signal can be written as

$$s(t) = A\left\{\left[\sum_{n=-\infty}^{\infty} I_n p(t-2nT)\right]\cos(2\pi f_c t) + \left[\sum_{n=-\infty}^{\infty} Q_n p(t-2nT-T)\right]\sin(2\pi f_c t)\right\} \tag{6.37}$$

where I_n and Q_n are the nth in-phase bit and nth quadrature bit, respectively, and $p(t)$ is the pulse-shaping function. This signal would be comparable with a QPSK signal with symbol time $2T$. Notice the half-symbol (T) delay of the quadrature signaling.

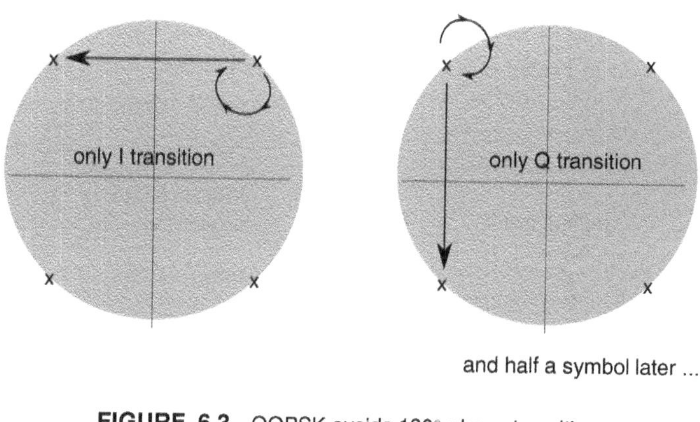

FIGURE 6.3 OQPSK avoids 180° phase transition.

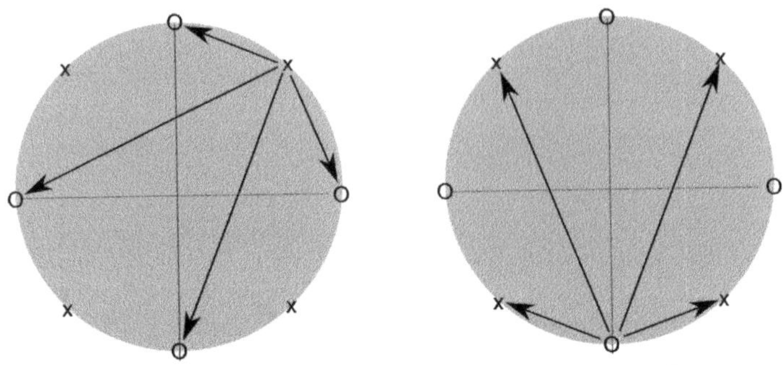

FIGURE 6.4 $\pi/4$ QPSK avoids 180° phase transition.

$\pi/4$ QPSK alternates between two different constellations that are rotated by 45° from each other. Thus, the maximum transition becomes 135°. See Figure 6.4 for an illustration of how this works.

6.2.3 Nonlinear Modulation: MSK

Unlike phase or amplitude modulation schemes where the phase or amplitude of the carrier is modulated by the data stream, with the FSK family it is the frequency of the carrier that is modulated by the data stream. Consider a binary FSK scheme where the carrier is shifted by $+f_\Delta$ or $-f_\Delta$, depending on the current bit being transmitted. Thus, the frequency may be $f_c + f_\Delta$ or $f_c - f_\Delta$. For good spectral properties, it is desirable to avoid an abrupt phase shifts, as explained in our discussion of QPSK. One way to use FSK while avoiding abrupt phase shifts is known as *minimum shift keying* (MSK), where the "minimum shift" in the name refers to minimum phase shifts. It is part of the family of *continuous-phase frequency shift keying* (CPFSK) modulation schemes, where the phase is continuous (without abrupt phase shifts). A mathematical analysis of MSK is out of scope of this book, but an intuitive view of MSK can be seen in Figure 6.5. Let T be the symbol period; then $f_\Delta = 1/4T$. So,

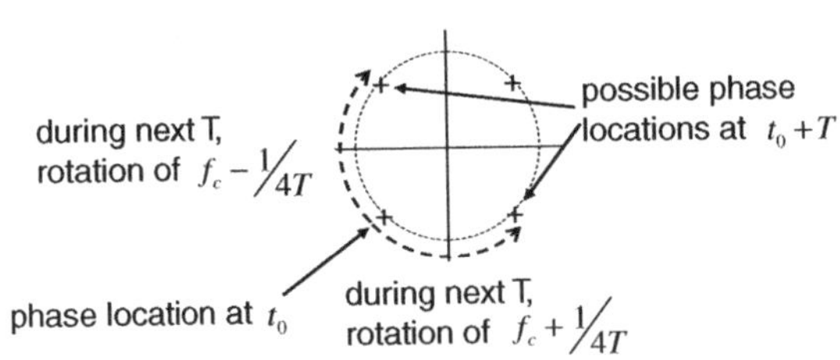

FIGURE 6.5 MSK.

viewing the phasor as a rotating phasor rotating at frequency f_c, the two frequencies $f_c + f_\Delta$ or $f_c - f_\Delta$ can be viewed as follows:

- $f_c + f_\Delta$ is a positive rotation (anticlockwise) representing an increase (accumulation) in phase within one symbol period given by $2\pi f_\Delta T = \pi/2$ radians.
- $f_c - f_\Delta$ is a negative rotation (clockwise) representing a decrease in phase within one symbol period given by $\pi/2$ radians.

So, if we start at a particular phase (e.g., $5\pi/4$, as shown in the diagram), with each bit we move clockwise or anticlockwise by $\pi/2$ radians.

For a given bit rate, MSK has a wider spectral main lobe than QPSK, but smaller side lobes. A popular variation of MSK is *Gaussian MSK* (GMSK), which is used in GSM. GMSK reduces side lobes even more than does MSK.

6.3 THE CELLULAR CONCEPT

A first attempt at building the wireless access part of a wireless personal communications system might go something like this:

1. License a suitable block (or suitable blocks, if not contiguous) of frequency spectrum.
2. Break up the block of spectrum into different frequency channels.†
3. Find a suitable location for a base station, preferably around the center of the region to be served by the system, and preferably high up (e.g., on a hill) to reduce the power needed for transmissions.
4. Install the base station and have it communicate with mobile devices anywhere in the service region, where different frequency channels are allocated to each of the mobile devices.

Let's suppose that the amount of spectrum licensed is B_{total} (kHz), and each channel uses B_{channel} (typically, on the order of 30 kHz for voice signals). This system has certain limitations:

1. The number of subscribers is limited. If the frequency channels were fixed, only $\lfloor B_{\text{total}}/B_{\text{channel}} \rfloor$ subscribers are allowed before capacity is reached. If the frequency channels were assigned dynamically to users as needed, the number of users could exceed $\lfloor B_{\text{total}}/B_{\text{channel}} \rfloor$, but the number of users communicating simultaneously is still $\lfloor B_{\text{total}}/B_{\text{channel}} \rfloor$.

† For now, we assume that channelization is by use of different frequency bands. Later we revisit some of the concepts discussed in this section, when other methods of channelization are introduced.

2. If the service region is large (e.g., a city area or city plus suburbs), the devices and base station would need to transmit at very high powers, to mitigate the high path losses in the wireless propagation environment.

Thus, the system would be very expensive and support very few users.

The cellular concept was a technological breakthrough, introducing many base stations to cover the service region, but with base stations covering only a small portion of the service region. Thus, the service region is broken down into *cells*, each of which is the coverage region of a base station. By increasing or decreasing the maximum transmitter power of each base station, the coverage of each cell could be increased or decreased accordingly. Furthermore, frequency channels (or, briefly, "frequencies") could be reused. This is because some distance away from a base station, the signal would be so weak that another base station could reuse the same frequency channels without there being excessive interference from the communications in one cell to the other, and vice versa. This concept of *frequency reuse* is at the heart of the cellular concept (Figure 6.6).

To reuse frequencies systematically, the available spectrum, B_{total}, is divided into N_s channel sets (where N_s is a positive integer), each of which has B_{total}/N_s kHz of spectrum, and $N_c = \lfloor B_{\text{total}}/N_s B_{\text{channel}} \rfloor$ channels. Each cell is assigned one of the channel sets and only uses the N_c channels within it. Cells that use frequencies in the same channel set are called *co-channel* cells. In terms of geometrical arrangement of the cells and co-channel cells, the neatest arrangement is when cells are considered hexagonal in shape. We allocate channel sets to cells with the principles that:

1. For each channel set, the co-channel cells should be far apart from one another, for a fixed N_s.
2. Each cell should belong to a channel set, and the channel set assignments should tessellate the entire area completely.

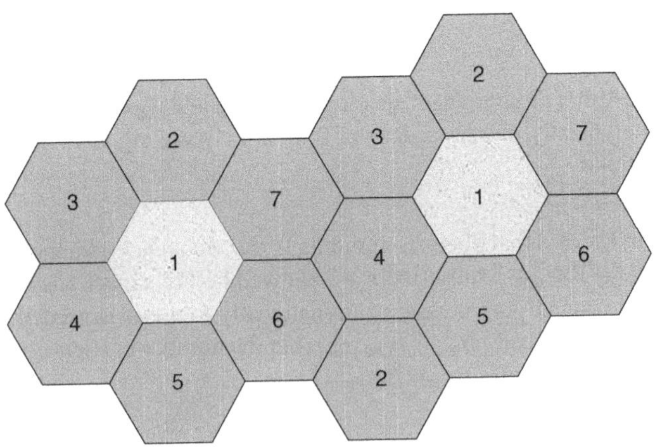

FIGURE 6.6 Example of frequency reuse, with reuse factor 7.

Then the co-channel cells for each channel set can be distributed in an elegant, regular pattern, while at the same time the hexagonal cells tessellate the entire area completely if and only if

$$N_s = i^2 + j^2 + ij \tag{6.38}$$

where i and j are both nonnegative integers. Equation (6.38) can be derived geometrically. In these cases, there will be six co-channel cells closest to any cell, and a reuse distance, D, can be defined as the distance between the center of a cell and the center of any of these six nearest co-channel cells. To locate any of these nearest co-channel cells geometrically, just take i "steps" in any of the six directions around the cell, turn 60° to the left or right, and then take j "steps." A *step* would be going from the center of a cell to the center of an adjacent cell. N_s is also called the *frequency reuse factor.*

6.3.1 Relating Frequency Reuse with S/I

What frequency reuse factor should be chosen for a cellular system? The smaller the frequency reuse factor, the more channels we would have available per cell. But the smaller the frequency reuse factor, the higher the (average) interference in the system. How do we quantify the relationship of the frequency reuse factor with S/I? One way is by the approximation we now derive. We start by assuming that the signal strength from the base station to the mobile drops off as d^n, so the signal power received is roughly P_0/d^n for some P_0 (see Section 5.1.1).

Consider a hexagonal idealization of a cellular system where D is the distance between the center of a cell and the center of its nearest co-channel cell as earlier defined, and R is the "radius" of the cell, the distance between its center and one of its vertices (again, by symmetry, it doesn't matter which one). We would expect the S/I requirements at the mobiles to impose tighter constraints on design choices than the S/I requirements at the base stations. So, looking at S/I at the mobiles, we can imagine the worst case, where the mobile is at one of the vertices of the hexagonal cell (given our assumed signal strength model). Then we have the interference power being the sum of interference powers from the co-channel cells:

$$S/I \approx \frac{P_0/R^n}{\sum_{\text{nearest interferers}} P_0/D^n} = \frac{(D/R)^n}{6} \tag{6.39}$$

where the approximation arises from assuming that (1) we need only consider interference from the nearest co-channel cells, since it would be much weaker from co-channel cells that are farther away; and (2) approximating the distance between the mobile and the base stations in those co-channel cells by D. Since there are six of these nearest co-channel cells, the result follows.

We need one more piece to complete the puzzle. It can be shown (see the exercises at the end of the chapter) that $D/R = \sqrt{3N_s}$. Thus, we have

$$S/I \approx (1/6)\left(\sqrt{3N_s}\right)^n \tag{6.40}$$

This imposes a lower limit on N_s, given a minimum required S/I.

6.3.2 Capacity Issues

We consider some ideas for increasing system capacity, focusing on the cell level.

Smaller Cells. Frequency reuse allows high numbers of users to be served by a cellular system, but (6.40) imposes a lower limit on N_S. However, even if we are down to the smallest N_s that we can use, there are ways to increase capacity at the cell level. One of these ways is the use of small cells, also known as *microcells* or *femtocells*.

Sectorization. An alternative way to increase capacity is the use of *sectorization* or *sectoring*. The basic idea is that instead of using omnidirectional antennas at the base stations, directional antennas are used to subdivide the coverage region of each base station into several sectors. This technique is illustrated in Figure 6.7 and has the benefit of reducing interference levels and thus allowing a smaller N_s to be used, resulting in increased capacity.

Other Parts of the System. Changes in other aspects of a wireless system (in addition to changes at the cell level) could also increase capacity. For example, more efficient speech and channel coding and other link-level techniques to reduce required bandwidth per user can also increase capacity. Such techniques might also be used to reduce required S/I, thus allowing a smaller N_s by (6.40). Some of the more established ideas presented in Chapter 7 and others, such as HARQ and multiple antenna techniques, may be found in Chapter 9.

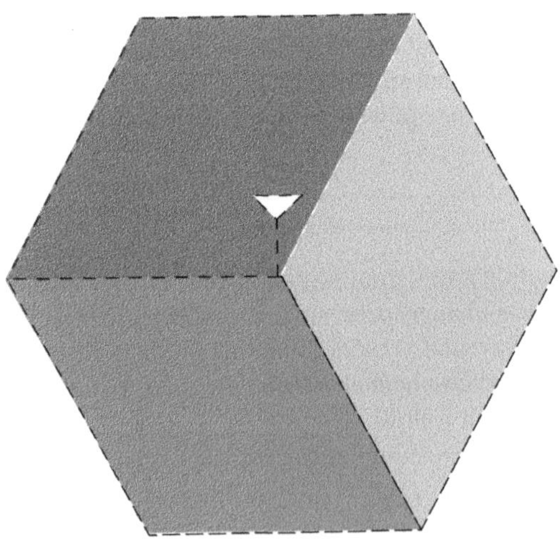

FIGURE 6.7 The use of sectoring helps to increase capacity in cellular systems. Here, a nominally hexagonally shaped cell is divided into three sectors.

6.4 SPREAD SPECTRUM

In this section we focus on the single-user case of spread spectrum with one transmitter and one receiver. While some of the benefits of spread spectrum are already apparent from such an inspection, we wait until Section 7.1 to discuss the multiuser aspects of spread spectrum, which allows us to get a more complete appreciation for the pros and cons of spread-spectrum techniques.

In conventional communications systems, the signal bandwidth of the modulated signal is usually on the order of the data rate. A lot of work has been done to make communications bandwidth efficient, to maximize use of the radio spectrum. With spread-spectrum communications, however, the signal bandwidth of the spread-spectrum modulated signal is much larger than the data rate (Figure 6.8). This may at first sound like a bad idea given the scarcity of usable bandwidth, but there are good reasons for using spread spectrum. The ratio between the signal bandwidth and the data bandwidth is known as the *processing gain*. One of the reasons for using spread spectrum is that in a multitransmitter environment it can be designed to be comparable to conventional systems in terms of bandwidth efficiency (Section 7.1).

The spread signal is obtained from a narrowband signal through a careful process whereby some controlled "randomness" is put in the process. The controlled randomness is typically from a pseudo-random number generator, and usually is in the form of a pseudo-random sequence of numbers. In the context of spread spectrum such pseudo-random number sequences are also called *PN sequences*. It is important that PN sequences be used rather than a truly random sequence, because the receiver has to be able to recover the signal transmitted, and it can do so only when the sequence is not truly random, so that the receiver can reproduce the sequence.

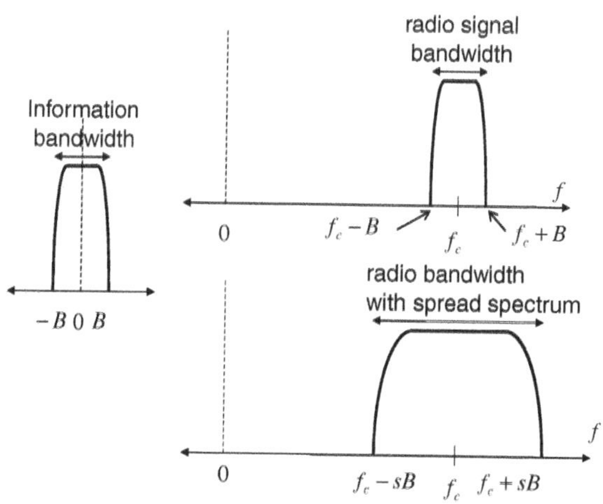

FIGURE 6.8 A signal with a given bandwidth (on the left) occupies a similar bandwidth at the carrier frequency (top right) with regular wireless transmission, but occupies a much larger bandwidth with spread spectrum (bottom right).

Given a source signal, $s(t)$, spread-spectrum signals may be generated in various ways, such as:

- *Direct sequence.* $s(t)$ is multiplied directly by the PN sequence.
- *Frequency hopping*. The signal $s(t)$ jumps around between different frequency channels.
- *Time hopping*. The signal $s(t)$ modulates the positions of pulses in a pulse train. In other words, a regular pulse train consists of a sequence of pulses spread evenly in time, whereas time hopping causes some displacement of pulses from their regular position.

We shall refer to these as *spreading schemes*. Some of the typical benefits of spread spectrum are:

- Low probability of intercept (sometimes called LPI)
- Low probability of detection (sometimes called LPD)
- Low interference to narrowband signals
- Interference rejection
- Mitigation of multipath delay spread effects

LPI, LPD, and the low interference to narrowband signals are benefits obtained because the signal power is spread so thinly over the wide bandwidth. Thus, it is difficult to detect that the signal is even there, and even if it is detected, it is not easy to intercept. Because it knows the PN sequence to use, the nature of the spread-spectrum receiver results in the interference rejection properties and the ability of spread spectrum to mitigate the dreaded multipath delay spread.

Within the family of spread-spectrum technologies, there is much scope for variations in the ways that the signal is spread, how the PN sequences are generated and used, what properties the PN sequences possess, and so on.

6.4.1 PN Sequences

In various applications in communications and computing, it is useful to generate sequences of bits that are "random." Pseudo-random noise sequences should look like random noise, such as white noise. As we have seen in Chapter 3, ideal white noise has a flat spectrum, being equally present at all frequencies. In other words, its ideal autocorrelation function is $R(\tau) = k\delta(\tau)$. We have also seen that a random binary wave looks almost like white noise (Section 1.3.5.5), with a broad spectrum and narrow autocorrelation function.

The specific properties of a PN sequence depends on how the sequence is generated. So we now discuss one way that a particularly useful type of PN sequence, the m-sequences, are generated, and then look at the autocorrelation properties of these m-sequences in Section 6.4.1.2 and some multiplication properties in Section 6.4.1.3. The reason for these examinations is so as to lay the foundations for our subsequent discussion of direct sequence spread spectrum (DSSS) in Section 6.4.2. DSSS is one of the fundamental technologies used in IS-95 CDMA systems and also in 3G wireless systems.

6.4.1.1 Use of LFSRs to Generate m-Sequences For pseudo-random noise, we want to generate a sequence that looks like white noise but is such that exactly the same sequence can be generated repeatedly. Furthermore, it needs to be relatively easy to generate.

A popular way to generate PN sequences that is easy to use and easy to generate repeatedly is by using *linear feedback shift registers* (LFSRs; see Figure 6.9). Moreover, the PN sequences thus generated look random. We call each unique set of 1's and 0's in the LFSR a *state*. Let the length of the LFSR be m (i.e., it contains m memory elements); then there are only $2^m - 1$ different nonzero states. The LFSR shifts every time it receives a clock signal, which is once every T_c seconds. Each output is one *chip* in the PN sequence (the components of the PN sequence in spread spectrum are typically called *chips*). Thus, the chip rate is $1/T_c$.

Suppose that the LFSR starts with some initial state. If this is the all-zero state, the LFSR will clearly remain in this state after subsequent shifts, so all zeros is not a useful initial state. So suppose that the initial state contains at least one 1. With each shift, the state of the LFSR changes. It may or may not cycle through all the $2^m - 1$ nonzero states before it returns to its original state. Sequences that cycle through all the $2^m - 1$ nonzero states are called *m-sequences*, and are preferred. These sequences are periodic with period $T_p = T_c(2^m - 1) = T_c\mathcal{P}$ seconds, where $\mathcal{P} = 2^m - 1$ is the sequence period in number of states. m-Sequences have certain highly desirable autocorrelation properties that we examine next. Figure 6.9 illustrates an LFSR with five registers that generates a $2^5 - 1$ m-sequence. One period (length 31) of the output is shown; afterword, the five registers would be back to 01000 and it would repeat. In systems such as IS-95 CDMA (Section 8.2), LFSRs with more shift registers are used, but the basic principle is the same.

To be precise in our terminology, over the next few subsections, we let $x(t)$ represent the output of the LFSR (in generating the PN sequence) as a continuous-time variable and let $y[n]$ represent the output of the LFSR as a sequence of digital values, and we then look at the autocorrelation function of both $x(t)$ and $y[n]$. In both cases we assume that binary antipodal signaling (as in Section 6.1.5.1), $x(t)$, comprises 0's

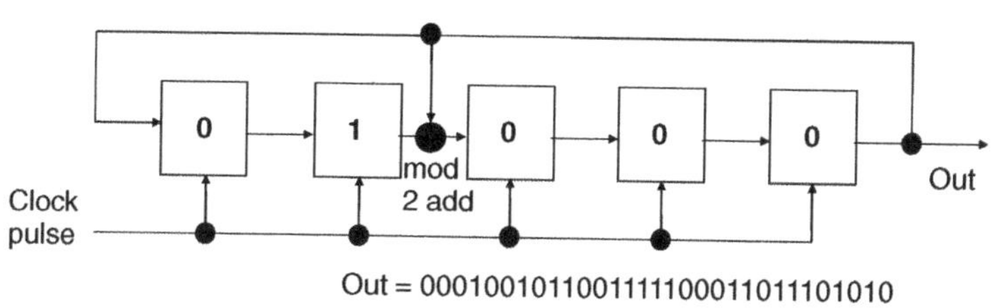

FIGURE 6.9 Linear feedback shift register.

mapped to a pulse function $p(t)$ and 1's mapped to $-p(t)$, whereas the values of $y[n]$ are 1 and -1, where $0 \to 1$ and $1 \to -1$.

6.4.1.2 *Autocorrelation Properties of m-Sequences* We can think of the autocorrelation properties of m-sequences either in continuous time or as the autocorrelation of a sequence. The two are essentially the same, but sometimes it may be easier to work with one or the other.

With $x(t)$ and $y[k]$ as just defined, it can be shown that $x(t)$ has normalized autocorrelation function

$$R_x(\tau) = (1 + 1/T_c)\Lambda(t/|T_c|) - 1/T_c \qquad \text{for } -T_p/2 \leq t \leq T_p/2 \tag{6.41}$$

which then repeats periodically. It is normalized, so it has a maximum value of 1, at $t = 0, t = \pm T_p$, and so on; we plot $R_x(\tau)$ in Figure 6.10. (NB: In the figure the vertical scaling has been exaggerated, for clarity of illustration.) However, normally, $-1/T_c$ is very small, and normally $T_p \gg T_c$, so the spikes or peaks are much farther apart than one might be led to thinking from just casually glancing at Figure 6.10.

Similarly, we can apply (6.26) to $y[n]$ to obtain

$$R_y[k] = \begin{cases} \mathcal{P} & \text{for } k = 0, \pm\mathcal{P}, \pm 2\mathcal{P}, \ldots \\ -1 & \text{otherwise} \end{cases} \tag{6.42}$$

which can also be normalized (Figure 6.11) for easier comparison with (6.41):

$$R'_y[k] = \begin{cases} 1 & \text{for } k = 0, \pm\mathcal{P}, \pm 2\mathcal{P}, \ldots \\ -1/\mathcal{P} & \text{otherwise} \end{cases} \tag{6.43}$$

The difference between $R_x(\tau)$ and (1.80) is that the former is periodic and has a finite width where it peaks, whereas the latter is the truly random case where the peak is a delta function and there is no periodicity. A similarly comparison can be made betwen $R_y[k]$ and (6.29).

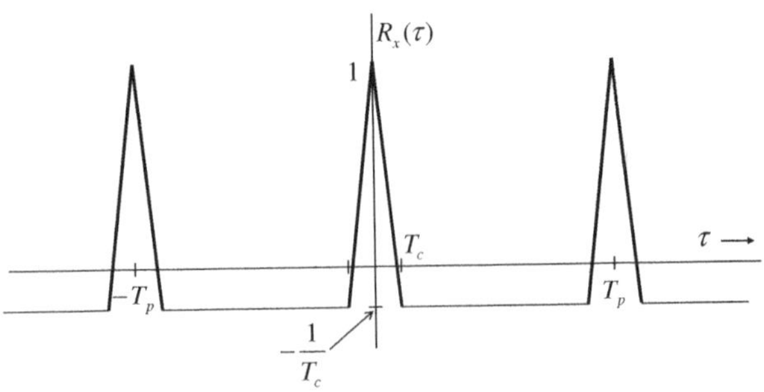

FIGURE 6.10 Autocorrelation of m-sequence.

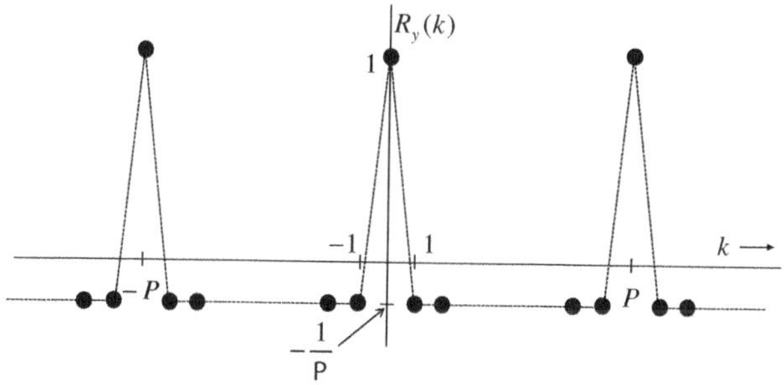

FIGURE 6.11 Discrete autocorrelation of an m-sequence (normalized).

6.4.1.3 Multiplication Properties of m-Sequences

Let us consider what happens when we multiply:

- An m-sequence by itself (0 offset)
- An m-sequence by a time-shifted version of itself (nonzero offset)
- An m-sequence by a low-rate sequence of bits
- The output of the previous multiplication, by the m-sequence (0 offset)

As seen in column (a) of Figure 6.12, when an m-sequence is multiplied by itself, a simple, flat output emerges. This is because $1 \times 1 = 1$ and $-1 \times -1 = 1$, so the output is constant at 1. This corresponds to $R_y[0]$, where the 1's are simply added up to yield $\mathcal{P}$. However, when an m-sequence is multiplied by a time-shifted version of

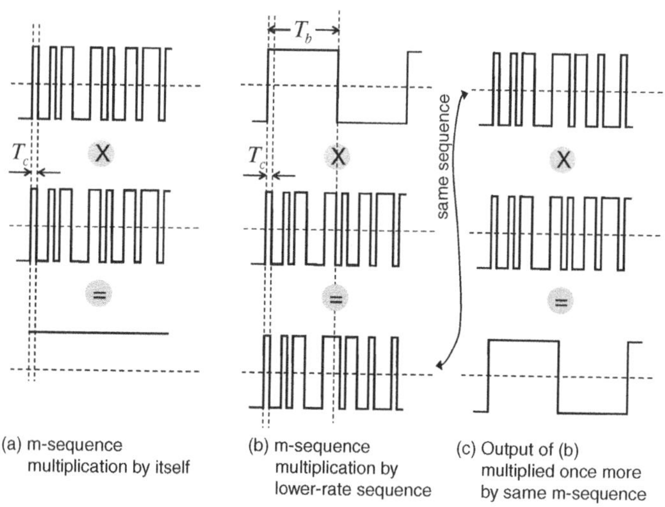

FIGURE 6.12 Various cases of multiplication of m-sequences.

itself, the result is an almost equal mix of 1's and −1's. It can be shown that for an m-sequence of length $2^n - 1$, the number of −1s would be 2^{n-1} and the number of 1's would be $2^{n-1} - 1$. (We cannot have an equal number of 1's and −1's since the period, $2^n - 1$, is odd). This agrees with (6.42), since the 1's and −1's would add up to −1.

When an m-sequence is multiplied by a low-rate (low relative to the m-sequence chip rate) sequence of bits, another random-looking chip-rate sequence emerges, as seen in column (b) in Figure 6.12. In the figure we have indicated the chip period, T_c, and the bit period of the low-rate sequence, T_b. Now, if the output sequence from column (b) is then multiplied again by the same m-sequence (with 0 offset), we get back the low-rate data sequence. Why is this? The result would be the low-rate sequence multiplied *twice* by the m-sequence (with 0 offset), so the m-sequence cancels out [exactly as what is seen in column (a)], and the low-rate data sequence reemerges. NB: If instead of multiplying by the same m-sequence with 0 offset, we multiply by the same m-sequence at a nonzero offset, the signal remains spread. So it has to be the *same* m-sequence at the *same* offset.

6.4.2 Direct Sequence

Direct sequence spread spectrum (DSSS) is illustrated in Figure 6.13. A bit sequence (information sequence), with bit period T_b, is multiplied by a higher-rate PN sequence,

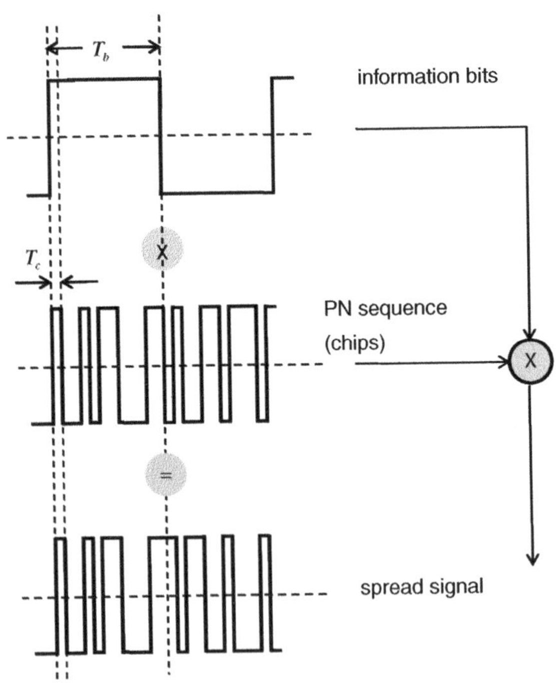

FIGURE 6.13 Direct sequence spreading.

with chip period or interval T_c. We represent each sequence by 1 and −1, as discussed in Section 6.1.5.1. The figure only shows the multiplication within a window of two bit periods. Notice that in the first bit period, the information bit is 1, so the output is just the PN sequence within that bit period. In the second bit period, the information bit is −1, so the output is the PN sequence "inverted" ($1 \rightarrow -1$ and $-1 \rightarrow 1$) within that bit period.

As can be seen in Figure 6.13, the output sequence is at the chip rate $1/T_c$. Thus, you multiply a PN sequence with an information sequence, and the result is a sequence with a bandwidth on the order of the PN sequence. This is the classic spectral characteristic that we find in spread-spectrum systems. Intuitively, we multiply two sequences in time, and hence we have convolution in the frequency domain, that is, the spectrum of the result is the convolution of their individual spectra (which are the narrowband spectrum of the information sequence and the wideband spectrum of the PN sequence).

Thus, we get what is often called a *scrambled signal* coming out of the multiplication (the information sequence is said to be scrambled by the PN sequence). This scrambled signal is wideband. It can be recovered by multiplication with the *same PN sequence* at the *same offset*. This follows from our discussion in Section 6.4.1.3. Multiplication by a different PN sequence, or the same PN sequence at a different offset, would not recover the information sequence.

Having seen what happens in direct sequence spread spectrum, we now revisit our list of benefits of spread spectrum to see how they are realized. Figure 6.14 (on the left side) shows that the scrambled signal often has its power spread out so much over a wide bandwidth that it is difficult to detect or intercept, hidden as it may be below the noise floor. Yet it is still there, and just needs the right PN sequence and the right offset to recover the narrowband information sequence (shown on the right side of the figure after recovery). The resulting gain in the spectral density is roughly the *processing gain*. The scrambled signal is spread out so much that it causes only a small amount of interference to any narrowband signal within its wide bandwidth. What about interference from narrowband interferers to the DSSS scrambled signal? A narrowband interferer is shown on the left side of the figure. However, after multiplication by the PN sequence in the receiver, the interferer becomes spread. This is again an application of what was illustrated in column (b) of Figure 6.12. Meanwhile,

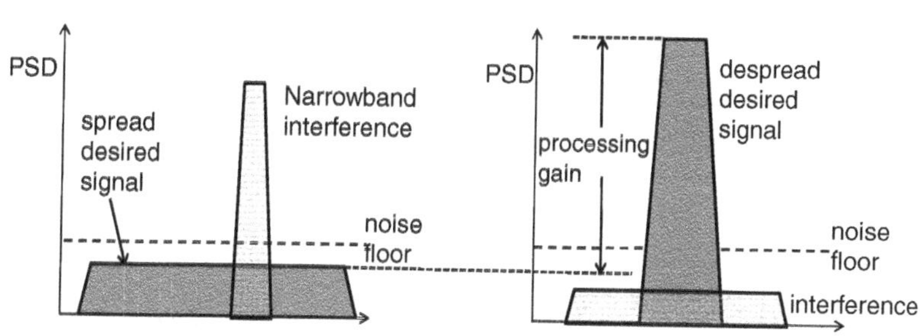

FIGURE 6.14 Interference rejection and other properties of spread spectrum.

the desired signal would become narrowband, and can be filtered, so most of the interference power is filtered off.

6.4.2.1 Rake Receiver Recall that we model our wireless channel as a linear time-invariant (LTI) system over a short period of time (comparable to the coherence time of the channel). The rake receiver (Figure 6.15) is a receiver that exploits:

- The linearity of the channel
- The autocorrelation properties of the PN sequence

to largely overcome the distortions of a frequency-selective fading channel. By linearity, the signal received can be written as a sum of time-delayed replicas of the same signal.

We have seen that the signal received, $r(t)$, can be written in (5.25) as a sum of multipath contributions. Here we rewrite (5.25) as

$$r(t) = \sum_{n=1}^{N} s_i(t) \tag{6.44}$$

where

$$s_i(t) = \alpha_n s(t - \tau_n) e^{-j2\pi f_c \tau_n} \tag{6.45}$$

to make it easier to talk about the signal arriving on each path, $s_i(t)$. Bear in mind that since the signal at the transmitter was multiplied by the PN sequence, all the replicas $s_i(t)$ also contain the data signal multiplied by the PN sequence (but at different time

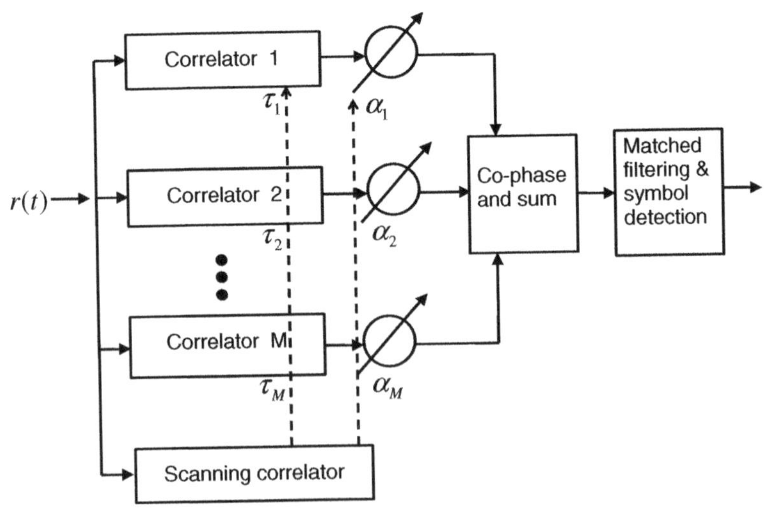

FIGURE 6.15 Rake receiver.

offsets). Hence, if we can correlate $r(t)$ with a PN sequence synchronized with $s_i(t)$, we would expect:

- The output to contain a large contribution from $s_i(t)$
- The contributions from $s_j(t)$, where $j \neq i$, to be very small, as they are suppressed by the low autocorrelation of the PN sequence

This is the main idea behind the rake receiver, where there are a few correlators (also called *fingers* of the rake receiver) that can be set to capture and extract the signal arriving at a few delays. The input signal to the rake receiver, $r(t)$, is as described in (6.44), a combination of contributions from multiple paths arriving at different times. In practice, $r(t)$ would already have been down-converted to a low IF frequency or to baseband, and the correlators would also be at the same low IF frequency or baseband. In a multipath delay spread channel, the best performance is obtained if the signal in the strongest arriving paths can be captured. How can the correlators or fingers be set to the best delays or offsets? After all, the wireless channel is time varying. The solution is that there is one more correlator that is constantly scanning for the best offsets (with the highest signal strengths), and then the other correlators can move to the best offsets.

How are the signals extracted through the various fingers combined? Maximal ratio combining is the best choice, since it has the best performance, and the signals can easily be co-phased, as the delays between them are set in the rake receiver and thus are known. Moreover, the scanning correlator also knows the relative SNRs of the correlators, so the maximal ratio weightage can be applied to the signals.

Worked Example. Suppose that the signal captured on three rake fingers has an SNR of 4, 5, and 7 dB, respectively. After combining with a maximal ratio combiner, what is the SNR of the signal coming out of the rake receiver?

Using (5.53), we just add to get $4 + 5 + 7 = 16$ dB.

6.5 OFDM

Orthogonal frequency-division multiplexing (OFDM) for wireless communications is not a new idea (OFDM for wireless communications had been proposed as far back as 1985 by Cimini [1], and the fundamentals of OFDM not specifically for wireless communications had been investigated even earlier). However, only within the last decade has it become commonplace in popular wireless standards and systems.[†]

As discussed in Section 5.3.3, we run into problems with frequency-selective fading when $\sigma \gg T_s$. If we could transform it to a flat fading channel, it would be an easier challenge for our wireless system. For a given wireless environment, σ cannot

[†] It should be noted that as early as 1995, OFDM was incorporated into the digital audio broadcast (DAB) system.

be changed, so we have more control over T_s than σ. However, if we wish to have high data rates in the usual way with single carrier communications, T_s needs to be small. For example, to achieve $R = 10$ mbit/s, we need $T_s = 0.1$ μs (assuming binary modulation). Even if we use higher-level modulation, we would need $T_s =$ 1 μs (4-level modulation), $T_s = 10$ μs (8-level modulation), $T_s = 100$ μs (16-level modulation), or $T_s = 0.001$ s (64-level modulation).

Suppose that we want to transmit data at a rate R but want to avoid the effects of frequency-selective fading. With OFDM, the data are transmitted over N multiple parallel channels, each of which need only transmit at rate R/N, so the overall data rate is still $N \times R/N = R$. Thus, the *symbol period* (also known as *symbol time*) in each of the parallel channels can be NT_s, an N-fold improvement. We denote the OFDM symbol period/time by $T_s' = NT_s$. The sampling interval, though, is still T_s, so there will still be N samples within the period T_s'.

With OFDM, and *multicarrier modulation* in general, the multiple parallel channels are formed by using multiple frequency carriers, each of which is a low-rate channel, that together provide a channel of the required data rate. The multiple carriers are often called *subcarriers* since they each carry only a fraction of the total data. Multicarrier modulation is not new, and can be found in such military wireless links as Link 11 (MIL-STD-6011), for example. However, OFDM has attractive advantages over other multicarrier modulation schemes:

- Among multicarrier schemes, OFDM provides the most compact spacing between adjacent subcarriers. If the subcarriers are any closer together, orthogonality would be lost.
- The multicarrier modulation is highly efficiently accomplished in the digital domain by the use of digital signal processing. In other words, instead of using N baseband filters and N subcarrier frequency generators, plus coherent demodulators for each one of them (resulting in much more hardware and cost), OFDM accomplishes the distribution of data into subcarriers through digital signal processing (i.e., through IDFT and DFT), and can still use one baseband filter with one carrier frequency generator and one coherent demodulator on the receiver side as though it were a single carrier system. Viewed another way, the subcarriers are generated virtually, using signal processing.
- The IDFT and DFT can be performed efficiently with IFFT and FFT.
- With the addition of a *cyclic prefix*, the effects of multipath delay spread are further mitigated.

With reference to the subcarrier spacing, it makes an interesting exercise to compare OFDM subcarrier spacing with non-OFDM multicarrier spacing, as in Exercise 6.5.

We let the difference in frequency between the adjacent subcarriers be such that within T_s', the higher-frequency subcarrier completes exactly one more cycle than the other one, as illustrated in Figure 6.16. [NB: This implies that the difference in frequency between adjacent subcarriers is $1/T_s'$, and by (6.1.8.3), this is the minimum spacing between adjacent subcarriers that preserves orthogonality.] At any given time

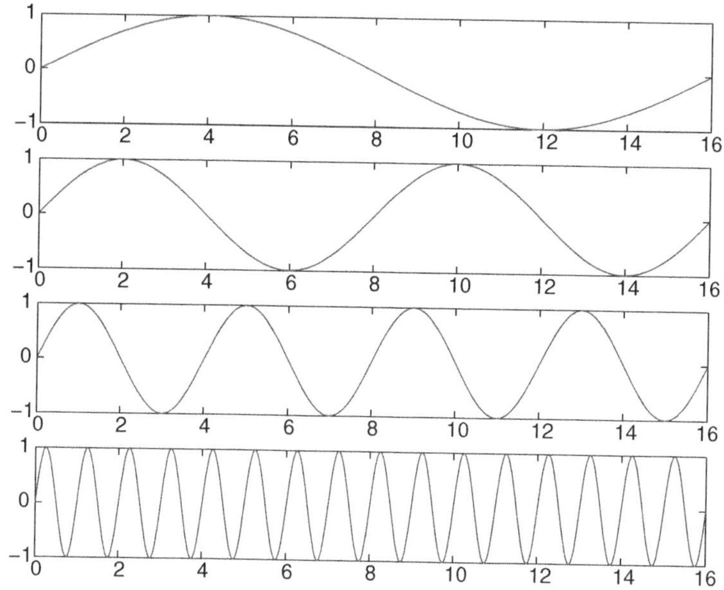

FIGURE 6.16 OFDM as transmission over multiple parallel subcarriers.

within the OFDM symbol period, $0 \leq t \leq T_s'$, the OFDM symbol is the sum of the signals on the N subcarriers:

$$x(t) = \sum_{n=0}^{N-1} X_n \exp\left(j2\pi \frac{nt}{NT_s}\right) \tag{6.46}$$

where X_n is the symbol being carried by the nth subcarrier. We can discretize time by writing $t = kT_s$ [i.e., by sampling the continuous-time signal (6.46)], and we have

$$x(kT_s) = \sum_{n=0}^{N-1} X_n \exp\left(j2\pi \frac{nkT_s}{NT_s}\right) \tag{6.47}$$

and now we can see how we get the DFT/IDFT in OFDM if we let $x_k = x(kT_s)$, and then we have

$$x_k = \sum_{n=0}^{N-1} X_n \exp\left(\frac{j2\pi nk}{N}\right) \tag{6.48}$$

which is the IDFT of the sequence $X_0, X_1, \ldots, X_{N-1}$.

Even though the effects of multipath delay spread are mitigated by the use of subcarriers, we can do even better. In OFDM, a *guard time* is inserted between OFDM symbols, and the length of the guard time might be chosen to be on the order of the RMS delay spread. What gets transmitted during the guard time? Rather than transmitting nothing or some random data during this time, in OFDM the last few samples of the symbol are copied to the guard time immediately before the symbol. This makes the samples during the guard time a *cyclic prefix* because it makes the entire transmission appear to be part of a cyclic signal.

With the addition of the cyclic period, a question arises: What is the OFDM symbol period? Is it still NT_s or is it NT_s+ guard time? Different definitions can be found in the literature. Sometimes, to be more precise, NT_s is called the length of the OFDM symbol, or the *useful symbol length*, and NT_s+ guard time may be called the *symbol interval*, which emphasizes that it is the time from the start of a symbol to the start of the next symbol. In IEEE 802.11a, NT_s is called the *FFT period* and NT_s+ guard time is called the *symbol interval.* It is helpful to think of all the subcarriers as a unit and to associate a reference carrier frequency to the whole (e.g., to up-convert or down-convert the entire multicarrier signal together).

We now revisit (6.46) and note that it applies only "at any given time within the OFDM symbol period, $0 \leq t \leq T_s'$," so we now rewrite the equation to be applicable for all t:

$$x(t) = \sum_{n=0}^{N-1} X_n \Pi \left(\frac{t - T_s'/2}{T_s'} \right) \exp \left(j2\pi \frac{nt}{NT_s} \right) \tag{6.49}$$

where the function Π is the square pulse function we introduced in Section 1.2.4. Here, it is centered around $T_s'/2$ and is T_s' wide, making everything zero outside the interval $0 \leq t \leq T_s'$. We can interpret (6.49) as meaning that the terms of the sum are each the product of a rectangular pulse $X_n \Pi[(t - T_s'/2)/T_s']$ and a sinusoid. Thus, in the frequency domain, we then have sinc functions translated by n/T_s'. These are plotted in Figure 6.17. We make some observations:

- The figure provides another way for us to see why the different subcarriers are orthogonal to each other. To recover the subcarrier at frequency offset n/T_s', what we are doing, in effect, is multiplying it by a sinusoid at the same frequency, n/T_s'. That is equivalent to sampling in the frequency domain at n/T_s'. From Figure 6.17 we see that every other sinc function (representing the other subcarriers) passes through zero at each point where $f = n/T_s'$, for $n = 1, \ldots, N$.
- From the figure we see at a glance that if the subcarriers were to be any closer to each other, they would no longer be orthogonal.
- We may not be happy with the use of rectangular pulses $X_n \Pi[(t - T_s'/2)/T_s']$, because it is not spectrally efficient. In other words, the sinc function (in frequency) drops off slowly compared with other possible functions, thus increasing the spectral occupancy of the signal.

6.5.1 Spectral Shaping and Guard Subcarriers

As just mentioned, the basic OFDM system with rectangular pulses in time has poor spectral occupancy characteristics, because the sinc function in frequency drops off slowly. Unless controlled this can cause a lot of adjacent channel interference. Thus,

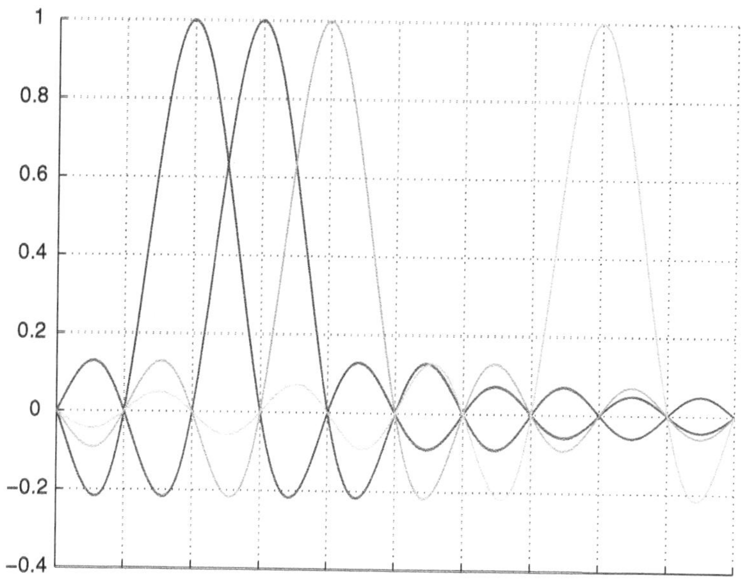

FIGURE 6.17 Orthogonality of the subcarriers.

typically, some spectral shaping is performed in OFDM systems (e.g., pulse shaping with a raised cosine pulse; see Section 1.4.2.1). Spectral shaping, however, affects the subcarriers at the edges most severely, so OFDM systems typically do not use the subcarriers at the edge. For example, with IEEE 802.11a, only 52 subcarriers are used out of the 64. The unused subcarriers are set to zero.

Viewed in the frequency domain, this results in considerable distortion of the subcarriers on both edges of the OFDM signal. Figure 6.18 shows a block diagram of an OFDM system. After some FEC coding, and possibly some coding specifically for PAPR reduction (see Section 6.5.2), the data symbols are modulated and then converted from serial to parallel so that N of them at a time can go into the IFFT. Coming out of the IFFT, the symbols are converted from parallel to serial (P-to-S), a cyclic prefix (CP) is added, and spectral shaping (e.g., pulse shaping) and digital-to-analog conversion (DAC) are performed. Subsequently, RF processing occurs, the signal is amplified and is transmitted.

On the receiver side, after down-converting, analog-to-digital conversion (ADC) and removal of the cyclic prefix in serial-to-parallel (S-to-P) conversion, the symbols go through an FFT before diversity combining. After parallel-to-serial (P-to-S) conversion, the symbols are decoded.

6.5.2 Peak-to-Average Power Ratio

In a multicarrier system such as OFDM, the data are independently modulated onto parallel subcarriers. At any moment in time, the resulting signal is a linear combination

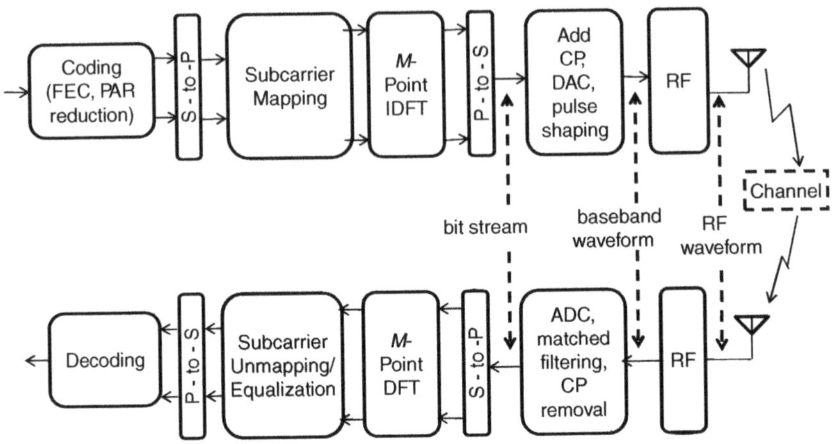

FIGURE 6.18 OFDM block diagram.

of the modulated signals on all the subcarriers. It can be pictured as a sum of random complex vectors. What if they happen to all add in phase at some point in time? We would have a peak power at that time that is much larger than the average power (averaged over many OFDM symbols). We can quantify the *peak-to-average-power ratio* (PAPR, or simply, PAR) as

$$\gamma_c = \max_t |x(t)|^2 \tag{6.50}$$

where $x(t)$ is given by (6.46).

γ_c is the largest instantaneous envelope peak power of the baseband signal. It is nontrivial to compute, so a computationally feasible alternative, the discrete-time PAPR, γ_d, is often used instead of γ_c and is defined as

$$\gamma_d = \max_{0 \leq k \leq LN-1} |x_k|^2 \tag{6.51}$$

where

$$x_k = \frac{1}{N} \sum_{n=0}^{N-1} X_n e^{j2\pi nk/LN} \tag{6.52}$$

with L being the oversampling rate, the idea of oversampling being that the more we oversample the analog signal (the larger the L), the closer γ_d would be to γ_c. However, it has been shown that once we get to around $L = 8$, γ_d is close enough to γ_c for most practical purposes [5].

Generally, high PAPR/PAR is undesirable for the following reasons:

- Amplifiers may be saturated, resulting in clipped signals.

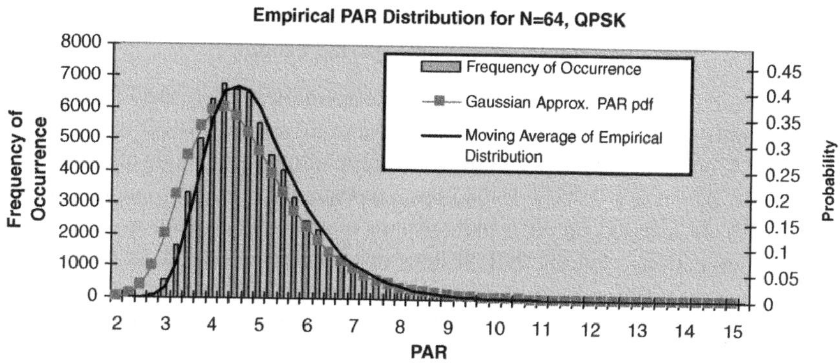

FIGURE 6.19 OFDM peak-to-average power ratio.

- Amplifiers have a limited range of operation over which they are linear. The wider the range of possible PAR, the less amplification can be performed if the PAR range is mapped to the linear operation range. Alternatively, the wider the range of PAR, the more nonlinear is the distortion of blocks with high PAR.

Therefore, it is desirable to reduce the PAR.

Figure 6.19 shows the PAPR/PAR distribution (over possible input blocks) for $N = 64$, which is a typical value for some OFDM-based systems such as IEEE 802.11a. An analytical, Gaussian approximation can also be derived easily and is also plotted in Figure 6.19. The approximation is

$$P(\gamma_d < R) \approx \left(1 - e^{-R}\right)^N \tag{6.53}$$

The empirical PAR probability distribution shows good agreement with the approximation. Also, the tails of the distribution have been found to be well approximated by the Gaussian approximation. Note that the expected PAPR/PAR of a given block increases with N.

Although much has been said about PAPR in the literature, in recent years another metric, the *cubic metric*, has emerged that better quantifies the impact on the power amplifier efficiency.

EXERCISES

6.1 Verify that the period of the discrete-time Fourier transform is 1.

6.2 Let D, R, and N_s be as defined in Section 6.3. Show that $D/R = \sqrt{3N_s}$. *Hint*: The law of cosines [(A.10) in Appendix A] may be useful here.

6.3 IEEE 802.11a has 20 millisymbol (msym)/s and 20 MHz channel spacing. Given the 20-msym/s signaling rate, what is the subcarrier spacing, Δf? What is the sampling interval? What is the OFDM symbol period? Adding the guard interval of 0.8 μs, what does the OFDM symbol period become?

6.4 Consider a multicarrier modulation system as follows: There are 15 data subcarriers and one pilot subcarrier. The subcarriers are spaced 110 Hz apart and uses DQPSK encoding. $T_s' = 13.33$ or 22 ms, depending on the mode of usage. Thus, the user data rate is 2.25 or 1.364 kbps, respectively, for the two modes of usage (verify). It is based on an actual system, the Link-11 system used in tactical communications systems. NB: It does not use the FFT/IFFT as with OFDM, so there are differences in the terms of data rates, subcarrier spacings, and so on. Let us compare it with OFDM. Suppose that we have an OFDM system with the same separation between adjacent data subcarriers. Suppose that the OFDM system has $N = 16$ but that one subcarrier is not data carrying, so it also has 15 data-carrying subcarriers. Suppose that it is also using DQPSK. What would be the OFDM symbol period and the sampling interval? What would be the signaling rate of that OFDM system? What is the data rate of the OFDM system? How does it compare with Link 11?

6.5 Again we compare OFDM with the multicarrier system described in Exercise 6.4. This time we take the same data rate (say, 2.25 kpbs on each subcarrier). What would be the signaling rate of the OFDM system, again with $N = 16$ and 15 subcarriers used for data with DQPSK? What would be the subcarrier spacing? How about the OFDM symbol period?

REFERENCES

1. L. Cimini. Analysis and simulation of a digital mobile channel using orthogonal frequency division multiplexing. *IEEE Transactions on Communications*, 33(7):665–675, July 1985.
2. B. G. Lee and S. Choi. *Broadband and Wireless Access and Local Networks*. Artech House, Norwood, MA, 2008.
3. J. S. Lee and L. E. Miller. *CDMA Systems Engineering Handbook*. Artech House, Norwood, MA, 1998.
4. A. V. Oppenheim and R. Schafer. *Discrete-Time Signal Processing*, 2nd ed. Prentice Hall, Upper Saddle River, NJ, 1999.
5. K. D. Wong, M. O. Pun, and H. V. Poor. The continuous-time peak-to-average power ratio of OFDM signals using complex modulation schemes. *IEEE Transactions on Communications*, 56(9), Sept. 2008.

7

COMPONENT TECHNOLOGIES

In this chapter we examine some component technologies that are part of wireless access. These technologies are typically needed to meet various requirements of the wireless access subsystem. For example, the wireless medium is a shared medium, so medium access control schemes are needed, and we cannot simply reuse wireline medium access schemes, as discussed in Section 7.1. In cellular-type wireless systems, the ability to move users from cell to cell in an efficient and timely way is a critical part of the cellular idea. We explore this in Section 7.2. In wireless access systems, interference control is very important, and one of the ways of controlling the amount of interference that any transmission causes to others is by controlling the transmit power levels, as discussed in Section 7.3. In section 7.4 we examine error control coding, which is especially crucial for wireless access since the raw error rates are higher than with wired communications.

7.1 MEDIUM ACCESS CONTROL

At a fundamental level, the wireless medium is a shared resource. If we examine a system and it appears that a particular wireless link is the exclusive property of a particular transmitter–receiver pair, with no interference from other transmitters, this can only be because something is happening at a lower layer to give it such an appearance. For example, a mobile phone application might be written as if it has an exclusive or dedicated link to a library of ebooks on a server somewhere in the wired network. In reality, when the bits are flowing from the ebook server to the application, the bits are actually going out on a shared medium. From the point of view of a layered

Fundamentals of Wireless Communication Engineering Technologies, First Edition. K. Daniel Wong.

communications model (see Section 10.1.1), it typically is the *medium access control* (MAC) sublayer of layer 2 of the protocol stack that deals with these challenges of having a shared medium. It may then present a point-to-point or point-to-multipoint service to the higher layers (for more on the concept of layering, see Section 10.1.1).

Medium access control refers to the schemes that are used by two or more devices to share the wireless medium. In the case of a licensed band, medium access control focuses on devices of the same type, whereas in the case of an unlicensed band, the system designers also have to consider transmissions from devices of other types (e.g., designers of Bluetooth systems have to consider possible interference from WiFi transmitters using the same band). We discuss licensed vs. unlicensed bands further in Section 17.4.1.

Multiple access is a subset of medium access control, as it refers to one type of medium access problem (i.e., the problem of multiple devices accessing a shared central point such as a base station or access point). In fact, strictly speaking, even in a cellular system based on TDMA or CDMA, for example, it is only the uplink that uses TDMA or CDMA. The downlink is based on *time-division multiplexing* (TDM) or *code-division multiplexing* (CDM). Both multiple access schemes and multiplexing schemes fall under the umbrella of medium access control. MAC schemes can be divided into those with a central controller and those without. We briefly examine schemes without a central controller in Section 7.1.1, and those with a central controller in Section 7.1.2.

7.1.1 Distributed-Control MAC Schemes

Through the decades, many schemes have been proposed for medium access control that are *distributed* (i.e., without a central controller). The famous *Aloha protocol* was one of the earliest, being deployed in AlohaNet in 1971. It works basically as follows:

- If there are data to send, just go ahead and send them (don't check if you are allowed to send, if the medium is free, etc.)
- If there is a collision of your message with another transmitter's message, try again later.

There can be different variations depending on when retransmissions are attempted, in the event of a collision, and so on. Aloha works well enough when there are not many transmissions so that the probability of collisions is low. As the number of attempts to transmit increase, and the probability of collisions also increases, the performance of Aloha tends to deteriorate. An improvement on Aloha is the *slotted Aloha protocol*. With slotted Aloha, time is broken up into discrete slots. All attempts are made only at the beginning of slots. It can be shown that this restriction decreases the probability of collisions. Slotted Aloha, perhaps surprisingly, has proven to be robust enough that it is used in GSM—not for every transmission, but at least on the random access channel (Section 8.1.1.3).

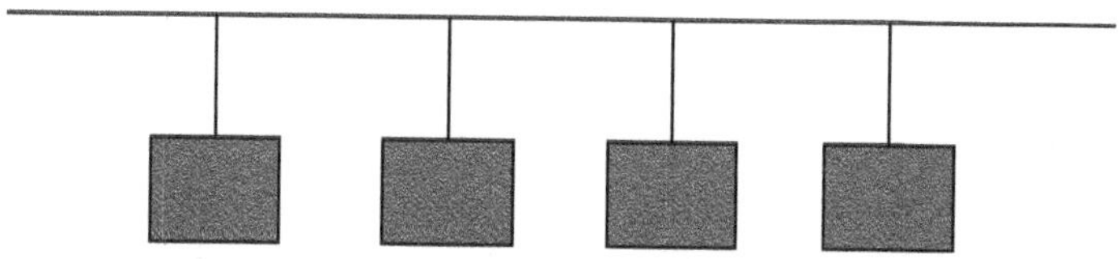

FIGURE 7.1 Bus topology as used by Ethernet, for example.

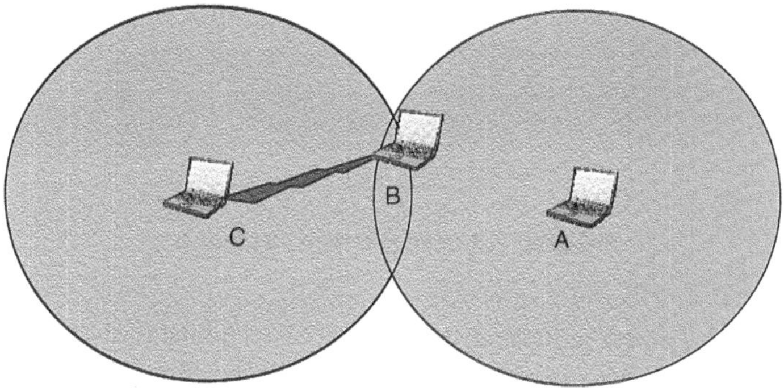

FIGURE 7.2 Hidden terminal problem.

7.1.1.1 Applying Lessons Learned from Ethernet Ethernet-based wired LANs are ubiquitous. Originally, an Ethernet-based LAN was a shared medium†, in some ways like the wireless medium but different in other ways. It was similar in the sense that multiple devices are connected to a shared *bus* (the bus topology is shown in Figure 7.1), so collisions were possible. Ethernet implements a carrier sensing scheme as well as a collision detection scheme. Thus, the Ethernet MAC protocol is called *carrier sensing multiple access with collision detection* (CSMA/CD). *Carrier sensing* means that before trying to transmit, a transmitting device will listen to the shared medium to sense if there is already a transmission (a carrier) ongoing. Although this reduces the probability of collisions, collisions can still occur (e.g., if two or more devices detect that the medium is free and then start transmitting at about the same time).

It is not possible to reuse CSMA/CD in a wireless LAN context, because of the limited range of wireless transmissions, as illustrated in the hidden terminal and exposed terminal problems. In the *hidden terminal problem*, illustrated in Figure 7.2, devices A and C are both within range of B but not within range of each other. Thus, they cannot hear each other so cannot sense when the other is transmitting. Therefore, if they both wish to transmit data to C (the "hidden terminal") at the same time, they will both proceed, and there will be a collision that neither of them can detect. The

† Today, with the pervasiveness of *switched Ethernet*, the situation is different, as each device is in its own *collision domain*.

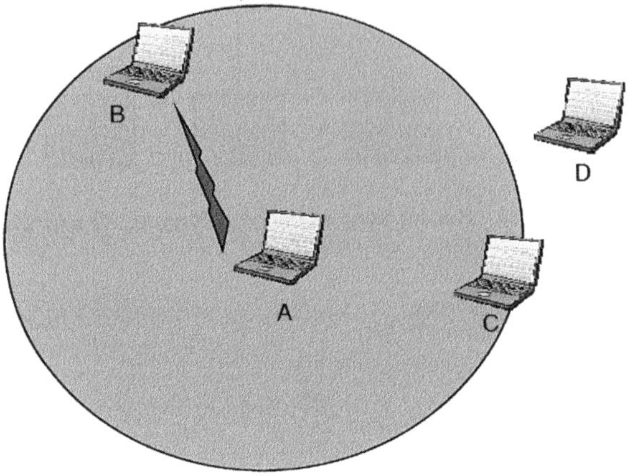

FIGURE 7.3 Exposed terminal problem.

exposed terminal problem, illustrated in Figure 7.3, is a kind of dual of the hidden terminal problem. In this case, C wants to transmit to D, and actually will succeed if it proceeds to transmit. However, A is transmitting to B, so C will sense the existing carrier and unnecessarily refrain from transmitting, even though transmission from A to B will actually not interfere, as D is out of range of A. In this case, C is the "exposed terminal."

Various solutions are possible. IEEE 802.11 uses *carrier sensing multiple access with collision avoidance*. We discuss this and other aspects of the 802.11 MAC in Section 8.3.2.

7.1.2 Central Controlled Multiple Access Schemes

Unlike in the preceding section, in this section there is a central point, such as a base station or access point, that controls and coordinates the multiple access attempts from various devices. How can transmissions from different devices to a base station or access point be separated into different channels? The two most obvious choices are different frequency bands or different time slots within the same frequency band: the *frequency-division multiple access* (FDMA) and *time-division multiple access* (TDMA) schemes. As their names suggest, FDMA is based on assigning different devices to different frequency bands (typically, all of the frequency bands would be of equal bandwidth), whereas TDMA is based on assigning different devices to different time slots [typically, all the time slots would be of the same length and would be arranged in a "frame" comprising some number of slots (e.g., eight time slots in a GSM frame; a transmitter would be assigned to a fixed time slot in every frame, e.g., the second time slot in every frame)]. The concepts of FDMA and TDMA are shown in Figure 7.4, where three users are divided by frequency or time. Notice the wasted spectrum or wasted time between adjacent bands or time slots. This is to minimize adjacent band or adjacent time slot interference due to imperfections, inaccuracies,

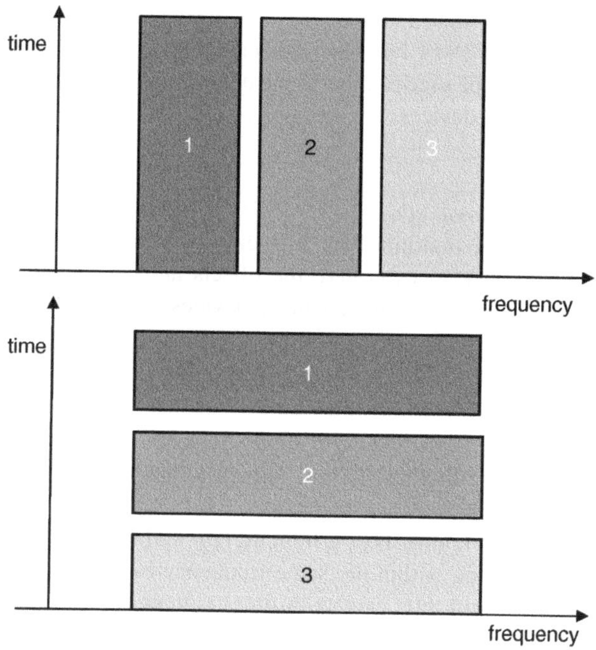

FIGURE 7.4 FDMA and TDMA: how frequency–time space is divided.

and so on, present at multiple levels in the systems. The gap between time slots is also known as *guard time*.

Other multiple access schemes are possible. *Code-division multiple access* (CDMA) is based on spread-spectrum principles that we have introduced in Section 6.4, where the devices are all using the same frequency band and transmitting at the same time. To contrast this with TDMA and FDMA, Figure 7.5 shows multiple

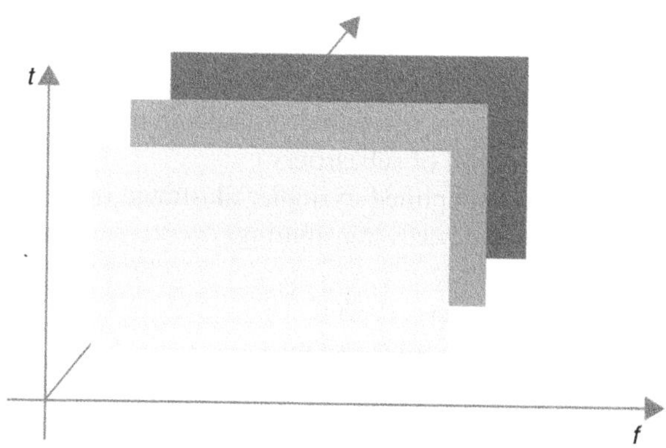

FIGURE 7.5 Code-division multiple access.

transmissions in the same time and frequency, but arranged along another dimension. Traditionally, the term *CDMA* has been associated with a direct sequence spread spectrum, but it can also be used to refer to the use of other types of spread-spectrum technologies to separate devices. For example, in a broader context, "CDMA" can refer to the following:

- *Direct sequence spread spectrum* (DSSS) *for multiple access*. The data from different devices are multiplied by different PN sequences, or different offsets of the same PN sequences (where the offsets are far enough apart to allow separation in a receiver); thus, all these devices could be transmitting at the same time in the same frequency band.
- *Frequency hopping for multiple access*. All the devices could be transmitting at the same time within the same frequency band, but using different hopping sequences (thus, they will be hopping to different sequences of frequency sub-bands within the overall frequency band, with a low probability of collisions).
- *Pulse position* (also known as time hopping) *for multiple access*. All the devices could be transmitting within the same frequency band, and transmitting very wide bandwidth (and thus narrow-in-time) signals, in the same overall time periods, but with different sequences of precise pulse positions/offsets, with a low probability of collisions.

Although DSSS may be the best known means of using spread spectrum for multiple access, frequency hopping can be found in some commercial systems (e.g., Bluetooth and one of the official physical layer specifications in IEEE 802.11 that did not become widely implemented unlike the DSSS physical layer of 802.11) and tactical data links such as Link 16. Pulse position modulation can be found in some ultrawideband (UWB) systems (see Section 17.4.2 for more on UWB).

OFDMA is a variation of OFDM that has been modified to allow for multiple access. The regular OFDM is just between one sender and one receiver. Instead of all the subcarriers being transmitted by the same transmitter, multiple devices transmit, but each transmits on a different set of subcarriers. The resulting signal that arrives at the receiver can be demodulated like an OFDM signal and then separated into the different signals from the different transmitters, based on the receiver knowing who was transmitting on which subset of subcarriers.

We illustrate with an oversimplified example, illustrated in Figure 7.6. Suppose that $N = 8$ and we have three devices transmitting.

- Let the subcarriers be $X_0, X_1, \ldots, X_7$.
- Let the three users be A, B, and C and the base station be D.
- A puts its data on X_0 and X_1 and 0 on other subcarriers and transmits it as a regular OFDM symbol.
- B puts its data on X_2 and X_3 and 0 on other subcarriers and transmits it as a regular OFDM symbol.

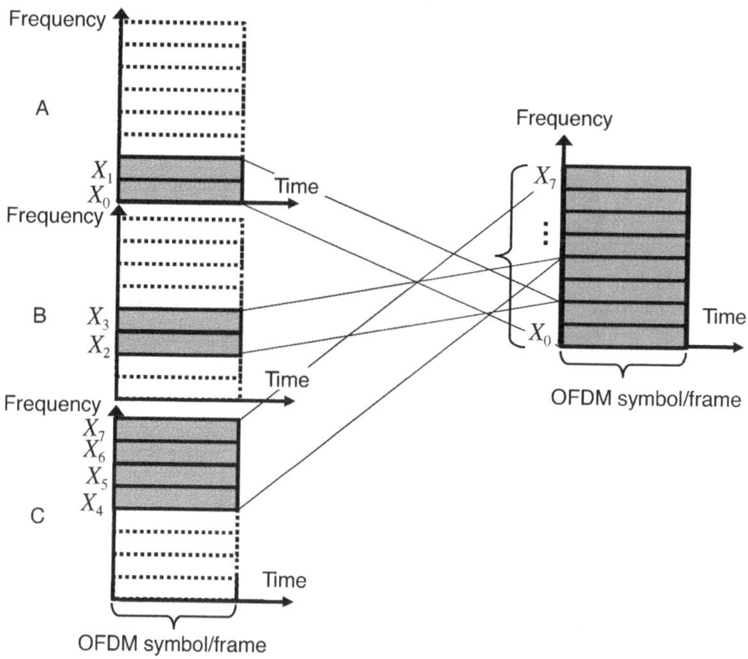

FIGURE 7.6 Simplified example of OFDMA.

- C puts its data on X_4, X_5, X_6, and X_7 and 0 on other subcarriers and transmits it as a regular OFDM symbol.
- A, B, and C transmit at the same time, and D receives all eight subcarriers and can demodulate it as a regular $N = 8$ OFDM symbol. NB: The linearity of the multiple access channel is assumed, so the signals are simply scaled and added at D. The signals may even arrive at different signal strengths at D, but that is OK because it looks like frequency-selective fading for the overall signal (but flat fading for each subcarrier).

In our simple example we had a very small N and only three devices transmitting. In a real system, N would be larger, as would be the number of transmitting devices. Also, a channel might be defined not just by a particular set of subcarriers in one OFDM symbol period, but by a set of subcarriers in two or more OFDM symbol periods. Some examples of what we mean will be seen when we look at OFDMA in a real system, WiMAX, in Section 9.4.

So, with OFDMA, allocations of subcarriers can change from OFDM symbol to OFDM symbol. Thus, this might invite comparisons with TDMA. In particular, one could imagine a TDMA/OFDM system in which transmitters take turns transmitting over the entire range of subcarriers of the OFDM symbol. It is important to realize that OFDMA is more flexible than this scheme, as can be seen in Figure 7.7. In this figure,

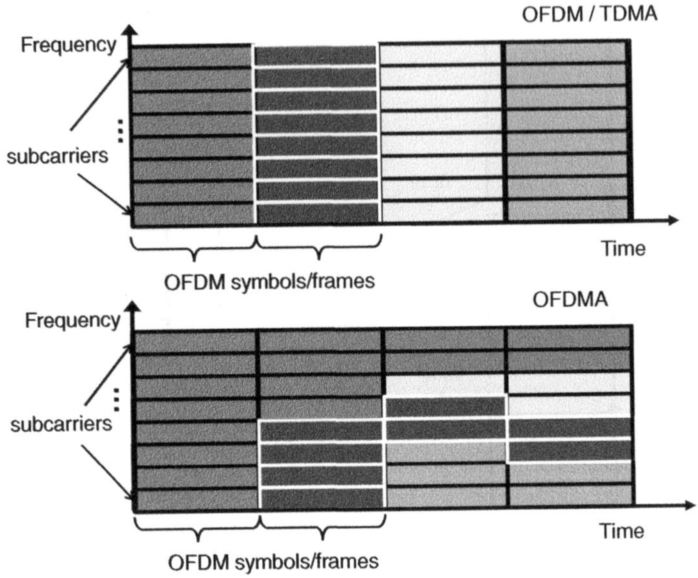

FIGURE 7.7 OFDMA is more flexible than OFDM/TDMA.

transmissions from different users are in different shades of gray. Clearly, OFDMA allows for better granularity of data rate for each user (more or fewer subcarriers can be allocated), and also for changing data rates (users need not be forced to use a fixed allocation of subcarriers, but the allocation can change with time as needed).

Also, in OFDMA, better subcarriers can use higher-rate modulation, and the receiver is simpler than: for example, CDMA. It is basically a normal OFDM receiver and does not require overly sophisticated signal processing. With OFDMA we also take advantage of *multiuser diversity*. Different devices would "see" different channels to and from the base station, and these channels would be changing with time: sometimes better, sometimes worse. Thus, subcarriers can be allocated to devices with the subcarrier quality in mind (e.g., if at a given time, a given device has good quality on a certain subset of the subcarriers, those could be allocated to it, and later, allocated to another device when the channels have changed and that other device has better quality on those subcarriers). If done well, this can result in better performance than that of CDMA, TDMA-OFDM, and so on, where the frequency-selective fading is in a sense averaged out. This is the first time that we are describing multiuser diversity, but we shall see in Section 9.3 that when we get to HSPA, and then EV-DO, that multiuser diversity is not the exclusive feature of OFDMA. Instead, it first rose to prominence in data-optimized air interfaces such as HSPA and EV-DO.

Other advantages of OFDMA over CDMA include:

- Interference cancellation is easier, since there are fewer interferers.
- The signal-to-interference ratio (SIR) is better in OFDMA because CDMA always has interference in both uplink and downlink, whereas OFDMA provides

orthogonality with much less interference in both uplink and downlink, at least from other transmissions within the same cell.

- It is easier to set up new connections with OFDMA. With CDMA, each new connection must be set up more carefully, to avoid causing too much negative impact on the interference environment.

It is worth noting that the 2G systems are dominated by TDMA/FDMA (as evidenced by the dominant position of GSM), the 3G systems are dominated by CDMA (WCDMA and cdma2000 being the main systems), and the 3.5G or 4G systems (WiMAX and LTE being prime examples) are dominated by OFDMA. In fact, rumor has it that even in the late 1990s, at one stage in the discussions on what air interface technology to use for the system that eventually became UMTS, there was a vote taken, and WCDMA beat OFDMA by only one vote. At that time, OFDMA was considered not mature enough. Perspectives have since changed.

7.1.2.1 Some Comparisons A problem with FDMA is that the spacing between adjacent channels is not as tight as it might be, and therefore some bandwidth is wasted. On the other hand, like OFDM, OFDMA uses the tightest intercarrier spacing that still allows the adjacent subcarriers to be orthogonal. TDMA, meanwhile, has the disadvantage of requiring a guard time between time slots. Both FDMA and TDMA do not exploit multiuser diversity, as the users are each assigned fixed frequency bands or time slots, whether or not there might be better channels at different frequencies or time slots. CDMA has the near–far problem, so power control is critical. OFDMA has problems with high peak-to-average power ratios (PAR or PAPR); however, a variation of OFDMA such as *single-carrier FDMA* can be used to reduce the PAR problem.

The differences between multiple access schemes (on the uplink) typically apply similarly on the downlink, although some of the issues may be less critical. For example, on the downlink, since transmissions are from a base station, guard times in TDM are less critical (the *same* base station is transmitting in consecutive time slots), and power control is less critical in CDM (see Section 7.3).

7.1.3 Duplexing

The concept of duplexing in wireless systems is related to that of multiple access and multiplexing, but is different in that it concerns only the link between a mobile device and a base station rather than a many-to-one or one-to-many situation as in the case of multiple access and multiplexing. Previously we have discussed multiple access and multiplexing without examining how a system would need to be handling both uplink and downlink. However, one or both sides may not be physically capable of transmitting and receiving simultaneously (this may be especially true for mobile devices, where the transmitting and receiving circuits may share certain elements to reduce costs, e.g., shared antennas). Also, it may be desirable for uplinks and downlinks to be in different frequency bands, to reduce interference between uplink and downlink.

In *time-division duplexing* (TDD), the uplink transmissions between a mobile device and a base station, and the downlink between the same base station and same mobile device, occur at different times. Thus, a single antenna at a mobile device can be used alternately for transmitting and receiving. In *frequency-division duplexing* (FDD), the uplink transmissions between a mobile device and a base station, and the downlink between the same base station and same mobile device, are in different frequency bands. Some systems (e.g., GSM) use FDD, but with a separation of three time slots between uplink transmissions and downlink receptions, so mobile devices need not transmit and receive simultaneously.

7.1.4 Beyond the Single Cell

Although multiple access schemes and multiplexing schemes are most easily understood at first in the context of a single cell, in many real systems (e.g., cellular systems) they are used in a multicell environment and so have to be designed with other-cell considerations in mind. In particular, how would our multiple access schemes handle interference from other cells (potentially from both the uplink and downlink transmissions in the surrounding cells)? The same question can be asked of our multiplexing schemes.

In the cases of FDMA/FDM and TDMA/TDM, there would be too much interference if the same frequencies are used in adjacent cells; hence, a frequency reuse factor must be introduced. We have already seen this discussed in Section 6.3. In the case of CDMA/CDM, we can exploit the spread-spectrum processing gain to use the same frequencies in every cell. Thus, CDMA/CDM is often said to use a frequency reuse factor of 1.

7.2 HANDOFF

Handoff refers to any change in wireless channel used. It is known alternatively as *handover*. Handoff is a key component of the cellular concept because (1) it is an integral part of the interference-controlling mechanism that is necessary for a cellular system to reuse frequencies and still be able to provide wireless links of reasonable quality; and (2) it allows a cellular system to support the mobility of its users (i.e., they are not restricted to the coverage region of any one serving base station). Without handoff, therefore, the careful cell planning that includes frequency reuse in strategically distributed cells would be severely affected. Recall that when talking about frequency reuse, a tacit assumption is that devices connect to the base station with the strongest signal (Section 6.3). Even if allowances could be made for users to stray far from their original serving base stations with certain probabilities (such allowances would be at the expense of precious capacity, since co-channel cells would have to be located farther apart), the system would be of limited use because calls would be dropped once the user strayed too far from his/her serving base station.

Hence, handoff is a foundational pillar of both the interference-controlling mechanism and the support of mobility in cellular systems.

We examine the costs of performing handoffs in Section 7.2.1, describe ways to categorize handoffs in Section 7.2.2, and then examine some of the issues that pertain to handoff algorithms in Section 7.2.3. We give an example in Section 7.2.4 and in Section 7.2.5 point to other sections in this book where other examples can be found.

7.2.1 What Does It Cost?

Every handoff requires network resources to proceed. This takes the form of over-the-air signaling, network signaling, database lookup, and network reconfiguration. Since handoff straddles the radio link and the network layers, it is only natural that it involves signaling across both the air interface and the network interface. Air interface signaling is between users and base stations; network interface signaling is between base stations and network entities such as mobile switching centers. Just as there is signaling on both sides, there are also costs on both sides. Control signaling such as handoff signaling consumes valuable radio bandwidth, whether in using dedicated control channels or in performing in-band signaling over traffic channels.

Although there is less of a bandwidth problem on the network interface side, the resources consumed are greater on that side. Control signaling performed between various network entities such as the base stations involved and the mobile switching center is only part of what happens. Database accesses for registration and authentication may also contribute to the cost of handoffs. There also needs to be network reconfiguration, in the form of setting up and tearing down of links between nodes, and bridging. These functions make the necessary internal adjustments to provide access for the user at the new base station and to stop providing access for the user at the old base station. In some systems, a short period of multicasting or buffering in strategic nodes is done to ensure seamless handoff without loss of data or voice.

7.2.2 Types of Handoff

One way of classifying handoffs is by the reason for handoff: to improve link quality, to reduce interference to others, and so on. A second way of classifying a handoff is by whether it is intercell or intracell. In *intercell handoffs* the previous and new serving base stations are different. In *intracell handoffs* the previous and new serving base stations are the same, but the channel used is different. Usually, when handoffs are discussed, intercell handoffs are the ones discussed. Intercell handoffs can be further subdivided into intraswitch and interswitch. On a larger scale, there may also be intersystem handoffs, as illustrated in Figure 7.8. These classifications are based on the supporting network topology.

An issue arises in the consideration of handoffs between sectors in the same cell, for systems in which sectoring (Section 6.3.2) is used. One could argue for treating handoffs between sectors in a cell as intracell handoffs. On the other hand, one could also argue that these should be treated as intercell handoffs. In our opinion, the latter is preferable, since the sectors can in many ways be treated as logically different cells

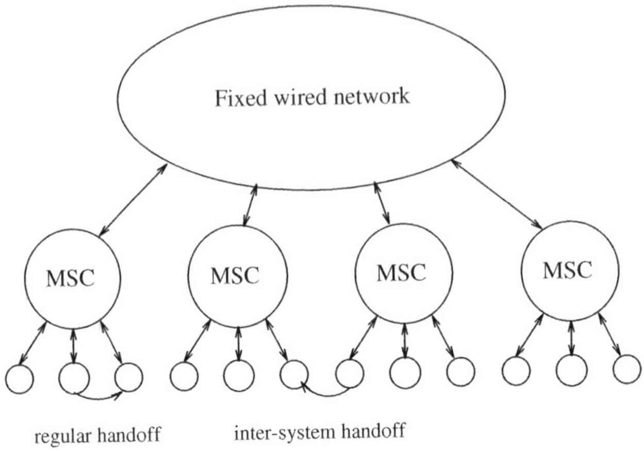

FIGURE 7.8 Hierarchical relationship between the fixed wired network, mobile switching centers (MSCs), and base stations. The difference between a regular intercell handoff and an intersystem handoff is illustrated.

that just happen to share a common base station location. Also, if a system has multiple frequency bands (e.g., GSM 900 and DCS 1800), even if the cells are co-located, an intercell handoff is considered to have occurred when a handoff is performed between them, since the frequency band is switched [3].

A third way of classification differentiates between hard and soft handoff. Handoffs may be classified as being hard or soft depending on what happens in the crucial period during handoff execution when there is communication between the user in question and more than one base station. With *hard handoff*, a *definite decision* is made on whether or not to hand off. On a positive decision, the handoff is initiated and executed without the user attempting to have simultaneous traffic channel communication with the two base stations. With *soft handoff*, a *conditional decision* is made on whether to hand off. Depending on the changes in pilot signal strength from the two or more base stations involved†, a definite decision will eventually be made to communicate with only one of the two base stations. This normally happens after it is clear that the signal from that base station is considerably stronger than those from the others. In the interim period, the user has simultaneous traffic channel communication with all candidate base stations.

A fourth classification of handoff is based on where the handoff decision control is located. In systems with *network-controlled handoff*, also known as base station-controlled handoff, handoff decision control is in the network of base stations. The users are passive and do not play a part in handoff decisions. In systems with *mobile-assisted handoff*, handoff decision control is still in the network of base stations, but the mobile devices play a part in the decisions. The mobile devices assist the base

† Soft handoff between two sectors of a sectorized cell is sometimes known as *softer handoff* [5].

stations by making measurements and relaying these measurements to the decision-making entities. In systems with *mobile-controlled handoff*, handoff decision control lies with the user terminal.

7.2.3 The Challenge of Making Handoff Decisions

We consider the case of a mobile moving between two base stations, BS 0 and BS 1, where the signal strength from the two base stations is S1 and S2, respectively. As the mobile moves between BS 0 and BS 1, the measurements of S1 and S2 are as shown in Figure 7.9. Suppose that at regular intervals of time, the following handoff decisions occur:

- If S2 – S1 > 0 and the serving base station is BS 0, hand off to BS 1. S2 – S1 is the relative signal strength of BS 1 with respect to BS 0.
- If S1 – S2 > 0 and the serving base station is BS 1, hand off to BS 0.
- Otherwise, do not hand off.

Using this handoff algorithm, the user will hand off three times back and forth between BS 0 and BS 1. At τ_1, the user hands off from BS 0 to BS 1, at τ_2 from BS 1 back to BS 0, and at τ_3 from BS 0 to BS 1 again, finally remaining with BS 1. This handing off back and forth several times between two base stations in a relatively short period of time is sometimes known as the *ping-pong effect*, analogous to the movement of the ball between the two ends of the table in a table tennis (a.k.a. ping pong) game. The problem is that each time a handoff is executed, network resources are consumed, as explained earlier in the chapter, and there is some probability of dropping the call every time a handoff is executed.

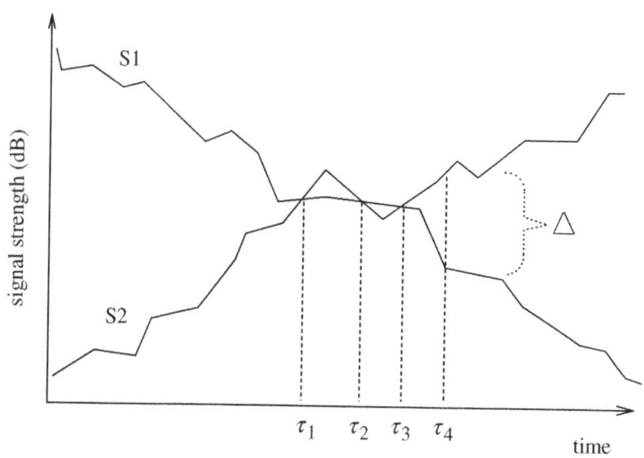

FIGURE 7.9 Signal strengths and handoffs: simplified picture.

To reduce the ping-pong effect, a standard feature of hard handoff algorithms is the incorporation of hysteresis. By this we mean that the basic algorithm is modified as follows:

- If S2 − S1 > Δ_0 and the serving base station is BS 0, hand off to BS 1.
- If S1 − S2 > Δ_1 and the serving base station is BS 1, hand off to BS 0.
- Otherwise, do not hand off.

Δ_0 and Δ_1 are hysteresis margins, and generally, $\Delta_0 = \Delta_1 = \Delta$. Hysteresis allows the system to wait before it performs a handoff until it is more certain that one should be performed, and it thus reduces the ping-pong effect. A variation of this algorithm is used in GSM. The value of using hysteresis can also be seen in Figure 7.9, where Δ is indicated on the right. If hysteresis is used, in this case, there would only have been one handoff performed from BS 0 to BS 1.

A disadvantage of using hysteresis is that the handoff decision is delayed on the average, and this delay increases with hysteresis margin. There is therefore a tradeoff between average delay and average number of unnecessary handoffs, which can be adjusted by changing the hysteresis threshold. The larger Δ is, the larger the average delay, but the smaller Δ is, the larger is the number of unnecessary handoffs.

Actually, Figure 7.9 shows a simplified picture of the signal strength. In reality, the signal strength measurements might look more like what is shown in Figure 7.10, due to all the fading (especially the small-scale fading) that occurs. Thus, this normally is smoothed out by sample averaging, so the size of the averaging window becomes another important parameter in handoff algorithms. In the case of a Rayleigh fading environment, other estimators have been shown to give better estimates of the average signal strength than the basic sample average estimator [6].

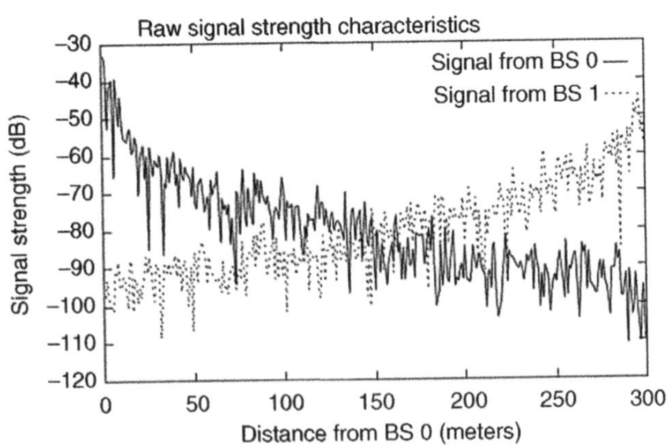

FIGURE 7.10 Signal strength and handoff.

7.2.4 Example: Handoff in AMPS

In the advanced mobile phone system (AMPS), a first generation cellular system, most of the handoff control is done by the cell site base stations and the mobile telephone switching office (MTSO), because handoff is of the network-controlled variety. Because AMPS uses frequency-division multiple access, the user terminal is tuned to one frequency channel during a call. To be able to monitor other channels associated with the serving base station or other base stations, the user terminal would therefore need another receiver. In the design of AMPS it was decided that additional hardware would not be desirable.

Whenever there is an active call, the serving base station monitors the signal strength of the uplink. If the signal strength drops below a certain threshhold, or when the supervisory audio tone† experiences too much interference, the MTSO instructs other groups of surrounding base stations to look for the user, and if a surrounding base station is found that can better serve the user, a decision is made to hand off to it. If such a base station is not found, the call is left undisturbed, with the possibility that the call might be dropped because of excessively poor link quality.

For the few handoffs between base stations served by different mobile switching centers (MSCs), (i.e., interswitch handoffs), certain protocols are used: for example, those specified in IS-41. For handoffs within the area served by the MSC (i.e., intraswitch handoffs), the handoff process has four steps:

- A cell-site trunk to the new serving cell site is found.
- The user is instructed to tune to the new cell site and is given the new supervisory audio tone frequency. Note that this information is transmitted over the present voice channel (in-band signaling in a "blank-and-burst" manner), which would be expected to cause an audible click in the voice channel.
- A new path is set up in the switching network.
- The old cell-site trunk is released.

The AMPS handoff procedures are relatively simple compared to those of some second-generation systems. Because of the engineering margin built into AMPS systems, users can go far (on the order of several times the diameter of a cell) from the serving base station before the system finds that it needs to hand off. The MTSO then has to find the user by trying to look for it at different groups of other base stations, which may take some time, and then it has to send vital signaling information crucial to the execution of the handoff over the (possibly rapidly) deteriorating channel.

7.2.5 Other Examples

Other examples of how handoffs are performed in cellular systems may be found in Sections 8.1.2 and 8.2.8.

† It allows some distinguishing of co-channel users from one another, being transmitted at one of three distinct frequencies.

7.3 POWER CONTROL

Power control is about adapting transmit power levels, both in mobiles and in base stations, in a controlled way, to optimize the interference environment in a way that address a specific problem, the near–far problem, which we consider in Section 7.3.1. In Section 7.3.2 we compare the need for power control in the uplink with power control in the downlink. Then, the difference between open- and closed-loop power control is explained in Section 7.3.3.

7.3.1 The Near–Far Problem

Interference rejection in spread-spectrum systems is actually interference suppression, whereby the strength of the interfering signals is reduced by a factor of about the processing gain. Therefore, it works best if the signal strength of all the arriving signals is about the same. Otherwise, the difference in signal strengths of the arriving signals can reduce the processing gain. This is an especially serious problem for a weak signal in the presence of one or more strong interferers.

If two or more transmitters (e.g., mobile devices) are transmitting to the same receiver (e.g., a base station) and one of them is near the base station and the other is far away, we would expect the path loss from the nearer transmitter to be significantly less than the path loss from the farther transmitter. Thus, the signals arrive unbalanced in strength. For best results (to maximize the capacity of a CDMA system), the base station tracks the signal strength from the various transmitters and sends instructions to the transmitters to step up or step down their transmitted power, expecting corresponding adjustments in received signal strength.

7.3.2 Uplink vs. Downlink

The need for power control is asymmetric. Power control is more critical on the uplink than on the downlink. This is because the transmissions are coming from different mobile stations (MSs), and power control would help tremendously the reception of signals from the weaker mobile stations. If done properly, power control can result in signals from all mobile stations arriving at the base station at about the same signal strength.

To discuss downlink power control more clearly, let's introduce two MSs, A and B, where A is close to the base station and generally receives a better signal from the base station, whereas B is farther away from the base station and generally receives a weaker signal from the base station. Unlike on the uplink, on the downlink the transmissions are all coming from the same source: the base station. What is the goal in the downlink case? Suppose that we try to follow the uplink example and cause the signals received by all mobile stations to arrive at each station at about the same power. The only way that this can be done is if the base station were to increase or decrease its transmitted power levels together for all the mobile stations (in lock step, as a group). Otherwise, if the base station were to increase the transmitted power for some channels (say, for MSs such as B) and decrease it for other channels (say, for

MSs such as A) at the same time, the former channels will be received at higher power than that of the latter channels. However, if the base station increases or decreases its transmitted power levels for all mobile stations together as a group, it must cater to the mobile stations that are receiving the weakest signals (e.g., B). The base station may then have to blast all the signals it is transmitting at the same high power so that mobile stations such as B would receive at acceptable signal levels. Most of the other mobile stations, including A, will then be receiving signals at power levels stronger than needed, while the base station might be causing higher levels of interference to communications in surrounding base stations.

Alternatively, the base station could attempt to reduce transmitted powers for mobile stations that are closer to it, and/or increase transmitted powers for mobile stations that are farther from it, using a type of "give more transmitter power to those who need it most" principle. This alternative is what in fact was chosen for IS-95. It leads to differences in received signal strength for different channels at each mobile station (higher for MSs such as B, lower for MSs such as A). This type of differential power control (different power levels for different channels) results in a double benefit for mobile stations such as B:

- The increase of transmitted signal power on the channel for MSs such as B results in higher received signal power on that channel at the mobile station.
- The simultaneous and corresponding decrease in the signal power on other channels for mobile stations who don't need the power boost (e.g., A), reduces the intrachannel interference at B (since the intrachannel interferers are weaker), which further helps reception at B.

At the same time, this double benefit for MSs such as B also doubly deteriorates the signal for MSs such as A. Not only does the power control result in their receiving less power, but there will be higher levels of interchannel interference (from channels for MSs such as B). This should be OK up to a point, since MSs such as A have some margin for this, but this is a reason why power control on the downlink cannot be as aggressive as power control on the uplink.

7.3.3 Open- and Closed-Loop Power Control

Power control, as used in CDMA systems, usually has a few components, as illustrated in Figure 7.11:

- *Open-loop power control*, where the mobile station sets it initial transmitter power without feedback from the base station. It is called "open loop" because of the lack of "closing" of the loop, which would be feedback from the base station. The power level is set at the mobile station based on the signal strength that it measures from the base station. Thus, if the measured signal strength is high, it may set its initial transmitter power level lower, whereas if the measured signal strength is low, it may set its initial transmitted power level higher. This

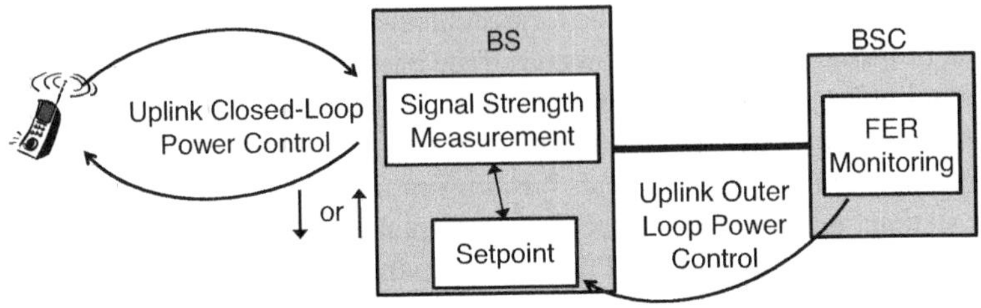

FIGURE 7.11 Power control loops in CDMA systems.

inverse proportionality to the received signal strength makes sense since the goal is to try to have the signals from all mobile stations arriving at about the same power level at the base station.

- *Closed-loop power control* addresses the issue of open-loop power control not being based directly on signal strength received *at the base station* but on an estimate based on the signal strength received *at the mobile station*. With closed-loop power control, the actual signal strength received at the base station of each mobile station is used as input to a decision process in the base station, which decides whether the mobile station should increase or decrease its transmitted signal power level.
- Longer-term adjustment of FER targets, leading to changing received power targets.

Specific numbers related to power control in specific systems, such as the power control step size, the frequency of the uplink closed-loop power control commands from the base station to the MS, and the dynamic range of power control, are presented in subsequent chapters in discussions of specific systems (e.g., Section 8.2.8.1).

7.4 ERROR CORRECTION CODES

Error correction codes are an essential component of digital communications systems. Since wireless communications systems generally have higher bit error rates than those of wireline communications systems, error correction codes are even more essential in wireless systems. Other terms for *error correction coding* are *error control coding* and *forward error correction*. The reason for the word "forward" is that the source adds the coding and it is then sent in the forward direction toward the destination. The receiver can then detect and correct errors (within limits), without necessarily sending anything backward to the sender, such as a retransmission request.

All error correction codes add some redundancy to the data bits that are to be sent from source to destination. An *encoder* adds redundancy on the source side, and a *decoder* tries to recover the original data bits on the destination side. The nature of the added redundancy depends on the code and the encoder (Figure 7.12).

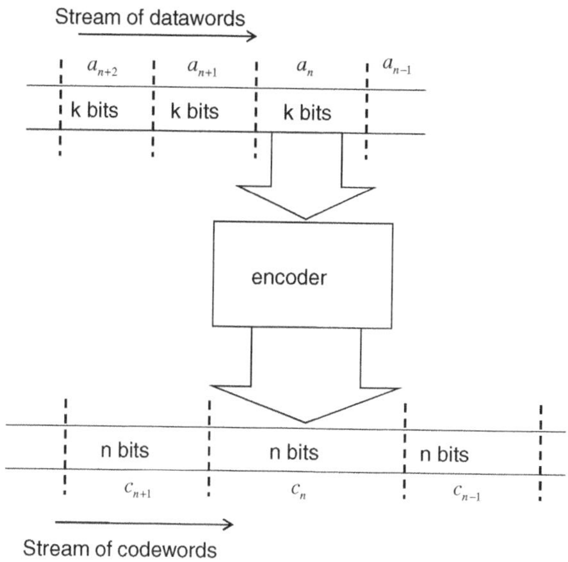

FIGURE 7.12 Block codes vs. convolutional codes in a common framework.

There are two basic types of codes: block codes and trellis codes. In *block codes*, the encoder works on blocks of k data bits at a time and produces *codewords* of n bits, where $n > k$ (hence, redundancy is added). The corresponding code is called an (n, k) *block code*. The *code rate* is $R = k/n$. If we call the blocks of k data bits *datawords*, then there are 2^k possible datawords, since there are 2^k possible combinations of k data bits, so 2^k codewords are needed to represent them. In a block code, the mapping from datawords to codewords is independent of earlier or later bits in the data bitstream. Therefore, block codes are said to be memoryless. They can be implemented by combinational logic.

In *trellis codes*, on the other hand, we can still think of segments of k data bits being mapped to n-bit codewords, however, the codewords are no longer constrained to being dependent merely on the present k data bits, but on previous data bits as well. A trellis code therefore has memory. It needs to be implemented by sequential logic. In addition to n and k, we need another parameter, m, that specifies the number of previous blocks of k data bits on which the present n-bit codeword is also dependent. We call the corresponding code an (n, k, m) *code*. m may be known as the memory order of the code. As with block codes, the code rate is $R = k/n$. A *convolutional code* is a very important type of trellis code. It is a trellis code that is time-invariant and linear.

We introduce block codes and convolutional codes in Sections 7.4.1 and 7.4.2, respectively. Common alterations of block or convolutional codes, and concatenation, will be seen in Section 7.4.3. We then briefly introduce the more recent turbo codes and LDPC codes in Sections 7.4.4 and 7.4.5, respectively.

7.4.1 Block Codes

It might at first appear that a block code must be defined by the set of codewords (all 2^k of them) used by the encoder as well as the mapping from datawords to codewords. Actually, however, if the datawords have certain statistical properties (all k-bit sequences are equally likely, and independent of all earlier and later bits), it doesn't matter what mapping is used from the datawords to codewords. The set of codewords, C, is all that we need to analyze the code's performance. The code is often considered to be C, and design of codes to meet performance objectives focuses on the selection of good sets of codewords.

The mapping from datawords to codewords, meanwhile, can be chosen to simplify implementation of the encoder. The mapping is often chosen such that the codeword contains the dataword plus some other bits. This is known as a *systematic* encoding of a code. In this case, it is common practice to speak of the bits in the codeword as being either *systematic bits* (the bits coming from the dataword) or *parity bits* (these being the bits added to the dataword to make the codeword). For example, a dataword 11000011 may be encoded to 11000011010 in an $(11, 8)$ code, where 010 are the parity bits added to the systematic bits 11000011 in this systematic encoding.

The *Hamming weight* of a codeword is the number of 1's in it. The *Hamming distance* between any two codewords is the number of places where they differ in a bit-by-bit comparison. The *minimum distance* of a code, $d_{\min}$, is the smallest Hamming distance between any two of its codewords. The error *detection* capabilities of a code may differ from its error *correction* capabilities. For example, the block received may be different from all codewords, so we know there are errors, but it may be the same Hamming distance from two or more codewords. Specifically, if a code has a particular $d_{\min}$, it can detect up to $\lfloor d_{\min}/2 \rfloor$ errors but only correct up to $\lfloor (d_{\min} - 1)/2 \rfloor$ errors. The spacing of the codewords, and the minimum distance, can be visualized as shown in Figure 7.13. NB: The code space is actually n-dimensional, so to visualize the code space accurately, we need to mentally extrapolate from the three dimensions implied by a diagram such as Figure 7.13.

7.4.1.1 *Linear Codes* A code C is said to be *linear* if every codeword in C is a linear combination of two or more other codewords. Then C forms a k-dimensional subspace of the vector space of all n-tuples, and encoding can be performed very efficiently by a $k \times n$ generator matrix, $\mathbf{G}$. Then each codeword $\mathbf{c}$ can be related to

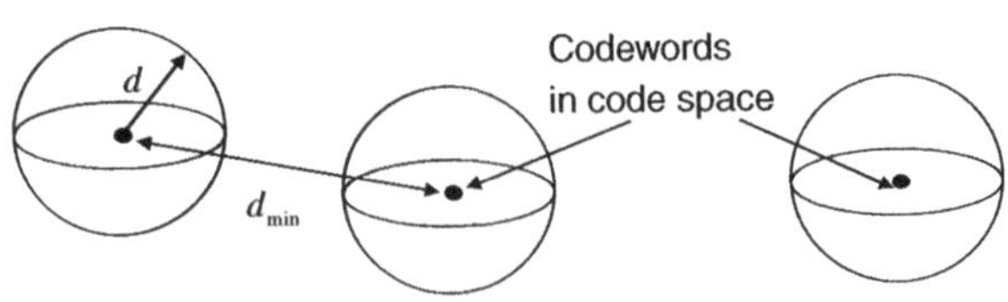

FIGURE 7.13 Codes in code space.

its associated dataword **a** by

$$\mathbf{c} = \mathbf{aG} \qquad \text{dimensions: } 1 \times n = (1 \times k)(k \times n) \tag{7.1}$$

For linear codes, a systematic encoding can always be found, where $\mathbf{G} = [\mathbf{I_k P}]$ or $\mathbf{G} = [\mathbf{PI_k}]$, depending on which convention is followed, $\mathbf{I_k}$ is the $k \times k$ identity matrix, and **P** is the encoding of the parity bits. Also, $d_{\min}$ can be related back to n and k by the *Singleton bound*:

$$d_{\min} \leq 1 + n - k \tag{7.2}$$

Furthermore, for a linear code, $d_{\min}$ is equal to the minimum weight of all its nonzero codewords.

Another important matrix related to a linear block code is its *parity check matrix*, **H**. For any $\mathbf{c} \in C$,

$$\mathbf{cH}^{\mathrm{T}} = \mathbf{0} \tag{7.3}$$

so $\mathbf{GH}^{\mathrm{T}} = \mathbf{0}$ also. Thus, **H** can be used to check whether or not a received vector is a codeword.

7.4.1.2 Hamming Codes The Hamming codes are among the earliest class of linear block codes devised. They exist for all integers $m > 3$ and with the parameters $n - k = m, n = 2^m - 1$, so $k = 2^m - m - 1$. For all Hamming codes, $d_{\min} = 3$. For example, the (7, 4) Hamming code is the smallest Hamming code, and larger Hamming codes include the (15, 11), (31, 26), and (63, 57) Hamming codes. An encode for the (7, 4) Hamming code is shown in Figure 7.14, and the corresponding decoder is shown in Figure 7.15.

7.4.1.3 Cyclic Codes Besides linearity, another structural property is found in some popular block codes: the cyclic property. A code is *cyclic* if for any codeword

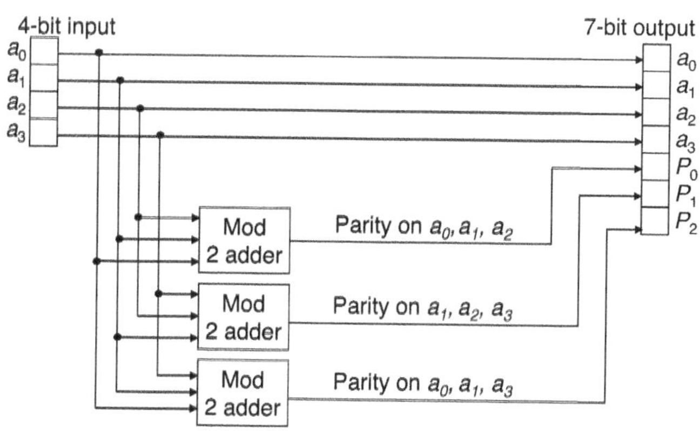

FIGURE 7.14 Encoder for a simple code.

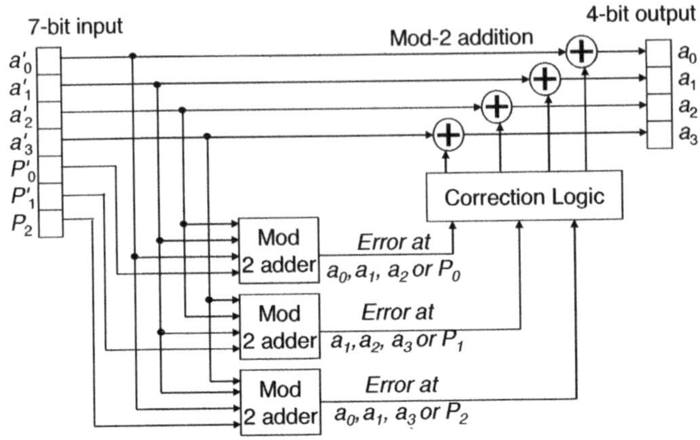

FIGURE 7.15 Decoder for a simple code.

c $\in C$, all cyclic shifts of c are also codewords in C. For example, the cyclic shifts of [10010], shifting leftward, are [00101], [01010], [10100], and [01001].

Cyclic codes are strongly related to a mathematical structure called finite fields, or Galois fields. They can be thought of as *ideals* of finite fields. Conveniently, the codewords can easily be generated by *linear feedback shift registers* in some configurations. Popular and practical codes such as BCH, Reed—Solomon, the Fire codes, and even some Hamming codes are cyclic.

7.4.2 Convolutional Codes

A *convolutional code* is a linear time-invariant trellis code with memory. Typically, therefore, it would be implemented with linear shift registers, for example, as shown in Figure 7.16. From the shift register structure of a convolutional code, we can derive the generator polynomials of the code.

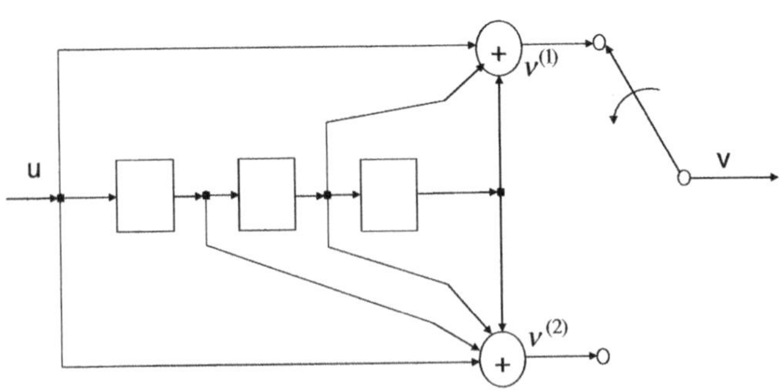

FIGURE 7.16 Example of a convolutional code.

An alternative view of a convolutional code is as a block code of infinite length. In this view, the entire infinite sequence of input bits is the dataword, and the entire infinite sequence of output bits is the codeword. Since there would be infinitely many datawords, the size of the code would also be infinite. This view naturally leads to the generator matrix **G** of a convolutional code as an infinite-dimensional matrix of the following form:

$$\mathbf{G} = \begin{bmatrix} \mathbf{G_0} & \mathbf{G_1} & \mathbf{G_2} & \mathbf{G_3} & \cdots & \mathbf{G_{m-1}} & \mathbf{G_m} & & & \\ & \mathbf{G_0} & \mathbf{G_1} & \mathbf{G_2} & \cdots & \mathbf{G_{m-2}} & \mathbf{G_{m-1}} & \mathbf{G_m} & & \\ & & \mathbf{G_0} & \mathbf{G_1} & \cdots & \mathbf{G_{m-3}} & \mathbf{G_{m-2}} & \mathbf{G_{m-1}} & \mathbf{G_m} & \\ & & & \ddots & & & & & & \ddots \end{bmatrix} \tag{7.4}$$

where each submatrix $\mathbf{G_i}$ is $k \times n$.

In practice, no infinite sequences are used. Instead, the input stream would have a finite length: say, kL. Convolutional codes can be represented by encoder state diagrams, trellis diagrams, and so on. Figure 7.17 shows an example of a trellis diagram for a convolutional code.

Corresponding to the minimum distance of block codes, we have the *minimum free distance.* Since convolutional codes have memory, past information bits can affect current bits. The number of past information bits that affect current bits is called the *constraint length.* The larger the constraint length, the more efficient the code but the higher the complexity of the decoder. The most popular decoder for convolutional codes is the *Viterbi decoder.*

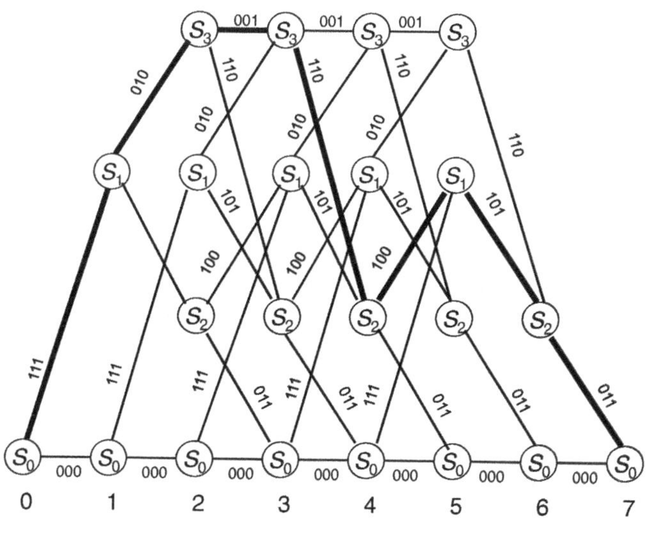

FIGURE 7.17 Example of a convolutional code trellis diagram.

7.4.3 Concatenation

Basic simple modifications that can be made to a linear code include:

1. *Extending*: increasing n while keeping k unchanged.
2. *Lengthening*: increasing both n and k by the same amount.
3. *Puncturing*: decreasing n while keeping k unchanged.
4. *Shortening*: decreasing both n and k by the same amount.
5. *Augmenting* increasing k while keeping n unchanged.
6. *Expurgating*: decreasing k while keeping n unchanged.

Two or more codes can also be used together, in back-to-back fashion. For example, on the sender side, the data can first be passed through a block encoder (a Reed–Solomon code is a popular choice for this purpose), and the output of the block encoder is then passed through a convolutional encoder. On the receiver side, the encoded bits are first passed through the corresponding convolutional decoder. The output of the convolutional decoder is then passed to the block decoder. Such an arrangement is known as *concatenation*, and the overall code is called a *concatenated code*. In the case of two concatenated codes, the first code in the encoder is called the *outer code* and the second is called the *inner code*. On the decoding side, the inner code is decoded first, followed by the outer code. The main reason for using concatenated codes is the lower complexity for similar bit error rate performance compared to using one code alone.

7.4.4 Turbo Codes

A turbo code can be viewed as a type of concatenated code with performance that is near the theoretically best achievable (called the *Shannon limits*, based on work by Claude Shannon). In a basic turbo encoder there are two identical recursive systematic convolutional encoders, one of which is fed the information bits in original sequence, with the other fed the information bits *after passing through an interleaver*. In general, the same convolutional encoder could be repeated multiple times, with the inputs to the different encoders coming out of different interleavers.

On the decoder side, an iterative procedure is commonly employed. Going back to the example with two component convolution encoders, what happens on the decoder side is that a corresponding two-component decoder is used. In the first iteration, each decoder takes as input the received bits corresponding to one of the encoders, respectively. The decoders then each produce *log likelihood ratio* (LLR) estimates of the bits and pass these estimates to each other to be used as a priori estimates for the next iteration of the decoders. NB: The decoders do not merely pass hard decisions on the bits (i.e., 0 or 1) to each other but also log likelihood ratio estimates in the form of probabilities, where the LLR for a bit k, written as $L(k)$, may be expressed as

$$L(k) = \ln \frac{P(k = 1)}{P(k = 0)} \tag{7.5}$$

Three key elements in turbo codes result in their good performance [2]:

- Use of the interleaver as part of the encoding and decoding process. The interleaver provides a random permutation element to turbo codes.
- Use of systematic convolutional codes.
- Use of soft inputs by the constituent decoders.

Turbo codes are sometimes called *convolutional turbo codes*, which emphasizes that the constituent codes are convolutional codes. This is in contrast to block turbo codes, where the constituent codes are block codes. Block turbo codes are not popular, for various reasons, so the turbo codes in use in most systems are convolutional turbo codes.

7.4.5 LDPC Codes

Low-density parity-check (LDPC) codes are another class of codes that have emerged in recent years as an alternative to traditional error correcting codes. LDPC codes have made their way into IEEE 802.16e (WiMAX) as one of the error correcting code options. They are characterized by a sparse parity check matrix $\mathbf{H}$, a parity check matrix with a low density of 1's. A sparse $\mathbf{H}$ does not necessarily give a good code, but the main idea behind the power of LDPC codes is that *iterative decoding* can be used. Iterative decoding of LDPC codes is made feasible by the sparseness of $\mathbf{H}$, and nearly maximum likelihood performance is possible (for a general $\mathbf{H}$, the complexity would be high to do maximum-likelihood decoding) [4].

7.4.6 ARQ

ARQ refers to a set of techniques whereby the receiver can request that the sender retransmit some data that may be corrupted or even be lost. Thus, it can be classified under error correction technologies [but not as part of forward error correction (FEC)]. However, it can also be considered as something that should go under network architecture, because it is often implemented at higher layers of the protocol stack, unlike FEC, which is most often implemented at the data link layer. In this book we discuss ARQ in Section 10.1.3.1. Hybrid ARQ, which is a hybrid of FEC and ARQ, is discussed in Section 9.2.1.

EXERCISES

7.1 Suppose that we have a medium access scheme in which everybody's clock is synchronized to a common reference time t_0. When any transmitter has something to transmits, it doesn't listen but just tries its luck at transmitting and hopes that there is no collision. The transmitters all agree only to try at times $t_0 + k\Delta$, where k is an integer and Δ is a finite time interval. What medium access scheme is this?

7.2 What is the difference between multiplexing and multiple access: for example, between TDM and TDMA, or CDM and CDMA? Which is more challenging?

7.3 Explain the near–far problem. Why does power control help?

7.4 Suppose that you are given a block error correcting code with $n = 424$ and $k = 300$. Without knowing anything more about the code, what is the largest possible number of errors that it could detect? How about the largest possible number of errors that it could correct?

7.5 To get a feel for the log likelihood ratio as used in turbo codes, compute a few values. For example, let $P(k = 1) = p$, so $P(k = 0) = 1 - p$, and compute $L(k)$ for $p = 1/2$, $p = 3/4$, and $p = 1/4$. Are there any symmetries? Can you express $\ln[P(k = 0)/P(k = 1)]$ in terms of $\ln[P(k = 1)/P(k = 0)]$?

REFERENCES

1. R. Blahut. *Algebraic Codes for Data Transmission*, 2nd ed. Cambridge University Press, New York, 2003.
2. A. Giulietti, B. Bougard, and L. van der Perre. *Turbo Codes: Desirable and Designable.* Springer-Verleg, New York, 2003.
3. L. Hanzo and J. Stefanov. The pan-European digital cellular mobile radio system—known as GSM. In R. Steele, editor, *Mobile Radio Communications*, chap. 8, pp. 677–765. IEEE Press, Piscatoway, NJ 1994.
4. W. E. Ryan and S. Lin. *Channel Codes: Classical and Modern.* Cambridge University Press, New York, 2009.
5. S.-W. Wang and I. Wang. Effects of soft handoff, frequency reuse and non-ideal antenna sectorization on CDMA system capacity. In *IEEE Vehicular Technology Conference*, pp. 850–854, Secaucus, NJ, May 1993.
6. D. Wong and D. Cox. Estimating local mean signal power level in a Rayleigh fading environment. *IEEE Transactions on Vehicular Technology*, 48(3):956–959, May 1999.

8

EXAMPLES OF AIR-INTERFACE STANDARDS: GSM, IS-95, WiFi

In this chapter we examine physical layer and link layer aspects of the most popular commercial wireless personal communications systems. We begin in Section 8.1 with GSM, the predominant TDMA-based second-generation system. We then discuss the second-generation IS-95 CDMA system in Section 8.2. The third standard we survey here is IEEE 802.11, in Section 8.3.

In a cellular system with multiple uplink transmissions (from mobiles to base stations) and downlink transmissions (from base stations to mobiles) occurring simultaneously, it is crucial in designing standards to carefully consider various separation problems. By *separation*, we mean avoiding or minimizing interference between signals from different transmissions. We consider five separation problems (Table 8.1):

1. Consider a base station that is serving multiple mobile stations. How are the signals it sends to these mobile stations separated from one another?
2. Again, consider a base station that is serving multiple mobile stations. How are the signals from these multiple mobile stations to this base station separated?
3. In the larger context, the cellular system would have multiple base stations. How are signals from these multiple base stations separated from those transmitted by other base stations?
4. How is a signal from a mobile station to its base station separated from the signals that other base stations are trying to receive from their mobile stations?
5. Between any transmitter–receiver pair in a multipath environment, a communication signal comes along multiple paths. How are these multiple copies of the signal separated from one another?

Fundamentals of Wireless Communication Engineering Technologies, First Edition. K. Daniel Wong.

TABLE 8.1 Overview of Achievement of Separations

Type of Separation	GSM	IS-95 CDMA
DL: between MSs in the same cell	Time–frequency channels	Walsh codes
UL: from MSs in the same cell	Time–frequency channels	Long code offsets
DL: from BSs in different cells	Distance (frequency reuse)	Short code offsets
UL: from MSs in different cells	Distance (frequency reuse)	Long code offsets
Between radio paths	Equalization	Rake receiver resolving multipath

8.1 GSM

GSM is by far the most widely deployed second-generation cellular system in the world. As such, it has lived up to its name (GSM: global system for mobile communications), going far beyond the boundaries of Europe, where it originated.

GSM uses TDMA and TDM on the uplink and downlink, respectively, with FDD. Although FDD is used, uplink and downlink signals are nevertheless offset by three time slots. This eases the hardware design requirements compared with a case where both uplink and downlink transmissions might have to be simultaneous, or one after the other without any time interval in between.

The signaling rate of each GSM carrier is 270.833 kHz, and the intercarrier spacing is 200 kHz. Each GSM carrier contains frames of eight time slots each. Usually, an operator would have multiple GSM carriers per base station (with a suitable frequency reuse plan, as in Section 6.3.1, to reduce co-channel interference to and from adjacent cells), so mobile devices are assigned a particular carrier and a particular time slot within that carrier. Hence, it is more correct in a sense to say that GSM uses a combination of TDMA and FDMA. A final wrinkle is that the assignment of a particular carrier to a mobile station is not fixed, but changes regularly. Since entire bursts are transmitted without changing frequency, it is a form of slow frequency hopping. The details of slow frequency hopping in GSM are beyond the scope of this book, but we note that it provides a form of channel diversity, especially for very slow moving mobile stations that might otherwise be stuck in a Rayleigh fade for a long time. Also, frequency hopping does not apply to the common channels (e.g., FCCH, SCH, BCCH, PCH/AGCH, RACH).

Each time slot is 0.577 ms long and comprises 148 bits, with 8.25 bits of guard time between time slots (Figure 8.1). The transmission within the time slot is called a *burst*, and a *normal burst* consists of the following (other types of bursts are discussed in Section 8.1.1):

- Two blocks of 3 *tail bits* each on each end of the time slot
- Two blocks of 57 *information bits* each next to each set of tail bits
- A block of 26 *sounding bits* in the middle of the time slot
- Two *stealing bits*, one on each end of the sounding bits

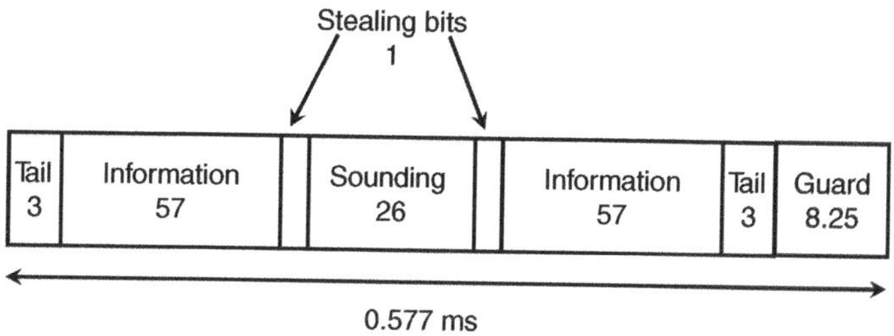

FIGURE 8.1 GSM time slot.

The *channel sounding bits* are pilot bits used for channel estimation (to train the equalizer) and for synchronization. They are in the middle of the time slot to minimize the time from when the sounding bits are transmitted to when bits at either extreme of the slot (the start and the end) are transmitted. If the sounding bits were at the start of the time slot, the channel might have changed significantly by the end of the time slot. Similarly, if the sounding bits were at the end of the time slot, the channel might be significantly different at the beginning of the time slot. Because of their position at the middle of the time slot, they are also known as the *midamble*, to contrast them with a *preamble* (a sequence of bits found in the beginning of a frame in some transmission formats).

The *stealing bits* are used to indicate whether the time slot is filled with user data or whether it is "stolen" for control signaling. The system sometimes needs to use traffic channel time slots for control signaling (e.g., when a handoff is occurring and more control signaling than normal is needed.

The *tail bits* allow the mobile transmitter to ramp up and ramp down, respectively, at the beginning and end of a burst. The three bits are all set to zero. They also help as additional guard time, even though there is 8.25 bits allocated for guard time between time slots. What does "8.25 bits" really mean? Can a quarter bit be sent? No, "8.25 bits" refers to the length of time, since no actual bits are transmitted during the guard time. Thus, the guard time is 8.25 times the symbol interval. A guard time is more crucial on the uplink (TDMA) than on the downlink (TDM, so the base station can synchronize its own transmissions to different mobile stations). Nevertheless, in both directions, the same length of time is used. The guard time helps to take care of timing inaccuracies and delay spread that might result in interference between adjacent time slots without a guard time or with insufficient guard time. In fact, the guard time might need to be even longer if it has to take into account differences in propagation times between one mobile station to the base station, and another mobile station to the same base station (where in extreme cases, one of the mobile stations could be very near the base station and the other could be very far from the base station). However, the *timing advance mechanism* helps to take care of that. This is a mechanism in which the propagation time from mobile to base station is estimated and then the transmission times of different mobile stations are adjusted accordingly to compensate for differences in propagation times, so that they would arrive roughly synchronized at their assigned time slot at the base station.

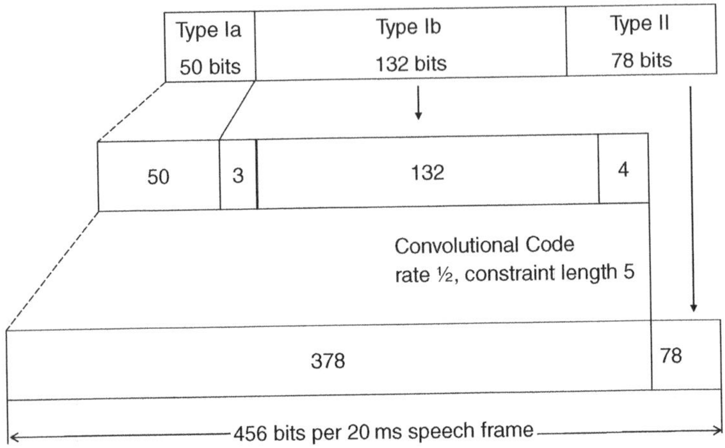

FIGURE 8.2 GSM use of error control coding.

The *information bits* come from the output of the voice codec as follows. GSM speech frames from the voice codec are each 20 ms long. At 13 kbps, that works out to 260 bits per speech frame. Error protection is not applied equally across all data bits of voice data. Instead, the bits are divided into three categories. The 50 most important bits receive the most protection, followed by the next 132 bits, and then the least important 78 bits. Three parity bits are computed from the 50 most important bits, and these 53 bits, together with the next 132 bits and 4 state-clearing bits, are entered into a rate-1/2 convolutional code, resulting in 378 coded bits. The least important 78 bits are added without protection, and this results in 456 bits (Figure 8.2). These 456 bits representing the 20 ms of speech are then distributed into eight 57-bit blocks in GSM time slots. As a form of interleaving, the two 57-bit information blocks in a time slot are taken from two different 20-ms speech frames.

GSM frames are also organized into larger units called multiframes, superframes, and hyperframes, as shown in Figure 8.3. There are two types of *multiframes*, “26 multiframes” and “51 multiframes.” Normally, user data are transmitted in the 26 multiframes (although some control signaling can also be found in these multiframes), and the 51 multiframes are for various control signaling. 51 multiframes are transmitted only from the base station (not the mobile stations). Figure 8.4 shows how a 51 multiframe is shared between control channels. Why are there so many types of control channels? What are the differences between them? It is perhaps best illustrated by example, which we will see in Section 8.1.1. We note at this time that these control channels are *common channels*, since they are not specific to any mobile station, whereas the traffic channels are dedicated channels. And how about the *superframe*? A superframe is 51 “26 multiframes” long, which is the same as 26 “51 multiframes.” It is the smallest unit that can contain either an integer number of 26 multiframes or an integer number of 51 multiframes, as 51×26 is the least common multiple of 51 and 26. The largest frame is the *hyperframe*, which is 2048 superframes long. This makes a hyperframe 3 hours 28 minutes 53.76 seconds long. It is the smallest time unit for certain functions of GSM related to frequency hopping and ciphering.

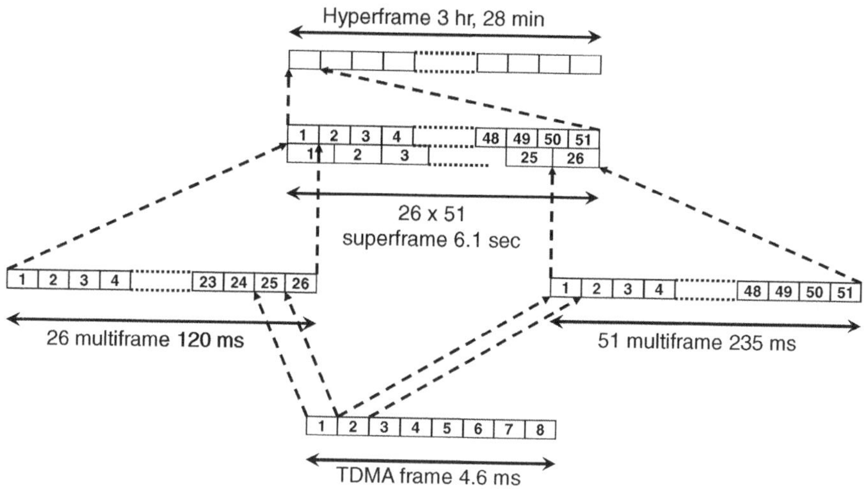

FIGURE 8.3 GSM frames.

8.1.1 Access Control

We now see how the mobile station can make use of various channels in the control multiframe to synchronize to a base station, obtain information about it, and possibly attempt to gain access. The base station may then assign a channel to the mobile station. We examine each of these steps in turn.

8.1.1.1 Synchronization The *frequency correction channel* (FCCH) and *synchronization channel* (SCH) are designed to help mobile stations synchronize to a base station. As can be seen in Figure 8.4, the FCCH is always transmitted in the time slot immediately before the SCH, and each subsequent FCCH comes eight time slots after the preceding SCH is transmitted.

The FCCH contains all 0's, resulting in a pure sine wave. Thus, a mobile station can scan through different frequencies at all times looking for FCCHs. When it hears an

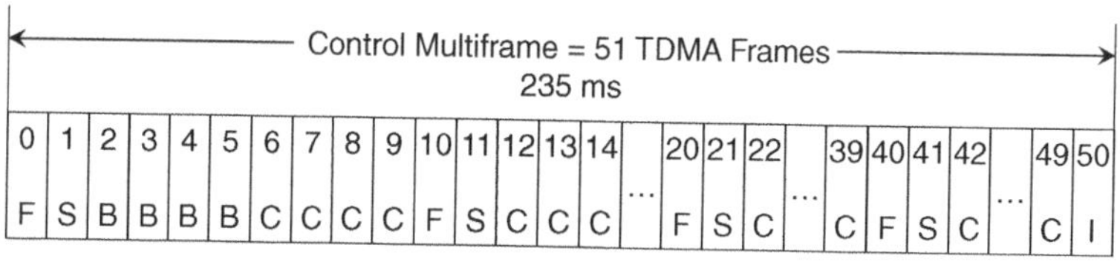

F : FCCH burst (BCH)
S : SCH burst (bch)
B : BCCH burst (bch)
C : PCH/AGCH burst (CCCH)
I : Idle

FIGURE 8.4 GSM control multiframe.

FCCH, it is able to make small adjustments in its internal frequency clock if necessary and also to have enough of an idea where the boundaries of the time slots are that it can then demodulate a subsequent SCH.

The SCH uses a special burst, different from a normal burst. It contains information that the mobile station can use so that the mobile station can fine-tune its knowledge of exactly where the slot boundaries are as well as where in the overall sequence of slots the base station is (as indicated in Figure 8.3, the current time slot could be anywhere within the $8 \times 26 \times 51 \times 2048$ time slots within the hyperframe, and the mobile station needs to know where it is; then it will know in which part of each cycle the base station currently is).

8.1.1.2 Acquiring Cell-Specific Information

Once synchronization is acquired, thanks to the FCCH and SCH, the mobile station is ready to move on to the BCCH. The base station regularly broadcasts information about itself on the *broadcast control channel* (BCCH). Such information includes:

- Information for cell selection, to assist mobile stations in selecting which cell(s) to access (or at least to try to access). Indeed, a mobile station may hear signals from multiple base stations and would apply certain criteria to decide which of these to attempt to access. Since the BCCH can be demodulated only after acquiring synchronization, the mobile station must first synchronize to each base station in turn.
- Information for idle mode functions. We discuss this in more detail in Section 11.1.4.
- Information for access. Information related to the scheduling of access attempts and repetitions are broadcast on the BCCH as part of what is sometimes called the *RACH control parameters*.
- Other miscellaneous information.

8.1.1.3 Accessing the Base Station on the Random Access Channel

After synchronization and acquiring the additional information it needs from the BCCH, the mobile station is finally ready to attempt access on the *random access channel* (RACH) if it so chooses. The frequency and location of the RACH can vary, and mobiles can obtain such information from the BCCH.

Like FCCH, SCH, BCCH, and PCH/AGCH, the RACH is a *common channel*. However, unlike these other channels, which are downlink common channels, the RACH is an uplink common channel. As the name implies, mobile stations are not assigned specific time slots but attempt to transmit on the RACH in a random access way. This makes sense, because the base station does not know when any particular mobile station may wish to access it. Certain time slots on certain frequency carriers are designated as RACH time slots and the mobile stations access RACH in a *slotted Aloha* manner (Section 7.1.1). It comes with a random backup upon occurrence of a collision, followed by attempts at retransmission.

8.1.1.4 Initial Channel Assignment If the mobile access request is successful and the base station (in conjunction with its BSC) admits the mobile, it indicates the initial channel assignment to the mobile station on the *paging channel/access grant channel* (PCH/AGCH). The PCH/AGCH could possibly be divided into different groups for groups of mobile stations; or a subchannel could be reserved only for channel assignment messages. Again, the mobile stations can determine such details from the BCCH.

8.1.2 Handoffs and Power Control

Intracell handoffs are performed in the BSC if it is a handoff between two base stations under the control of the same BSC, and in the MSC otherwise. Handoffs are of the mobile-assisted handoff variety, and measurements are transmitted by the users to their serving base stations every half-second.

The relevant measurements are the received signal strength (RXLEV), the received signal quality (RXQUAL), and the absolute distance between base station and user (DISTANCE). They are measured as follows:

- RXLEV is measured by the user terminal. What it measures is the signal strength on the broadcast control channel (BCCH) carrier, which is transmitted continuously by the base station on all time slots and without variations of the power level. This is measured by the user terminal from the serving cell and from the base stations in all adjacent cells (by tuning and listening to their BCCH carriers, too). The measurements are averaged over 15 s and quantized into 64 levels. The base stations are identified by base station identity codes (BSIC) on the BCCH. The monitoring of BCCH of adjacent base stations can be performed by an MS in a different time slot from either when it is transmitting or when it is receiving, as shown in Figure 8.5. RXLEV of user transmissions are also measured by base stations.

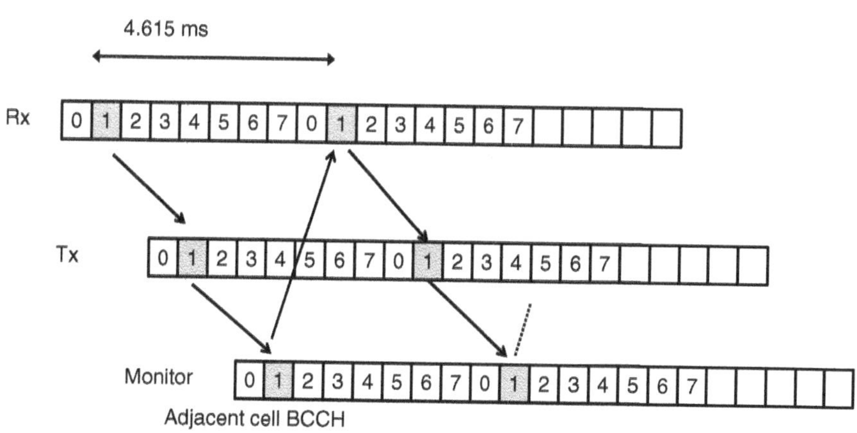

FIGURE 8.5 Different time slots for transmitting, receiving, and listening to the BCCH.

- RXQUAL is obtained from estimation of the chip error rate, which is the BER before channel decoding, using information from the Viterbi channel equalizer or convolutional decoder. It is quantized into eight levels. It is measured by both the base station and the user on their communication link.
- DISTANCE is measured by looking at the "timing advance" parameter (can measure 0 to 70 km with an accuracy of 1 km). It is measured by the base station.

Power control in GSM uses only RXLEV (the exact strategy is determined by the network operator), whereas handoff uses all three. A parameter known as the *power budget parameter* allows the system to take into account power control levels, user maximum power capability, and so on.

8.1.3 Physical Layer Aspects

GSM uses Gaussian minimum shift keying (GMSK; Section 6.2.3) for its modulation scheme. GMSK is more spectrally efficient than MSK. As mentioned earlier, the signaling rate is 270.8333 kHz. This comes from the use of the symbol period $T = 48/13\ \mu s$.

From the RF perspective, there is significant overlap between adjacent frequency channels (that are only separated from each other by 200 kHz), with the modulated carrier amplitude down by only about 40 dB at the center frequency of an immediately adjacent carrier. However, adjacent frequency channels are not used in the same cell or even in immediately adjacent cells (because of the frequency reuse factor).

GPRS (see Section 12.2) uses the same physical layer as GSM, and GPRS data are transported on the *packet data channel* (PDCH [2]), which is mapped to time slots as follows: In each 52-multiframe, there are 12 *radio blocks*. Each radio block consists of four time slots in four consecutive TDMA frames (e.g., time slot 3 in each of four consecutive TDMA frames). There are also four idle frames. The radio blocks can be allocated dynamically to mobile stations, on a block-by-block basis. Thus, multiple mobile stations can share the same PDCH (each is allocated different blocks in the PDCH). Higher rates can be achieved by allocating multiple time slots to a user. *Enhanced data rate for GSM evolution* (EDGE, a GPRS enhancement), on the other hand, uses a new modulation scheme, 8-PSK, instead of GMSK, in order to achieve higher data rates than GPRS. Like GPRS, EDGE reuses most of the GSM framework, including the 200-kHz radio carriers, and the same network architecture. Only over the air is the difference seen, and mostly in the modulation. There are a few small changes in the RLC and MAC layers as well. The constellation for 8-PSK can be seen on the right side of Figure 1.9.

8.2 IS-95 CDMA

It must be noted that IS-95 CDMA systems reuse frequencies in every cell [i.e., the channel reuse is 1 (the most aggressive possible)], nor are the signals separated in time (as in a TDMA system). Thus, the proper use of codes for separating the various signals is essential for the operation of a CDMA system. In Section 6.4.2.1 we have seen how the autocorrelation properties of the PN sequences can be exploited to distinguish signals that are arriving at different time offsets.

IS-95 can be confusing at first to newcomers, because it uses two types of codes: *channelization codes* and *scrambling codes*. Keeping Table 8.1 in mind, IS-95 uses orthogonal Walsh codes for channelization (the downlink separation of mobile stations in the same cell), and scrambling codes (PN sequences of the sort we were introduced to in Section 6.4.1) for the other four separation problems. In particular, two PN sequences are used:

- The *long PN sequence*, whose period is $2^{42} - 1$ chips
- The *short PN sequence*, whose period is $2^{15} - 1$ chips

Different MSs (whether in the same cell or different cells) use different offsets of the long PN sequence to provide inter-MS separation within a cell and across cells for uplink transmissions. Different base stations use different offsets of the short PN sequence to provide inter-BS separation in the downlink.

8.2.1 Downlink Separation of Base Stations

We consider the downlink, where code-division multiplexing (CDM) is used. In the case of signals arriving from different base stations, a mobile can exploit the autocorrelation property of PN sequences to track its desired base station and simultaneously to suppress the interference from other base stations. In particular, all base stations share the same *short PN sequence*, a $2^{15} - 1$ chip sequence, and they transmit at different offsets of this sequence. A minimum separation of 64 chips between allowable offsets provides 512 different offsets that base stations can use. Thus, in a network with more than 512 base stations, offsets will eventually have to be reused. However, the reuse distance would be so large that no problems result in practice. Figure 8.6

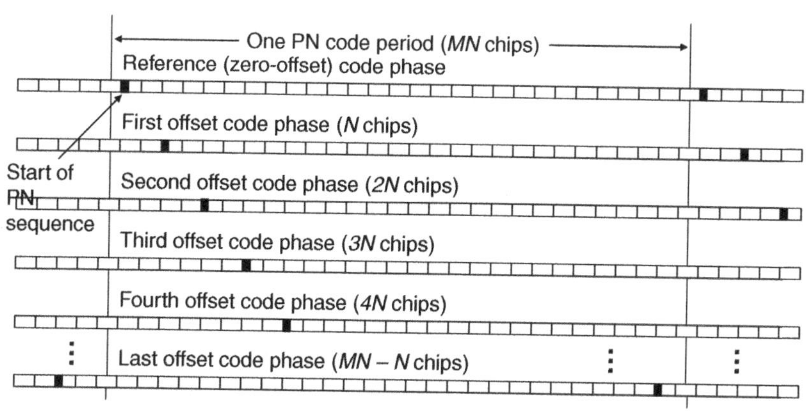

FIGURE 8.6 IS-95 base station offsets.

shows how different base station offsets are positioned relative to a particular zero-offset reference. Now, imagine if the base stations are not precisely synchronized (i.e., if their clocks are not precisely in step with one another). The minimum separation of 64 chips might then be reduced, resulting in higher error rates. Thus, it is very important in IS-95 systems that the base stations be precisely synchronized with one another. This is often accomplished through the use of GPS.

8.2.2 Single Base Station Downlink to Multiple Mobile Stations

We "zoom in" to the communication between a single base station and its mobile stations (i.e., the MSs communicating with it). The base station transmits signals meant for specific mobile stations as well as common signals meant for all its mobile stations (an example of such a common signal would be the pilot signal that all mobile stations can use). From the base station, all these signals are transmitted at the same time on the same frequency. The concept of separating these different signals is known as *channelization*; that is, the base station creates different *channels* for each signal, so mobile stations can separate out the desired signals from the combination of signals they receive. In particular, we are talking about *downlink channelization*. Moreover, since all these signals are transmitted from the same base station, they are using the same offset of the short PN sequence, so another mechanism is needed to separate out the different signals in this bundle of transmissions from the base station (i.e., another mechanism is needed for channelization). Indeed, there is another set of codes that can be (and is) used for this purpose. These are the Walsh codes.

Walsh codes are a kind of orthogonal code that can be constructed easily. Usually, the Walsh codes used in a system (e.g., IS-95) are all the same length, K (a power of 2). For example, the Walsh codes used in IS-95 are all 64-element binary sequences (here we use the terms *Walsh code* and *Walsh sequence* interchangeably). There are K Walsh sequences of length K, and we may write them as W_i^K, where $i = 0, \ldots, K-1$, where i is the number of zero crossings of the Walsh code (e.g., W_3^K would have three zero crossings). Any two of the codes W_i^K are orthogonal; that is, if we represent the Walsh code binary sequences as sequences of 1 and -1, the inner product of any two Walsh codes of length K is 0, except where the two codes are the same code. Thus, letting $W_i^K(k)$ represent the kth element of the Walsh sequence W_i^K and applying (6.32), we have

$$\sum_{k=0}^{K-1} W_i^K(k) W_j^K(k) = \begin{cases} 0 & \text{if } i \neq j \\ K & \text{if } i = j \end{cases} \tag{8.1}$$

Table 8.2 lists the codes for $K = 8$. For convenience, rather than representing the binary values as 1 and -1, we map $1 \rightarrow 0$ and $-1 \rightarrow 1$, so we can simply represent them as sequences of 0's and 1's (see the discussion in Section 6.1.5.1). In a real system implementation, to get the orthogonality by multiplication, of course, the Walsh codes should take opposite values, such as 1 and -1. The orthogonality of any pair of codes may be verified for Table 8.2.

TABLE 8.2 Walsh Codes of Length 8

Designation	Sequence
W_0^8	00000000
W_1^8	00001111
W_2^8	00111100
W_3^8	00110011
W_4^8	01100110
W_5^8	01101001
W_6^8	01011010
W_7^8	01010101

Thus, ideally, the channelization of different signals by the use of orthogonal Walsh codes is perfect, whereas the separation of different signals through the use of different offsets of a PN sequence is imperfect and still suffers from a small amount of interference, albeit suppressed by the processing gain. What do we mean by "ideally"? Notice that in (8.1) the Walsh codes need to be synchronized. The inner product of two Walsh codes that are not synchronized is, in general, not equal to zero. However, since the signals are all transmitted by the same base station, they can be transmitted by the base station at the exactly the same time. The signal received at the mobile device is a linear sum of the multiple signals, but the desired signal can be extracted by taking the inner product (correlating) with the specific Walsh code. So, by the orthogonality relationship, as expressed in (8.1), the other signals are "killed," leaving the desired signal only. In practice, though, there is multipath, so there will still be some small amount of interference because the multipath removes the ideal case of perfectly aligned Walsh codes.

Why is channelization done by using Walsh codes, whereas the separation between signals from different base stations is done using different offsets of the short PN code? First, orthogonal separation is better than the imperfect separation obtained from using different offsets of the same PN sequence. Second, because the signals are all coming from the same transmitter (the base station), it can transmit them at the same time, and most will be received at the same time (except when there is multipath, as discussed earlier). In the case of separation of base stations, even if base stations coordinate their transmissions, the signals are coming from multiple sources, so depending on where the receiver (mobile station) is, the signals will arrive at different times, making it challenging, if not practically impossible, to exploit the orthogonality properties of Walsh codes. Figure 8.7 shows multiple downlink channels that are separated orthogonally in the BS in the downlink, using different Walsh codes (and they are also spread up to 1.2288 MHz in the I/Q modulations, with the appropriate BS-specific short PN code offset).

8.2.3 Downlink Channels

The downlink traffic channel is processed as shown in Figure 8.8. Compared to the GSM case (as shown in Figure 8.2), the data bits in IS-95 are all treated equally

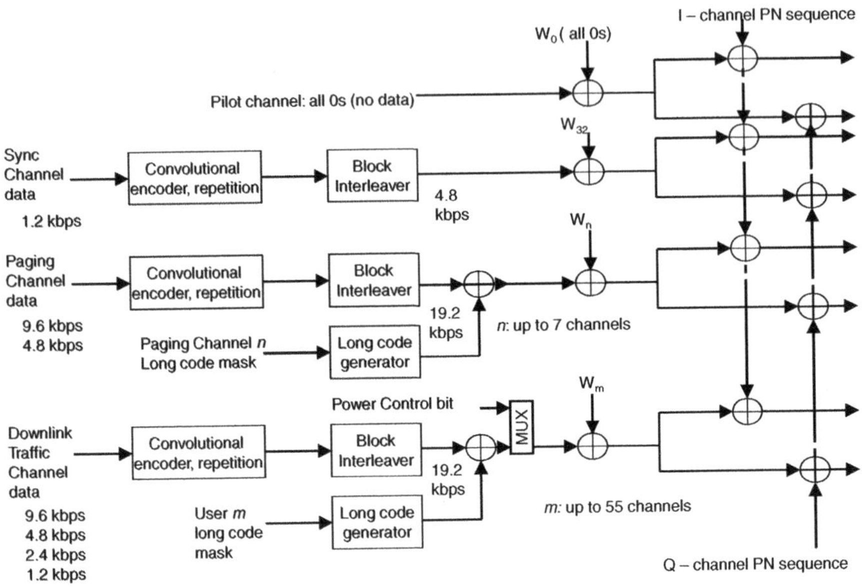

FIGURE 8.7 IS-95 downlink channels separated by Walsh codes.

as far as application of FEC is concerned. They are all passed through a rate-1/2 convolutional code of constraint length 9. Even though the transmitter may start with one of four rates (0.8 to 8 kbps, which go up slightly to 1.2 to 9.6 kbps, with the addition of CRC and 8 "tail bits" for the convolutional encoder), a simple repetition code follows the convolutional code to bring the rate up to 19.2 kbps no matter which

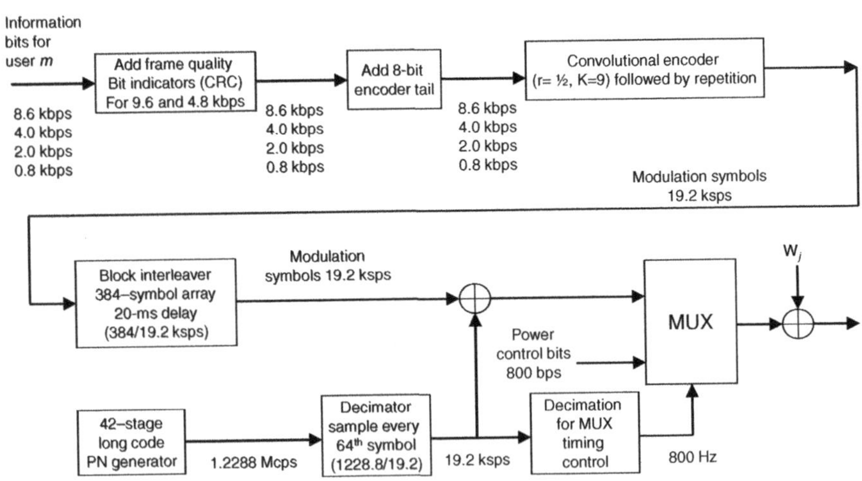

FIGURE 8.8 IS-95 downlink traffic channel.

rate it started with. This is followed by an interleaver, which operates on 384 symbols at a time. At 19.2 kbps, it takes 20 ms to gather 384 symbols, so the interleaver delay is 20 ms.

In addition to the downlink traffic channels, there are some control channels, as can be seen in Figure 8.7, where we see the pilot channel, sync channel, and paging channels. Since there can be up to seven paging channels, making a total of 9 control channels, each using one Walsh code, that leaves $64 - 9 = 55$ Walsh codes for downlink traffic channels. We discuss the use of the pilot, sync, and paging channels in Section 8.2.7. W_0 (on the sync channel) is basically all zeros, transmitted at higher power than the other channels. Thus, it is basically an unmodulated channel that is very helpful for the rake receiver in the mobile station in finding and tracking the strongest-arriving paths from the base station. Even though there is an in-phase/quadrature modulation at the end, the same PN sequence, at the same offset, is used for both I and Q, providing a form of quadrature diversity.

8.2.4 Uplink Separation of Mobile Stations

Each mobile station uses a different offset of the long PN sequence (Figure 8.9). The long code "mask," which corresponds to the offset, is a function of the mobile identity, making it unique. This allows the base station to extract the mobile station's signals from the rest of the mobile stations' signals, using the autocorrelation properties of the long PN sequence. When we say "the rest of the mobile stations' signals," we include both the other mobile stations in the same cell, and other mobiles in adjacent cells that are transmitting to the other base stations.

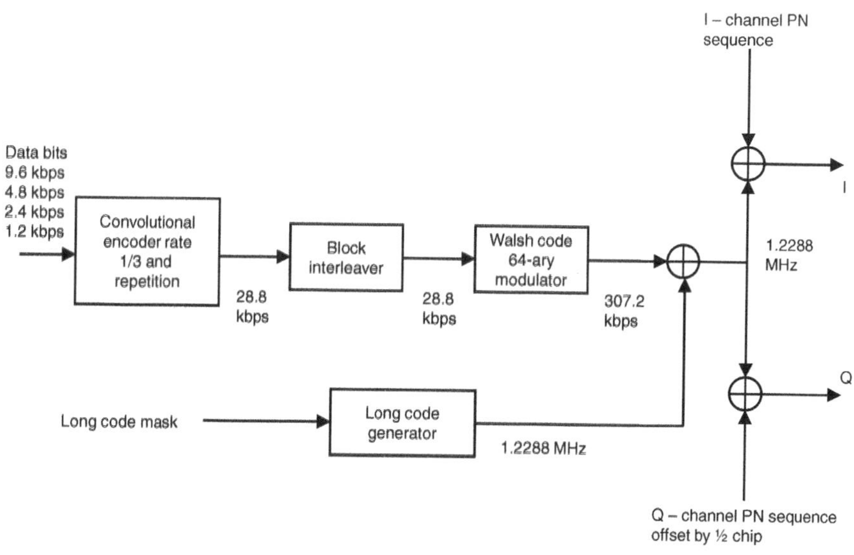

FIGURE 8.9 IS-95 uplink traffic.

Note that in the I/Q modulator, the quadrature chip sequence is delayed by half a chip, providing a form of OQPSK. However, as in the downlink, the same PN sequence is for use with both I and Q (with just the 1/2-chip offset).

8.2.5 Uplink Traffic Channel

The uplink traffic channel does not include a pilot signal, which is one reason that it needs a stronger convolutional code. Also, Walsh codes are used on the uplink, but for a completely different purpose than in the downlink. Here, they are used as a form of 64-ary orthogonal modulation (i.e., each unique sequence of 6 bits out of the interleaver is mapped to a different one of length $2^6 = 64$ Walsh sequences). Thus, the rate goes up from 28.8 kbps to $28.8 \times 64/6 = 307$ kbps.

8.2.6 Separation of the Multipath

Last but not least, we come to the fifth of our separation problems. Considering a single link between MS and BS or the other way around, we can again use the autocorrelation properties of PN sequences to resolve the multipath, by using a rake receiver, as discussed in Section 6.4.2.1. In IS-95, the mobile station implements a rake receiver with three fingers, whereas the base station implements a rake receiver with four fingers. In addition to the fingers tuned to each strong path for the arriving signal, one finger is used to search for new positions of strong arriving signals.

8.2.7 Access Control

8.2.7.1 Synchronization and Acquiring Cell-Specific Information Synchronization is a two-stage process:

- Pilot PN sequence synchronization
- Obtaining additional synchronization-related information from the sync channel

The pilot channel, shown in Figure 8.8, is a sequence of all zeros that is modulated by the 0 Walsh function (also consisting of all zeros). This is then multiplied by the short PN code. Thus, the pilot channel can be said to be an *unmodulated channel*, which simply gives the short PN code offset. Furthermore, it is not power controlled (it is transmitted at a constant power). Since the pilot channel is unmodulated, the mobile station can use its rake receiver to receiver the pilot channel relatively easily. After tuning a finger of the rake receiver to the pilot channel, the mobile station would have information about the short PN code offset used, plus timing and phase information on the chips so that it can perform coherent demodulation. Furthermore, in a multipath environment, the tuning finger of the rake receiver will tune to a few strongest paths, allowing other fingers to tune to these paths, and preparing the rake receiver for further reception of signals from the base station via the multiple paths.

The E_c/I_0 of the pilot channel received on the multiple paths is used for soft handoff purposes (see Section 8.2.8).

Once the mobile station has acquired the pilot, as just discussed, the sync channel can be acquired. This is quite straightforward, as the sync channel is aligned with the pilot PN sequence. Like the pilot channel, the sync channel is a broadcast channel. However, it carries messages, including information that mobile stations need for the rest of the synchronization process, and more. These include:

- Identification information (of the base station and network)
- Pilot PN sequence offset
- Various timing information, including the long code timing information
- Paging channel data rate

With the base station and network IDs, the mobile station can decide whether it should use this network or look for another network. The various timing information can be used for additional synchronization (e.g., of frames and slots, which are timing units on some channels), and long code PN sychronization. After obtaining information from the sync channel, the mobile station is then ready to monitor the paging channel, since it has the paging channel data rate from the sync channel. It obtains additional cell-specific information from the paging channel. Although there may be multiple paging channels, the mobile station monitors only one. The monitored paging channel is selected through the use of a hashing algorithm based on the mobile ID number. Various information is obtained from the paging channel: for example, broadcast information such as:

- System parameters such as the number of paging channels
- Access channel parameters: needed for accessing the base station on the random access channel
- List of neighboring base stations' pilot short PN code offset

The paging channel is also used for mobile-specific messages, such as paging when there is an incoming call, and channel assignment.

8.2.7.2 Accessing the Base Station on the Random Access Channel

The random access channel (RACH) is used by a mobile station to access a base station when there are no dedicated resources for the mobile station. It is a contention-based channel, and IS-95 specifies a procedure for sending *access probes* in sequences of increasing power (until the base station can receive the RACH message and acknowledge it). The values of *initial power* and *power increments*, if very large, would reduce the number of probes needed, but would result in additional interference with other uplink transmissions. If initial power and power increments are very small, however, more probes would be needed, each of which causes interference to other uplink transmissions (and also, increases the latency).

8.2.8 Soft Handoffs and Power Control

We introduced the difference between hard and soft handoffs in Section 7.2.2. IS-95 introduces the concept of *channel sets* for soft handoff purposes [4]. Each mobile station maintains lists of channels (each channel being specified by PN offset and frequencies) including the *active set*, the *candidate set*, the *neighbor set*, and the *remaining set*. The active set includes the currently used channel or channels and contains two or more members during a soft handoff and one member at all other times. The candidate set includes channels that are good candidates to be promoted to the active set. Thus, new channels are chosen for soft handoff from the candidate set. The neighbor set is the set of channels that are reasonably strong but that do not meet the criteria to be included in the active and candidate sets. The remaining set includes all other channels. There are rules by which channels may move from one set to another.

Three handoff procedures are possible in the traffic channel state (regular communication link established):

- *CDMA-to-CDMA soft handoff*. This is between CDMA channels on identical frequencies.
- *CDMA-to-CDMA hard handoff*. This is between CDMA channels on different frequencies.
- *CDMA-to-analog hard handoff*. This is when the handoff is to one of the coexisting analog channels.

Both the CDMA-to-CDMA soft and hard handoffs are normally initiated by the user. The user makes the measurements on the pilot channel without changing frequencies. (Each base station transmits a pilot channel continuously for every CDMA frequency it supports. There could be pilot channels at other frequencies.) The user searches for usable multipath components in ranges of PN offsets (known as *search windows*) specified by the base station. When a user detects a pilot of sufficient strength, not one of its current downlink channels (and there could be several; the user provides diversity combining in that case), the measurement is sent to the serving base station, which can then make the appropriate decision, perhaps to hand off. The information sent to the base station by the user includes the following:

- *Strength of pilot*. This is computed by adding the ratios of the pilot energy received per chip to the total interference-plus-noise spectral density received from at most k usable multipath components, where the number k is the number of correlators that is implemented by the user for demodulating:

$$\text{strength of pilot} = \left(\frac{E_C}{I_0}\right)_1 + \cdots + \left(\frac{E_C}{I_0}\right)_k$$

- *Handoff drop timer*. The user maintains a handoff drop timer for each pilot of the channels it is using and good alternatives. The timer is started when the pilot strength drops below a threshold. The timer is reset when the signal strength rises above the threshold.

- *PN phase measurements*. These might be used by the base station to make an estimate of propagation delay to the user, for faster uplink traffic channel acquisition time (by more intelligent setting of its correlator delays, for example).

8.2.8.1 Power Control We introduced the general issues and principles of power control in Section 7.3. In IS-95, the rate at which the uplink power control directives are issued by the base station to each MS is 800 Hz (i.e., 800 times per second). The power control step size is typically 1 dB. The dynamic range of power control on the uplink may be up to 70 dB, whereas it may be only about 20 dB on the downlink.

8.2.8.2 Power Control in a Soft Handoff What happens to the various control functions in the system during a soft handoff? What happens to power control is of particular interest. Normally, when not in soft handoff, a user communicates with one base station, and that is the base station that gives it power control instructions, to increase or decrease transmitted power. In soft handoff, one possibility is that the different base stations transmit power control instructions independently, requiring that the user arbitrate between the instructions if they are different. Another possibility is that they all agree upon and transmit the same power control instructions.

In IS-95, something in between the aforementioned possibilities is used. All downlink traffic channels associated with pilots in the user's active set carry the same modulation symbols, except for the power control subchannel. Sets of downlink traffic channels carry identical power control information, but each set could be different from the others. The user performs diversity combining on each set (because the information is identical), and looks at all the resulting power control bits obtained from all the sets. If even one of them instructs the user to decrease the power transmitted, it obeys and decreases it. Otherwise, it increases it. This is because if there is at least one instruction to decrease power, there is at least one base station that is able to provide coverage.

8.2.8.3 Idle-Mode Handoff Even when idle, the mobile device is always associated with a base station. After the mobile device is turned on, it acquires the pilot signal and timing from the best base station around. It then goes into a user idle state, during which it monitors pilot channel signals continuously. If the user detects a sufficiently strong pilot channel signal different from the current base station's, an idle handoff occurs. In the idle state the mobile station can communicate with a base station over a paging channel.

8.3 IEEE 802.11 WiFi

The current revision of IEEE 802.11 is 802.11-2007, published in 2007. However, the IEEE 802.11 family of systems used to consist of the 1999 revision (of the

TABLE 8.3 Some Members of the IEEEE 802.11 Family

Family Member	Features
802.11a	OFDM, up to 54 mbps, 5-GHz band
802.11b	CCK modulation, up to 11 mbps, 2.4-GHz band; cheap, popular, especially before 802.11g
802.11e	QoS extensions
802.11f	Access point interoperability
802.11g	OFDM, up to 54 mbps, 2.4.-GHz band, backwardly compatible with 802.11b
802.11i	Much better security than WEP
802.11n	OFDM, MIMO, up to 600 mbps
802.11s	Mesh networks

original baseline standard from 1997) and some amendments: 802.11a, 802.11b, and so on (Table 8.3; for an explanation on revisions, amendments, etc., in the standards, see Section 17.2.6, and Section 17.2.6.1 specifically for revisions and amendments pertaining to 802.11). Many of these amendments are now part of 802.11-2007. Nevertheless, people still popularly refer to aspects of 802.11 covered by these amendments by their amendment names [e.g., 802.11a rather than the relevant section(s) in 802.11-2007].

Because 802.11 is designed to operate in unlicensed spectrum, the physical layer needs to use spread spectrum for its interference reduction properties and to spread out the transmitted signal energy to meet regulatory requirements (unlicensed spectrum usage rules typically impose a maximum EIRP per hertz; since the constraint is in EIRP per hertz, more power can be transmitted with spread-spectrum transmission). However, 802.11 is a good example of a system where spread spectrum is used just for these purposes and *not* for medium access, unlike the CDMA cellular systems.

IEEE 802.11 specifies only the link layer and below (Figure 8.10). In a sense, 802.11 is therefore more purely a wireless access technology than cellular systems such as GSM, which are more completely specified systems. The mobile devices are called *mobile stations*.

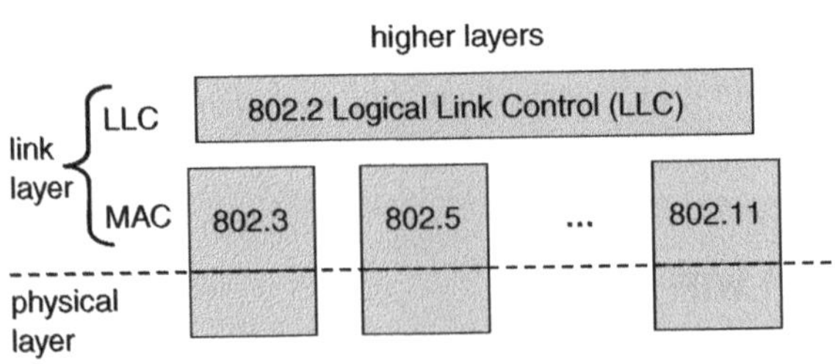

FIGURE 8.10 IEEE 802.11 specifies only lower layers of the protocol stack.

8.3.1 LAN Concepts

A fundamental building block in 802.11-based networks is the *basic service set* (BSS) concept. There are actually two kinds of BSSs, so in a sense there are two fundamental building blocks.

- The *independent BSS* (IBSS), also known as the *ad hoc network* of 802.11, comprises a group of stations that communicate with each other such that each is within radio range of all of the others and can transmit and receive packets directly to/from each of the others without needing the assistance of a central controller. The IBSS is also independent of other networks (i.e., it is not connected to other networks, such as the Internet). It may be useful, for example, where a small group of devices that wish to communicate with one another are out in the desert where there is no fixed infrastructure. This ad hoc network should not be confused with the concept of *mobile ad hoc networks* discussed in Section 13.3.
- The *infrastructure BSS* comprises a group of stations together with a special station called an *access point* (AP). Generally, even if two non-AP stations in an infrastructure BSS are within radio range of each other, they are not allowed to talk to each other directly, but always go through the AP. One benefit of this is that it helps with power savings, since stations can go to sleep and the AP can buffer packets for them. The AP performs this "layer 2 forwarding" function, also known as *bridging*.

In both cases, all the stations (as well as the AP in the case of the infrastructure BSS) transmit and receive in the same frequency band. In the case where integration with another network (such as a wired network, or the Internet) is desired, the connection or integration point to the other network is called a *portal*.

IEEE 802.11 allows for three concepts of a local area network (see Section 10.2.1 for more on LANs):

- The independent basic service set (IBSS) concept, where a group of mobiles independently form a BSS. All mobiles are peers. This is shown at the top left of Figure 8.11.
- Basic infrastructure BSS with one AP, where the AP plays the role of AP and portal, as shown at the bottom left of Figure 8.11. This is a common scenario in home WiFi networks, where the single AP is also the portal to the wired Internet. (In this common scenario, the same device often acts as AP, portal, switch, IP router, and DSL modem or cable modem!)
- The extended service set (ESS) concept, where there are multiple infrastructure BSSs and a *distribution system* (DS) that connects the BSSs. The DS is an abstract concept that in practice is often implemented as a wired network, (e.g., Ethernet), but a wireless DS based on 802.11 is also posible. The multiple BSSs and the DS in an ESS together form one big LAN. Thus, if a router is present in the LAN, each mobile station in the ESS is just one hop away from the router as

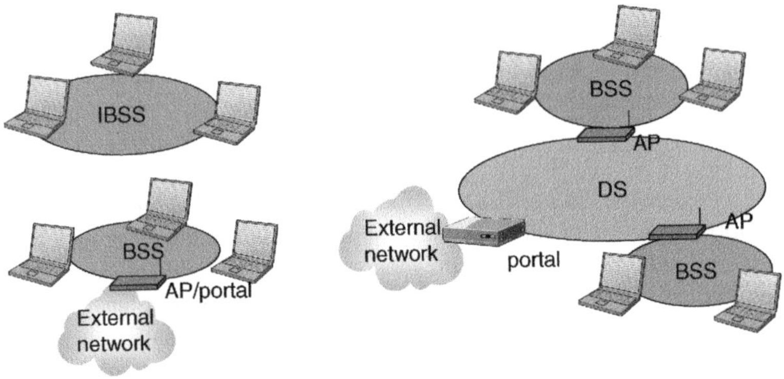

FIGURE 8.11 LAN concepts in IEEE 802.11.

far as IP is concerned. Movement within the ESS is transparent to IP, with no IP address change needed (as is the case when there is *layer 3 mobility*). Instead, the WLAN handles any necessary rerouting of packets internally. Hence, movement within the ESS is also known as *layer 2 mobility*. An ESS may or may not connect to other networks, such as the Internet. If the ESS is to be connected to other networks, the integration point is the portal. The ESS is shown on the right of Figure 8.11.

The infrastructure BSS with one AP can be viewed as a special case of the ESS where there is only one AP, the DS has been collapsed, and the single AP and portal are merged into one device. As for the regular ESS, 802.11 specifies the functions, frame structures, and messages that allow it to behave as one big LAN even though it comprises multiple BSSs connected by a DS. An interesting case occurs when the DS is also implemented using wireless (802.11). Then, in the case of communications from a mobile station in one BSS to a mobile station in a different BSS (both within the same ESS), there are four relevant MAC addresses: the address of the wireless interface of each mobile station and that of each of the two APs involved, as shown in Figure 8.12. IEEE 802.11 provides for the specification of all four MAC addresses, along with two flags, "From DS" and "To DS." Notice how the flags are set and which MAC addresses are specified as the frame moves from the sending mobile station to the receiving mobile station.

8.3.2 IEEE 802.11 MAC

As discussed in Section 7.1.1.1, there are differences between the wireless and wired environments that preclude the use of CSMA/CD (used in Ethernet) in 802.11. Instead, 802.11's MAC is based on *carrier sense multiple access with collision avoidance* (CSMA/CA). Built on the foundation of CSMA/CA are two modes of the MAC protocol, *distributed coordination function* (DCF) and *point coordination function* (PCF).

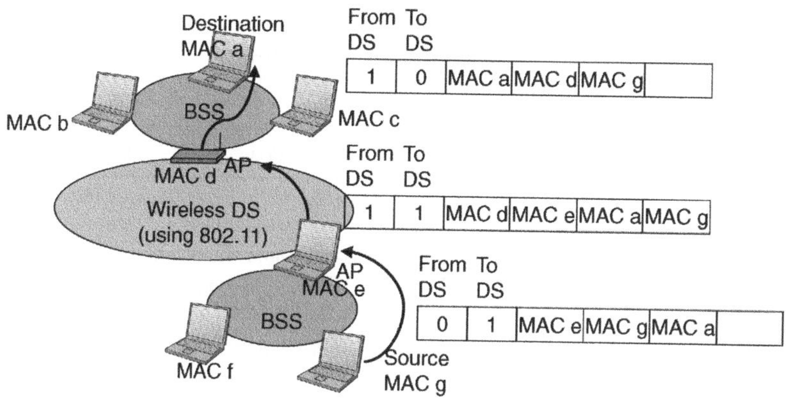

FIGURE 8.12 Up to four MAC addresses may be found in an 802.11 frame.

Since there is a chance of collisions, and because of the relative high error rates on wireless channels, 802.11 requires that an acknowledgment (ACK) frame be sent after each (unicast) frame is received. But to begin sending, it first needs to access the network. It uses its distributed MAC protocol (local decisions are made at each mobile rather than coordinated through a central point such as an AP) for channel access. To understand the channel access scheme, we first introduce some of the key variables used in the scheme and then discuss the scheme in Section 8.3.2.1.

Network Allocation Vector. For the distributed MAC protocol to work, the mobile needs continually to be aware of whether the medium is idle or busy. This might waste battery power if the mobiles need to be listening to the medium frequently to detect if it is idle or busy. Instead, the *network allocation vector* (NAV) is used in this case to help save power. The NAV is an estimate of the amount of time it will take for the current frame to be transmitted. Under some conditions, the *duration/ID* field of an 802.11 packet contains a duration value that other mobile stations can use in setting their NAV. When a mobile station receives an 802.11 packet whose *duration/ID* field satisfies these conditions (so the mobile station knows that the field contains a duration value), it updates the NAV. The duration/ID field is part of the 802.11 header of every transmission, so when idle mobiles are monitoring the medium, if they can hear the duration/ID field of packets in ongoing transmissions and these indicate a duration, they know that the medium will be busy for at least that amount of time, and they can stop listening during that time.

Contention Window and Backoff Counter. The contention window is a key parameter for the exponential backup procedure that is a subprocedure of the channel access procedure. The backoff counter is another key variable for the exponential backup procedure.

8.3.2.1 Channel Access Procedure

Now, when a packet arrives at the transmitter in a mobile, it goes through the following procedure (Figure 8.13):

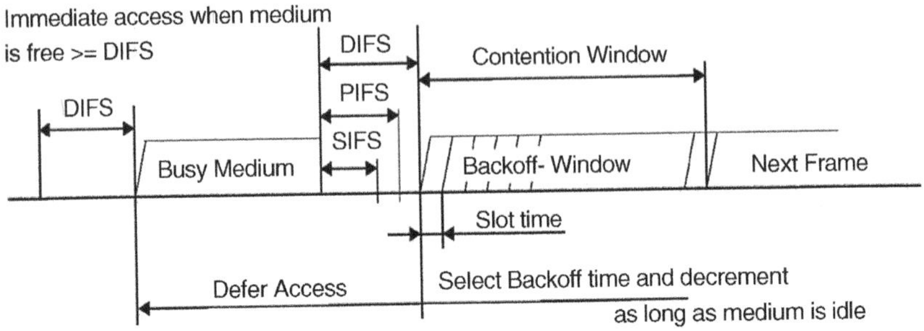

FIGURE 8.13 Base MAC protocol of 802.11. (From IEEE 802.11-2007 [3]; copyright © 2007 by IEEE, reprinted with permission.)

1. If the medium has been idle for at least the required IFS (this has to do with the prioritization mechanism that we discuss later), it proceeds to step 2. Otherwise, the medium is busy or it is idle, but for a shorter time than the required IFS; the mobile proceeds to step 3.
2. It transmits immediately. After transmitting, if this was a unicast transmission, it waits for the ACK from the receiving station. If one is received, it is done. If one is not received, it needs to try again, and proceeds to step 3.
3. The mobile increases the size of its contention window according to a predetermined amount (i.e., not a random increase) specified in the standard. The mobile needs to wait for the medium to be idle for the required IFS, but then instead of transmitting immediately (as was the case in step 2), it waits a random amount of time (from the backoff procedure). This random amount of time is given by the backoff counter, which is drawn from a uniform distribution between zero and the size of the contention window (thus, it would tend to be larger when the contention window is larger). Only when this additional time is over and the channel is idle the entire time does it go to step 5. Otherwise, if the channel becomes busy during the wait, the mobile goes to step 4. A main reason for the random wait is to avoid a situation in which multiple stations are all waiting to transmit, they all wait for the required IFS, and then all "pounce" and begin transmitting at the same time.
4. The backoff procedure is suspended, to be continued with the same backoff after the channel becomes idle again for another DIFS interval.
5. It transmits immediately. After transmitting, if this was a unicast transmission, it waits for the ACK from the receiving station. If one is received, we move to step 6. Otherwise, it needs to retransmit, and goes back to step 3. Notice the difference between this step (step 5) and step 2 earlier: If step 5 is successful, we proceed to step 6, whereas if step 2 is successful, we are done.
6. The contention window is reset to the minimum value. Then, another random backoff is performed, for fairness, so a station that has just transmitted something backs off and gives others a chance to transmit. Of course, once this backoff is

completed, the station resets, so the next packet to be transmitted gets handled as in step 1 again.

We see the need for ACKs in steps 2 and 5 above. Typically, ACKs have high priority (with the shortest IFS; see Section 8.3.2.2). The ACK may be a separate control frame from the receiver, but often it is piggybacked on a data frame for reduced overhead.

CSMA/CA tries further to avoid collisions using the following methods:

- A RTS/CTS scheme may be used as a handshake, allowing a transmitter–receiver pair to agree that the transmitter will be sending a data packet. The request-to-send (RTS) control message is sent when the transmitter is ready to transmit some data. Only if the receiver replies with a clear-to-send (CTS) identifying that particular sender does the sender transmit the data. We will see an example of the use of RTS and CTS in Section 8.3.2.3.
- The *network allocation vector* (NAV): Each station maintains a timer, called the NAV, that indicates when it will not transmit data, whether or not it senses that the wireless medium is busy at that time. The NAV is set based on information that the mobile station may receive (e.g., from an access point seizing control of the medium during the PCF) from another mobile station transmitting an RTS, or from another mobile station transmitting a CTS. Having the NAV in each mobile station reduces the probability of collisions occurring.

Since RTS/CTS is overhead that is especially significant for smaller packets, it is typically used only for packets larger than a specified threshold, if RTS/CTS is even turned on at all.

The DCF takes CSMA/CA and adds a little prioritization. PCF then sits on top of DCF in the sense that it makes use of features of DCF to allow a point coordinator to "seize control of the medium." First, we consider DCF and then briefly describe PCF.

8.3.2.2 Interframe Spacings The prioritization in DCF enabled by the interframe spacing concept. The main idea is that once the wireless medium is free, different priority traffic would have to wait different lengths of time before attempting to transmit. Thus, higher-priority traffic would have a shorter wait time. The wait time is known as *interframe spacing* (IFS), and there are a few IFSs defined in 802.11:

- *Short IFS (SIFS)*. These shortest IFSs are used when a frame has been sent and an immediate response is expected (e.g., ACK, or a frame has been unicast, or CTS, if an RTS has been sent).
- *PCF IFS (PIFS)*. The PIFS is the second shortest IFS. It is used by the point coordinator (when the AP takes that role) to "seize control of the medium." Except for outstanding responses that use SIFS, no other transmissions will wait for less time than the point coordinator waiting for PIFS, so it can transmit

before them. Once it starts transmitting, the other mobiles cannot transmit freely (because they wait for DIFS), except when allowed to by the point coordinator.

- *DCF IFS (DIFS)*. The DIFS is the normal wait time for mobiles trying to transmit in DCF mode.
- *AIFS*. Introduced with 802.11e, we discuss AIFS briefly in Section 11.3.3.1.

Pointed Coordinated Function (PCF). The 802.11 WLAN can alternate between periods of using just DCF and periods of using PCF. In a sense, DCF is never turned off, but the access point, as the point coordinator, implements PCF on top of DCF. It does this by using the PIFS to seize control of the medium and thus switch to PCF and control the transmissions while in the PCF mode. In the PCF mode, the point coordinator will use polling to find out which mobile stations have data to transmit, and then coordinate the transmissions. PCF has not been implemented widely [1].

8.3.2.3 Example: Timing of Transmissions with RTS and CTS and SIFS

Figure 8.14 shows a case where RTS and CTS are used, and where the SIFS is used. We note the following points:

- The wait before the RTS is DIFS, not SIFS. As mentioned in Section 8.3.2.2, the SIFS is used for a "CTS if an RTS has been sent." This does not mean that the RTS enjoys the same privilege. This makes sense because the sending of the RTS is not as time-critical as the sending of a CTS once a corresponding RTS has been sent.
- After the CTS, the source only needs to wait for the SIFS before transmitting data. So, once an RTS has been transmitted and received successfully, that

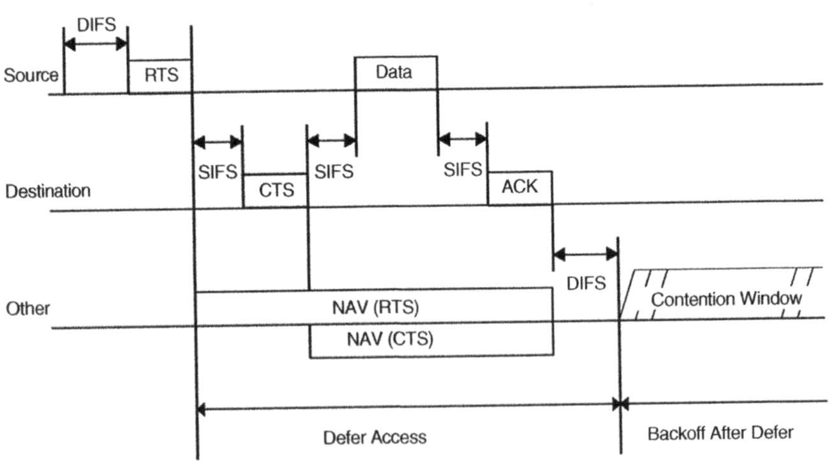

FIGURE 8.14 Timing diagram of RTS and CTS and SIFS. (From IEEE 802.11-2007 [3]; copyright © 2007 by IEEE, reprinted with permission.)

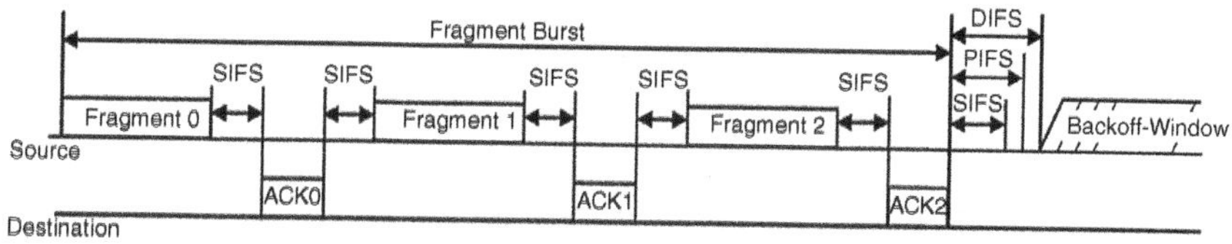

FIGURE 8.15 Timing diagram when multiple fragments are transmitted. (From IEEE 802.11-2007 [3]; copyright © 2007 by IEEE, reprinted with permission.)

transmitter–receiver pair has priority over the other mobile stations, to complete the transmission and reception of those data.

- We also notice from the figure how the NAV is set. For another mobile station hearing the RTS, it sets the NAV until the end of the ACK. If the mobile station doesn' hear the RTS, but hears the CTS, it also sets the NAV until the end of the ACK, but of course, the corresponding value is smaller than if it had heard the RTS.

8.3.2.4 Example: Timing of Transmissions with Multiple Fragments Another scenario where SIFSs are used is when a transmission is broken up into multiple fragments. After each fragment there is a corresponding ACK. After the ACK, the next fragment is allowed to be sent after a wait of only SIFS. This is to minimize delays in reception of subsequent fragments once the first fragment has been sent and received (Figure 8.15).

8.3.2.5 Control, Administrative, and Management Aspects In an Ethernet-based wired LAN, membership in the LAN is straightforward. Stations that are physically connected (by cables) to the LAN are part of the LAN.† With WLAN, however, just being within radio range of other mobiles does not mean that a mobile is part of the LAN. DCF, PCF, and CSMA/CA describe what happens once a mobile station is part of a BSS. We also need to consider how a mobile station joins the BSS or leaves it in the first place. When there is an AP (infrastructure BSS), a mobile station needs to *associate* with the ESS to join. Upon leaving, it may *disassociate*. As it moves from one AP to another within the same ESS, it *reassociates*.

Management frames are defined for such purposes. NB: These are different from *control frames* such as RTS/CTS, polling, and ACK that we have seen so far, and they are also have a different name: *management frames*. The management frames are for such functions as:

† By default at least; in modern LAN switches, the switch can administratively allow only certain ports, certain MAC addresses, and so on, to be part of the LAN, and it may also break the physical LAN into multiple virtual LANs.

- *Association (request and response).* To join a BSS, a mobile station needs to be associated with it; this is initiated by the sending of an association request to the AP.
- *Reassociation (request and response).* To move from one AP to another AP within the *same* ESS, reassociation is used.
- *Probe (request and response).* This is used to obtain information about a BSS.
- *Beacon.* Beacon frames are broadcast by the AP (or distributed among members in an IBSS) regularly. It includes system information such as the direct sequence spread-spectrum parameter set, SSID, beacon interval, and traffic indication map.
- *Disassociation.* This is the opposite of association.
- *Authentication.* Authentication is of either the open system or the shared key variety (see Chapter 15 for more details).
- *Deauthentication.* This is used to terminate an authentication.

The beacon broadcasts valuable information for mobile stations wishing to join a BSS. These include the SSID (service set ID), used for identification of the BSS, as well as key physical layer parameters. The *traffic indication map* (TIM) helps mobile stations save power since they can go to sleep during times when there are ongoing transmissions that do not concern them.

Rather than waiting for a beacon, a mobile station may send a probe request to obtain a probe response from an AP. In fact, if the mobile station is just scanning for APs, it will probably want to send probe requests at different frequencies to elicit whatever probe responses it can get rather than waiting patiently for the beacons. Probe responses are very similar to beacons, but without some information, such as the TIM.

8.3.2.6 Variations in Implementation Although the specifications are very detailed, they leave room for variations in implementation. The MAC functions, for example, need not be all implemented in an AP. Instead, a "split MAC" implementation may divide the MAC functions between individual APs and another element, such as a WLAN controller (Figure 8.16). There may be dozens or even hundreds of APs per WLAN controller. This allows the reduced APs to be cheaper, so the overall cost (compared to having the same number of APs, but full-featured ones and without a WLAN controller) can be reduced. Besides potential cost savings, a

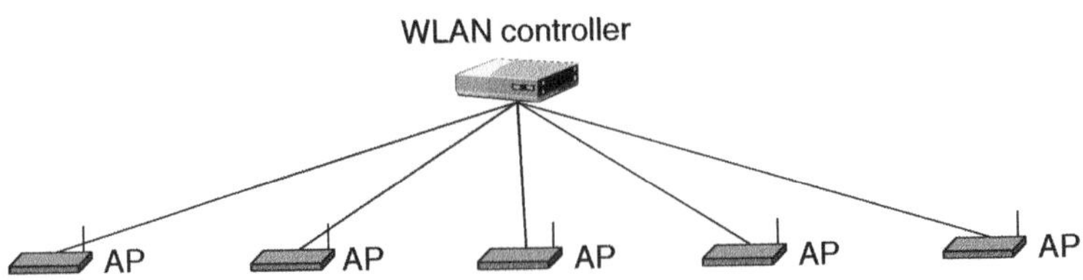

FIGURE 8.16 "Split MAC" implementation of 802.11.

split MAC implementation also allows the vendor to incorporate elements of central coordination not part of 802.11. For example, frequency coordination between APs is not part of 802.11. A WLAN controller could implement algorithms for optimizing frequency assignments to various APs. It could also coordinate security settings, QoS settings, and so on.

8.3.3 A Plethora of Physical Layers

The initial 802.11 (in 1997) contained just three options for the physical layer: direct sequence spread spectrum (DSSS), frequency hopping spread spectrum (FHSS), and infrared (IR), each of which supported data rates of either 1 and 2 Mbps. Actual implementations have largely neglected the FHSS and IR options. Meanwhile, additional physical layer options have been added in the past decade, based on OFDM, CCK modulation, and so on.

The original DSSS was accomplished by multiplying each symbol with an 11-chip Barker sequence: $1, -1, 1, 1, -1, 1, 1, 1, -1, -1, -1$. The difference between the 1- and 2-Mbps transmissions is that DBPSK is used in the former and DQPSK is used in the latter.

IEEE 802.11b adds 5.5 and 11 Mbps through the use of CCK modulation, a form of m-ary orthogonal modulation. IEEE 802.11a is the first amendment that uses OFDM and that uses the 5-GHz band (the 802.11 baseline DSSS and 802.11b are specified for 2.4 GHz). It can support up to 54 Mbps. IEEE 802.11g, like 802.11a, uses OFDM and can support up to 54 Mbps. Unlike 802.11a, though, it operates in the 2.4-GHz band and so is more compatible with the baseline 802.11 and 802.11b. IEEE 802.11n uses OFDM and also adds MIMO, to support rates of up to 600 Mbps. It can be used in both 2.4 and 5 GHz.

8.3.3.1 A Closer Look at IEEE 802.11a

IEEE 802.11a is an amendment to IEEE 802.11-1999 that introduces the first physical layer for 802.11-based wireless LANs that is based on OFDM. It uses the following system parameters:

- 20 MHz channel spacing
- A 20-Msps sampling rate
- $N = 64$ FFT
- 48 data subcarriers and four pilot subcarriers
- Modulation on each subcarrier (BPSK, QPSK, 16-QAM, or 64-QAM)
- Pilot-assisted coherent detection
- Convolutional coding with rate-$1/2$ mother code
- A 0.8-μs guard period (16 time samples)

Depending on the channel quality, 802.11a will adaptively change its transmission rate. In the best cases it can transmit at 54 mbps, with 64-QAM and only rate-3/4

TABLE 8.4 IEEE 802.11a: Possible Data Rates

Transmission Rate (mbps)	Modulation	FEC Rate	Coded Bits per OFDM Symbol	Data Bits per OFDM Symbol
6	BPSK	1/2	48	24
9	BPSK	3/4	48	36
12	QPSK	1/2	96	48
18	QPSK	3/4	96	72
24	16-QAM	1/2	192	96
36	16-QAM	3/4	192	144
48	64-QAM	2/3	288	192
54	64-QAM	3/4	288	216

FEC, but in many cases the channel will not be good enough and thus would need to scale back to a lower rate. The possibilities are indicated in Table 8.4.

EXERCISES

8.1 Given the GSM signaling rate and the time allocated to each time slot, verify that the number of bits in the time slot, as shown in Figure 8.1, actually fits in the time slot. Also, compute the length of time of the guard period and compare it with the delay spread that might be expected in a cell.

8.2 What is the average FEC rate of GSM; that is, how many bits get mapped to the 456 bits per 20-ms speech frame? Why is the GSM solution better than just using a code at the average rate for all the bits?

8.3 The minimum separation between base station offsets in the downlink of an IS-95 system is 64 chips. How long is 1 chip (in time)? How long, therefore, is 64 chips? (This is to give us an idea of how critical it is for the base stations to be synchronized, because 64 chips is not a very long time.) Compare this with typical RMS delay spread for a cellular system.

8.4 The power control rate in IS-95 is 800 Hz. What is the corresponding time interval between power control directives?

8.5 Can an 802.11 IBSS connect to the Internet?

REFERENCES

1. B. G. Lee and S. Choi. *Broadband and Wireless Access and Local Networks*. Artech House, Norwood, MA, 2008.
2. E. Seurre, P. Savelli, and P.-J. Pietri. *GPRS for Mobile Internet*. Artech House, Norwood, MA, 2003.

3. IEEE Computer Society. IEEE standard for information technology—telecommunications and information exchange between systems—local and metropolitan area networks—specific requirements: 11. Wireless LAN medium access control (MAC) and physical layer (PHY) specifications. IEEE 802.11-2007 (revision of 802.11-1999), June 2007. Sponsored by the LAN/MAN Standards Committee.
4. D. Wong and T. J. Lim. Soft handoffs in CDMA mobile systems. *IEEE Personal Communications*, pp. 6–17, Dec. 1997.

9

RECENT TRENDS AND DEVELOPMENTS

There have been many new developments in the past decade as wireless technologies have continued to improve rapidly and have been providing higher data rates, optimizations for packet data traffic, and so on. Since this chapter is one of the chapters on wireless access technologies, we focus here on recent trends and developments in wireless access technologies. Recent trends and developments in other aspects (e.g., in the network) are covered in other chapters, especially Chapter 12, and in service architectures in Chapter 13.

In this chapter we begin in Section 9.1 with the third-generation systems WCDMA and cdma2000. Then in Section 9.2 we explore some emerging technologies in wireless, such as HARQ and multiple-antenna techniques. While WCDMA and cdma2000 can be said to be the last wireless air interfaces in the more voice- and circuit-centric traditional mold, a variety of enhancements and improvements to these systems were soon developed to support high-speed packet data more efficiently. These, called by various names, including HSPA and EV-DO, are discussed in Section 9.3. We wrap up the chapter by looking at aspects of WiMAX (Section 9.4) and LTE (Section 9.5).

9.1 THIRD-GENERATION CDMA-BASED SYSTEMS

There are two major third-generation (3G) systems: WCDMA (Section 9.1.1) and cdma2000 (Section 9.1.2). In general, these systems could be called "3G" because they addressed the ITU requirements for higher data rates, variable data rates, and so on (see Section 17.2.3.1 for more on the IMT-2000 process of ITU for 3G systems). The 3G systems also saw turbo codes and beamforming coming into the mainstream,

Fundamentals of Wireless Communication Engineering Technologies, First Edition. K. Daniel Wong.

with support for such technologies being included in the standards (e.g., specification of turbo codes to be used, and additional pilots that could be used with beamforming; see Section 9.2.2.3 for more on beamforming). The more political and standards development aspects of WCDMA and cdma2000 are deferred until Section 17.2, and we just discuss some technical features of these systems here.

9.1.1 WCDMA

The *wideband CDMA* (WCDMA) air interface was created as part of *universal mobile telecommunications system* (UMTS), the system that would replace GSM. The air interface was a complete redesign from that of GSM, replacing TDMA with CDMA for multiple access. It was called "wideband" CDMA because it was designed to operate in 5 MHz of spectrum, which was several times wider than the 1.25-MHz bandwidth occupied by IS-95 channels and much wider than the 200-kHz channels of GSM. Correspondingly, the chip rate is 3.84 Mbps.

Notable innovations in WCDMA compared to IS-95 include:

- WCDMA supports higher data rates than do 2G systems, taking advantage of the wider bandwidth to do so.
- WCDMA supports variable-rate channels. IS-95, being a 2G system, focuses primarily on voice channels. WCDMA supports a range of data rates.
- WCDMA does not require that the base stations be synchronized with each other, whereas IS-95 uses GPS to ensure that base stations are synchronized with each other.

Additionally:

- WCDMA closed-loop power control is at a rate of 1.5 kHz, compared to 800 Hz in IS-95.
- Instead of one pilot per cell on the downlink (in IS-95), there are multiple pilots, in both downlink and uplink, some of which are specific to individual links (this supports coherent detection on the uplink, antenna beamforming, etc.). In particular, for beamforming to direct a signal in a particular direction toward a mobile, a dedicated pilot is sent on the same beam so that the mobile can perform coherent demodulation. See Section 9.2.2.3 for further discussion of beamforming.

To support variable-rate channels, WCDMA uses *orthogonal variable spreading factor* (OVSF) *codes* in place of same-rate Walsh codes on all channels. OVSF codes are a generalization of same-rate Walsh codes and can be thought of as a collection of Walsh codes of different rates. Why do variable-rate channels need variable spreading factors? Considering that the final chip rate coming out of the transmitter is 3.84 Mchip/s, and this rate is the same for all channels, higher-data-rate channels need less spreading; for example, a channel with the rate 0.96 Mbps needs spreading

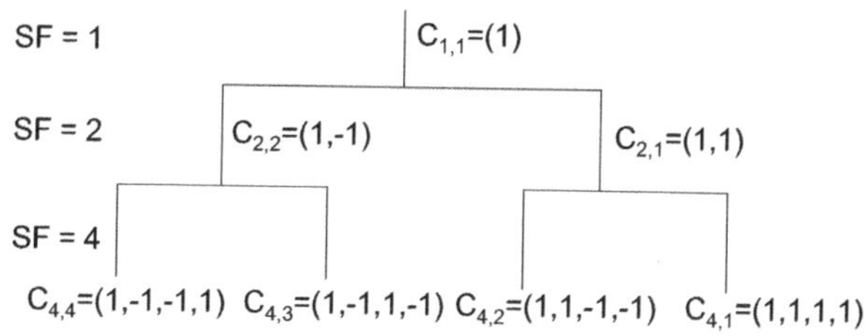

FIGURE 9.1 OVSF codes.

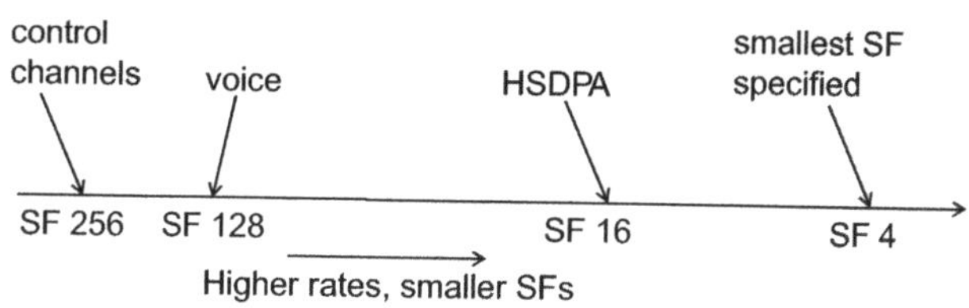

FIGURE 9.2 Typical use of some OVSF code spreading factors.

only by a factor of $3.84/0.96 = 4$ to reach 3.84 Mchip/s, whereas a channel with a rate of 0.24 Mbps needs spreading by a factor of $3.84/0.24 = 16$.

We can conceive of OVSF codes by constructing them as a tree, with "parents" and "children" branching out as in Figure 9.1, each child code derived from its parent code and having a higher spreading factor than its parent. Figure 9.1 shows only the beginning of the tree, and there would usually be more levels, extending the figure downward. For example, in WCDMA, codes of up to spreading factor 256 are used. Like Walsh codes, any two OVSF codes of the same rate are orthogonal to each other. Some OVSF codes of different rates may be orthogonal to each other. However, any two OVSF codes of different rates *need not* be orthogonal to each other. Thus, care needs to be taken to ensure that the set of OVSF codes being used at any time is always an orthogonal set. Given a set of codes, a convenient way to see which codes can or cannot be used together is to check if any of them is a parent (not necessarily an immediate parent but even an "ancestor," i.e., parent of parent, etc.) of any of the other codes in the set. If such parent–child relationships exist in the set, those codes are not orthogonal. Equivalently, if we can trace the parent path of each code back to the root of the tree without encountering another code from the set, the codes in the set are all orthogonal to each other.

Figure 9.2 shows some of the uses of codes of various spreading factors. Additionally (not shown in the figure), for various rates of data traffic (in contrast to voice or control traffic), various spreading factors would be used.

9.1.2 cdma2000

cdma2000 refers to the evolution of the IS-95 system. It was originally designed as a wideband CDMA system similar to WCDMA. Since it was, however, created as an

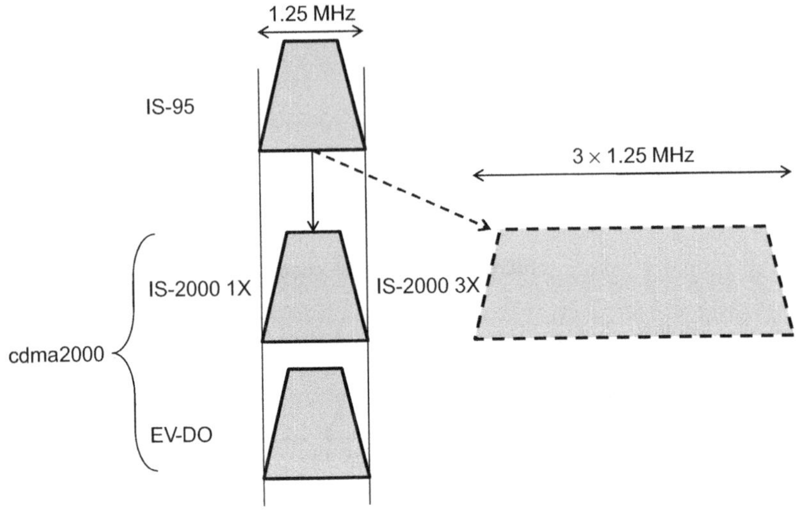

FIGURE 9.3 Evolution of IS-95 to cdma2000.

evolution of the IS-95 system, there were some differences between cdma2000 and WCDMA. Since cdma2000 was evolving from an existing CDMA-based system, IS-95, that used 1.25-MHz channels, cdma2000 was designed to work with certain integer multiples of 1.25-MHz channels (i.e., 1.25, 3.75, 6.25 MHz, etc.) denoted by 1X, 3X, 5X, and so on (thus, 1X means one times the IS-95 1.25-MHz bandwidth, 3X means three times 1.25 MHz, etc.). Initially, a lot of attention was focused on the 3X version of cdma2000, which at 3.75 MHz was comparable to the 5-MHz channel of WCDMA. In fact, the two air interfaces were the main competitors to be selected by ITU as *the* single global air interface standard for 3G in the late 1990s (as part of the IMT-2000 framework; see Section 17.2.3.1 for more details).

However, most operators who were already operating CDMA networks based on IS-95 opted to keep using their 1.25-MHz channels. The systems were upgraded from IS-95 to IS-2000, also known as cdma2000 1X. cdma2000 1X added various improvements to IS-95, as we discuss in Section 9.1.2.1. Besides the upgrades to cdma2000 1X, interest also shifted toward high-speed wireless data, as 1×EV-DO emerged (see Section 9.3) as a more data-optimized technology that utilized the original 1.25-MHz channels. A graphical summary of the evolution from IS-95 to cdma2000 is shown in Figure 9.3.

9.1.2.1 Improvements of IS-2000 over IS-95 To meet the ITU requirements for 3G, IS-2000 was designed to support 144-kbps data rates, requiring improvements in both signaling and transmission aspects [13]. On the signaling side, the focus was on signaling to support rapid acquisition and release of radio resources, so high-rate packet-switched data could be supported efficiently. This included the addition of new control channels and the specification of shorter frames for some control channels.

On the transmission side, new channels were added to support high-rate user data. Unlike IS-95, which has a pilot channel only on the downlink and not the uplink (so coherent demodulation can be performed only on the downlink), IS-2000 has a pilot channel on the uplink as well. Unlike WCDMA, which has dedicated downlink pilots for each user, IS-2000 has optional *auxiliary pilot channels* that can be used with beamforming, but these are limited to four in number, so only four beams can be used simultaneously. One of the most notable improvement was an increase in physical layer capacity as follows:

- On the *downlink*. Recall that IS-95 uses QPSK but that *the same* symbols are transmitted on both the in-phase and quadrature carriers. This is a form of transmit diversity, but it is not bandwidth efficient. With IS-2000, QPSK is also used, but different symbols are transmitted on the in-phase and quadrature carriers, thus allowing the data rate to double. Another way of thinking of this is that IS-95 is practically using BPSK, since there is only one chip per symbol (duplicated on in-phase and quadrature), whereas IS-2000 uses two chips per symbol. In conjunction with this capacity gain, the Walsh code length was increased from 64 in IS-95 to 128 in IS-2000, making more Walsh codes available.
- On the *uplink*. Recall that IS-95 uses 64-ary orthogonal modulation on the uplink and does not have an uplink pilot. With the addition of the pilot in the uplink of IS-2000, coherent demodulation of the uplink becomes possible, which in general increases the capacity. Furthermore, addition of the pilot allows Walsh codes to be used for channelization, so a mobile station can transmit multiple channels to the base station simultaneously (with different Walsh codes for each).

Does cdma2000 make use of OVSF codes as W-CDMA does? Yes and no. Yes, in the sense that the same idea is used and one can map the OVSF codes to the codes in cdma2000 in a one-to-one way. No, in the sense that the codes are just called Walsh codes, as before, albeit Walsh codes of different lengths. Thus, in the downlink, Walsh code lengths of between 4 and 128 (in powers of 2) could be used, with the shortest Walsh codes being used for the highest-rate channels.

9.1.3 Summary

We summarize some major differences between 3G CDMA wireless access technologies such as WCDMA and cdma2000, and IS-95, in Table 9.1.

9.2 EMERGING TECHNOLOGIES FOR WIRELESS ACCESS

Among the key innovations to emerge in the wireless access portion of wireless systems since UMTS and cdma2000 were introduced are the use of hybrid ARQ (HARQ), multiple-antenna technologies, and OFDMA. We introduced OFDMA in Section 7.1.2 and will see how it is implemented in standards in Sections 9.4 and 9.5.

TABLE 9.1 Comparisons among WCDMA, cdma2000, and IS-95 CDMA

	IS-95 CDMA	WCDMA	cdma2000
Chip rate	1.2288 Mchip/s	3.84 Mchip/s	1.2288 Mchip/s
Turbo code rates	N/A	1/3	1/2, 1/3, 1/4, 1/5
DL channelization codes	Walsh	OVSF	Variable-length Walsh
UL channelization codes	N/A	OVSF	Variable-length Walsh
DL scrambling codes	PN code offsets	Gold codes	PN code offsets
UL scrambling codes	Long PN code	Gold codes or short codes	Long PN code
UL power control rate	800 Hz	1.5 kHz	800 Hz
DL power control rate	50 Hz	1.5 kHz	800 Hz
UL pilot	N/A	Multiple	Multiple, common only
DL pilot	One per cell	Multiple	Multiple

In this section, then, we focus on HARQ (Section 9.2.1) and multiple-antenna techniques (Section 9.2.2).

9.2.1 Hybrid ARQ

Hybrid ARQ is a hybrid of traditional ARQ (as introduced in Section 10.1.3.1) and forward error correction (FEC, Section 7.4). There are several ways of combining ARQ and FEC, and these include *type I* and *II HARQ*.

9.2.1.1 Type I HARQ: Chase Combining In type I HARQ, also known as *chase combining*, only a subset of the FEC-encoded bits is sent (so it can be viewed as a punctured version of the original code; we discussed code puncturing in Section 7.4.3). This subset is sufficient for the receiver to detect and correct a limited number of errors, but in general it is a less "powerful" set of encoded bits than the complete set. However, in the case of good channel conditions, this subset may be sufficient for decoding, and so bandwidth is saved. If the channel is not as good, the *same* subset of FEC-encoded bits is resent, up to a specified limit. Each time the bits are resent, the reliability of the bit values in the receiver increases, since we have a form of time diversity in receiving multiple versions of the same bits. In particular, the multiple versions of each can be combined using maximal ratio combining. This is sometimes described as *soft combining*, because the combining happens before the bit decisions, rather than *hard combining*, which would be after the bit decisions are made, in which case valuable information about what was actually received has been lost and the best that can be done is just to take a "majority vote" decision.

9.2.1.2 Type II HARQ: Incremental Redundancy In type II HARQ, also known as *incremental redundancy*, the first transmission is similar to that of type I HARQ: only a subset of the FEC-encoded bits is sent, and these are sufficient for detecting and correcting a limited number of errors. Under good channel conditions,

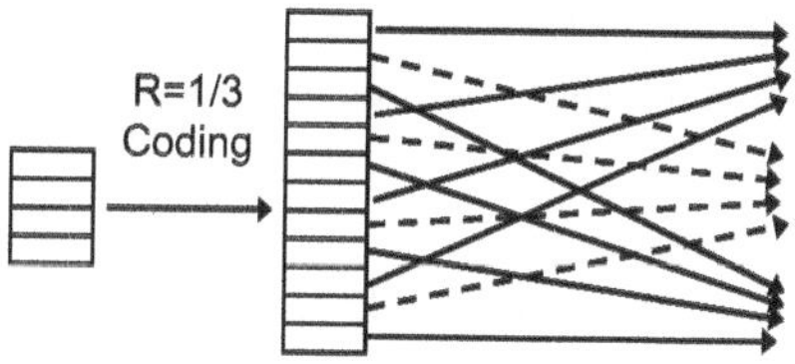

FIGURE 9.4 Hybrid ARQ.

this first transmission might be sufficient for decoding, saving bandwidth. The difference from type I HARQ is what happens when retransmissions are needed. Instead of sending the same subset of bits with each retransmissions as with type I HARQ, in type II HARQ, different encoded bits are sent each time. This has the effect of lowering the code rate (from a high rate, less powerful code, to a lower rate, more powerful code) with each retransmission. For example, Figure 9.4 shows an example where 4 data bits are encoded into 12 coded bits using an $R = 1/3$ code. We suppose that it is a systematic code; that is, the coded bits include a complete copy of the original data bits. These four original data bits are then sent in the first transmission, resulting in an essentially uncoded transmission (i.e., one without FEC protection, with rate $R = 1$). This may be OK under good channel conditions, but if a retransmission is needed, 4 of the other bits from the 12 coded bits can be sent, effectively making it an $R = 1/2$ code. If these are still not enough, another retransmission can be done. The remaining 4 bits are sent, bringing it back to the original $R = 1/3$ code. NB: Our simple example does not show how the receiver might detect whether or not the first transmission contains errors, since it is an uncoded transmission. In practice, there must be some way for the receiver to detect this so that it can decide whether or not retransmission is necessary. However, we did not show such a way in this example, for purposes of clarity and simplicity. We discuss this issue further in Section 9.2.1.4. Type I HARQ could even be thought of as a special case of type II HARQ, where the code is a repetition code (and hence, the same bits are sent with each retransmission).

9.2.1.3 Worked Example: Effective FEC Rate Suppose that we are using an $R = 1/4$ code, where $1/4$ of the bits are sent each time, making effective code rates of 1, 1/2, 1/3, and 1/4, respectively. Suppose that 60% of the transmissions need no retransmissions, 20% need one retransmission, 10% need two retransmissions, and 10% need three retransmissions. What is the effective FEC rate of this HARQ scheme in this case? The effective rate is

$$0.6 \times 1 + 0.2 \times \frac{1}{2} + 0.1 \times \frac{1}{3} + 0.1 \times \frac{1}{4} = 0.75833$$

which is $R \approx 3/4$. What if we had just started with a rate-$3/4$ code instead of using HARQ? Bandwidth utilization would have been about the same. However, the rate-$3/4$ code might not be powerful enough for the 20% of cases where the second or third HARQ retransmission was required (corresponding to code rates $1/3$ and $1/4$), and it might not be good enough for some of the 20% of cases where one retransmission was required (corresponding to code rate 1/2). Thus, there might have been severe performance degradation in the use of a flat-rate 3/4 code compared to using HARQ.

9.2.1.4 *Further Discussion of HARQ*

How would the receiver know when the bits received are sufficiently reliable or when it needs to ask for a retransmission? Typically, some kind of error detection check (e.g., a CRC check) might be carried out. The advantage of such error detection codes is that they typically comprise many fewer bits than FEC (whose numbers would be on the order of the number of data bits), and so contribute very little overhead.

It might be expected that type II HARQ would perform better than type I, since there might be an effect of diminishing marginal gains with type I HARQ, when just the same bits are retransmitted, whereas new information is sent with each transmission in type II HARQ. Indeed, research studies have found that type II HARQ performs better. However, it is also more computationally complex than type I HARQ.

9.2.2 Multiple-Antenna Techniques

Since the late 1990s, there has been a flurry of R&D in multiple-antenna techniques. The newcomer may be confused hearing the variety of different terms used: *multiple input, multiple output* (MIMO), spatial multiplexing, space-time coding, Alamouti scheme, transmit diversity, and so on. Furthermore, is transmit diversity new or is it a variation of older receiver diversity techniques that have been known and used for many years (and which were discussed in Section 5.3.5)?

Having multiple antennas on both the transmitting and receiving sides allows a dimension of freedom whose gains can be realized in different ways. We can classify the multiple antenna techniques into four types, based on how the gains are realized, as follows:

- *Spatial multiplexing*: tries for higher data rates by simultaneous transmission of multiple independent data streams.
- *Spatial diversity*: tries for lower error rates through space-time coding, receiver antenna diversity, etc.
- *Smart antennas*, also known as *adaptive antenna arrays*: beamforming for antenna gains or *array gains*, providing better SNR and reducing co-channel interference
- *Hybrid techniques*: combines two or more of the above

The term *MIMO* can be used to describe the general setup: that is, having multiple antennas on both sides (at the input and output of the wireless medium). However, it is sometimes also used to refer specifically to spatial multiplexing. In this section we follow the second convention, and use MIMO synonymously with spatial multiplexing. Which approach should be taken for a given situation? In cases of high SNR, spatial multiplexing makes a good choice provided that the wireless environment supports it. For cases of low SNR, spatial diversity or beamforming techniques may be a better choice.

9.2.2.1 Spatial Multiplexing for Higher Bit Rates The seminal work of Foschini and Gans, and of Telatar, in the late 1990s has inspired intense R&D interest in MIMO ever since, and MIMO techniques have now found their way into recent commercial wireless systems as a crucial component for systems to deliver the highest data rates. Suppose that m and n antennas are used in the transmitter and receiver, respectively (traditionally called *transmit* and *receive antennas*, respectively). Then, there can be up to m different independent channels, one for each transmit antenna (indeed, with spatial multiplexing, each antenna would be transmitting a different stream of data). It can be shown that the capacity grows roughly linearly with the minimum of m and n [5], where bandwidth and total power are kept constant.

However, this is for the ideal case of statistically independent channels between the transmit and receive antennas, which is more likely to be found in a *rich fading environment* where a lot of reflection, refraction, and so on, happens between transmitter and receiver, so the combination of effects is different for each pair of transmit and receive antennas. Thus, in practice, wireless systems can realize capacity gains with a small number of antennas [e.g., 2×2 (two transmit antennas and two receive antennas), or even up to four, provided that the propagation environment supports it]. Adding antennas beyond a certain number (depending on the propagation environment) does not allow higher throughput, and it is sometimes recommended that in that case, the additional antennas could be used for other purposes, such as beamforming.

Given the theoretical capacity, a range of options are available for realizing data rates close to capacity at low error rates. The best performance tends to come from using maximum likelihood detectors in the receiver, at the expense of high complexity. Various lower-complexity alternatives to maximum likelihood detectors can be used. Some are based on successive interference cancellation principles. The idea here is that since multiple streams are transmitted within the same bandwidth and at the same time to recover any given stream at the receiver, the other streams would appear as interference that would need to be canceled or removed. One of the earliest schemes, BLAST from Bell Labs, was based on this type of approach, recovering each stream "layer" by "layer."

9.2.2.2 Spatial Diversity In spatial diversity schemes, the aim is not so much to go all-out for higher data rates by putting independent data streams on different antennas, but to take advantage of spatial diversity to achieve *diversity gains* and, in some schemes, also to achieve *coding gains* through the use of advanced coding schemes.

Receiver diversity techniques such as selection diversity, equal gain combining and maximal ratio combining have been known for many years, as discussed in Section 5.3.5, and were analyzed extensively in the classic book edited by Jakes [8].

More recently, transmit diversity schemes have become popular, where techniques on the transmitter side are used either for diversity gains only or for a combination of diversity and coding gains. Two milestone papers that showed the value of such techniques were Alamouti's scheme for two transmit antennas [1] and Tarokh et al.'s paper on space-time codes [12]. Subsequently, numerous generalizations and extensions to such space-time codes have been created [14].

9.2.2.3 *Smart Antennas and Beamforming*

We discussed antenna arrays in Chapter 4. Beamforming is about using antenna arrays to create antenna patterns that have beams pointing in certain directions and/or have nulls in certain directions. As we saw in Section 4.3, small changes in the relative phase of the signal between different antennas in the array could result in drastic changes in the antenna pattern. The more antennas in the array, the more "degrees of freedom," which could be used to create beams or nulls. In particular, an N-element array would have $N - 1$ degrees of freedom. Such antenna arrays are sometimes called *smart antennas*. They would typically be found on base stations, since it is more realistic for them to be at a base station than at a mobile. However, in theory, smart antennas could be used at both transmitters and receivers. Often, especially in traditional beamforming, the antennas are close together and highly correlated. This is in contrast to diversity schemes where low correlation is essential.

The beams are used to improve the radio link for specific mobiles or groups of mobiles, by pointing the beams toward them. The nulls can be used in directions where interfering devices are located. Thus, interference can be reduced. As a practical matter of interest, when beamforming is used, the channel from base station to mobile through a beam could look very different from the main, omnidirectional channel from the base station to the mobile. For example, the channel through the beam could be essentially a line-of-sight channel, whereas the omnidirectional channel could include a number of non-line-of-sight paths. Thus, if only a common pilot channel is transmitted by the base station (on the omnidirectional channel), that pilot channel would be of limited value for channel estimation and coherent demodulation of the signal transmitted through the beam. Hence, additional pilots are needed for each beam, and this is one of the trade-offs that comes with using beamforming.

9.3 HSPA AND HRPD

High-speed packet access (HSPA) actually encompasses two technologies: *high-speed downlink packet access* (HSDPA) and *high-speed uplink packet access* (HSUPA). HSDPA came before HSUPA, and there are some differences in the way they work, although there are also common elements. HSDPA was one of the major additions to UMTS in Release 5, and HSUPA was one of the major additions to UMTS in Release 6 (see Chapter 12 for a description of the evolution of UMTS, including Release 5 in Section 12.3.3 and Release 6 in Section 12.3.4). *High-rate packet data* (HRPD) is also known as 1×EV-DO, the 1× referring to the bandwidth of the system (1 times the IS-95 bandwidth), the EV referring to evolution and the DO referring to "data optimized."

9.3.1 HSDPA

One way to view HSDPA is as a natural step in the continuing evolution from 2G to 3G systems, in the quest to better support data traffic. The 2G systems are very voice-centric and circuit-switched. So-called 2.5-G technologies such as GPRS add some flexibility for data traffic but are still based on the GSM time slots. With the first UMTS systems in 3G, the new dedicated channel (DCH) supports variable-rate traffic. However, it can be viewed as a variable-rate *circuit* rather than a true *packet-switched* channel. A true packet-switched channel that is more optimized for variable-rate data traffic to and from multiple mobile devices should be something that can switch quickly and flexibly between devices. Since data are bursty, at one moment device A might need a huge amount of bandwidth to receive an urgent transmission and all the other mobile devices could afford to wait awhile, whereas at another moment, devices B and C could share the available bandwidth and other devices could afford to wait awhile. Wouldn't it be great if there were a scheme that could provide such flexibility?

Indeed, there is, and one of the key ideas in HSDPA is that it uses what is sometimes informally called "one big channel" that it could completely allocate to device A for a short period of time, and perhaps later share between B and C for another short period of time. This is illustrated in Figure 9.5. The *y*-axis label is "spreading codes." What happens with HSDPA is that a group of 1 to 15 OVSF codes (with spreading factor 16) is allocated for HSDPA transmission, out of the total pool of OVSF codes available at a base station. The other codes may be used as usual for circuit-switched services, control signaling, and so on. Of the codes allocated to HSDPA, at any time from zero to all of them could be allocated to any given mobile, and some of the codes could be distributed to two or more mobiles, as shown in the figure.

Besides allocating channel resources based on changing demand, another factor in the allocation decision is the changing channel conditions between the base station and each of the mobile stations. Thus, the allocation of channel resources to mobile devices can give preference to those mobile stations that have a better signal. Then,

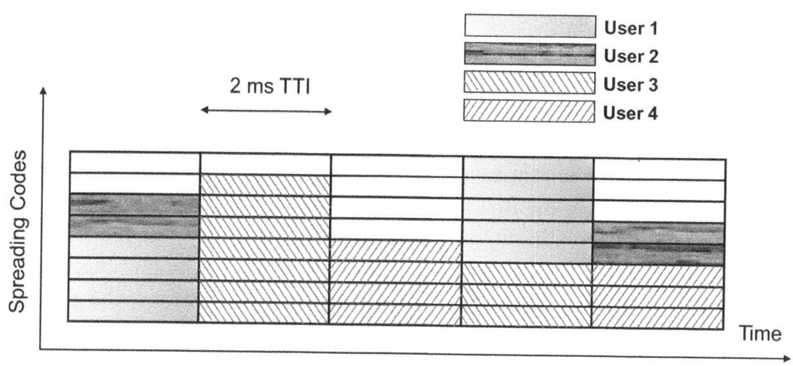

FIGURE 9.5 HSPA: one big channel that can be allocated flexibly to different users at different times.

when those mobile stations are in fades, other mobile stations may have a better signal, and preference could be given to those. This is a form of *multiuser diversity*. It is sometimes called *channel aware scheduling* or *channel-dependent scheduling*. So we see that there are multiple inputs for the design process of the scheduler. There is the changing bandwidth demand from the mobiles on the one hand, and the changing channel conditions on the other. Moreover, there might be different QoS requirements for different traffic to the mobiles. If the scheduler optimizes transmissions based mostly on channel conditions, this may result in too much unfairness of allocations to mobiles (e.g., if one mobile tends to have a weaker channel, it may get starved of bandwidth as resources are allocated to other mobiles most of the time). If the scheduler optimizes based mostly on the bandwidth and quality of service needs of the mobiles, the resulting overall capacity may be decreased significantly (it may be scheduling transmissions on weaker channels more often). Thus, there is a trade-off between throughput and fairness.

Furthermore, for channel-aware scheduling to work well, the base station would need to be able to rapidly schedule allocation of the big channel as needed, and the allocation should be for short intervals, to provide a considerable degree of flexibility. In fact, the fast allocations are another of the innovations that come with HSDPA. Whereas previously, channel allocation was handled by the BSC, this introduces too much delay. Thus, in HSDPA, the base station itself handles the allocation and with very short *transmission time intervals* (TTI) of 2 ms, which provides a lot of flexibility to switch allocations as needed.

As we have seen in Section 7.3, power control is critical on the uplink of CDMA systems. However, it is not so critical on the downlink. In IS-95 systems, the dynamic range in transmitter power resulting from uplink power control is 70 dB, whereas it is only about 20 dB for downlink power control. Instead of doing power control on the downlink and reducing the transmitter power when there is a strong-signal channel to a mobile device, the power level is kept constant in HSDPA, and the better channel is exploited through the use of higher-level modulation. Thus, either QPSK or 16-QAM could be used with HSDPA (initially; later, the possibility of using 64-QAM was added to HSPA+ in UMTS Release 7). A related ability to adapt is the ability to adapt the FEC code rate, also depending on channel conditions.

Yet another feature of HSDPA that results in higher data rates is the use of *hybrid ARQ* (HARQ). We introduced HARQ in Section 9.2.1. HSDPA allows for both chase combining and incremental redundancy. The FEC coding is with a punctured rate-1/3 turbo code. There are potentially up to three transmissions (the first transmission, followed by one or two retransmissions). For each transmission or retransmission, the code is punctured (for the first transmission, the puncturing removes only some parity bits, so the systematic bits plus some remaining parity bits are transmitted; for the retransmissions, for incremental redundancy, the systematic bits are punctured and only parity bits are sent). For incremental redundancy, the puncturing patterns are different for each transmission or retransmission, whereas for chase combining, the same pattern would be used on the retransmissions as on the first transmission.

9.3.2 HSUPA

HSUPA is the uplink counterpart to HSDPA. Unfortunately, some of the features of HSDPA cannot simply be imported for use also in HSUPA. For example, we have explained how adaptive modulation on the downlink is done instead of power control. On the uplink, however, it is not possible to avoid power control, so that still has to be used. Thus, adaptive modulation is not a feature of HSUPA. Furthermore, unlike the downlink, there is no common shared channel in HSUPA. This is because the transmissions are coming from different mobiles (unlike the downlink, where they are all from the base station), and the uplink does not use orthogonal channels. So the uplink is more similar to the original uplink channel, but like HSDPA, it still uses fast scheduling and fast HARQ in the physical layer.

9.3.3 1×EV-DO

1×EV-DO (HRPD, part of cdma2000) is very similar to HSPA, and has the features:

- There is one big "pipe" that can quickly be scheduled for multiuser diversity, with short frames.
- Adaptive modulation is used for higher data rates.
- Physical layer HARQ with incremental redundancy is used. This also supports higher data rates.

More recently, cdma2000 also includes another high-speed packet optimized evolution, known as 1×EV-DV or cdma2000 Release D. 1×EV-DO is for data only, whereas 1×EV-DV is designed to be backward compatible with earlier releases of cdma2000, and so can be used in the same systems that also support the voice and low-rate traffic channels of earlier releases. DV stands for "data and voice."

9.3.4 Continuing Enhancements

HSDPA was introduced before HSUPA. However, after they had both been introduced, they were and are often referred to simply as HSPA. HSPA+ is a more recent enhancement of HSPA that includes:

- Adaptive modulation up to 64 QAM
- MIMO

It brings the maximum data rate from 14.4 Mbps up to 21.1 Mbps (with 64 QAM but without MIMO), and up to 42.2 Mbps with MIMO.

More recently, along with LTE in UMTS Release 8, *dual-cell HSPA* was introduced. Also known as *dual-carrier HSPA*, it allows for the use of two carriers and joint resource allocation and load balancing across both carriers, thus doubling the size of the "big channel." Initially, dual-cell HSPA was introduced without MIMO, so

TABLE 9.2 Summary of Features of HSPA and 1×EV-DO

Feature	DCH	HDSPA	HSUPA	1×EV-DO
Variable spreading factor	Yes	No	Yes	No
Fast power control	Yes	No	Yes	No
Soft handoff	Yes	No	Yes	"Virtual"
Adaptive modulation	No	Yes	Yes	Yes
Channel-dependent scheduling	No	Yes	Yes	Yes
L1 HARQ	No	Yes	Yes	Yes
TTI (in ms)	10, 20, 40, 80	2	2, 10	1.6

the maximum data rate is also 42.2 Mbps, but more recently, combinations of dual-cell HSPA and MIMO have been specified, pushing the maximum data rate up to 84.4 Mbps. Table 9.2 provides a summary of the features of HSDPA, HSUPA, and 1×EV-DO compared to the DCH of UMTS.

9.4 IEEE 802.16 WiMAX

IEEE 802.11 for wireless LANs (Section 8.3) quickly became popular after it was introduced, but it was designed for wireless LANs. As such, the range is limited to up to about a few hundred meters, and the MAC protocol is best for short-range communications. For wider areas such as *metropolitan area networks* (MANs; see Section 10.2.1 for the differences between LAN, MAN, PAN, etc.), a new design was needed. Thus, in 1998, a year after the initial version of 802.11 came out, work began on IEEE 802.16. Whereas IEEE designed 802.11 for wireless LAN, 802.16 was designed for point-to-multipoint MAN coverage. The system popularly known as *WiMAX* is based on the IEEE 802.16 standard. Actually, 802.16 is an entire family of standards with many options, but the main choices these days are:

- Options designed for fixed broadband wireless applications, popularly known as *fixed WiMAX*, and specified in the revision IEEE 802.16-2004 (often also known as 802.16d, which is, strictly speaking, an incorrect designation for 802.16-2004, since it is a revision and not an amendment)
- Options designed for mobile broadband wireless applications, popularly known as *mobile WiMAX*, and specified as amendment 802.16e-2005 (often referred to as just 802.16e), which is an amendment to 802.16-2004 that adds mobility support.

We discuss the development of the 802.16 standards further in Section 17.2.2. There, we also discuss the relationship between the actual IEEE standards and the WiMAX forum, in terms of how they both affect what options get deployed. Features in WiMAX include:

- HARQ (both types I and II)
- Multiple-antenna support
- OFDMA

9.4.1 Use of HARQ

The use of HARQ is optional in WiMAX.

9.4.2 Use of OFDMA

In the most general case, any combination of subcarriers could be assigned to any mobile device for transmitting or receiving. In 802.16, however, only groups of subcarriers can be assigned. These groups are called *slots* and *bursts*. The smallest allocation is the slot. In general, a slot might contain subcarriers distributed both in frequency (subcarriers with different center frequencies) and time (subcarriers in different OFDMA symbols). The exact combination of subcarriers in a slot depends on various factors, including whether it is for the uplink or the downlink, and on the *subchannelization* scheme.

9.4.2.1 Subchannelization Schemes IEEE 802.16 defines a number of subchannelization schemes. These are schemes for the groupings of subchannels into units like slots. There are two main approaches to subchannelization:

- Distributed schemes, which group together subcarriers that are far apart; these include the *downlink full usage of subchannels* (DL FUSC), *downlink partial usage of subchannels* (DL PUSC), and *uplink partial usage of subchannels* (UL PUSC) schemes.
- Adjacent subcarrier schemes, which group together subcarriers that are adjacent to each other; these include the *downlink adaptive modulation and coding* (AMC) and the *uplink adaptive modulation and coding* (AMC) schemes.

The idea behind the distributed schemes is that they can best take advantage of frequency diversity. The idea behind the adjacent subcarrier schemes is that they can best exploit good channels (the good subcarriers would tend to be clustered together in frequency rather than spread out) to perform adaptive modulation and coding.

DL FUSC. In the DL FUSC, there are two types of pilot subcarriers (Figure 9.6):

- *Constant set pilot subcarrier*: fixed subcarriers in consecutive OFDMA symbols
- *Variable set pilot subcarrier*: variably located subcarriers in consecutive OFDMA symbols

In addition, there are also guard subcarriers on the left and right.

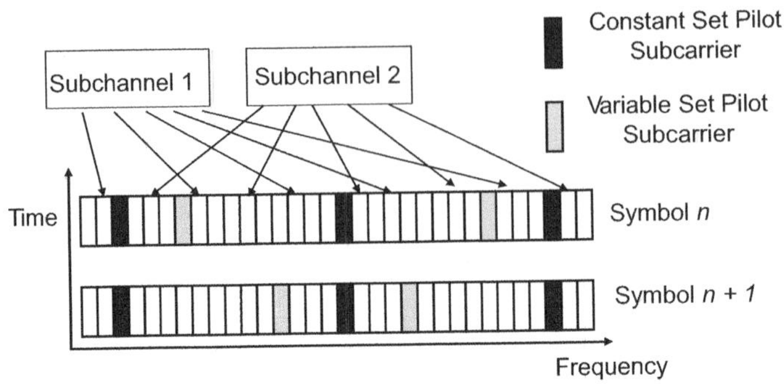

FIGURE 9.6 WiMAX FUSC scheme.

For example, for the $n = 1024$ OFDMA, there are 173 guard subcarriers, 82 pilot subcarriers, and 1 DC subcarrier, leaving 768 data subcarriers. These 768 data subcarriers are divided evenly into 16 subchannels, with 48 subcarriers each. The 48 subcarriers are distributed over the full range of the data subcarriers in the OFDMA frame.

DL PUSC. In downlink PUSC, the entire set of pilot and data subcarriers are grouped into *clusters*. Each cluster is 14 adjacent subcarriers in two consecutive OFDMA symbols, making a total of 28 subcarriers (the same 14 adjacent subcarriers in the two OFDMA symbols). Of the 28 subcarriers, 24 are data subcarriers and 4 are pilot subcarriers. Each cluster is pseudo-randomly distributed into one of six groups. A subchannel is formed from two clusters in the same group. Thus, there are 56 subcarriers per subchannel (14 adjacent subcarriers in two consecutive OFDMA symbols, plus another 14 adjacent subcarriers in two consecutive OFDMA symbols), and because of the pseudo-random way the groups are formed, the clusters within the subchannel could be close together or far apart in frequency. The downlink PUSC scheme is shown in Figure 9.7.

NB: The cluster size is always the same (the same 14 adjacent subcarriers in two consecutive OFDMA symbols) even for different values of N. Different values of N just result in more clusters or fewer clusters. Why bother with the six groups? Why not just define the subchannels directly as combinations of *any* two clusters? The distribution of clusters into groups allows the network operator the option of using all the groups, or just a subset of them, with a particular transmitter. For example, a base station may be sectorized and allocate two groups to each of its three sectors even while using the same frequency band for all the sectors. Thus, interference between the base station transmissions in different sectors is reduced.

UL PUSC. On the uplink there is also a PUSC scheme, but it is different from the downlink PUSC. Instead of clusters, it uses *tiles*, which we can think of as similar to clusters but a different size. In particular, whereas a cluster is 14 adjacent subcarriers over two OFDMA symbols, a tile is four adjacent subcarriers over three consecutive

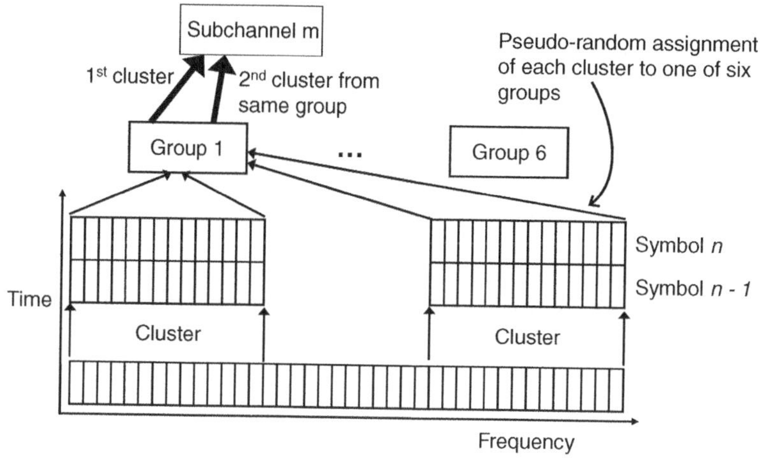

FIGURE 9.7 WiMAX downlink PUSC scheme.

OFDMA symbols. Eight of the subcarriers are data subcarriers and four are pilot subcarriers. There is also another option for three adjacent subcarriers over three OFDMA symbols, with eight data subcarriers and one pilot subcarrier, which may be useful in applications of WiMAX where channel tracking is easy.

As with DL PUSC, the tiles are then randomly distributed into six groups. Whereas with DL PUSC, one subchannel is two clusters from the same group, with UL PUSC one subchannel is six tiles from the same group, for a total of 72 subcarriers. UL PUSC is illustrated in Figure 9.8.

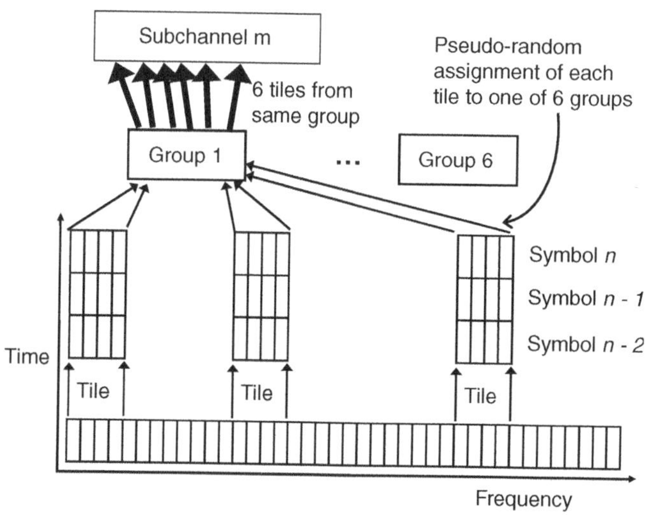

FIGURE 9.8 WiMAX uplink PUSC scheme.

NB: There is an option to use, on the downlink, the same structure as UL PUSC, and this option is sometimes called *tile usage of subcarriers* (TUSC).

Band AMC. WiMAX provides operators with a lot of flexibility to deploy their systems for their particular requirements. Thus, it was recognized that there are two very different strategies that could be used in the allocation of subcarriers to channels in an OFDMA-based system:

- Spread the subcarriers out to achieve frequency diversity. We would expect that most of the time, we would have a range of subcarrier quality, so we would get an averaging effect, which is more stable compared to the case where we allocate only adjacent subcarriers to a channel and then sometimes have to deal with a situation where all of the subcarriers in the subchannel are bad at the same time.
- Allocate only adjacent subcarriers to each subchannel. We then expect that the subcarrier quality will tend to go up and down together as a group. However, this doesn't have to be a bad thing, if we can adaptively change the modulation on the subcarriers. Thus, the modulation can be a low-rate modulation when the group of subcarriers is bad, and it can be a high-rate modulation when the group is good.

FUSC and PUSC follow the first strategy, whereas *band adaptive modulation and coding* (band AMC) follows the second strategy. Band AMC allocates subcarriers adjacent to each other to a subchannel (Figure 9.9). Thus, it loses the potential benefits of frequency diversity. However, it is more convenient to adapt the modulation on the subcarriers as a group, or band, than when they are spread out. Furthermore, when used in conjunction with multiuser diversity, it could achieve pretty good performance

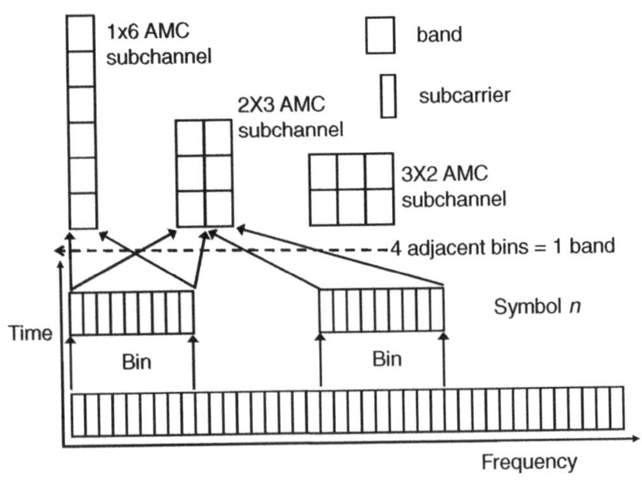

FIGURE 9.9 WiMAX band AMC scheme.

by scheduling transmissions to take advantage of when subchannels for different users are good. At those times, transmission is allocated to those users, and their subchannels can as a band or group use a higher-level modulation so that the average data rate can be high.

The specific grouping of subcarriers into bands and subchannels works as follows:

- Nine adjacent subcarriers (eight data subcarriers and one pilot subcarrier) form a *bin*.
- Four adjacent bins form a *band*.
- Six contiguous bands form a *subchannel*.

When we talk about subcarriers (in each bin) or bins (in each band) being adjacent, we mean adjacent in frequency. However, when we say that the bands in a subchannel are contiguous, they don't have to be adjacent in frequency, and in fact, they cannot be six bands adjacent to each other in frequency. Instead, a subchannel is either:

- Six consecutive bands in time (i.e., the same one band over six OFDMA symbols)
- Two adjacent bands over three OFDMA symbols
- Three adjacent bands over two OFDMA symbols

9.4.2.2 Slots, Bursts, and Frames Now that we have seen various options for forming subchannels, we move on to the related concepts of slots, bursts, and frames. A *slot* is comparable to a subchannel. A slot can be thought of as the data subcarriers (and just the data subcarriers, not the pilot subcarriers) within a subchannel. In fact, it is the slot, rather than the subchannel, that is the smallest unit of physical layer resources that can be allocated to a mobile device. A slot is always 48 data subcarriers (but not necessarily in the same OFDMA symbol). In fact, the reader might have noticed in Section 9.4.2.1 that even though the number of subcarriers per subchannel varies from subchannelization scheme to subchannelization scheme, in all cases, once we subtracted the pilot subcarriers we had 48 data subcarriers per subchannel. The interested reader may verify this by doing Exercise 9.4.

A *burst* is multiple slots that are contiguous in time and in *logical subchannel number*. When we say that slots are contiguous in logical subchannel number, we mean as in Figure 9.10. Thus, once the FUSC (or PUSC, etc.) mapping is performed, the actual physical subcarriers in a burst could be spread out over frequency.

The frame structure in OFDMA is usually based on time-division duplexing (TDD), so a paired spectrum is not needed. It is a sequence of OFDMA symbols on the downlink, followed by a *Tx/Rx transition gap* (TTG), and then followed by a sequence of OFDMA symbols on the uplink. There is then an *Rx/Tx transition gap* (RTG) before the next frame begins. The TTG and RTG are guard times whose values (e.g., around 80 μs in some cases) are chosen depending on the variations in transmission delay between mobile stations far from, and near to, the base station.

Figure 9.10 shows an example of an OFDMA frame used in a WiMAX system. It is important to note that the vertical axis is a subchannel logical number, so when this

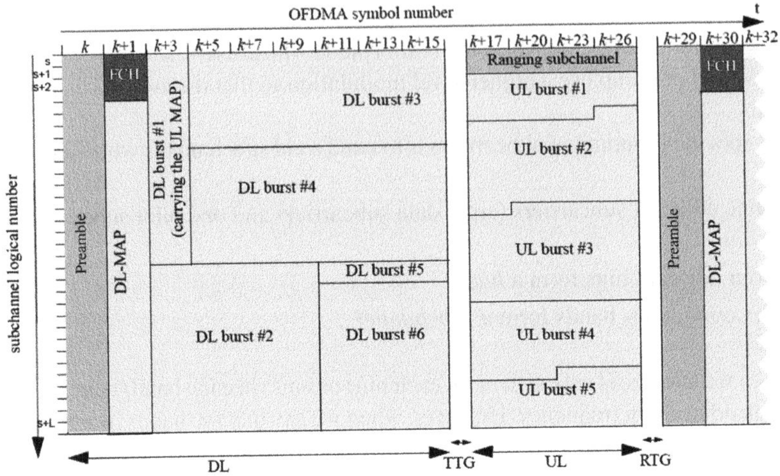

FIGURE 9.10 WiMAX OFDMA frame. (From IEEE 802.16-2009; copyright © 2009 by IEEE, reprinted with permission.)

gets mapped to actual subcarriers, they could be spread out (in frequency) all over the OFDMA symbol.

The downlink frame begins with a *preamble*, which is one OFDMA symbol long. The next OFDMA symbols in the frame are divided between a *frame control header* (FCH) and the DL-MAP. These are then followed by data bursts. The first downlink data burst, in additional to carrying user data, also contains the UL-MAP. We briefly explain the preamble, FCH, DL-MAP, and UL-MAP now.

Preamble. The preamble is a set of known sequences and helps with synchronization and equalization in the receivers.

FCH. The FCH provides crucial information like the channel coding used in the DL-MAP and the length of the DL-MAP.

DL-MAP. The DL-MAP, as its name suggests, contains information on the allocation of downlink subchannels for each mobile station. In addition to these mappings, it also contains general information such as the base station identification and the number of OFDMA symbols in the downlink subframe of the current frame.

UL-MAP. The UL-MAP is similar to the DL-MAP, but for the uplink.

The preamble is fixed, and the FCH and DL-MAP must use the downlink PUSC. After that, the subchannelization on the downlink bursts can be any of the subchannelization schemes we discussed in Section 9.4.2.1. This information is also provided in the DL-MAP.

As for the uplink frame, it begins with three OFDMA symbols that together contain the *ranging channel*, *channel quality information* (CQI) channel, and the

ACK channel. It can then be followed immediately by the uplink bursts, since the base station would have already provided the downlink and uplink maps in the downlink subframe.

Ranging Channel. The ranging channel comprises six adjacent UL PUSC subchannels (adjacent in the sense of having consecutive logical subchannel numbers). It is used by the mobile stations for procedures such as *initial ranging* and *periodic ranging*. Ranging is about the MS acquiring a suitable timing offset and making the necessary adjustments to transmission power. It is a type of random access channel (with ranging slots provided by the BS in the UL-MAP), where a ranging slot is chosen randomly by the MS, and a CDMA ranging code is sent to the BS during that time. The BS can respond with a timing offset correction and power-level correction.

CQI Channel. The CQI channel is used to provide to the BS, channel quality information about the downlink transmission.

ACK Channel. The ACK channel is used to provide feedback for the HARQ on the downlink.

9.4.3 Other Aspects

Power control in WiMAX is, as in other systems, more critical on the uplink than on the downlink. Different subchannels of an OFDMA system are arriving from different mobiles at the base station, and the base station needs to coordinate with each of the mobiles by issuing power control commands. Since variable rates are supported, there is the question of whether the total power transmitted by the mobile is controlled or the density (power per subchannel). In WiMAX, it is the power per subchannel that is power controlled. Hence, as the number of subchannels transmitted by the mobile increases or decreases, the total power it transmits goes up or down, respectively, in proportion. For the initial transmit power settings from the mobile, a procedure called *ranging* is used. In ranging, a special physical layer is sent from the mobile and the base station uses it to estimate the channel response, and so on. It then indicates the initial transmit power and timing offset for the MS to use.

WiMAX systems following 802.16e must support hard handoff. However, they can optionally use *macro diversity handoff* (MDHO) or *fast base station switching* (FBSS). These are both forms of soft handoff. Thus, they both share the concept of an active set with traditional soft handoffs. The major difference between MDHO and FBSS is that with MDHO there are simultaneous communications with all the base stations in the active set (as in traditional soft handoffs), whereas with FBSS, the mobile is communicating with only one of the base stations in the active set (that base station is called the *anchor base station*), while it only monitors the rest of the base stations in the active set and performs ranging and maintains a valid connection ID with each of them. It can therefore very quickly switch the anchor BS from one base station to another without needing to perform handoff signaling.

The basic error control code in 802.16e is a rate-1/2 convolutional code of constraint length 7. Depending on the mode and options chosen for a given system, Reed–Solomon codes, block turbo codes, convolutional turbo codes, and LDPC codes have been specified and may be found in WiMAX systems.

Multiple antennas are supported in WiMAX for spatial multiplexing, spatial diversity and beamforming, and combinations thereof. They are an integral part of 802.16e, and the standard specifies how to divide the pilot subcarriers in each OFDMA symbol between the antennas (the channel from each transmit antenna is different and must be estimated at the receiver, so each transmit antenna needs pilot subcarriers). A number of feedback options are also specified, allowing closed-loop MIMO to be used.

9.5 LTE

Features in LTE include:

- HARQ
- Multiple-antenna support
- OFDMA

There are various channels in LTE, grouped into *logical channels* that make use of the services of underlying *traffic channels*, which are then mapped to underlying *physical channels*. A discussion of all the channels is out of our scope, but we briefly introduce a few that will be relevant to the discussion in this section.

Uplink Channels. Most of the uplink traffic goes on the *physical uplink shared channel* (PUSCH), except for control signals such as HARQ acknowledgments, channel status reports, and so on, that are sent on the *physical uplink control channel* (PUCCH). The transport channel *uplink shared channel* (UL-SCH) is mapped to the PUSCH.

Downlink Channels. Most of the downlink traffic goes on the *physical downlink shared channel* (PDSCH). Two transport channels are mapped to the PDSCH. They are the *paging channel* (PCH) and the *downlink shared channel* (DL-SCH). There are also other transport channels, such as the *broadcast channel* (BCH) and the *multicast channel* (MCH), that are not mapped to the PDSCH. BCH is used to broadcast certain system information, and MCH is used to support MBMS (Section 13.2.4).

9.5.1 Use of HARQ

HARQ in LTE is part of the MAC layer, although aspects of it, like soft combining (of the original transmission and the retransmissions), are done in the physical layer. Additionally, at the radio link control (RLC) layer above the MAC layer, LTE has traditional ARQ. Why ARQ at the RLC as well as HARQ at the MAC and physical layer? Since HARQ relies on feedback from the receiver, there can be errors that go uncorrected by HARQ. The use of ARQ at the RLC can significantly reduce the number of these remaining errors and present a very low error service to higher layers, albeit at the cost of additional latency.

It doesn't make sense to use HARQ for broadcast channels such as the BCH or multicast channels such as the MCH, so HARQ is used only for the DL-SCH and UL-SCH.

9.5.2 Use of OFDMA on Downlink

The basic unit of time–frequency resource that can be allocated is a *resource block*. Each resource block consists of 12 contiguous subcarriers over seven time slots (seven OFDM symbols), thus containing 84 resource elements (if we consider one subcarrier over one time slot as a resource element). Such grouping helps reduce signaling overhead while giving up on the finest level of control (if individual subcarriers could be allocated), and a similar method is used in WiMAX, as seen earlier. It may seem at first that the use of OFDMA in LTE sacrifices frequency diversity (since 12 contiguous subcarriers are grouped together in each resource block), whereas we have seen how schemes such as FUSC in WiMAX can provide a lot of frequency diversity. However, the resource blocks in LTE are shared (as in HSPA), not dedicated (as in WCDMA). So one or more resource blocks across the band (hence, providing frequency diversity) can be assigned to a mobile, and the TTI is 1 ms, so the resource blocks can be switched rapidly between users depending on need, channel response to each mobile, and so on. If some resource blocks are received poorly (in a fade) at one mobile, and others are received better, the stronger ones can be allocated to that mobile.

9.5.3 SC-FDMA or DFTS-OFDM on Uplink

LTE uses a variation of OFDMA on the uplink. It is called *single-carrier FDMA* (SC-FDMA; Figure 9.11). A main reason it is used on the uplink is that SC-FDMA has been shown [10] to suffer less from the PAPR problem (Section 6.5.2). This is especially important on the uplink because lower PAPR means more efficient use of the power amplifier. On the uplink, that means that the mobile station can transmit with higher power efficiency with SC-FDMA than OFDMA, thus extending battery life, which is very important on the mobile station side.

SC-FDMA can be thought of as OFDMA with the addition of a DFT before the subcarrier mapping on the transmitter side, and the addition of a corresponding IDFT in the receiver side. As a result of the DFT, what is mapped to each subcarrier is not just 1, 2, 4, or 6 user bits (depending on the modulation level) after FEC and/or interleaving, but is a function of the entire block of bits. Therefore, it is sometimes called *DFT-spread OFDM* (or DFTS-OFDM), because of this "spreading" effect of the DFT. It can be thought of as a single-carrier FDMA system in the sense that the added DFT transforms from the time domain to frequency domain, and then the IFFT transforms it back to the time domain, back to single carrier.

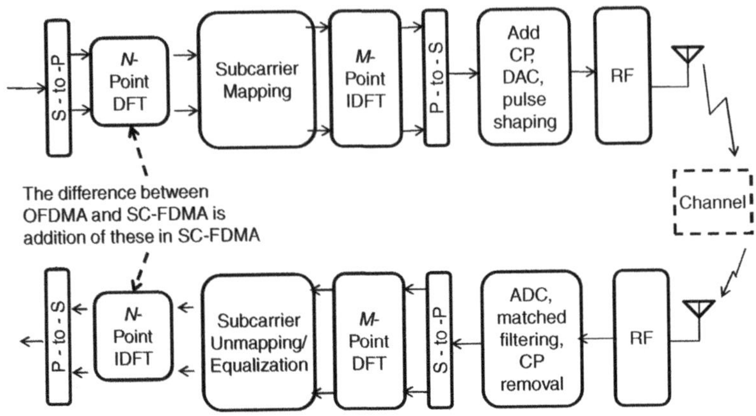

FIGURE 9.11 SC-FDMA block diagram.

In general, though, the size of the added DFT/IDFT would be different from the size of the standard IFFT/FFT in OFDMA. Otherwise, if the two are both N-point transforms, they would just cancel each other out. Here, we use M for the standard IFFT/FFT and use N to denote the size of the added DFT/IDFT. Then, with SC-FDMA, $M > N$, and a basic point about multiple access with SC-FDMA is that different transmitters can be assigned different subsets of the M subchannels. This is very similar to OFDMA. The main difference is that instead of the input to the IFFT being the data symbols (at the assigned subsets of the subchannels and zero everywhere else), they are Fourier-transformed combinations of the data symbols.

As with OFDMA, there is the question of how to map the N symbols into a subset of the M SC-FDMA subchannels. In theory, the N symbols from each transmitter might be clustered together, or they might be spread out completely with zeros in between subchannels so that at the receiver, the subchannels from different transmitters are "interleaved," or some other way might be chosen to map the symbols to subchannels. Each mapping scheme has its pros and cons. For LTE, the N symbols from each transmitter are clustered together, as a group contiguous in frequency [3].

Of course, the N-point DFT/IDFT can be implemented with FFT and IFFT, for efficiency. But perhaps to avoid confusion, they are written customarily as DFT/IDFT.

9.5.4 Other Aspects

Like many other wireless systems, LTE uses open- and closed-loop power control. The focus, as usual, is on the uplink power control. Power control commands are sent by the base station for the mobile to increase or decrease it transmit power level. Unlike UMTS or HSPA, LTE uses a 2-bit power control command, thus providing for four possibilities for the power control step: -1 dB, 0 dB, 1 dB, or 3 dB. In contrast, earlier systems, with 1-bit power control, could only request an increase or decrease. Additionally, unlike UMTS and HSPA, where all the channels are scaled up or down together, in LTE the power control commands could be different, and independent, for the physical uplink control channel (PUCCH) and for the physical uplink shared channel (PUSCH).

Since LTE is not CDMA-based, it does not need to use soft handoffs, so it uses hard handoffs. LTE uses the same convolutional Turbo codes as UMTS and HSPA [3], but with an improved interleaver.

9.6 WHAT'S NEXT?

New systems continue to be specified, such as WiMAX 2 and LTE-Advanced (3GPP Release 10). ITU-R has evaluated both WiMAX and LTE and declared neither to be worthy of the label "4G," whereas WiMAX 2 and LTE-Advanced would qualify. However, ITU-R has subsequently relented and allowed operators to call their networks "4G" for marketing purposes even if they do not meet ITU-R's original technical requirements for 4G.

EXERCISES

9.1 For a similar multipath environment, which of the following spatial multiplexing configurations could potentially yield the highest data rates? (a) 2×2; (b) 3×3; (c) 2×4; (d) 6×2?

9.2 A wireless system uses HARQ with chase combining. Suppose that 40% of the time no retransmission is necessary, 30% of the time one retransmission is necessary, 20% of the time two retransmissions are necessary, and 10% of the time three retransmissions are necessary. For the single transmission, the code rate is $1/2$. What is the effective code rate?

9.3 Why is the data rate on HSUPA not as good as on HSDPA?

9.4 In DL FUSC it is clear that each subchannel has 48 subcarriers. Check the following:

- DL PUSC: How many subcarriers are there per subchannel? How many pilot subcarriers are there in each subchannel? So how many data subcarriers are there in each subchannel?
- UL PUSC: How many subcarriers are there per subchannel? How many pilot subcarriers are there in each subchannel? So how many data subcarriers are there in each subchannel?
- Band AMC: How many subcarriers are there per subchannel? How many pilot subcarriers are there in each subchannel? So how many data subcarriers are there in each subchannel?

9.5 In SC-FDMA, why does the N-point DFT before the subcarrier mapping work? Why wouldn't it "cancel out" the IFFT in the OFDM processing, giving us a simple time-domain transmission without the frequency-division aspects?

REFERENCES

1. S. Alamouti. A simple transmit diversity technique for wireless communications. *IEEE Journal on Selected Areas in Comunications*, 16(8):1451–1458, Oct. 1998.
2. J. Andrews, A. Ghosh, and R. Muhamed. *Fundamentals of WiMAX*. Prentice Hall, Upper Saddle River, NJ, 2007.
3. E. Dahlman, S. Parkvall, J. Sköld, and P. Beming. *3G Evolution: HSPA and LTE for Mobile Broadband*, 2nd ed. Academic Press, San Diego, CA, 2008.
4. K. Etemad. *cdma2000 Evolution: System Concepts and Design Principles*. Wiley, Hoboken, NJ, 2004.
5. G. Foschini and M. Gans. On limits of wireless communications in a fading environment when using multiple antennas. *Kluwer Wireless Personal Communications*, 6:311–335, Mar. 1998.
6. H. Holma and A. Toskala, editors. *HSDPA/HSUPA for UMTS*. Wiley, Hoboken, NJ, 2006.
7. H. Holma and A. Toskala. *WCDMA for UMTS: HSPA Evolution and LTE*, 4th ed. Wiley, Hoboken, NJ, 2007.
8. W. C. Jakes, editor. *Microwave Mobile Communications*. 2nd ed. Wiley-IEEE Press, New York, 1994.
9. J. Mietzner, R. Schober, L. Lampe, W. Gerstacker, and P. Hoeher. Multiple-antenna techniques for wireless communications—a comprehensive literature survey. *IEEE Communications Surveys and Tutorials*, 11(2):87–105, 2009.
10. H. G. Myung, J. Lim, and D. J. Goodman. Single carrier FDMA for uplink wireless transmission. *IEEE Vehicular Technology*, 1(3):30–38, Sep. 2006.
11. H. Schulze and C. Lüders. *Theory and Applications of OFDM and CDMA: Wideband Wireless Communications*. Wiley, Hoboken, NJ, 2005.
12. V. Tarokh, N. Seshadri, and A. Calderbank. Space-time codes for high data rate wireless communication: performance criteria and code construction. *IEEE Transactions on Information Theory*, 44(2):744–765, Mar. 1998.
13. S. Yang. *3G CDMA 2000*. Artech House, Norwood, MA, 2004.
14. C. Yuen, Y. L. Guan, and T. T. Tjhung. *Quasi-Orthogonal Space-Time Block Code*. Imperial College Press, London, 2007.

IV

NETWORK AND SERVICE ARCHITECTURES

10

INTRODUCTION TO NETWORK AND SERVICE ARCHITECTURES

In this part of the book, Chapters 10 to 13, we discuss network and service architectures for wireless networks. We begin in this chapter by laying some foundations focused primarily on general networking concepts and IP networking, keeping in mind the move toward "all-IP" networking in wireless networks. (This can be viewed as a convergence toward all-IP wireless networks from two starting points: the traditional wireless cellular networks and the traditional IP-based data networks.) We then spend two chapters tracing the evolution of wireless networks and network architectures from GSM to LTE and the gradual shift from voice- and circuit-centric networks toward packet-centric networks. In the first of these chapters, Chapter 11, we discuss the GSM network. In the same chapter we introduce steps taken to expand the capabilities of IP to support voice over IP (VoIP) and quality of service (QoS), both of which are important building blocks needed for IP networks to take a more prominent role, such as in wireless all-IP networks. We continue in Chapter 12 to examine more steps taken to fit IP networks with capabilities to be used in wireless networks, such as the addition of mobility support. We also see how, on the other hand, wireless networks started adding packet-carrying capabilities in the form of GPRS. The chapter continues by examining the evolutionary steps through different releases of the 3G network (UMTS) until LTE is reached. The last of the chapters in this part of the book, Chapter 13, examines service architectures as well as alternative network architectures such as mobile ad hoc networks.

The rest of the present chapter contains a review of fundamental general networking concepts in Section 10.1, before moving on specifically to network architectures and related concepts in Section 10.2. IP networking, including IPv6, is then reviewed in Section 10.3, and we introduce teletraffic analysis in Section 10.4.

Fundamentals of Wireless Communication Engineering Technologies, First Edition. K. Daniel Wong.

10.1 REVIEW OF FUNDAMENTAL NETWORKING CONCEPTS

Just as on a road transportation network, the cars and other vehicles are collectively called *traffic*, so in communications networks the user data that move around the networks are called traffic. Communication networks are *complex*, *distributed*, and *have limited resources*. Since communication networks are complex systems with many functions, we try to design them to help network engineers and architects to better understand, design, implement, and troubleshoot communication networks. To achieve this objective, the system complexity must be managed. One of the ways that this happens in most communications networks is through the use of modules called *layers*. We discuss the concept of layering in Section 10.1.1.

For the distributed elements to work together to implement certain functions, they need to communicate among themselves to negotiate communications parameters, reserve resources, and so on. These communications are internal to the communications network and are necessary in order for the system to work; however, they are distinct from the communications of the users of the network. Hence, we call these *control signaling* or *control traffic*, in contrast to *data traffic* or *user traffic*, which refers to user data traffic. Control signaling may be *in-band* or *out-of-band* (see Section 10.1.2.1 for more details).

The two basic ways to do switching in a network have to do with how the distributed elements handle resource allocation. With *circuit switching*, resources are prearranged for communication paths from one end user to another end user, whereas the resources are not prearranged with *packet switching*. We elaborate on circuit switching vs. packet switching in Section 10.1.2.

10.1.1 Layering

Modular design is one way to tackle the complexities of large, complex systems by breaking them up into subsystems or modules, each of which is responsible for a subset of the functions of the entire system. Thus, it can be thought of as a divide-and-conquer approach and is a well-respected and useful approach. Modular design has benefits in system design, implementation, and operations (including network management and troubleshooting). In communications, the most prominent example of modular design can be seen in the concept of *layering*. Functions and protocols are distributed between layers, making the design, implementation, and operation of each layer more manageable than if there were no layers. Layering is a form of modular design with a hierarchical flavor: Higher layer functions are constructed with the assistance of lower layer functions. The layers are often arranged in a vertical manner consistent with this hierarchical flavor, and the vertical arrangement is called a *protocol stack*.

A famous reference model for how a communication system can be built in a layered way is the *open systems interconnection* (OSI) reference model, shown in Figure 10.1. It comes from the International Standards Organization (ISO). The name suggests that it meant for interconnection of systems that are open for communication with other systems. To provide some structure to how those communications might be organized, the OSI reference model breaks the functions down into seven layers,

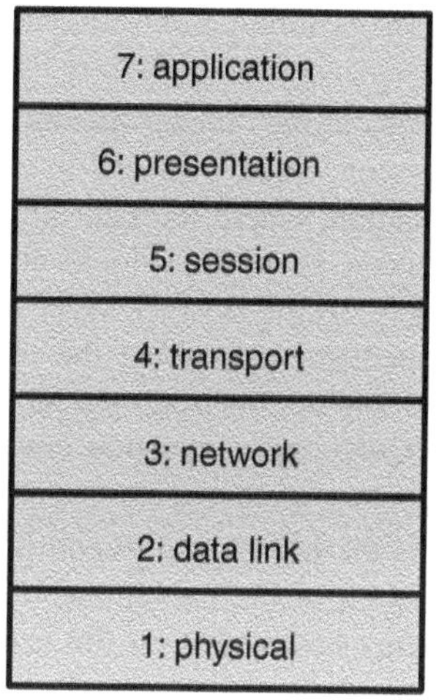

FIGURE 10.1 OSI model with a seven-layer protocol stack.

where the choice of layers and what functions go into each layer is based on such principles as:

- The functions of each layer should be well defined.
- The functions of each layer should be such that they can be standardized by international standards organizations.
- Each layer should present a different abstraction to higher layers.
- The groupings of functions to each layer should be such that information flow is minimized across the interfaces.
- There should not be too many or too few layers.

These principles also apply more generally to the modular design of systems in a more general context (with modules replacing layers in that case; in the case of modular design, other principles can be suggested in addition to these, but we don't discuss them since they are outside our scope).

The seven layers of the OSI reference model (from lowest to highest) are:

1. *Physical.* This layer contains functions related to the physical medium, how raw bits are transmitted over it, what type of processing should be done in the transmitters and receivers to minimize bit error rates, and so on.
2. *Data link.* This layer uses the services of the physical layer for the transmission of bits. While the physical layer would usually be a link with errors (even if the bit error rate is low), the data link layer tries to present "a line that appears free of undetected transmission errors" [6] as its service to the network layer. This layer is also the lowest layer in which flow control can, and sometimes is, handled. For *broadcast networks*, the data link layer also handles

access to the shared network through its sublayer, the *medium access control* (MAC) sublayer.

3. *Network.* This layer looks at the "bigger picture" than the physical and data link layers, in that it is concerned with end-to-end communications, from a source to a final destination, where the path might need to go through one or more intermediate nodes. However, it looks at the end-to-end communications in a distributed way, so the network-layer peer of a source node might be just the next intermediate node along the path to the destination (only from the transport layer and up, the peers are the true endpoints). Issues such as quality of service (QoS) and congestion control are also handled in the network layer.
4. *Transport.* This layer is the lowest end-to-end layer where the peers are the true source and destination of the communications. It may provide reliable data transfer to the higher layers, even if the network layer provides a best-effort service. Tunneling protocols (which we discuss in Section 10.2.6) can be said to carry data at the transport layer, because they deliver data from one endpoint to another endpoint.
5. *Session.* This layer manages dialogs and connections between the endpoints.
6. *Presentation.* This layer translates between data in different syntax or semantics, to present to the application layer, for example; if the application layer on the two ends uses different syntax, the presentation layer would need to do the necessary translation for the two applications to talk to each other.
7. *Application.* This layer is where communications applications reside. Examples include the hypertext transfer protocol (HTTP) and the file transfer protocol (FTP).

The physical and data link layers are the only two layers where the two peers are usually connected directly to the same medium. The session, presentation, and application layers may sometimes be said to be merged into one layer, for example, in TCP/IP networks, where it may be said that there are just five layers of the protocol stack, with an application layer sitting on top on the transport layer.

It should be noted that services and protocols are distinct concepts. As Tanenbaum puts it succinctly [6]:

> A *service* is a set of primitives (operations) that a layer provides to the layer above it. The service defines what operations the layer is prepared to perform on behalf of its users, but it says nothing at all about how these operations are implemented. A service relates to an interface between two layers, with the lower layer being the service provider and the upper layer being the service user.
>
> A *protocol*, in contrast, is a set of rules governing the format and meaning of the packets, or messages that are exchanged by the peer entities within a layer. Entities use protocols to implement their service definitions. They are free to change their protocols at will, provided they do not change the service visible to their users. In this way, the service and the protocol are completely decoupled.

10.1.2 Packet Switching vs. Circuit Switching

Traditionally, the telephone network has been a circuit-switched network. Data can be carried on circuit-switched networks, too, but arguably not very efficiently. With the emergence of packet data networks, we have two major paradigms for switching of traffic in communications networks. In this section our discussion focuses on examples of circuit and packet switching. The traditional telephone network is given as an example of a circuit-switched network (Section 10.1.2.1), IP networks (like the Internet) are given as an example of a packet-switched network (Section 10.1.2.2), and ATM is given as an example of a type of hybrid between the two (Section 10.1.2.3). Later, the issues of packet switching and circuit switching from a broader architectural perspective are discussed. In Section 10.2.7 we consider the movement toward *convergence*, one of the main ideas of which is to carry all kinds of traffic (even those that have traditionally been carried on circuit-switched networks) over next-generation packet-switched networks.

10.1.2.1 Traditional Phone Network We sometimes use the terms *calling party* and *called party* to refer to the two ends of a voice session, such as a phone call. The difference between calling party and called party is that the *calling party* is the one that initiates the call or session, whereas the *called party* is the one that responds to the initiation by the other party. To be precise, if we need to differentiate between the human user and the device (such as a phone) that they are using, we could refer to the human user as the calling party or called party, and to their device as the calling party device or called party device. However, it should usually be clear from the context to what we are referring, so for convenience we generally just say calling party or called party. The process of trying to initiate a call (by the calling party) is called *call initiation*. The process in the network, during call initiation, of finding the called party and alerting it is known as *call delivery*. Although these terms were originally used to discuss calls made with a traditional phone, they are generic enough that they are also used for VoIP calls (using SIP, for example, as we will see in a subsequent chapter).

The traditional phone network is also known as the *public switched telephone network* (PSTN). In the PSTN, circuits are set up using SS7 signaling (as we will shortly explain) between switches. The switches are located in *central offices*. *Class 4* switches or *tandem switches* are those that connect other switches together. *Class 5* switches are connected to subscriber telephone lines. They need to provide dial tone and handle the subscriber lines. Thus, there would be hundreds or thousands of lines going out to subscriber lines from a class 5 switch, the number of which would be planned according to teletraffic analysis principles (Section 10.4) to provide high availability.

Signaling. Early signaling schemes in telephone networks were of the *in-band* variety (also known as *channel-associated signaling*), whereas later schemes such as SS7 are of the *out-of-band* variety, also known as *common channel signaling*. With in-band signaling, as the name implies, signaling is carried on the same channel on which speech is carried. Thus, signaling bandwidth may be limited, and it may be easier to perpetrate fraud (using tone generators, etc., as users have easier access to the signaling channel than in a case of out-of-band signaling). With out-of-band signaling, the signaling is carried on a separate network that can be provisioned to meet the bandwidth requirements and other requirements of the signaling, which makes it more difficult for hackers to mess with. Earlier signaling systems, up to SS5, used in-band signaling, but from SS6 onward, out-of-band signaling was used. In 1980, ITU-T (Section 17.2.3) introduced SS7, with improvements over SS6, which is better suited for digital systems.

SS7 is not needed for intraoffice (also known as intraexchange) calls, where both the calling party and called party are connected directly to the same central office (i.e., the same exchange). For all other calls, multiple switches are involved, and SS7 signaling is used between them. The SS7 signaling traverses from one signaling endpoint (a switch) to another (another switch) either through *signaling transfer points* (STPs) or directly. The use of STPs is popular in North America but is not common in Europe. Each approach has its pros and cons.

The lower layers of the SS7 protocol stack up to the network layer are handled by the *message transfer part* (MTP)—in particular, MTP levels 1, 2, and 3. For some higher layers, there is also the *signaling connection control part* (SCCP), which completes the network layer functionality, whereas some higher layers bypass SCCP and use the services of MTP directly. For example, on top of MTP, the *integrated services digital network* (ISDN) *user part* (ISUP [3]) is used for call-related messages between telephone switches. It occupies a central role and is used every day in the setting up, maintenance, and tearing down of circuits for millions of calls around the world. Later, the emergence of wireless cellular networks brought requirements for mobility handling to the phone network. With the introduction of these mobility requirements, additional messages and protocols were needed. The GSM *mobility application part* (MAP), for example, is used alongside ISUP in GSM networks. Another example of an application part is the *intelligent network application part* (INAP) for an intelligent network (see Section 13.2.5). These parts, whether called "*x* user part" or "*x* application part" (where *x* is ISDN, mobility, etc.), can be considered to be application layer parts of SS7. Thus, we could consider SS7 to just have the lower three layers (up to network layer) followed by the application layer on top of them. An example of ISUP signaling to set up a telephone call is shown in Figure 10.2.

More recently, *bearer independent call control* (BICC, ITU-T Q.1901) has been introduced and is gradually expected to replace ISUP. BICC is not tied to SS7 and can be transported over other networks, such as IP or ATM. When BICC is used with SS7, though, both ISUP and BICC use MTP services.

10.1.2.2 IP Network Networks based on the *Internet protocol* (IP) are ubiquitous these days. Even though IP was not designed with wireless and mobility in mind, IP has evolved to the point where it is essential to have a good understanding of IP networking in order to work with wireless networks. Although wireless networks were originally designed as traditional circuit-switched networks based on SS7, along with suitable

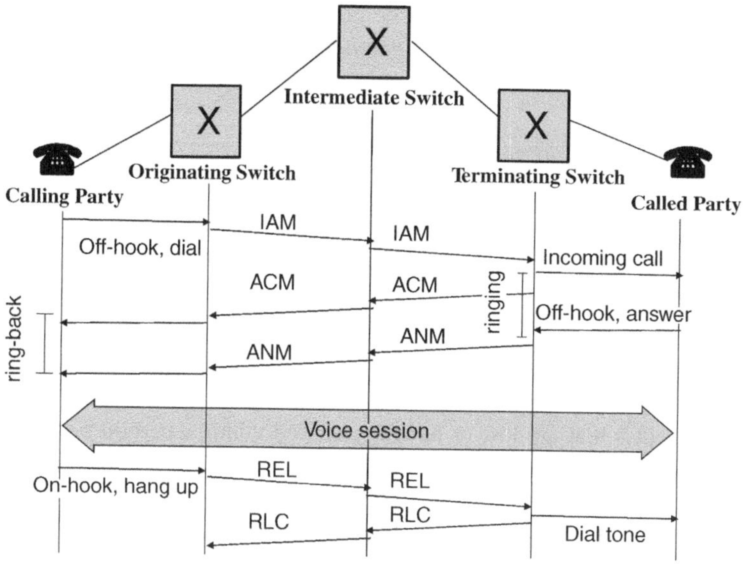

FIGURE 10.2 ISUP signaling in the PSTN.

extensions to support mobility, wireless networks have been moving toward an "all IP" architecture for over a decade. We devote Section 10.3 to IP networking.

10.1.2.3 ATM Network The *asynchronous transfer mode* (ATM) is a hybrid of packet and circuit switching. It was designed to provide good support for both voice and data in very high-speed switched networks. Since it was designed to support data efficiently, it does not use traditional circuits. To support voice and other constant-bit-rate applications, however, ATM provides for *virtual circuits*. The virtual circuits, as well as all other ATM traffic, are transported by a common carrier, the *cell*. Each ATM cell is a fixed-size "packet" consisting of a fixed-size 5-byte header followed by 48 bytes of data.

10.1.3 Reliability

In planning a reliable network, there are issues regarding reliability of the equipment (routers, switches, lines, etc.), how long they last, how often they fail, and son on. These are quantified by metrics such as *mean time to failure* (MTTF) and *mean time to repair* (MTTR). Redundant links and redundant routers or switches can help to reduce downtime further due to failures.

Then, there are also the dynamic operational reliability issues related to how the network handles changing traffic conditions. One advantage of circuit switching is that it is relatively more reliable than packet switching. With circuit switching, circuits are allocated for each communication between endpoints. With packet switching, dynamic factors can result in network congestion, queues getting full, and packets being dropped. To the newcomer it may be surprising that IP, which is so widely used for so many applications, is actually an *unreliable* protocol. For example, IP does not provide any guarantees:

- That a packet will arrive at the intended destination
- That a packet will arrive within a certain period of time
- That the delay variation in arrival times (also called *jitter*) will be bounded
- That packets sent in a certain sequence will arrive at the destination in the same sequence

However, all is not lost! For applications that need more reliability, TCP is available, and TCP provides reliability by using an ARQ mechanism (see Section 10.1.3.1) to retransmit packets that are lost or have not arrived within a certain period of time. For other applications, such as voice/video streaming, TCP may not be such a good idea. We discuss why, in Section 10.3.2, and provide more details on the use of alternative transport protocols such as UDP and RTP. For now, though, we note the power of modularity and layering that is evident here. IP was intentionally designed as unreliable, with a retransmission mechanism. This allows it to be used with a transport protocol such as TCP for applications that need reliable transport, whereas it can be used with UDP/RTP for applications such as voice and video that may not need reliable transport. This divide-and-conquer, mix-and-match approach has served the Internet well as it has grown rapidly and taken on all kinds of new applications that the original designers did not conceive of.

10.1.3.1 ARQ *Automatic repeat request* (ARQ) refers to a family of related techniques to support a reliable transport service over an underlying unreliable network or communication service. It can be considered a form of error control as well. The basic techniques are:

- *Stop and wait*: the simplest technique, in which the sender sends a packet and then stops and waits for an acknowledgment before sending the next one; clearly, this is inefficient because of the waiting, but the sender only needs to store the current packet.
- *Go back N*: in this technique, the receiver can request that the sender go back and resend packets from up to *N* packets back from the latest packet. Unlike stop and wait, the sender can have a "window" of up to *N* packets for which it has not yet received delivery confirmation from the receiver (through ACKs). It must be prepared to roll back in time and resend from up to *N* packets back from the latest packet (if it doesn't receive an ACK for that packet or receives a negative ACK, indicating that the receiver has not received it). Thus, it needs to store the *N* latest packets. The receiver, on the other hand, doesn't need to store any packets since the sender in go back *N* will just retransmit all the packets, starting from up to *N* packets back.

- *Selective repeat*: the receiver also has a buffer, so the receiver can simply afford to request a specific packet (up to N back) and the sender doesn't have to transmit $N, N-1$, and so on, up to the present packet, because the receiver is maintaining its own window of packets and can wait for the selective repeat transmission to arrive, whereupon it can insert the packet into the proper sequence.

10.2 ARCHITECTURES

Network architecture has to do with how networks are organized, classified, and structured to perform various functions. We begin by introducing popular classifications of networks according to size (Section 10.2.1), then distinguish between network areas: core, distribution, and access (Section 10.2.2). Next we introduce topics related to network topology (Section 10.2.3) and communication paradigms (Section 10.2.4), and briefly introduce one of the design philosophical areas of debate within the networking community: namely, how much intelligence should be put into the network (Section 10.2.5). Last but not least, we revisit the concept of layering (Section 10.2.6) and introduce network convergence concepts (Section 10.2.7).

10.2.1 Network Sizes

Not all networks are the same. Often, the geographical size of a network is a useful indicator of its requirements and of similarities with other networks of the same size. A useful classification of networks, therefore, is by their area:

- *Local area network* (LAN): a network in a limited geographical area, such as home or office, that may be hundreds of meters in diameter or less.
- *Wide area network* (WAN): the complement of LANs; can include networks that span the width of a city as well as networks that span the globe.
- *Metropolitan area network* (MAN): a network on the order of the area of a city. Although the terms LAN and WAN pretty much cover the entire range of network sizes, when we wish to refer to a more specific type of WAN, we can call it a MAN.
- *Personal area network* (PAN): a network with a smaller coverage area than a LAN. Bluetooth is a good example of a PAN wireless technology.

10.2.2 Core, Distribution, and Access

At a high level, networks can be divided into *edge devices* or *end users* and *infrastructure devices*. Edge devices are those at the edge of a network (e.g., a phone or an ADSL modem that connects to the network), and infrastructure devices are switches, routers, and so on, that help get traffic from place to place. Beyond this division, the infrastructure part of networks, especially large networks, are sometimes divided into different parts, based on how far the parts are from the edge devices.

- *Core*. The core has the highest volume of traffic to deal with and should focus on moving packets around as quickly as possible. Thus, computation-intensive decisions (related to routing, QoS, etc.) are moved to the distribution.
- *Distribution* (sometimes known as *aggregation*). Policies, access control lists, and so on, are applied in the distribution network. Various computation-intensive decisions are off-loaded from the core by the distribution network so that packets sent to the core can be moved around as quickly as possible.
- *Access*. Responsible for connecting end devices to the network, access deals with the particular challenges of different access technologies. In wired networks, high port densities are normally found in the access network.

The core is sometimes also called the *backbone*. Networks are sometimes divided into core and access only, omitting the distribution part. GSM networks, with the *radio access network* and *core network*, are an example of such networks.

10.2.3 Topology

The network topology has to do with the arrangement of the network devices: how they connect to each other. As such, it makes a difference whether the devices are connected to a *shared medium* (also known as a *broadcast medium*) or with *point-to-point links* (also known as a *non-broadcast medium*).

- *Hub-and-spoke*. Only one of the devices is connected directly to all the other devices. This device is called the *hub*, whereas all the other devices are *spokes*. Spoke devices that want to send data to other spoke devices need to send the data through the hub device. A hub-and-spoke topology utilizes $n - 1$ links, which is the smallest number of links between devices in which there is a connected path between any two of the devices.
- *Point-to-multipoint*. This is another name for hub-and-spoke.
- *Mesh*. Every device is connected to every other device, and there are $n(n - 1)$ links; sometimes the distinction is made between *full mesh* and *partial mesh*. A partial mesh is somewhere in between a mesh arrangement and hub-and-spoke.

10.2.4 Communication Paradigm

Similar to topology in some ways, but different, are the basic communication paradigms that may be adopted in sending particular data, namely:

- *Unicast*: one to one
- *Multicast*: one to many
- *Broadcast*: one to all

Other communication paradigms are possible, too: for example, *anycast*, as found in IPv6, but which is outside the scope of this book.

Broadcasting is a paradigm where all possible recipients are supposed to receive the broadcast. If random packets were broadcast to the entire Internet, it would consume and waste a tremendous amount of network resources. Thus, broadcasts are usually restricted to some defined area (e.g., to within a LAN).

The traditional argument for multicast services is that sometimes, the same data need to be sent to more than one recipient. One alternative is to separately unicast the same data to each of them individually. However, the paths taken between the sender and these recipients may overlap partially, at least from the sender up to some point, before the paths diverge. In the event that multiple recipients are in the same LAN, the entire path from sender to recipient could be the same, all the way to the destination LAN. In view of the partial or complete overlaps of paths, why send the same data along the same path, or partial path, multiple times? Multicasting is a way to remove this inefficiency.

Although multicasting can remove the inefficiency in unicasting of the same data along the same paths or partial paths, it does come with some costs. It requires some overhead to set up and manage the multicasting [there are different ways that it could be done; for TCP/IP, there are protocols such as *Internet group management protocol* (IGMP)]. Thus, it typically makes most sense to use multicasting either in situations where the savings from multicasting are considerable, or in special cases where little setup is required, if any. An example of the former case is how multicasting is a popular paradigm for distribution of multimedia (video, sound, etc.), where the savings can be considerable because of the bandwidths involved. An example of the latter case is special IPv6 multicast addresses such as the solicited-node multicast address (see Section 10.3.6.3).

10.2.5 Stupid vs. Intelligent Networks

A long-lasting debate in the networking community concerns the amount of "intelligence" in the network. The network needs to provide at least basic connectivity for communications between endpoints. However, over and above that, what else should the network provide? A more "intelligent" network might provide, for example, error correction, location information, and security services such as encryption, for its users, whereas a more "stupid" network might provide as few of these as possible, and leave it to the end users to provide these services where needed. Both approaches have their pros and cons. Most networks lie in between the two extremes.

10.2.6 Layering Revisited

Layering in communications networks is a powerful concept. One may not realize the flexibility and power that could be obtained from a layered approach to networking until one encounters such situations as the following:

- The link layer is often split into multiple *sublayers*, such as the logical link control (LLC) and medium access control (MAC).

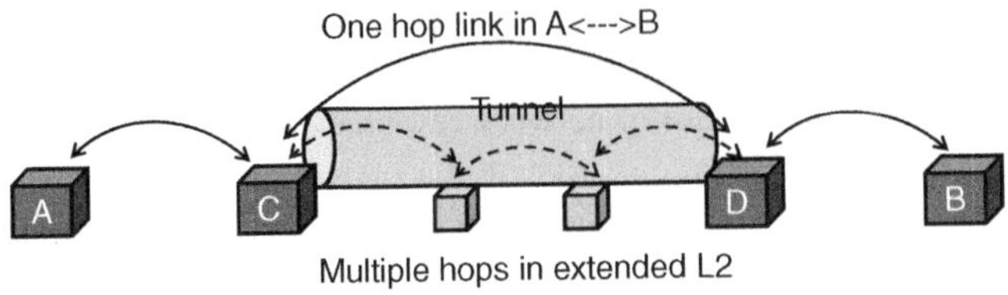

FIGURE 10.3 What layer are we on?

- The "link" at layer 2 is often not a simple shared physical medium; instead, the "link" may actually but an entire network connecting the two layer 2 peers together. This type of extended link layer is often seen: for example, in GPRS and in the ESS of WiFi networks. Millions of people also use *point-to-point protocol* (PPP) with access technologies such as DSL, where the two ends of the PPP connection appear as a layer 2 link to the higher layers, but the PPP connection traverses long distances and multiple pieces of equipment.
- Tunnels between two points in a network are often found: for example, in mobile IP and for security purposes (IPsec tunnels). They can also be viewed as a form of extended link layer.

These situations could cause confusion in our understanding of the layers unless one realizes that the layers are not hard and fast rules but a conceptual framework to help guide our thinking. Also, what layer something is at could depend on the context. Thus in the case of the tunnel in Figure 10.3, as far as the communications from A to B is concerned, the tunnel between C and D is one hop, that is, a direct link between C and D. However, the tunnel between C and D is actually made up of multiple hops over intermediate equipment such as routers. It provides a transport service (some might even say a transport layer service) to other communications (such as between A and B), whereas from the perspective of the IP/network layer of the communications between A and B, it is similar to the link layer.

10.2.7 Network Convergence

Various communications systems have arisen out of different backgrounds, each with their own requirements, design considerations, and applications, and each with their own resulting architecture. Other factors that might have gone into the architecture of a network might be the state of technological knowledge, or even philosophical trends (e.g., stupid vs. intelligent network), during the design of the architecture. Some are wired networks, some are wireless. There may be differences in requirements for bandwidth, and requirements arising from the types of traffic that the network might be expected to carry. Sometimes, design changes may be required in network architectures and protocols as the range of applications and other factors change (see Section 10.3.6 on the movement of IP to IPv6, for example). For example, we have the telephone networks, the cable TV networks, the cellular networks, the Internet,

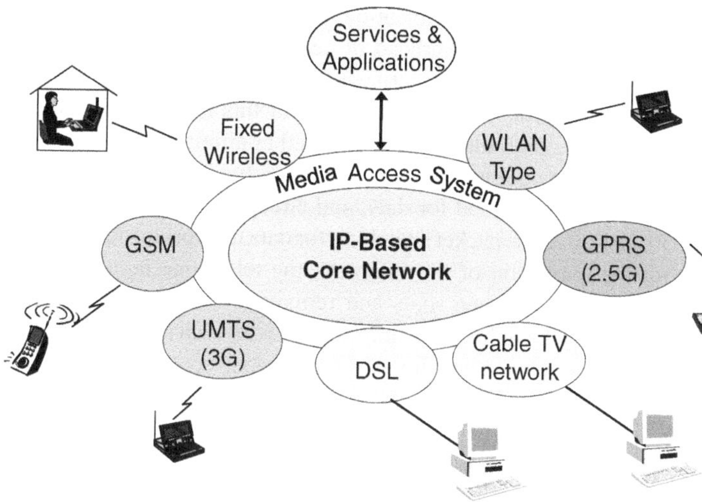

FIGURE 10.4 Vision of a converged network.

and various other networks. Where it comes to access networks, e.g., access networks providing broadband access to the Internet, the list increases (we have DSL, satellite, cable, WiFi, GPRS, etc.).

Thus, we have a situation where there are different systems and different networks, each of them complete in themselves, serving their respective applications. However, in many cases, they do not talk to each other, and integration requires effort. Thus, the different systems/networks are said to be "separate silos," *vertically integrated* (in terms of each having their respective protocol stacks) but not horizontally integrated, and so on. Therefore, one vision for a converged future network is as shown in Figure 10.4. In this figure we see that there are various access networks that connect to the same core network, and then services are provided through the common core network. One of the big challenges in realizing such a vision is the ability to share the same core network. We have seen that some traffic, like voice, used to be mostly circuit switched, and data are mostly packet switched. How can we converge these networks, then, especially in the core? We now turn to discussing this challenge.

10.2.7.1 Convergence of the Underlying Transport Mechanism Circuit-switched communications works best when the traffic is relatively constant in volume, because the circuits need to be set up, and resources allocated, for a specified bit rate. Voice traffic is at a constant bit rate (relatively speaking—of course, during periods of silence, fewer bits need to be transmitted, but compared to most other kinds of traffic, like Web browsing, voice is at a roughly constant bit rate). Thus, circuit-switched communications is more suited than packet-switched communications to carrying voice traffic. The circuits are set up, and resources reserved, and then voice as a type of constant bit rate traffic is carried efficiently over the circuits.

However, packet-switched communications is more suited to carrying most other types of data traffic. Despite the overhead of needing a header for every packet, in packet-switched communications, packet-switched communications is more efficient for most kinds of data traffic. A primary reason for this is that most data traffic is of variable bit rate type, such that circuit-switched communications would be very wasteful, as the circuits would not be utilized efficiently.

Thus, packet switching is good for data, and circuit switching is good for voice. Over the past 50 years, even as packet switching for data networking has grown rapidly (especially with the explosion of the Internet), the telephone network has largely stayed true to its circuit-switched roots and remained a circuit-switched network. However, various forces have been pushing for *network convergence*. These drivers include the following perceived benefits of convergence:

- Rather than operating and maintaining two parallel networks, we have just one network to operate and maintain. This streamlines the network operation and can result in cost savings and better optimization of the one network.
- Convergence can help facilitate *computer telephony integration* (CTI), resulting in new conveniences and services.
- Convergence can provide better statistical multiplexing of traffic on the one network.

10.3 IP NETWORKING

There are many books on IP networking, such as Comer's [1]. Here we just provide a brief overview, especially including material that will be needed as background for discussions in later chapters. We start with some basic features of IP (Section 10.3.1), move on to transport protocols (Section 10.3.2), and then to fundamental protocols such as DNS and DHCP (Section 10.3.3). We then consider some matters having to do with the style of IP protocols (Section 10.3.4), how IP interacts with lower layers (Section 10.3.5), and round off with an introduction to IPv6 (Section 10.3.6).

10.3.1 Features of IP

At a basic level, an IP network consists of hosts and routers. *Hosts* are the end devices that connect to the IP network, and *routers* are intermediate machines (typically, with multiple interfaces) that receive packets and forward them on toward their destination. Because IP is a packet-switched technology, all IP packets have an IP header that allows the intermediate routers to know how to process it (Figure 10.5). For example, the basic thing an IP router does when an IP packet arrives (which has not been filtered out by lower layers, e.g., the Ethernet layer) is decide if the packet is addressed to it (the router), and if not, it decides where to forward it (i.e., it decides on an outgoing interface). In Section 10.3.1.1 we discuss how a *routing table* is used to make forwarding decisions.

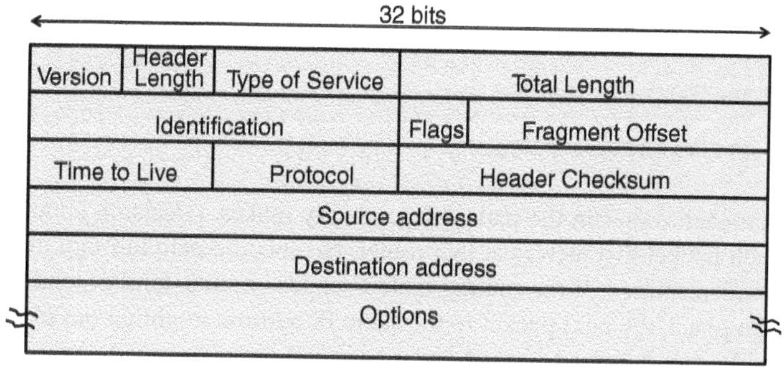

FIGURE 10.5 IP header.

10.3.1.1 Forwarding and the Routing Table Every IP-enabled device would have an engine (either in software or hardware, or some combination of software and hardware) that decides what to do with incoming IP packets. Incoming IP packets could come from either:

- The device itself
- Some other source, through one of the IP interfaces of the device

We highlight some essentials of what type of processing the engine does for every incoming IP packet. First, the engine checks if the destination IP address of the IP packet is an IP address belonging to the device itself. If it is, the packet has arrived at its destination, and the packet does not need to travel any further. It can be passed to the appropriate higher-layer handler for further processing.

Second, if the destination IP address of the IP packet is *not* an IP address belonging to the device itself, the device would need to figure out what to do with it. If the IP packet did not come from the device itself but from some other source through one of the IP interfaces of the device, the device will *silently discard* the packet *unless* IP forwarding is turned on. If the packet is a regular packet, and not using source routing (see Section 10.3.1.2), the device will then consult its *IP routing table* to figure out where to send the IP packet. Here is an example of a typical routing table that might be found on a home PC.

Network Destination	Netmask	Gateway	Interface	Metric
0.0.0.0	0.0.0.0	192.168.1.254	192.168.1.64	20
10.0.2.0	255.255.255.0	10.0.3.1	10.0.3.2	31
10.0.3.0	255.255.255.0	On-link	10.0.3.2	286
10.0.3.2	255.255.255.255	On-link	10.0.3.2	286
10.0.3.255	255.255.255.255	On-link	10.0.3.2	286
127.0.0.0	255.0.0.0	On-link	127.0.0.1	306
127.0.0.1	255.255.255.255	On-link	127.0.0.1	306
127.255.255.255	255.255.255.255	On-link	127.0.0.1	306
192.168.1.0	255.255.255.0	On-link	192.168.1.64	276
192.168.1.64	255.255.255.255	On-link	192.168.1.64	276
192.168.1.255	255.255.255.255	On-link	192.168.1.64	276

```
        224.0.0.0          240.0.0.0         On-link          127.0.0.1    306
        224.0.0.0          240.0.0.0         On-link           10.0.3.2    286
        224.0.0.0          240.0.0.0         On-link       192.168.1.64    276
  255.255.255.255    255.255.255.255         On-link          127.0.0.1    306
  255.255.255.255    255.255.255.255         On-link           10.0.3.2    286
  255.255.255.255    255.255.255.255         On-link       192.168.1.64    276
```

Note that each router in the path independently makes a decision about where to forward each packet that arrives at that router. No specific path through the network has been predetermined. If the routing table changes or some kind of load balancing scheme is in place, the next packet to the same IP address might go out of the router through a different interface.

10.3.1.2 Source Routing Usually, the IP network is free to deliver an IP packet to the specified destination address along any path it pleases. Sometimes, the source may wish to have more control over the path taken (e.g., to specify that it should go through certain specific routers). IPv4 provides this capability, called *source routing*. However, it is not implemented uniformly, so not all routers will process source routes properly even when they are specified in IP packets, and therefore source routing is not normally used. What is supposed to happen when source routing is used is that the packet needs to travel through each of the destinations in the source route list in turn, before arriving at the final destination. Hence, in consulting the forwarding table, the router would need to check for the next hop toward the next element in the source route list, rather than toward the actual final destination.

10.3.2 Transport Protocols

As mentioned in Section 10.1.3, the *transmission control protocol* (TCP) provides reliable transport over the unreliable IP, IP being a "best effort" protocol. TCP is the dominant transport protocol used over IP and provides the features shown in Table 10.1. For transmissions that do not need the services of TCP, the *universal datagram protocol* (UDP) is a more lightweight transport protocol that can be used instead of TCP. It needs less processing and can afford to be stateless (unlike TCP), since it doesn't have to be concerned about sequence numbers or retransmissions.

Various applications will pass packets down to the TCP or UDP layer. To keep the applications separate, TCP and UDP utilize the concept of *ports*. For example,

TABLE 10.1 Comparison of TCP and UDP

Feature	TCP	UDP
Reliability	ACK, retransmissions	No
Ordering	Sequence number	No
Guaranteed delay	No	No
Jitter control	No	No
Integrity	Checksum	Checksum (optional)
Efficiency		Less processing, stateless

TABLE 10.2 Some Well-Known and/or Important TCP Ports

Port Number	Protocol	Reference
20	FTP: file transfer protocol	
21	FTP (control)	
22	SSH: secure shell	
23	telnet	
25	SMTP: simple mail transfer protocol	
50	IPSec ESP	Section 15.3.1
51	IPSec AH	Section 15.3.1
80	http	
110	POP3: post office protocol 3	
443	SSL	Section 15.3

TABLE 10.3 Some Well-Known and/or Important UDP Ports

Port Number	Protocol	Multicast Address	Reference
161	SNMP requests and responses		Section 14.3
162	SNMP traps		Section 14.3.5
Variable	RTP		Section 10.3.2.1
5060	SIP		Section 11.2.2
434	Mobile IP		Section 12.1.1
520	RIP: routing information protocol	224.0.0.9	

file transfer protocol (FTP) is assigned TCP ports 21 and 22, telnet is assigned port 23, and so on; when TCP is handling an incoming packet (at the receiving end), it knows which application to pass the data to, based on the port number in the TCP header. Some well-known TCP ports are shown in Table 10.2. Similarly, some well-known UDP ports are shown in Table 10.3. Note that SIP and mobile IP messages can be sent over TCP as well, but UDP is often a better choice.

Although TCP and UDP are adequate for many kinds of traffic over IP networks, they are not good matches for the transport requirements of certain other kinds of traffic. For real-time traffic such as voice and video, RTP has been created as a more suitable transport protocol (Section 10.3.2.1), and for SS7 signaling over IP, SCTP has been created as a more suitable transport protocol (Section 10.3.2.2).

10.3.2.1 Transport of Voice and Video: RTP Two of the biggest challenges to providing sufficient QoS for voice and video over IP are related to delay: The end-to-end delay needs to be less than 400 ms, and the delay variance needs to be small as well. The delay variance is often called *jitter*, referring to the small (hopefully!) fluctuations in arrival time of packets at the destination. Unlike some other kinds of traffic, though, voice and video traffic do not need the occasional dropped packets to be retransmitted (this is partly by design of the voice and video codecs, where occasional lost packets can be tolerated; also, the playback would have moved on by the time the retransmitted packet arrives, so it would no longer be useful). This is quite different from the requirements for file transfer, for example, which is more flexible as far as delay tolerance is concerned, but where all packets must eventually arrive at their destination.

Therefore, traditional transport layer protocols such as TCP and UDP are not well suited for voice and video traffic. Instead, *real-time transport protocol* (RTP) was created for transport of real-time traffic such as voice and video. Since UDP already existed, RTP adds functionality to UDP. So RTP is typically used together with UDP rather than stand-alone. A VoIP packet would have an RTP header added by RTP, and the resulting RTP packet would then be passed to UDP.

10.3.2.2 Transport of PSTN Signaling: SCTP In the movement toward "all-IP" networks and convergence, the need to transport SS7 (Section 10.1.2.1) over IP networks naturally arises. It turns out that TCP, UDP, and RTP are not good candidates for this role, so the SIGTRAN working group in IETF created a new transport protocol for this purpose. The *stream control transmission protocol* (SCTP [5]) was therefore designed as a transport protocol to transport SS7 messages over an IP network.

As stated in RFC 4960 [5], SCTP provides the following services:

- Acknowledged error-free nonduplicated transfer of user data
- Data fragmentation to conform to discovered path MTU size
- Sequenced delivery of user messages within multiple streams, with an option for order-of-arrival delivery of individual user messages
- Optional bundling of multiple user messages into a single SCTP packet
- Network-level fault tolerance through support of multihoming at either or both ends of an association

10.3.3 Related Protocols and Systems

Some protocols and systems are so fundamental and used so broadly in IP networks that they merit at least a brief introduction.

10.3.3.1 Domain Name System Most IP-based networks rely heavily on the *domain name system* (DNS) for various name translation services. For example, given a human-friendly name such as `www.google.com`, a DNS server can be queried for the IP address(es) corresponding to that name. In some cases, such as with `www.google.com`, multiple IP addresses are returned, to facilitate load balancing. Thus, a query to a DNS server (also known as a *DNS query*) may return the IP addresses 74.125.224.17, 74.125.224.19, 74.125.224.16, 74.125.224.18, and 74.125.224.20. A *reverse lookup* can return a name, given an IP address. Other queries are possible: for example, for IPv6 addresses.

10.3.3.2 Dynamic Host Configuration Protocol *Dynamic host configuration protocol* (DHCP) is a protocol used on IP networks for automatic configuring (*autoconfiguring*) devices connected to IP networks. It allows a device to obtain configuration information, such as an IP address it can use, from a *DHCP server* in the network. The DHCP server functionality might be colocated with a router. Besides the IP address, other information, such as a gateway IP router address, can be provided by the DHCP server.

10.3.4 Style

IP and its related protocols have certain stylistic traits. Often, roles are defined (e.g., client/server, manager/agent) that are like hats a person can wear; network elements can, and often do, play multiple roles. In certain cases, a given network element can play multiple roles simultaneously (e.g., SIP server and SIP client). However, the protocols make clear separations between the different roles.

IP-style protocols tend to be text-based, lightweight, and focused on particular tasks. Multiple protocols are often needed to accomplish more complex tasks (e.g., to do voice over IP, to provide security services at the network layer, to manage a network). This type of design, sometimes called *modular*, is a basic and useful system engineering principle, allowing appropriate pieces to be selected for particular tasks. For example, for VoIP, there are SIP and related protocols to handle session control, RTP/UDP and related protocols to handle transport of the voice traffic, and voice coders such as G.729 that are specified outside the IETF. For security services at the network layer, IPSec can be used, with TCP for transport, and cryptographic algorithms such as 3DES, which are specified outside the IETF, can be used. For network management, SNMP and related protocols specify and support exchange of the appropriate information, and UDP is used for transport of the SNMP messages.

In the last paragraph, we used the phrase "and related protocols" a few times, for a reason. SIP uses SDP to describe sessions, RTP needs to be used with RTCP for control purposes; IPsec is actually a suite of protocols, an important component of which is IKE, which itself can be broken down into ISAKMP and Oakley; SNMP uses MIBs that are described with ASN.1 and an object-naming scheme from ISO.

10.3.5 Interactions with Lower Layers

When IP is used over Ethernet, both the IP and Ethernet layers have their own addresses. We now consider a common situation where Ethernet addresses of other devices in a LAN would need to be obtained in order for IP networking to work over Ethernet. As is common practice, we refer to the Ethernet addresses as MAC addresses.

Typically, a router provides forwarding services for the hosts in a LAN, such as an Ethernet-based LAN. Most traffic from hosts will typically need to go outside the LAN, and thus need to go through the router. The router interface on the LAN is often set as the default route for the hosts in the LAN. Similarly, most traffic for hosts would typically be coming from outside the LAN, through the router. When a host wants to send a packet to its router, it may only have the final destination IP address, as well as the IP address of the router (perhaps entered manually as a static route). When it puts the IP packet in an Ethernet frame for transmission over the Ethernet, it needs the destination MAC address, which is the router interface's MAC address. It knows only the IP address of the router interface on the LAN. How can it find the MAC address that corresponds to that IP address?

The *address resolution protocol* (ARP) is the signaling protocol that is used in IP networks to allow the host to find the router's MAC address given that it has the router's IP address. In general, it is used not just for hosts to find MAC addresses of routers, but for any two devices on the same Ethernet, for a source to find a target device's MAC address. Suppose that the target device's IP and MAC addresses are a.b.c.d and u.v.w.x.y.z, respectively. Then, the source broadcasts a "who has IP address a.b.c.d?" query message out, and the target device hears the broadcast and replies (unicast, because it would have the MAC address of the source from the source MAC address of the ARP broadcast message that was just sent) "a.b.c.d is at u.v.w.x.y.z."

ARP is used all the time. There are also variations of ARP that are less frequently used, but helpful in specific situations. Reverse ARP and inverse ARP are for obtaining an IP address given that the sender knows the MAC address already. Proxy ARP is where another device can respond to ARP queries on behalf of the actual target device and thus can provide its own MAC address and "capture" packets meant to be sent to the target device. It may appear that proxy ARP can be abused, so a malicious node can respond to ARP queries for a victim node by giving its (the malicious node's) MAC address and so causing packets meant for the victim node to go to the malicious node instead. Indeed, such abuse of ARP is possible, but is outside the scope of our present discussion. Instead, we note that proxy ARP can be helpful, for example, in cases where the target device is unable to respond to ARP queries for its MAC address, or it may even not be present to respond (the analogy with proxy voting comes to mind). In Section 12.1.1.1 we will see an example of how this is useful in mobile IP.

10.3.6 IPv6

The version of IP that is most widely deployed is IP version 4 (IPv4). IPv4 (or IP, for short) was designed in the 1970s when the Internet was a very different network from what it is today. Back then, most of the users were researchers in universities and research labs, and applications such as file transfer and email were all the network had to handle. IPv4 did not have to deal with the wide range of traffic that is found over today's Internet. Some of the types of traffic are more demanding in terms of service quality, or security requirements, or mobility support requirements. A few decades ago, a 32-bit address space was thought to be adequate for many, many years, so IPv4 addresses were limited to 32 bits. It was not realized that the Internet would take off and grow exponentially, such that within a few decades, the IPv4 address space would be nearing exhaustion. In short, by the 1990s, the requirements for the Internet protocol had evolved to a different, more exacting set than the set of requirements of a few decades earlier. A new base protocol was needed for the Internet. IPv6 [2] is the new, enhanced version of the Internet protocol designed to better address the ever-evolving requirements for the Internet protocol.

10.3.6.1 IPv6 Addresses IPv6 addresses are much longer than IPv4 addresses. Hence, they are often written with hexadecimal digits, and there are shorthand ways to write the long addresses with less effort. Before introducing the shorthand notation, let's look at the basic IPv6 address. The 128 bits are divided into eight groups of 16 bits each, and separated by semicolons. Each group is written as four hexadecimal digits. For example, we may have fe80:1234:0000:0000:abcd:00ff:e0f4:0001. The shorthand rules are:

- In each group of 16 bits, leading zeros can be dropped. Thus, our example address becomes fe80:1234:0:0:abcd:ff:e0f4:1.
- Long strings of zeros may be replaced by "::", but only once in the address. Thus, our example address becomes fe80:1234::abcd:ff:e0f4:1.

With IPv4, usually each interface has only one IP address. With IPv6, they will usually have more than one IPv6 address. Part of the reason is that the different IPv6 addresses may be different in scope (discussed next).

Address Scopes. IPv6 introduces a variety of address scopes that make it more flexible than IPv4 (Figure 10.6). The address scopes are:

- *Global.* Like global IPv4 addresses, these are meant to be globally routable, and thus globally unique.
- *Site local.* Site-local addresses are IPv6 addresses that are routable only within a site/network. Hence, different networks can use the same site-local addresses and not have a conflict.

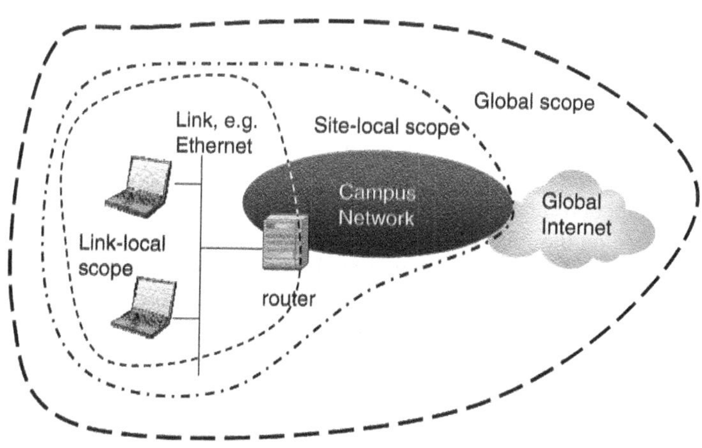

FIGURE 10.6 IPv6 address scopes.

- *Link local*. These are applicable only on local links, so will not be forwarded by any routers to go beyond the LAN.

In IPv6 terminology, a *link* is defined as "a communication facility or medium over which nodes can communicate at the link layer, i.e., the layer immediately below IPv6" [2].

10.3.6.2 Autoconfiguration A major feature of IPv6 is its enhanced autoconfiguration capabilities. These capabilities are designed to ease the burden on network administrators to configure and maintain a network. Address autoconfiguration may be *stateless* or *stateful*. With stateful autoconfiguration, a server, such as a DHCPv6 server, gives out, and keeps track of, IPv6 addresses to clients. With stateless autoconfiguration, an IPv6 node configures its own address(es) without such assistance. Stateless autoconfiguration [7] takes place in two stages. First, the node acquires a link-local address for communications only on the link. Second, the node acquires other addresses for internetwork communications (e.g., a site-local and a global address).

The large address space of IPv6 allowed the designers to make some interesting choices trading off efficiency (in usage of the addresses) for convenience and other benefits (one may compare it with spread spectrum, where the large amount of bandwidth is not used efficiently, in the sense that it is not used to obtain a maximum data rate; instead, it is traded off for such benefits as interference suppression). One example of such a design choice is how it is very easy for a node to autoconfigure a link local address for itself. It may do this by using its "unique" MAC address (we put the word *unique* in quotation marks because in theory MAC addresses are supposed to be unique, but in practice, some device's MAC address can be changed, and so can result in duplicates), processing it slightly, and then using the result as the 64-bit *interface identifier* for its link local address (it is concatenated with a fixed 64-bit prefix to form the link local address).

There is a small chance that an autoconfigured link local IP address is already in use by another device on the same link. Thus, the autoconfigured link local address begins its life as a *candidate link-local address*. It needs to make sure that no other node on the link happens to be using it. The *duplicate address detection* procedure is used for this purpose. This procedure is part of the neighbor discovery functions in IPv6 (Section 10.3.6.3).

After the autoconfiguration of the link local address, the hosts need to figure out if any routers are on the link. If routers are present, they would provide information to the host in their router advertisements, such as whether to use stateless or stateful autoconfiguration, and prefix information that hosts can use for generating site-local and global addresses. Once the relevant information is obtained from router advertisements, the host can complete the second stage by autoconfiguring site-local and/or global addresses. Since these addresses would also be formed by concatenation of the host's link identifier with various prefixes, it is not necessary to test again for uniqueness, as uniqueness has been tested in stage 1, with the link-local address using the same link identifier.

As for stateful address autoconfiguration, DHCPv6 has been standardized as one means of carrying it out. DHCPv6 relies on the host to already have a link-local address before it interacts with DHCPv6 servers. Unlike in DHCP for IPv4, where the request message is broadcast, DHCPv6 uses a special multicast address, the All_DHCP_Relay_Agents_and_Servers address, to which the host sends a request for stateful autoconfiguration, over UDP. In one of the modes of usage, a suitable DHCPv6 server responds with a DHCPv6 reply. It provides the host with the addresses requested, and also with other configuration information such as the address of DNS servers. It is also perfectly legitimate for a node to obtain its addresses through stateless autoconfiguration and still contact a DHCPv6 server to obtain other configuration information about DNS servers, and so on. When would stateless autoconfiguration be preferred to stateful autoconfiguration, and vice versa? Stateless autoconfiguration is preferred when a site cares less about the exact addresses that hosts use, except that they are routable and unique, whereas the stateful approach is used when more control is desired. The stateless approach is more convenient in some ways, because DHCPv6 servers are not required. As a network size grows, it would gradually make more sense to use stateful autoconfiguration for more control over address management.

10.3.6.3 Neighbor Discovery Neighbor discovery [4] is a group of related functions that are very important for the smooth functioning of IPv6. Neighbor discovery lets a node discover information about a link and neighbors on the link, where by *neighbors* we mean other nodes on the same link. Neighbor discovery is depended on for the following:

- *Router discovery*. Hosts can use router discovery to find routers on a link to which they are attached.
- *Discovery of other link information*. The link prefix, link MTU, and so on, can be discovered by router discovery.
- *Address resolution*. IPv4 needs ARP for address resolution; ARP is not needed in IPv6, since address resolution is part of neighbor discovery.
- *Neighbor unreachability discovery*. Sometimes, a neighbor can be reachable earlier on and then become unreachable later. Neighbor unreachability discovery is about finding out which neighbors are no longer reachable.
- *Duplicate address detection*. To find out if a candidate address (for use by a node) is already in use by another node.
- *Autoconfiguration*. In the process of autoconfiguration, neighbor discovery is used a number of times.
  - To autoconfigure a global address, router discovery is needed, followed by obtaining link information from the router. Routers can also specify if hosts should use stateful or stateless address autoconfiguration.
  - When the node has autoconfigured a candidate IPv6 address, it needs to make sure that it is unique (duplicate address detection)

Router advertisements, router solicitations, neighbor advertisements, and neighbor solicitations are critical building blocks of neighbor discovery. One cluster of the neighbor discovery functions use router advertisements and router solications, while a second cluster of neighbor discovery functions use neighbor advertisements and neighbor solicitations.

Router advertisements from routers provide the information needed for router discovery, prefix discovery (the prefix, or prefixes in general, that are considered on-link and hence can be used for autoconfiguration and next-hop determination), and parameter discovery (parameters such as MTU). To avoid bandwidth wastage, router advertisements are broadcast only once every few minutes. Hosts may send out router solicitations (soliciting for router advertisements), however, in order to speed up receipt of router advertisements. The router discovery and next-hop determination procedures can be used as an alternative to having to manually configure a default route on a node.

Neighbor solicitations and neighbor advertisements are very versatile in that they are used for address resolution, duplicate address detection, and neighbor unreachability detection. These messages contain an IPv6 *destination address* and an IPv6 *target address*. Figure 10.7 shows these three uses of neighbor solicitations and neighbor advertisements, indicating the destination address by “d” and the target address by “t.” For short, the solicited-node multicast address of address x is indicated by m[x]. The solicited-node multicast address of a unicast address x is simply the last 24 bits of x, with the prefix ff02:0:0:0:0:1:ff00::/104. This is more efficient to use than a broadcast, since only nodes with the same last 24 bits of their unicast address would receive and process such multicast packets.

As seen in Figure 10.7(a), address resolution is performed by setting the target address field to the IPv6 address in question, and then multicasting a neighbor

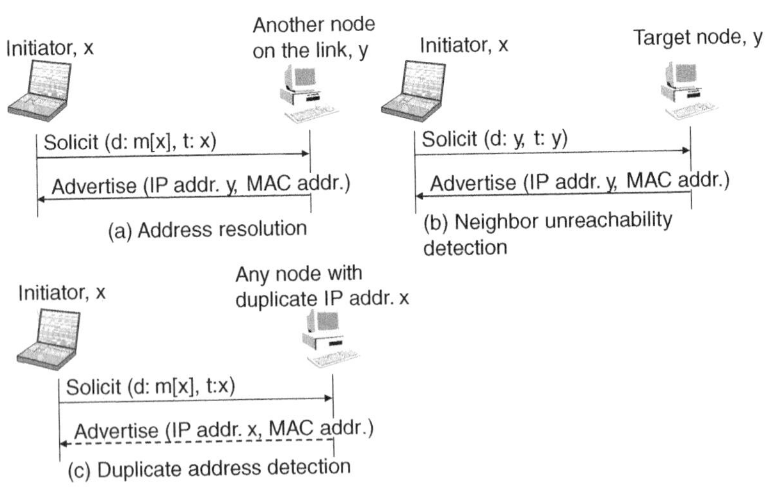

FIGURE 10.7 IPv6 neighbor discovery.

solicitation to the solicited-node multicast address of the target address. The target node then returns its link layer address by unicasting a neighbor advertisement back. As seen in Figure 10.7(b), neighbor unreachability detection is similar to address resolution, except that it is unicast directly to the target address. If there is no reply, the neighbor is considered unreachable. Duplicate address detection, shown in Figure 10.7(c), is similar to address resolution except that the target address is the node's own candidate address. This is multicasted to the solicited-node multicast address. Receiving a reply is a bad sign; it means that the address is already in use by another node.

10.4 TELETRAFFIC ANALYSIS

How many channels do we need per cell? How many users can we support in the system if we have a certain number of channels per cell? What happens when all the channels are in use and a new user tries to make a call? The answers to questions like these lie in the domain of teletraffic analysis.

It usually is more cost-effective to share costly resources than to have enough of the resources available so that they are available to anybody who wants them on demand. For example, people pay less for a flight on a commercial airline or for owning a time-share apartment than for a flight on a private jet or for owning a regular apartment. In exchange for paying less, people take the risk that sometimes the shared resource might not be available when they want it: for example, all seats on a commercial flight might be occupied by others, or another of the time-share apartment's owners might have reserved it for the time desired. Similarly, in telecommunications, network resources are costly and often shared rather than assigned exclusively.

10.4.1 Roots in the Old Phone Network

The field of teletraffic analysis has its roots in the old analog phone system. Thus, the basic analysis was carried out before the emergence of cellular systems. It has, however, been found to be useful in the design of cellular systems as well. In this section we first explain teletraffic analysis in the original context of the analog phone system, including the meaning and usage of the Erlang B formula. We then discuss how the analysis can be extended to cellular systems.

Consider an analog phone network where there is a separate physical line/wire from each phone to the central office. This will allow all the phones in that area to be used at the same time. However, it is costly for each phone to have a dedicated physical line/wire in this way. Instead, physical lines/wires are aggregated at one or more points between the phones and the central office, where typically at the aggregation points, there are more lines in the downstream direction (toward the phones) than in the upstream direction (toward the central office). For example, there might be 100 lines going out downstream and only 90 lines going out upstream. Whenever 90 or fewer of the downstream lines are active, they can be switched/connected to appropriate upstream lines in a one-to-one manner. Only in the very rare event that there are already 90 calls, and then another user tries to make a call on a different line, do we have a problem. In this case, there is no upstream line to which to switch/connect the new call, so it is *blocked* (i.e., it cannot be completed).

The *blocking probability*, P_b, is the probability that a new call gets blocked (i.e., it cannot be completed because all resources are busy). Generally, the telephone network is planned so that we have a finite but small P_b rather than $P_b = 0$, which is prohibitively expensive (all dedicated lines). Clearly, P_b cannot be too large either, or customers will get annoyed.

It is important to estimate P_b. An engineer named Erlang derived a few formulas for P_b based on different sets of assumptions. The *Erlang B* formula is the best known of these, and it is given by

$$P_b = \frac{(\lambda/\mu)^C/C!}{\sum_{k=0}^{C}(\lambda/\mu)^k/k!} \tag{10.1}$$

The model assumes the following:

- There are C "servers" (upstream telephone lines in our example).
- Each of the upstream lines can be either available or busy. When it is busy, it is being used by one of the customers, and the length of time it is used is distributed exponentially with mean $1/\mu$. This distribution is independent of the state of the system, which customer is using it, and so on. The length of use and the "service rate" are inversely proportional to each other (the shorter the calls, the more customers can use the line, over time, on average). Thus, μ is the service rate (e.g., in customers per second).
- There is a constant arrival rate, λ, where the interarrival times are distributed exponentially and independent of the state of the system. In our example, λ represents the arrival rate of new calls.
- Blocked calls are "cleared"; that is, they leave the system without automatic backoff and retry (as might be the case with Ethernet medium access, for example) or other such schemes. Blocked calls cleared is a good model for the telephone network, so the Erlang B model is very popular.

NB: There is an asymmetry between λ and μ. Whereas λ is the single arrival rate to the entire system, μ is the per-line (or per "server") departure rate. The system departure rate is $k\mu$, where k is the number of lines busy (thus the system departure rate is variable). The point of there not being enough lines is shown on the left in Figure 10.8. The setup and some of the assumptions for the Erlang B model are shown on the right of the figure. Notice that the arrival rate is a constant λ, where if more of the C servers are in use, the departure rate goes up, since each server serves independently at rate μ. Notice also that blocked calls (with a rate λP_b) are cleared rather than being queued up for retrying.

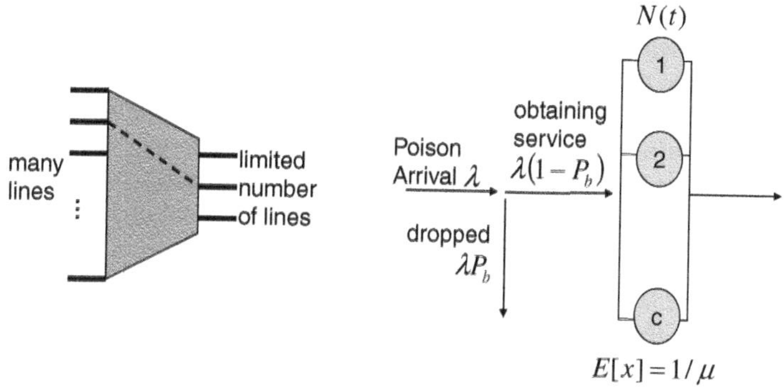

FIGURE 10.8 Teletraffic model used for Erlang B.

10.4.2 Queuing Theory Perspective

The Erlang B model, and other such models, can be expressed in the context of *queuing theory*, so we can draw on the knowledge we have of various systems that have been studied by queuing theorists to deepen our understanding and to apply similar ideas to related situations such as wireless systems. We discuss application to wireless in Section 10.4.2.1, but first discuss some basics of queuing theory.

We consider systems where there are m servers and a flow of customers into and out of the system. Queuing systems can be characterized by the nature of the arrival process of the customers, the nature of the service, the number of servers, and sometimes also the storage capacity of the system. Thus, queuing systems are described as $M/M/1$, $M/M/m/m$, and so on: a sequence of up to four values separated by a forward slash "/". The four values in the shorthand notation are as follows:

1. *Arrival process* [i.e., the inter-arrival time distribution (of the time between customer arrivals)]: M for exponential, D for deterministic, G for general, or other letters for other processes (but M is the most common)
2. *Service time distribution*: same notation as for the arrival process
3. *Number of servers*: a positive integer, denoted conventionally by the letter m if it is variable, or by a numerical value, for a specific system
4. *Storage space*: the number of customers that can be in the system at any given time, a positive integer denoted conventionally by the letter m

The storage space is often ∞, in which case the fourth value is often omitted. So an $M/D/3$ queuing system is one where the inter-arrival times are distributed exponentially, the service times are deterministic (i.e, not random), and there are three servers in the system and infinite-length queues. Other queuing systems, including queues with a finite total customer population, networks of queues, and so on, are outside our scope here.

$M/M/1$ is usually the first queuing encountered in studies of queuing systems, because the exponential distribution has some fascinating properties that make analysis more tractable than in other cases. With the $M/M/1$ queue, the arrival process is a *Poisson process*, so the interarrival times are independent, identically distributed exponentially distributed random variables. The service times are also independent, identically distributed exponentially distributed random variables. NB: With $M/M/1$, there is only one server, so if the server is busy and additional users arrive, a queue forms of users waiting to be served. Closely related is the $M/M/n/n$ queue, where there are n servers (rather than one, as in $M/M/1$), and the system can hold only n users at a time. Hence, additional attempts to use the system are dropped, exactly as in the Erlang B model. Thus, the Erlang B formula applies to the $M/M/n/n$ queue.

10.4.2.1 Application to Wireless The Erlang B model can be abstracted to a more general form and thus can apply to a wider range of phenomena, beyond traditional telephony. Rather than dealing with the availability of upstream phone lines, it can deal with the availability of some shared resources. In the case of cellular systems, then, the shared resources are the channels (or channel pairs, including both uplink and downlink). If a cell has a finite number of channels available to support mobiles, a new call is blocked once the available channels are all occupied.

The primary difference in the case of wireless is that we can distinguish between pure blocking, when a new call fails because of the lack of a channel, and another kind of blocking, when a handoff fails because of the lack of a channel in the destination cell. We call this new kind of blocking, *dropping*. We denote *dropping probability* by P_d. Dropping is generally considered to be more annoying to customers than blocking, since an existing conversation gets dropped. Therefore, schemes have been proposed to reduce P_d at the expense of P_b (e.g., the reservation of some channels only for handoff calls, not for new calls) (Figure 10.9).

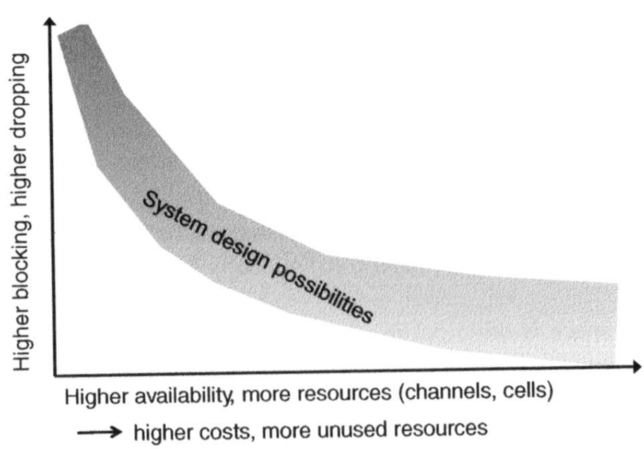

FIGURE 10.9 Fundamental trade-off in teletraffic engineering.

10.4.2.2 Variations and Alternative Assumptions There are many variations that can be conceived. For example, instead of exponential distributions for both arrival and service times, we may have other distributions for one or both of these. In traditional telephony, the exponential distribution is a reasonable assumption, and it has been found that the average length of a phone call is 3 minutes. With the increasing use of data traffic over phone lines and over wireless, different distributions might be more appropriate (and possibly more difficult to analyze). Furthermore, even in cellular systems with just voice traffic, it is unclear how close handoff traffic is to the exponential model.

Also, the constant λ assumption is valid only for a large, practically infinite number of users. It becomes less and less justifiable as we go to smaller and smaller pools of users, because we would expect the arrival rate to drop as more channels are being used (since there are fewer mobiles that are not busy). There are other models that account for a finite population. Yet another variation would be if instead of "blocked calls cleared," as in Erlang B, blocked calls wait in a special queue for the next available channel.

Other variations can be constructed, and in many cases, exact analytical expressions are not available, so computer simulations are used to estimate P_b and P_d.

EXERCISES

10.1 Isn't layering inefficient? You have multiple headers being added to the data as they move through the layers, and there might be duplication of functions at multiple layers. Why use it?

10.2 What are the differences among core, distribution, and access networks?

10.3 The following routing table is a typical one for an MS Windows PC. Referring to the routing table, specify the outgoing interface (e.g., eth0, eth1) if the IP address is:

- 210.78.150.130
- 210.78.150.133

```
Kernel IP routing table
Destination     Gateway Genmask           Flags   MSS Window   irtt Iface

210.78.150.177 *        255.255.255.255 UH        0 0           0 eth2
210.78.150.179 *        255.255.255.255 UH        0 0           0 eth2
210.78.150.133 *        255.255.255.255 UH        0 0           0 eth2
210.78.150.141 *        255.255.255.255 UH        0 0           0 eth2
210.78.150.140 *        255.255.255.255 UH        0 0           0 eth2
210.78.150.128 *        255.255.255.128 U         0 0           0 eth0
192.168.3.0    *        255.255.255.0   U         0 0           0 eth3
192.168.2.0    *        255.255.255.0   U         0 0           0 eth2
192.168.1.0    *        255.255.255.0   U         0 0           0 eth1
loopback       *        255.0.0.0       U         0 0           0 lo
default 210.78.150.129  0.0.0.0         UG        0 0           0 eth0
```

10.4 Expand the following IPv6 address written in shorthand notation: fe80:4:3333::a:15.

10.5 Compute the blocking probability for an $M/M/m/m$ queue where $\lambda/\mu = 20$ and $C = 20$. Try $C = 10$ and $C = 30$. How does P_b change?

REFERENCES

1. D. Comer. *Internetworking with TCP/IP, Vol. 1, Principles, Protocols, and Architecture,* 5th ed. Prentice Hall, Upper Saddle River, NJ, 2006.
2. S. Deering and R. Hinden. Internet protocol, version 6 (IPv6) specification. RFC 2460, Dec. 1998.
3. ITU-T. Signalling system no. 7 – ISDN user part functional description. ITU-T Recommendation Q.761, Dec. 1999.
4. T. Narten, E. Nordmark, W. Simpson, and H. Soliman. Neighbor discovery for IP version 6 (IPv6). RFC 4861, Sept. 2007.
5. R. Stewart. Stream control transmission protocol. RFC 4960, Sept. 2007.
6. A. S. Tanenbaum. *Computer Networks,* 4th ed. Prentice Hall, Upper Saddle River, NJ, 2003.
7. S. Thomson, T. Narten, and T. Jinmei. IPv6 stateless address autoconfiguration. RFC 4862, Sept. 2007.

11

GSM AND IP: INGREDIENTS OF CONVERGENCE

The big picture in this chapter and the next is to trace the development of the latest wireless networks from two main starting points:

- The old telephone network, extended to wireless phones with first-generation systems such as AMPS (the advanced mobile phone system), and later, second-generation systems such as GSM
- The data networking world, which is becoming predominantly about TCP/IP networking

The two starting points are quite some distant apart; for example, the old telephone network, AMPS, and GSM use circuit switching and related protocols, whereas TCP/IP is a packet-switched protocol. We will see how additions and changes to wireless networks such as GSM have led to increasing capability for these networks to go beyond mainly supporting mobile voice services, to providing data support as well. The addition of GPRS is one example of such an addition. We will also see how additions and changes to TCP/IP, to help it to better support voice, QoS, mobility, and so on, are ways that the data networking world has been moving to provide better wireless support. Such trends have been leading to convergence toward an "all-IP" wireless network. Recent systems such as WiMAX, and perhaps more so, LTE, can be seen as instantiations of wireless all-IP networks.

We begin this chapter with an introduction to the network aspects of a second-generation wireless cellular system, as typified by GSM. We introduced IP networking in Chapter 10, and introduced the concept of convergence in Section 10.2.7. In the

Fundamentals of Wireless Communication Engineering Technologies, First Edition. K. Daniel Wong.

rest of the chapter we discuss some important building blocks needed in a converged wireless network that supports both voice and data over IP. In particular, we discuss how IP has been extended from its original designs to support:

- VoIP (Section 11.2). Voice turns out to be a challenging form of traffic to transport over IP networks.
- QoS (Section 11.3). A converged wireless network carrying all kinds of traffic with a wide range of required service quality needs to support differentiated quality of service (QoS).

In Chapter 12 we discuss how more recent developments have continued to move in the direction of convergence and toward the "all-IP" wireless network.

11.1 GSM

11.1.1 Some Preliminary Concepts

The GSM network architecture extends the old telephone network architecture to support wireless and mobility management. This is no trivial matter. To support proper wireless operations, the ability of the network to find mobile phones (e.g., to deliver incoming calls, incoming packets, and short messages), *registration*, and *local updates* are crucial. In the telephone network, phones are static, so the route to the phone is always the same. In wireless networks, mobile phones can move around, so it is significantly more challenging for the network to find phones for delivery of incoming calls, for example; completely new procedures, such as registration and location updates, are needed in the wireless network to support this mobility capability that were not needed in the phone network. Furthermore, a mobile phone is designed to be reachable not just anywhere within its home operator's network, but potentially even in another operator's network on the other side of the world. This concept is called *roaming*.

The word *roaming* is sometimes used to refer to movement of a subscriber away from the home network, at a visited network. It is sometimes also used to refer to any movement of the subscriber at the network level, even if the movement is just within the subscriber's home network. With this broader use of the word *roaming*, one has to distinguish between intraoperator roaming and interoperator roaming. In this book we use the word *roaming* in the narrower sense, to refer to interoperator roaming only. This is in harmony with the following definition of roaming from the GSM Association [8]:

> Roaming is defined as the ability for a cellular customer to automatically make and receive voice calls, send and receive data, or access other services when travelling outside the geographical coverage area of the home network, by means of using a visited network. Roaming is technically supported by mobility management, authentication and

billing procedures. Establishing roaming between network operators is based on—and the commercial terms are contained in—Roaming Agreements. If the visited network is in the same country as the home network, this is known as National Roaming. If the visited network is outside the home country, this is known as International Roaming (the term Global Roaming has also been used). If the visited network operates on a different technical standard than the home network, this is known as Inter-standard Roaming.

Another difference between the old telephone network and the wireless networks has to be with more refined and subtle concepts of identity in the wireless networks. In the old telephone network, the phone line and the human subscriber are considered to be one and the same thing, whereas the phone itself is secondary, in the sense that when your phone number is dialed, the network just cares about delivering the call to the line that goes to your phone. It doesn't matter if the phone is replaced by another physical phone and connected to the same line. In the mind of the telephone company (and its billing system!) all calls associated with that line are billed to the human subscriber (according to the phone company's billing policy). In GSM, there is no longer another physical line that the operator can associate with the identity of the human subscriber. Instead, it is important to make a distinction between the *mobile station* (MS), *mobile terminal* (MT), and *subscriber identity module* (SIM). The MS is made up of a MT together with a SIM. Just like the old telephone subscription is tied to the line rather than to the phone, the GSM subscription is tied to the SIM card rather than to the mobile terminal. Each active SIM card has a unique *international mobile subscriber identity* (IMSI) associated with it. Each MT has a unique *international mobile equipment identity* (IMEI) associated with it. So the IMSI is seen and discussed more often than the IMEI. However, there are occasions where the network wishes to keep track of certain MTs, and in that case the IMEIs come into the picture, as we will see when we discuss the EIR in Section 11.1.2.

In light of the challenges in meeting the new requirements for wireless networks that were not present with the old phone network, new network protocols were designed to handle the challenges. In designing the network protocols to support the various requirements, some guiding principles included:

- To preserve battery power to lengthen the time between recharging of the mobile
- To use network resources efficiently

11.1.2 Network Elements

We introduce here briefly some of the main "players" (i.e., network elements) in a GSM network (Figure 11.1), before giving examples of how they work together to perform various essential procedures.

Base Transceiver System. The *base transceiver system* (BTS) is also popularly known as the *base station*.

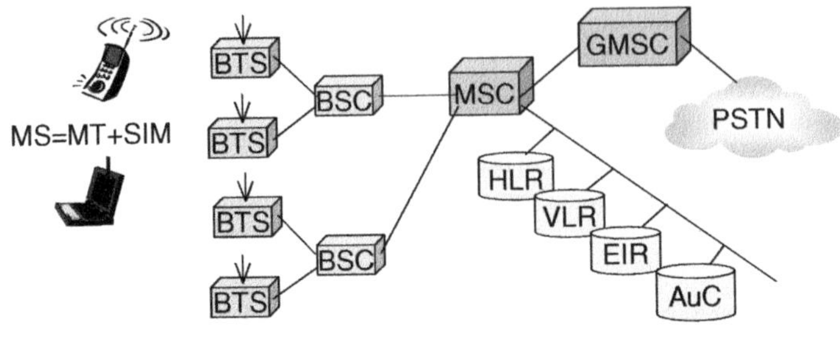

FIGURE 11.1 GSM network architecture.

Base Station Controller. The *base station controller* (BSC) controls several base stations. BSCs and BTSs together make up the *base station subsystem* (BSS).

Mobile Switching Center. The *mobile switching center* (MSC) is like a normal switch in the phone network, but with enhancements to support mobility. As a switch in the phone network, it uses SS7 signaling.

Gateway MSC. The *gateway MSC* is an MSC with the additional responsibility of being an entry point to the mobile operator's network from the external world.

Mobile Station. The *mobile station* (MS) is a *mobile terminal* together with a *subscriber identity module* (SIM) card.

HLR. The *home location register* (HLR) is a database that contains all kinds of important information about the subscribers of that network. Such information includes:

- Subscription-related information such as the services to which a user has subscribed
- (Partial) information on the current location of the MS (we will be more precise about this in Section 11.1.4).

There is usually one HLR per mobile operator (thus, this would be a very huge database, and could be implemented in several ways, e.g., as a distributed database, but as far as the GSM procedures are concerned, information for all subscribers is contained in *the* HLR), and usually is implemented separately from MSCs (contrast this to VLRs, described next).

VLR. The *visitor location register* (VLR) is a database that contains information related to subscribers who are currently in the network associated with the VLR. In other words, it contains information related to roaming users from other networks, and as such, the information is temporary. The VLR is often implemented internal to an MSC, but it doesn't have to be. A common misconception is that the VLR is used only when an MS is roaming in another operator's network.[†] Actually, the VLR is used even when an MS is in its home network. So, whether or not an MS is roaming in another operator's network or is in its home network, there should be information about it in the VLR associated with the MS's current-serving MSC. A key difference between the HLR and VLR is that the HLR handles tasks and information that are *not* related to the subscriber's current location, with the exception that it would usually have a pointer (in the form of an MSRN, as we will see in Section 11.1.4) to the current VLR that contains more specific current information about the MS's location. The VLR, on the other hand, is related to a specific geographical area (coverage area of one or more MSCs), and temporarily contains relevant information that allows the serving MSC to perform mobility-related functions.

EIR. The *equipment identity register* (EIR) is a database that contains information related to the equipment used to access the network (i.e., the mobile terminals).

AuC. The *authentication center* (AuC) is for security-related procedures. For example, it is one of two places where a user's secret key K_i is stored (the other being the SIM; see Section 15.4.1).

11.1.3 Procedures

In a complex system such as GSM, many procedures are needed to allow the system to provide various services and to handle different situations. We only highlight some of the most basic and important procedures here.

11.1.3.1 Outgoing Call An outgoing call is one that is originated by a GSM MS, to whatever destination (e.g., to another GSM phone, a non-GSM mobile phone, a landline phone) it is trying to reach. The signaling flow is straightforward, as shown in Figure 11.2.

11.1.4 Location Management

Whenever a mobile station is on, it may be in one of two general states:

- Idle mode
- Dedicated mode

Dedicated mode corresponds to the times when a mobile station is in active communications, such as when a call is in progress. When the mobile station is in such a

[†] Perhaps the misconception may be due to the word *visitor* in the name of the VLR, and common explanations of VLR that use the word *roaming* in the broad sense. It is correct to say that the VLR is used for roaming MSs, where *roaming* is meant in the broad sense, but incorrect to say it if roaming is meant in the narrower sense of roaming in another operator's network

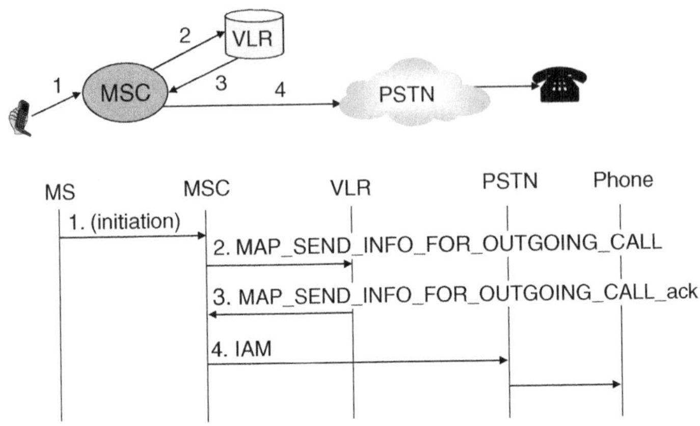

FIGURE 11.2 How GSM handles outgoing calls.

mode, dedicated radio resources are allocated to it to facilitate communications. *Idle mode*, on the other hand, is when the mobile station is not in active comunications, but is nevertheless powered on.

We have seen in Section 8.1.2 how handoffs are done. Handoffs are applicable only during the dedicated mode. In such cases, the network needs to know (and does know) which base station the mobile is attached to, and traffic is redirected promptly when the mobile hands off to another base station. Even in idle mode, however, the network needs to know where the mobile is located. Otherwise, how can the network deliver an incoming call to the mobile? In fact, whenever the mobile is in idle mode, the network tracks the location of the mobile, in a procedure known as *location management*.

Unlike with handoffs, location management does not need to track the location of mobiles as accurately; moreover, people calling the mobile phone can tolerate a small delay in the call delivery process. Thus, as long as the network has a rough idea where the mobile is, whenever there is a call delivery, it can quickly locate the mobile (through a process called *paging* that we will explain shortly). Given this characteristic of location management, the location management procedures are designed to preserve mobile battery power and conserve network resources through the following:

- When in idle mode, the mobile is mostly listening to signals from the base stations around it, but not connecting to any particular base station and handing off between base stations in the same way as during a call (this would use up battery power and consume network resources).
- The mobile only transmits to update the network when it moves between large areas called *location areas*. Each location area includes the coverage area of multiple base stations. The network thus only knows the location of the mobile to the granularity of a location area, not the finer granularity (smaller area) of a base station. In exchange for this loss of precision (compared to the case of regular handoffs), the mobile saves battery power by transmitting less often,

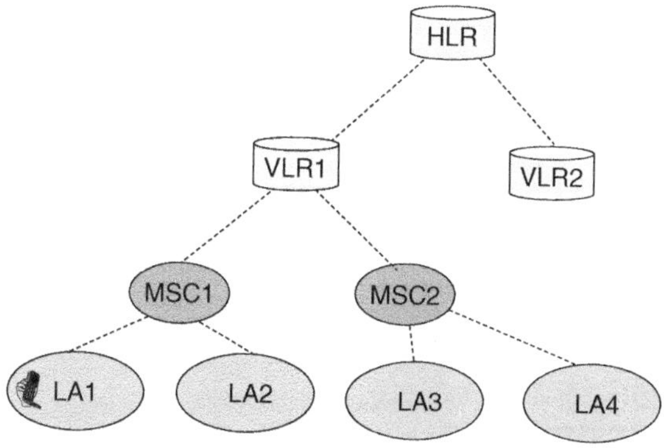

FIGURE 11.3 GSM location information hierarchy.

and network resources are saved by not having to track the mobile's location as finely.

So the use of large location areas helps to save network resources and battery power while a mobile is in idle mode, but how does this affect call delivery? When there is an incoming call, the network needs to figure out which of the cells within the location area is the one that the mobile is in. It does this through a process called *paging*. During this process, all the base stations within the location area will broadcast a special signal, the paging signal, to find the mobile. The mobile will respond through one of these base stations, and thus be found.

So far, we have been talking about "the network" as though it were a monolithic entity (e.g., as in "the network" keeping track of the mobile). In reality, the network comprises different network elements that cooperate to carry out various network functions. For location management, the location information of a mobile is not stored solely in one network element, but is distributed. In particular, the information is distributed as follows (see Figure 11.3 for a picture of this):

- The HLR in the MS's home network contains at least the address of the current MSC/VLR (whether it is one in the home network or in a visited network), and may also contain a *mobile station roaming number* (MSRN), which allows routing to that MSC/VLR for facilitating incoming call delivery.
- If MSCs and VLRs are not colocated, the HLR may just know the address of the appropriate VLR, and the VLR would have the MSRN to route to the right MSC.
- The MSC/VLR currently serving the MS would know which location area the MS is in.

As in the case that the mobile is in its home network, the distributed location information gets into place at first as part of the *IMSI attach* procedure, followed by a location update if necessary. Or the MS could just proceed with a location update (and not need to perform an IMSI attach) if it finds out that it is in a new location area

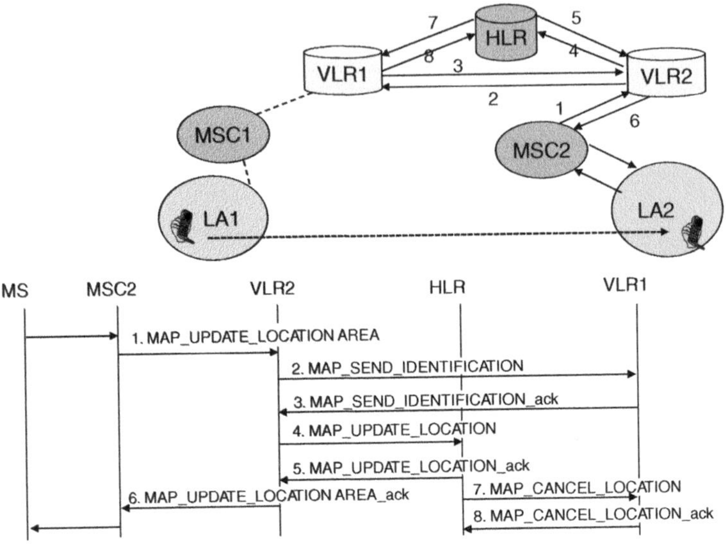

FIGURE 11.4 GSM location area update.

from where it was when it was turned off previously. In either case, subsequently, the location information is then updated as part of the location update procedure as the mobile moves around in idle mode.

11.1.4.1 Example: Location Area Update In Figure 11.4, signaling for a location area update is shown. In particular, it is for the case in which a mobile station moves from one LA to another LA under a different MSC. Thus, two MSCs are involved, labeled MSC1 and MSC2 in the diagram. We show the typical case where this would involve two different VLRs, labeled VLR1 and VLR2. The procedure begins when the mobile station enters LA2 and informs MSC2 about its entry into LA2. The MAP_SEND_IDENTIFICATION and its response, between the VLRs, allows the transfer of the IMSI internal to the network without unnecessary exposure of the IMSI over the air (see Section 15.4.1.3 for more details on GSM anonymity). After having obtained the IMSI of the mobile station, the VLR2 can communicate with the HLR so that the HLR will update its records and point to VLR2 now instead of VLR1. The HLR will also instruct VLR1 to "cancel" the location information for the mobile station in its database, since the mobile station has moved on. NB: This call flow applies whether or not the MS is in its home network or another operator's network (as mentioned earlier, the VLRs are used even when an MS is in its home network).

11.1.4.2 Example: IMSI Attach and Detach To conserve network and radio resources when a mobile station is turned off, the *IMSI attach* and *IMSI detach* functions are defined. Suppose that a mobile station is idle and has performed one or more location area update procedures; then the user turns off the phone. Before the phone goes off, it will perform an IMSI detach procedure so the network will know not to waste resources trying to locate the mobile station (e.g., if there is an incoming call). Later, if the mobile station is turned on again *within the same location area*, it performs an IMSI attach procedure to let the MSC/VLR know that it is back. Otherwise, if the mobile station has moved to a different location area when it is turned on again, it does not need to perform an IMSI attach procedure, but proceeds with location area updating in the new location area as normal.

The IMSI detach simply involves the sending of one message from the mobile station to the base station, and no acknowledgment is required. The mobile station is being turned off, anyway, and the consequences of the message being lost are not that serious anyway.

The IMSI attach is very similar to a location area update. It is not necessary that an IMSI attach must only follow a successful IMSI detach. The mobile station does not keep track of the attach/detach state anyway, and the IMSI detach might have failed, for whatever reason.

11.1.4.3 Example: Call Delivery

Call delivery is also known as *mobile termination*. Similarly, *call origination* is also known as *mobile origination*. Suppose there is an incoming call for a mobile station, which may be in its home network or roaming in another operator's network. The first part of the call delivery procedure is that based on the phone number (the MSISDN), and a partial circuit is set up all the way from the calling party to the *gateway MSC* in the home network of the MS. The gateway MSC will query the HLR to obtain information about the MS, such as its location. The HLR doesn't know where the MS is, but it does know the current VLR, so it queries the VLR (which may be in the same network or in another operator's network), to obtain an MSRN that corresponds to the MSC associated with the current location of the MS. The HLR returns a send routing information message to the MSC with the MSRN.

Once the MSRN is obtained, the GMSC can then begin to establish the next segment of the circuit with the MSC/VLR. The MSC/VLR will then check its information about the MS and initiate paging in the appropriate location area. If the MSC and VLR are not colocated, the MSC sends the MAP *send information* message to the VLR to obtain the MS-specific information. The VLR replies with a page message that contains the location area and TMSI of the MS. Figure 11.5 shows the signaling flow. Although the figure shows the case when the MS is roaming in another operator's network, the signaling is the same when the MS is in its home network, the difference being that the MSC and VLR would then be in the same network as the GMSC.

11.2 VoIP

In going from circuit-switched voice to packet-switched voice, several characteristics of circuit-switched voice are given up, such as:

- Guaranteed bandwidth is reserved for use by the circuit.
- Timing information is implicit. Bits will arrive at regular intervals.

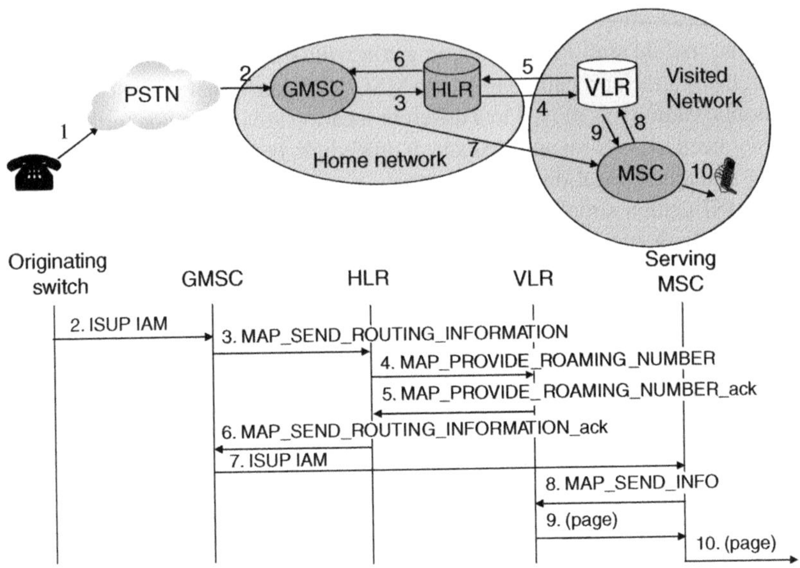

FIGURE 11.5 GSM call delivery while roaming.

- Source and destination is implicit. Once the circuit is set up, the source and destination are fixed and known.

Part of the solution is to have a new transport protocol more suited for voice and video, and real-time data in general: namely, the real-time transport protocol (RTP [6]). We discussed RTP in Section 10.3.2.1. Besides what RTP provides, other pieces of the solution are provided by buffering, and so on.

Moreover, besides the transport-related issues, there is also the question of control and control signaling. The control signaling protocols used for traditional circuit-switched communications are designed *for* circuit-switched communications and handles the setting up and tearing down of circuits, and so on. In circuit-switched communications, the transmissions in the session all go along the same circuit (i.e., the same path). Packet-switched networks, however, are more flexible, and all packets in a session do not necessarily take the same path between source and destination. Thus, session control protocols designed for packet-switched networks, IP networks in particular, can be adapted to the characteristics of these networks. For example, once the called party is found and initial signaling between the calling party and the called party is complete, the two parties will have the IP address of their peer, and could exchange voice packets directly without needing to send all those packets through the same path as that taken by the initial signaling. A popular solution for control signaling in VoIP is SIP (Section 11.2.2).

11.2.1 Other Parts of the VoIP Solution

The wireless link is usually where bandwidth is most scarce, in the end-to-end path between two communicating nodes (where the rest of the end-to-end path may be mostly wired). Therefore, especially over the wireless link, it is good system design to eliminate or reduce as many inefficiencies as possible. One such inefficiency is packet header overhead (also referred to simply as *header overhead*). We illustrate the problem, and solutions, in Section 12.1.2.

11.2.2 Session Control: SIP

As seen in Chapter 10, there's data traffic and there's control/signaling traffic. In traditional circuit-switched voice communications, what is the control signaling optimized for? Circuit-switched voice communications, of course! The signaling is called *signaling system 7* (SS7). When we move toward VoIP, it is still possible to use SS7 and carry SS7 signaling over IP. However, there are advantages to using an IP-style protocol for session control. Advantages include:

- It can be optimized for packet-based sessions, and in particular, IP-based sessions.
- It can be designed for easy integration with other IP-based protocols, such as http.
- The plain text headers are easy to understand and debug.

The leading protocol for this purpose is *session initiation protocol* (SIP [5]). It was designed from the ground up as a lightweight "IP-style" protocol (i.e., it doesn't try to do everything, but reuses other IP protocols for QoS, real-time transport, etc.). Thus, it just focuses on session initiation and control, and on doing it well. In fact, SIP has been constantly evolving and has shown great flexibility in the ability to take new features while retaining its lightweight feel and avoiding bloatedness.

SIP is text-based and is an "IP-style" protocol that has many similarities and uniformities with other IP-style protocols, such as http. As an IP-style protocol, SIP makes it easier to integrate telephony with other IP-based services, such as http. This integration is called *computer telephony integration* (CTI). For example, hypertext links on a business's web page could invoke SIP functionality (e.g., to initiate a phone call to the customer service department). This can bring the customer in more immediate contact with a service representative who could answer specific questions.

Doubtless, certain session establishment procedures in SIP are similar to procedures used in the PSTN signaling (Table 11.1). However, the entire structure and design of SIP is very different; from being a text-based http-like query-response protocol to the modular nature of the protocol design, it has the flavor of an IP-style protocol.

VoIP devices that allow users to make and receive calls over an IP-based network such as the Internet, are called soft phones or IP phones. An IP phone that uses may also be called an SIP phone.

TABLE 11.1 Phone Session Control: Comparison Between PSTN/SS7 and VoIP/SIP

Requirements	PSTN with SS7	VoIP with SIP
Phone A initiates the conversation: it needs to inform the other party	Phone A goes "off hook" and digits are transmitted to the PSTN	Phone A sends SIP INVITE message to phone B with session description in SDP
Terms, conditions, and other parameters are negotiated	No negotiations, as there is no choice of codec	Information in SDP is used to negotiate codec, etc.
Phone B may decline, or simply not accept	Phone B rings for awhile and is not picked up	Phone B rings for awhile and is not picked up, or negotiations fail
Phone B may accept	Phone B rings for awhile and is then picked up	Phone B rings for awhile and is then picked up

11.2.2.1 Session Description Protocol As part of session initiation, the calling and called parties need to negotiate session parameters. Thus, a language is needed for describing these session parameters. *Session description protocol* (SDP [3]) is the language used by SIP. Although both SIP and SDP are protocols, they are different types of protocols. SDP can be thought of as more of a linguistical framework and syntax for describing sessions (including the various parameters, such as codecs to be used, endpoint IP addresses and port numbers, etc.), whereas SIP involves messages and flows of messages between various entities in order to accomplish certain functions (e.g., registration with a SIP registrar, call initiation, call termination). SDP is not used exclusively by SIP; it is employed by other IP protocols as well. Like SIP it is in plain text.

11.2.2.2 SIP Call Flow Example We first examine a case where the calling party and called party know each other's IP address or DNS name, so the calling party can find the called party directly, without needing assistance from any SIP server. Then, setting up the session is very simple. Let the calling party and called party devices be denoted by A and B, respectively. Then it begins by A directly sending B a SIP INVITE message. The header of the INVITE message indicates that it is an INVITE message, and also includes such fields as a unique call ID, and the SIP names of the calling party and called party. The SIP names are in the form user@host. The body of the message contains a list of session parameters that the calling party is willing to use for the session. The session parameters include a list of codecs that the calling party is willing to use for the session.

B will respond with a SIP 200 OK message. [There are different categories of *response codes*, each beginning with its unique number, similar to http server responses that most people might be familiar with from our web browsing experiences; for example, the *provisional, successful* and *redirection* response codes start with 1, 2 and 3, respectively]. The 200 OK message is sent after the user has been alerted (through ringing) and has responded. The body of the 200 OK message contains a subset of the parameters proposed by A in A's INVITE message. This subset

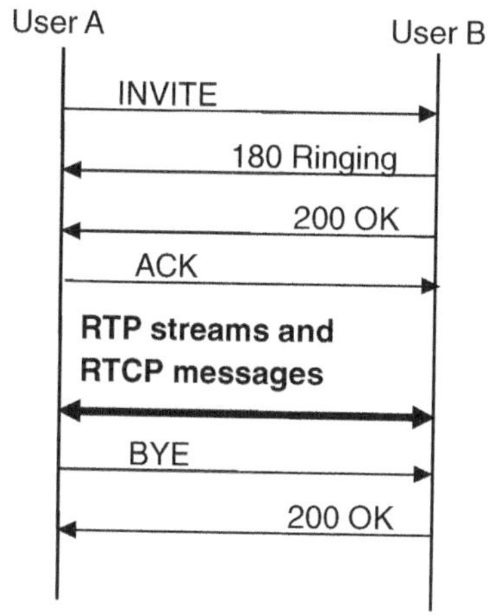

FIGURE 11.6 SIP signaling.

indicates the parameters (e.g., codecs) that B has chosen and is also specified in SDP. This completes the *offer and answer* model of negotiating parameters. The 200 OK message is an example of a *final response* (in contrast to the *provisional responses* such as 180 ringing, which we will soon discuss) to the INVITE message. A final response must be acknowledged. Therefore, A sends a SIP ACK message to B to complete the session initiation. And that's it! The session is initiated, and A and B begin exchanging RTP packets.

What if the human using B takes awhile to respond to the incoming call? SIP provides B a way to let A know that it is trying to alert the user. It can send a *180 ringing message* to A while waiting for the human to respond. The 180 ringing message is one of a set of *provisional messages* that do not have to be acknowledged (if they are lost, so be it!), and which can be recognized by the range of numbers in which they lie (i.e., 100 to 199).

That is the beginning of the VoIP session. As for how it ends, one party hangs up; the hang-up indication is a BYE message that is sent to the other party. In the example depicted in Figure 11.6, it so happens that the initiator of the session is also the one who wants to end it. It could also be the other party that sends the BYE message. In either case, the recipient of the BYE message then sends a 200 OK response. This indicates its cooperation in shutting down the session. NB: Unlike in the case of INVITE and 200 OK, an ACK is not needed after the exchange of a BYE and 200 OK.

An example of a SIP header follows. It is for an INVITE message from daniel@danielwireless.com to maeli@tee.sg

```
INVITE sip:maeli@tee.sg SIP/2.0
Via: SIP/2.0/UDP sip.danielwireless.com;branch=zlkjfdslkg89Ug3
Max-Forwards: 70
From: "Brother" <sip:daniel@danielwireless.com>;tag=mich
```

```
To: "Sister" <sip:maeli@tee.sg>;tag=schwester
Call-ID: hl3f432fklj3@sip.danielwireless.com
CSeq: 225 INVITE
Contact: <sip:daniel@danielwireless.com>
Subject: the Lord bless you and keep you
Content-Type: application/sdp
Content-Length: 142
```

11.2.2.3 Adding Scalability: SIP Proxy and Redirect Servers Although SIP can work purely peer to peer, as shown in Section 11.2.2.2, this model does not scale well. As the number of potential destinations increases, would the SIP agent in the end host have to keep track of reachability information for all of them (e.g., IP addresses)? In general, the calling party would not know the called party's IP address or DNS name, but would only know the SIP address/name of the called party. This would be of the form: `sip:daniel@danielwireless.com`. To make SIP scalable and not burden the SIP phones unduly, the task of finding the called party is distributed. One or more *SIP proxies* (also known as *proxy servers* or *forwarding proxies*) or redirect servers may assist in finding the called party. These are SIP servers that have at least partial information on how to arrive at various SIP destinations. There can be multiple SIP servers in the path between the calling party and called party. When a SIP message (such as an INVITE) arrives at a SIP proxy, the SIP proxy will forward the message closer to the destination. Figure 11.7 shows a SIP proxy in action. It is the same scenario as shown in Figure 11.6 except that we have added a SIP proxy in the middle.

A redirect server, on the other hand, upon receipt of a SIP message, will send a message back to the node from which it got the SIP message, which should provide information that helps get the message closer to the destination. Thus, it redirects the previous node closer to the destination. Most people who have used web browsers

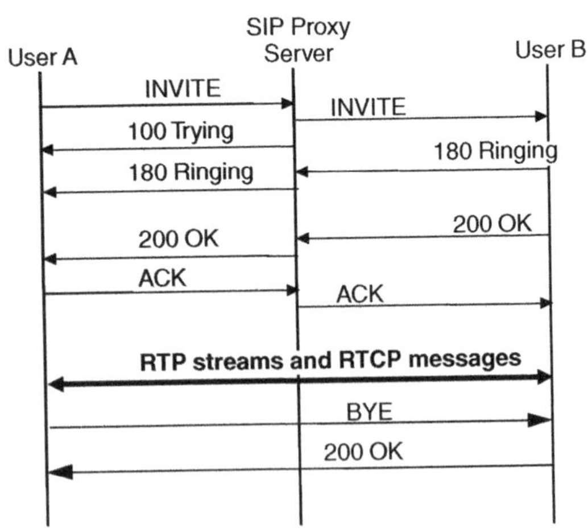

FIGURE 11.7 SIP proxy.

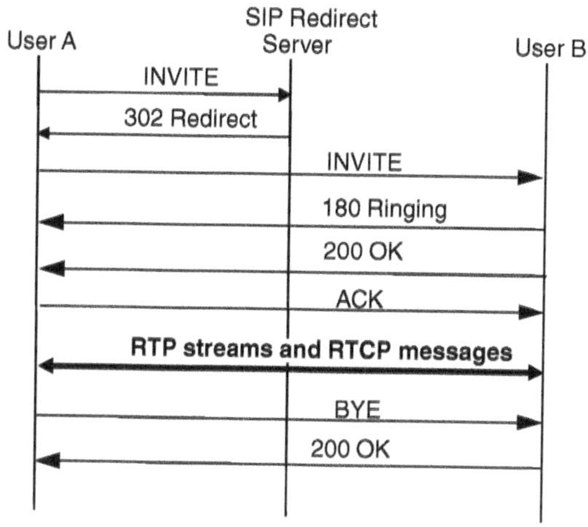

FIGURE 11.8 Redirect server.

would have seen occasional redirect responses from some web servers when a web page has moved. SIP redirect is similar to http redirect. Instead of forwarding the SIP message, it will help the previous server or SIP user agent to get closer to the destination by providing redirection information. Figure 11.8 shows a SIP redirect server in action. SIP proxy servers and redirect servers can be mixed within a call flow, as shown in Figure 11.9. In fact, there could be multiple SIP proxy servers and/or multiple SIP redirect servers within any given flow.

Notice how, in Figure 11.7, the SIP proxy is only involved in the initial dialog (`INVITE` and the corresponding `200 OK` and `ACK`) and that subsequent signaling

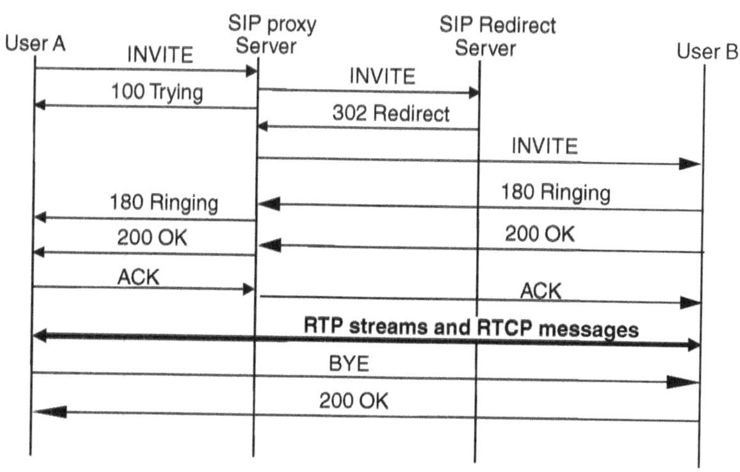

FIGURE 11.9 Basic SIP flow with both SIP proxy and redirect server.

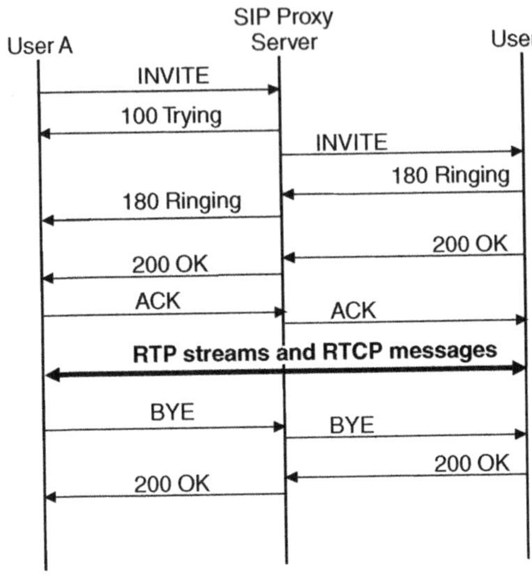

FIGURE 11.10 Use of record-route to keep a SIP proxy in the signaling path.

can go directly end to end since the two user agents, A and B, know each other's IP addresses by then. Thus, the `BYE` and its corresponding `200 OK` and `ACK` go directly between A and B. Sometimes, however, it may be desirable for one or more SIP proxies to remain in the call flow so that all the SIP messages pass through them. For example, in a service provider environment (such as with IMS, as we will see in Section 12.4), the SIP servers may need to remain in the flow so that various functions (including accounting functions such as collecting call records for accurate billing) can be handled properly. The standard way that SIP servers can remain in the flow is by means of the `Record-Route` and `Route` headers. A SIP server that wants to remain in the flow will insert itself into the `Record-Route` header of the `INVITE` message. The user agents at A and B will then add all such SIP servers to the SIP forwarding path by listing them in the `Route` header of subsequent messages. Notice how Figure 11.7 is modified to Figure 11.10 (and notice how the BYE goes through the proxy server).

11.2.2.4 Extending SIP The original SIP has some limitations, such that it cannot be used unmodified in the wide variety of scenarios and frameworks that is being demanded of it or envisioned for it. However, it has proven to be flexible enough in being able to grow to incorporate various requirements.

A good example of how the original SIP has been extended to meet additional requirements is how SIP how been embraced by service providers, such as wireless service providers, and SIP occupies a central place in the IMS (Section 12.4). Typically, the use of SIP in a service provider environment imposes additional requirements on SIP that are not found or needed in the use of SIP by hobbyists/enthusiasts. These additional requirements include:

- All the control signaling might need to go through some specific network elements, so that the necessary details can be kept and proper accounting can be done, allowing the provider to bill subscribers accurately.
- QoS and other preconditions must be supported. In a wireless network, it is not guaranteed that the necessary resources will be available to support the desired SIP call. Thus, some time must be allowed to try to arrange for the needed resources, before even alerting the called party, for example.
- Originally, it was envisioned that provisional response (e.g., `180 Ringing`) would not need to be reliable, so they would not need to be acknowledged. However, in service provider environments, it is usually required that these be reliable.

For the first requirement, we have already discussed how the `Record-Route` and `Route` headers can be used to keep certain SIP servers in the signaling path. QoS-related or other preconditions are handled by extending SDP to allow it to specify certain parameters as mandatory, and the `UPDATE` method can be sent from a user agent to the other side after the necessary QoS reservations have been made. Reliability of provisional responses [4] refers to a scheme whereby SIP is extended to provide reliability for the provisional responses. Basically, a new acknowledgment message, `PRACK`, is introduced, whereby a provisional response will get resent until a `PRACK` is received.

11.3 QoS

Quality of service (QoS) refers to the quality of the communications service provided by a network to various traffic. Without designing or operating the network for QoS, the default QoS could be very poor, especially when traffic loads are heavy. In addition to the QoS experienced by traffic in general, it is often the case that the network operator wishes to provide *differentiated QoS*, treating different kinds of traffic differently. This is because different kinds of traffic have different QoS requirements. Table 11.2 shows QoS requirements for different types of broad classes of traffic, as classified by 3GPP. The most stringent requirements are for *conversational traffic* (e.g., VoIP or video sessions). Streaming does not need low jitter since a buffer can be used to compensate for certain amounts of jitter and then to play back the stream

TABLE 11.2 QoS Requirements for Broad Classes of Traffic

Category	Constant Rate Needed	Low Delay Needed	Low Jitter	Low Delay Preferred
Conversational	Yes	Yes	Yes	
Streaming	Yes	Yes		
Interactive	—	—	—	Yes
Background	—	—	—	—

smoothly. However, with conversational traffic, no time is available to buffer large amounts of the traffic.

Various schemes have been designed to provide QoS and differentiated QoS. Some common terms used in talking about such schemes include:

- *Classification*: grouping packets into classes so that different classes can be handled differently.
- *Marking*: labeling packets (typically, setting a particular field to specific values), to identify their class.
- (traffic) *Shaping*: changing the overall characteristics of a flow of traffic to conform to a certain profile (e.g., a certain rate for some period of time, but profiles can include specifications about maximum rate, average rate, etc.).
- (traffic) *Policing*: similar to shaping; sometimes one of the two terms is used to refer to the case where packets can be dropped in order to get the traffic to conform to the profile, with the other term used to refer to the case where packets are delayed rather than dropped. It is always best to check the definition used in any particular context.

Terms such as *token bucket* and *leaky bucket* are sometimes used, describing mechanisms used for shaping and policing. Rather than being applied in an ad hoc manner, classification, marking, shaping, and so on, can be applied systematically in the context of QoS frameworks such as DiffServ and IntServ, as we will see in Section 11.3.1. Some of the QoS mechanisms that can be applied in these contexts or otherwise are discussed in Section 11.3.2. Then, in Section 11.3.3 we discuss QoS briefly specifically in a wireless context.

11.3.1 Frameworks

There are two classical frameworks for Internet QoS from the IETF: *integrated services* (IntServ) and *differentiated services* (DiffServ).

Integrated Services (IntServ). The IntServ framework for QoS [1] was the earlier of the two. It attempts to provide service for different types of non-real-time and real-time traffic on the same network, in an integrated way. It recognizes that different types of traffic have different requirements, so service differentiation is needed. IntServ uses admission control, rate control mechanisms, and resource reservation mechanisms (specifically, RSVP, which we discuss in Section 11.3.2.1.

Originally, IntServ did not specify in detail what service classes could be supported by IntServ. Subsequently, RFCs were written on supporting various service classes using the IntServ model. These RFCs include the controlled load service class and the guaranteed QoS service class. Each of them has its own QoS requirements.

Differentiated Services (DiffServ). IntServ is stateful and explicit control signaling is done for each flow to reserve the appropriate resources. This model does not scale well, so an alternative model, DiffServ, was introduced. DiffServ [2] is more stateless (so routers don't have to be burdened with maintaining state information about resource reservations). With DiffServ, explicit control signaling is not needed for each separate flow. Instead, each router performs the QoS-related functions based on concepts of *per hop behavior* (PHB) and *behavior aggregate* (BA) rather than based on treatment of the end-to-end flow that needs to be prearranged, as in the case of IntServ. We explain the meaning of BA and PHB next.

Instead of dealing with individual flows (of which there could be many thousands), DiffServ groups packets into classes known as behavior aggregates (BAs). Each packet in a BA is treated in the same way as every other packet in that BA. The treatments are called per hop behaviors (PHBs). The various BAs are distinguished by DSCP value (to be defined shortly). A region of the Internet where DiffServ is used may be called a *DiffServ domain*. The core networks of Internet service providers (ISPs), and the interconnections between them, often are DiffServ domains. The great attraction of DiffServ is that most of the computationally intensive QoS-related activities (e.g., classifying, marking, shaping) are performed at edge routers of a DiffServ domain. In the interior routers of a DiffServ domain, packets are already classified, so the routers just need to inspect the DSCP and apply the appropriate PHB.

The IETF DiffServ working group has redefined the IPv4 header type of service (TOS) octet as the DS field (for IPv6, the traffic class octet maps to the DS field). Packets are marked in the *differentiated services code point* (DSCP) field of the DS field, which is 6 bits long. The rest of the TOS octet is unused. The packets in a behavior aggregate (BA) all have the same DSCP value (and this is used to distinguish between different BAs).

Marking packets is a way to reduce the classification burden on routers. Deep inspection of the packets is needed only at the edge routers, where they are marked, and interior routers do not need to perform deep inspection to handle the packets appropriately, because they can just observe the markings. Thus, interior routers, which typically handle the heaviest loads, can have less "work" to do to support QoS in DiffServ, and can process packets more efficiently. Additional processing of the packets (e.g., *policing* and *shaping*) can be performed at the edges. At the edges, packets are marked according to the appropriate class of service. DiffServ keeps the forwarding path simple, and pushes complexity to the network edges as much as possible.

The PHB is the treatment applied at a DiffServ-compliant router to a BA. The edge routers classify and/or mark packets coming into and going out of the DiffServ domain, and the PHBs are the treatments they receive. Two groups of PHBs that have been proposed are *assured forwarding* (AF) and *expedited forwarding* (EF). AF is a means for a service provider DiffServ domain to offer different levels of forwarding assurances for IP packets received from a customer DiffServ domain. AF is a type of PHB group with three members, each with four instances, for a total of 12 possible treatments. Each AF class is allocated some bandwidth and buffer space in each DiffServ node. Meanwhile, EF configures nodes so that a BA has a minimum departure rate that is independent of the intensity of other traffic at the router. Together with appropriate handling of the BAs at the boundary routers, EF may be used to provide the "Premium" service.

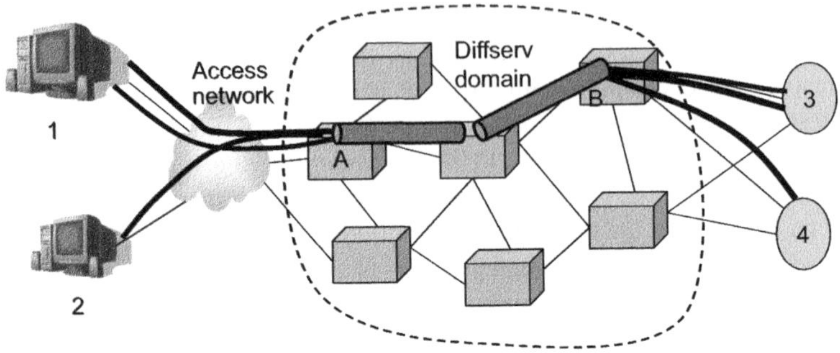

FIGURE 11.11 DiffServ.

A DiffServ domain is illustrated in Figure 11.11. There are multiple flows between nodes 1 and 2 on one side and nodes 3 and 4 on the other side. However, at the DiffServ edge routers, labeled A and B, we suppose in this example that the packets in these multiple flows are all marked the same (e.g., EF). Then, the interior routers can treat all these packets in the same way as far as QoS treatment is concerned, without having to process complicated criteria. As the figure indicates, we do not have a tunnel between A and B. The packets are still processed in the interior routers, albeit more efficiently than without DiffServ.

11.3.2 QoS Mechanisms

A number of different QoS mechanisms can be deployed in an IP network. Here, we focus on two aspects:

- Controlling the use of network resources before the network even gets congested. This is discussed in Section 11.3.2.1.
- Queuing and prioritization schemes. This is discussed in Section 11.3.2.2.

11.3.2.1 Designing and Planning for Adequate Resources Examples of controlling the use of network resources before the network gets congested include admission control and resource reservation schemes such as RSVP.

Admission Control. The idea behind admission control is to limit the rate of admission (of packets, or of flows) into the network so that the volume of traffic is manageable and it would be unlikely that the network becomes congested. It is analogous to admission control for road transportation networks, where during rush hour, the rate of entry into the freeways might be controlled through the use of alternating red and green lights. Various schemes fall under admission control. For example, at the edge of a DiffServ domain, the entry rate can be controlled by methods such as *token bucket*. The *bandwidth broker* concept can also be considered to be useful for admission control. More details are available, for example, in reference [7].

Resource Reservation Protocol. *Resource reservation protocol* (RSVP) is used for reserving resources across a network. Resources are reserved in one direction, from source to destination only (i.e., it is unidirectional), but it is easy for resources to be reserved in both directions, where each endpoint can initiate its own RSVP signaling. RSVP-capable routers in the path of the traffic flow will reserve appropriate resources after the RSVP signaling is completed. These reservations needs to be refreshed periodically. Interestingly, reservation requests can be shared between traffic from more than one sender, or just for traffic from one sender. The receiver is in a better position than the sender to know what resources are needed, from one or possibly more senders. Thus, the *receiver*, rather than the sender, is responsible for making reservation requests.

The reservations are made as follows:

- The source node sends *Path* messages toward the destination node. However, resources are neither requested nor reserved at this time.
- The destination node sends back *Resv* (reservation request) messages toward the source, using the same path but in the reverse direction. It is at this time that the resource reservations are actually requested. As the *Resv* traverses each RSVP-capable router, the resources are reserved.

We see from the above that RSVP is designed for the receiver to request reservations rather than the sender. Nevertheless, the Path messages serve a couple of purposes:

- To set up the soft state information (about the path back to the sender) at each intermediate router
- To inform the receiver about the kind of traffic that would be sent so that it can decide appropriately as to resource reservations.

RSVP messages are special transport layer packets, such as Internet control message protocol (ICMP) packets, which are processor intensive for routers. Hence, the use of RSVP does not scale well. RSVP is shown in Figure 11.12.

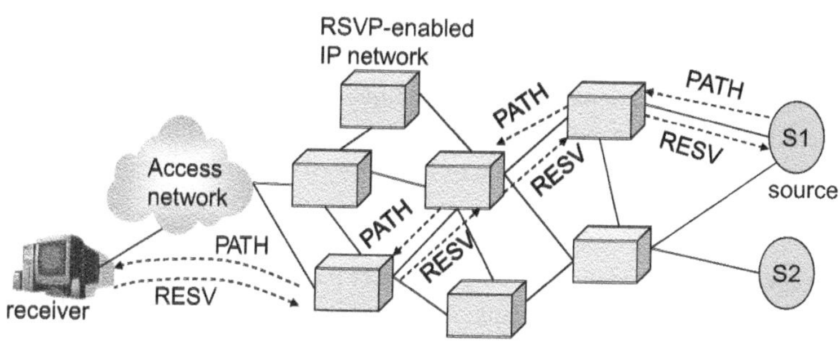

FIGURE 11.12 RSVP.

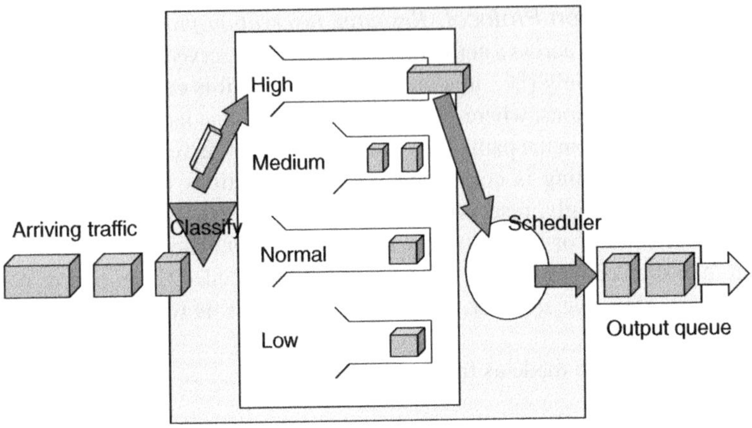

FIGURE 11.13 Queuing framework in routers.

11.3.2.2 Queuing and Other Prioritization Schemes Queuing is a fundamental tool in QoS. Each router has one or more input queues and one or more output queues, and many routers have numerous options for configuring *queuing disciplines* on these queues (Figure 11.13). By *queuing discipline* we mean how the packets in the queue are handled; whether some that arrive later can be sent out before others that arrived earlier (and if so, based on what criteria); and also (in the case of multiple queues) whether some queues have priority over others; and so on.

FIFO Queuing. FIFO (first in, first out) queuing is the simplest queuing discipline. All packets are treated equally without preference, except that earlier-arriving packets leave before later-arriving packets. Advantages of FIFO queuing include:

- It is the most straightforward method to implement.
- For lightly loaded networks, the queuing just needs to smooth intermittent traffic bursts. FIFO is adequate and efficient for this.

Disadvantages of FIFO queuing include:

- FIFO does not have a mechanism for giving any type of priority to any packets.
- For heavily loaded networks, when the queues are full and packets have to be discarded, FIFO does not distinguish between higher- and lower-priority traffic, so there may be a greater chance of high-priority traffic being discarded than in some other queuing discipline.

Priority Queuing. The router takes high-priority packets and places them in an output queue ahead of lower-priority packets. Advantages of priority quening (PQ) include:

- Unlike FIFO, traffic can be prioritized, so high-priority packets get through with less delay.

- The number of priority levels, and the granularity of each group of packets, is flexible.

Disadvantages of PQ include:

- If there are too many priority levels, the computational overhead might be too high and might affect packet-forwarding performance.
- Notice that we did not qualify our statement that the router takes high-priority packets and places them in an output queue ahead of lower-priority packets. This happens even if some lower-priority packets have been waiting a long time. As long as there is higher-priority traffic, it gets served first. Therefore, when higher-priority traffic volume is high, lower-priority traffic may be waiting for very long and/or mostly be dropped. It is said that they are starved of service.

The second disadvantage listed above is commonly known as *buffer starvation*. It may be better to provide reduced levels of service to the low-priority traffic instead of allowing buffer starvation to occur. That is one of the ideas behind "fair queuing" schemes.

Fair Queuing (FQ), WFQ, and CBQ. Different traffic flows [we can think of a flow as a set of packets moving in the network that have some kind of association with one another (e.g., from the same source to the same destination, the same kind of data, etc.); this is vague, because flows can be specified in many ways] may present different rates of incoming packets to a router. Fair queuing tries to balance traffic flow volume at the output queue across different input flows that may present different rates. It does the balancing using per-flow queues and interleaving between the queues. As a result, fair queuing favors low-volume traffic flows. This is not necessarily a good thing.

Weighted fair queuing (WFQ) is a variation of FQ that weights the per-flow queues and does not treat them all equally as FQ does. The weighting may be based on the IP type of service (TOS) field or on other criteria. Since WFQ is still a variation of FQ, it avoids buffer starvation just as FQ does, but at the same time it gives preference to higher-priority traffic.

Advantages of WFQ include:

- There is prioritization of packets *without* buffer starvation.
- Any misbehaving flow (e.g., from a rogue TCP session) is prevented from taking up too much output bandwidth at the expense of other flows.

Disadvantages of WFQ include:

- Even though there is fair queuing at the packet level (i.e., the number of packets from each flow that go to the output queue may be rate limited), there may still be problems with flows that present very large packets (so the rate in bits per second may be much higher for such flows).

- It does not scale well to many flows.

There are other variations [e.g., *class-based queuing* (CBQ)], or schemes that go by other names, but where different vendors and different people use the same name for different concepts or different names for the same concepts. So we just list some of the ideas that go under such names as CBQ and *low-latency queuing*:

- One definition of the basic WFQ has the router automatically dividing incoming packets into flows (which can lead to thousands of flows that can cause performance problems); the packets are divided into classes by some user-specified criteria (sometimes called a *policy*). It is obvious why such a scheme may be called CBQ.
- To avoid cases where flows with very large packets may get away with higher rates than other flows, (bps) rate limiters can be installed for each flow.
- A hybrid of PQ and CBQ may be used (and this is sometimes called *low-latency queuing*) where a special rate-limited queue is set aside for delay-sensitive traffic such as voice and video, and CBQ is used for the rest of the traffic. The special queue needs to be rate-limited to avoid buffer starvation.

11.3.3 Wireless QoS

In wireless systems, the wireless link(s) is often the critical part of the end-to-end path in terms of QoS. Since the bandwidth might be most severely limited in the wireless link, and would have higher error rates than wired portions of the network, care has to be taken, especially with the wireless link, in coming up with an end-to-end QoS solution.

We have already seen in Section 8.3.2.1 how 802.11's basic MAC channel access protocol includes some prioritization, albeit crude. It does this by using different values for DIFS, SIFS, and so on, and enforcing the waits for these different lengths of time.

11.3.3.1 IEEE 802.11e The basic 802.11 MAC incorporates a priority scheme using different interframe spacings, as we have seen in Section 8.3.2.1. However, it is restricted to giving priority to certain control frames. All data frames have the same priority, which is unacceptable for certain applications, such as voice and/or video over WLAN. Therefore, IEEE 802.11e introduces QoS mechanisms for WLAN. It has new coordination functions, the *enhanced distributed coordination function* (EDCF) and the *hybrid coordination function* (HCF), to support eight traffic classes. EDCF is designed as an enhancement of DCF built on top of the existing DCF mechanisms. Therefore, non 802.11e-enabled MSs can coexist with 802.11e-enabled ones. EDCF has both:

- Traffic class–dependent interframe spacings
- Traffic class–dependent minimum initial collision window sizes

A new interframe spacing, *arbitration interframe space* (AIFS), has been introduced which can be different for different traffic priorities. Unlike in DCF, in EDCF, the minimum size of the collision window can differ depending on the traffic class, so as to favor higher-priority traffic.

The HCF is like a QoS-enhanced version of the PCF. Although it may seem perfectly placed to control the polling sequence, and thus to control traffic and QoS carefully in the WLAN, the polling sequencing is not specified in 802.11. It may be up to the vendor to decide on polling strategy. There are also other reasons why the PCF fails to meet the more stringent requirements for delay-sensitive traffic such as voice.

- The point coordinator does not necessarily know what kind of traffic each station wishes to communicate.
- The point coordinator does not know the queue lengths in each station, and in the base 802.11 standard, data traffic is not placed in different queues depending on QoS priority.
- The point coordinator has control of the medium for only part of the total time.
- When a station is polled, it has the right to transmit as large a packet as it wants, up to the maximum packet size of 2304 bytes that applies for any packet passed to the 802.11 MAC. This may cause too much delay for other stations.

HCF tries to exploit the polling capabilities of the point coordinator (called the *hybrid coordinator*) while addressing the concerns just listed. HCF enhances the 802.11 MAC so that careful polling helps provide QoS priorities. It deals with the problems listed above. A new QoS control field allows information on the traffic from each station to be provided by the station to the hybrid coordinator. The hybrid coordinator can also specify a limit to the size of the packet that may be transmitted in response to each poll. Furthermore, the hybrid coordinator can initiate HCF access even while the DCF (not the PCF) is in use, so streaming or conversational traffic can be transmitted regularly with less delay.

EXERCISES

11.1 Why is it more difficult to deliver a call (GSM) or a packet (GPRS) to a mobile station than it is to originate a call or a packet?

11.2 What is a location area in GSM? Is it one or more cells? Why is it useful? What are one advantage and one disadvantage of having large location areas? *Hint*: Think in terms of frequency of location updates and size of paging areas.

11.3 What is the difference between a SIP proxy and a SIP redirect server?

11.4 How can a SIP server ensure that it stays in the signaling path?

11.5 What is buffer starvation? How can it be avoided?

REFERENCES

1. R. Braden, D. Clark, and S. Shenker. Integrated services in the Internet architecture: an overview. RFC 1633, June 1994.
2. K. Nichols, V. Jacobson, and L. Zhang. A two-bit differentiated services architecture for the Internet. RFC 2638, July 1999.
3. J. Rosenberg and Schulzrinne H. An offer/answer model with the session description protocol (SDP). RFC 3264, June 2002.
4. J. Rosenberg and H. Schulzrinne. Reliability of provisional responses in the session initiation protocol (SIP). RFC 3262, June 2002.
5. J. Rosenberg, H. Schulzrinne, G. Camarillo, A. Johnston, J. Peterson, R. Sparks, M. Handley, and E. Schooler. SIP: Session initiation protocol. RFC 3261, June 2002.
6. H. Schulzrinne, S. Casner, R. Frederick, and V. Jacobsen. RTP: a transport protocol for real-time applications. RFC 3550, July 2003.
7. K. D. Wong. *Wireless Internet Telecommunications*. Artech House, Norwood, MA, 2005.
8. GSM World. GSM roaming. http://www.gsmworld.com/technology/roaming/, 2011. Retrieved Mar. 2, 2011.

12

TOWARD AN ALL-IP CORE NETWORK

In this chapter we continue the story from Chapter 11 of the convergence of wireless and IP networks toward an all-IP wireless networks. We have seen in Chapter 11 how IP has been retrofitted to support voice (using such technologies as RTP and SIP) with QoS control. In this chapter we proceed in Section 12.1 to show how IP has also been retrofitted to support other aspects of using it in wireless networks, including mobility support and more limited bandwidth than in wired networks. We then discuss in Section 12.2 how GSM has evolved to add more support for packet data networking, with the addition of GPRS. Moving beyond GPRS, in Section 12.3 we trace the continued evolution of wireless networks, up to LTE. An important development is the addition of the IP multimedia subsystem (IMS) in UMTS/LTE, so we explain IMS concepts in Section 12.4. Finally, in Section 12.5 we look briefly at how other networks (from other tracks of development besides UMTS) have been moving in the same direction—toward convergence with IP networks.

12.1 MAKING IP WORK WITH WIRELESS

One of the beauties of IP routing is the hierarchical addressing scheme. It allows *aggregation* of addresses; that is, an entire block of contiguous addresses (including very large blocks of addresses) can be referenced by a single network address. All the addresses in the block share the same network prefix, and thus that shared network prefix is used as the single network address to represent them. This ability to aggregate addresses helps make IP scalable, in the following sense: Without address aggregation, routing tables will grow linearly with the number of IP addresses. Even if the concept

Fundamentals of Wireless Communication Engineering Technologies, First Edition. K. Daniel Wong.

of default route is used so that most hosts only need to have a limited number of entries in their routing tables, there will still be core routers in the Internet that will need to have multiple millions of addresses (and in theory, on the order of 2^{32}), one for *every* separate IP address.

With address aggregation, the routing tables in the core routers in the Internet are still large, but managable. Forwarding of packets toward an entire range of contiguous addresses is done by matching an the network prefix (which is the same for all those addresses). For example, the network prefix 18.0.0.0 famously belongs to MIT. Although this is an elegant and scalable scheme, it tends to restrict IP addresses that share the same network prefix to the same geographical region. Generally, the more specific the network address, the smaller the geographical region. More specifically, it is not so much about geographical location as about the location of a node in the network topology. In Section 12.1.1 we will see that the hierarchical addressing scheme creates a challenge for handling mobility, and we will see how the mobile IP scheme addresses the challenge.

Another challenge in making IP work with wireless is the relatively higher cost of bandwidth over wireless links. Thus, in wired networks, IP-related protocols can have relatively higher overheads (e.g., larger headers) without causing as much of a problem as in wireless links. In Section 12.1.2 we will see that *header compression* schemes may be needed over wireless links, to reduce the size of packet headers.

12.1.1 Mobile IP

When a wireless device moves and changes its point of attachment to the network, the movement can be within a LAN or it can be between LANs. In the former case, we have *layer 2 mobility*, and there is no mobility as far as the network protocol (e.g., IP) is concerned. In the latter case, we have *layer 3 mobility*, and the rerouting of packets needs to be handled by the network layer protocol. An example of layer 2 mobility is movement between access points in a WiFi network, where the two access points are part of the same ESS. An example of layer 3 mobility is movement between two access points in a WiFi network where the two access points are part of two different ESSs.

In the case of layer 3 mobility, should the mobile use the same or a different IP address at the new network? Some activities, such as much web browsing activities, do not require that the mobile keep the same IP address as it moves between networks. However, some activities, such as file transfer, require that the mobile keep the same IP address as it moves. If the mobile is playing the role of a server (e.g., an email server), it may also need to keep its IP address as it moves. We can call this the requirement for uninterrupted communications.

However, given the association of each IP address with its particular position in the Internet topology, a mobile cannot keep its IP address as it moves. Thus, the problem that mobile IP was introduced to solve:

- The mobile wants to keep an unchanging IP address as it moves, for uninterrupted communications.

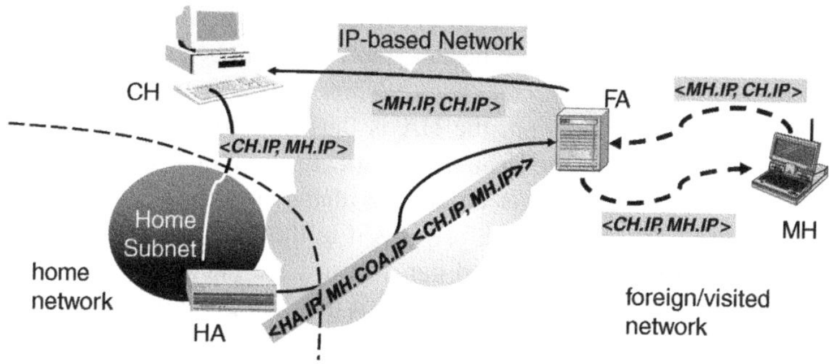

FIGURE 12.1 Mobile IP.

- The mobile needs to use a local IP address in the network it is visiting, in order for the regular Internet routing to work and packets to be delivered to it.

Mobile IP [3] (Figure 12.1) is the way that the IETF has chosen to solve the problem. We will go into more detail shortly, but at a high level what happens is this: The *mobile node* (MN), also known as *mobile host* (MH), has an unchanging *home address* that allows it uninterrupted communications while it moves. The home address is topologically part of the mobile node's *home network*. The mobile also acquires a *care-of address* in the visited network, which is a local IP address in that network whenever it moves into a visited network. Packets from any *correspondent node* (CN), also known as *correspondent host* (CH), traverse the Internet and are delivered (as usual) to the home network. In the home network, a *home agent* intercepts the packet and forwards it to the care-of address of the mobile. In the visited network, a *foreign agent* receives the forwarded packet and delivers it to the roaming mobile.

12.1.1.1 Delivery of the Packet from Home Agent to Mobile Node Let's revisit the journey of an IP packet from the correspondent node (CN) to the mobile node (MN). The packet will travel using normal IP routing to the home network of the MN, since its destination address is the home address of the MN, which is an IP address from its home network. This always happens whether or not the MN is at home or it is visiting another network. If it is at home, it receives the packet as usual, and that's the end of the story. If it is visiting another network, the packet will be lost in its home network but for the actions of its home agent (HA). The HA intercepts packets for the MN; for example, if the home network is an Ethernet-based LAN, the HA can use proxy ARP (Section 10.3.5) to capture the packets meant for the MN.

Once the HA is in possession of the packet, how does it get it to the MN in the visited network? The HA would have the care-of address of the MN in the visited network (in Section 12.1.1.2 we will find out how it obtains this information). It *encapsulates* the packet in a new packet. The new IP packet has the HA IP address as the source address, and the care-of address as the destination address, and the entire original packet, including the original header, just goes into the payload portion of the new packet. [This is called IP-in-IP encapsulation; it is simple but adds at least 20 bytes of overhead with the new header; more efficient schemes with less added overhead are possible (e.g., *minimal encapsulation* that can bring the overhead down to 8 bytes).] Using encapsulation, the HA is thus able to *tunnel* the packet from the MN's home network to the visited network. In particular, it arrives at the FA there, since the care-of address is an address that is routable to the FA. The FA then unencapsulates the packet and forwards it to the MN. Usually, the FA would be on the same LAN as the MN and can forward the packet to the MN *without* the need for IP forwarding. (Otherwise, since it is the original packet and no longer encapsulated, the packet would get forwarded all the way back to the home network of the MN!)

As for packets from the MN to the CN, they can be sent directly from the MN to the CN without going through the home network.† The HA and FA need to be prepped to perform these roles for the MN, and the setup happens during the registration procedure, which we discuss next.

12.1.1.2 Registration For mobile IP to work, the HA needs to hear from the MN about its new care-of address and the FA also needs to be aware that the MN is present in the visited network. This is accomplished by the registration procedure (Figure 12.2). In the registration procedure, the MN first discovers the presence of an FA. (It may simply wait until it hears the *agent advertisement* message that the FA periodically broadcasts, or it may reduce the wait by inciting the FA to send an agent advertisement outside its periodic schedule; it can do this by sending an *agent solicitation* message.) The agent advertisement message contains the IP address of the FA that the MN can use as its care-of address.

Armed with this potential care-of address, the MN then sends a registration message to the HA via the FA. The HA replies, also via the FA, and stores the *binding*

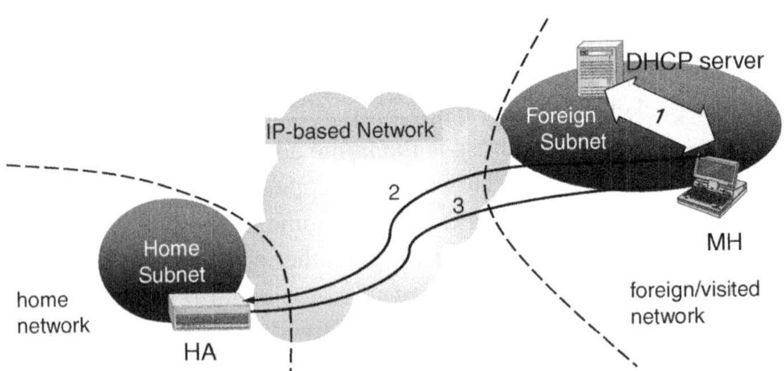

FIGURE 12.2 Mobile IP registration procedure.

† This may cause problems if some firewalls are in use in the visited network (e.g., if the firewalls do not let traffic out of the visited network whose source address is not part of the visited network); there are ways around this, and other such problems, but they are beyond the scope of this book.

of the home address and care-of address, for use when the next packets for the MN arrive in its home network.

You may wonder what is to prevent a malicious node from sending a fake registration message and thus diverting all traffic for an MN from its home network to an arbitrary network, using the HA as an unwitting accomplice. This is prevented through an authentication mechanism, whereby a mandatory mobile-home authentication extension is attached to the registation message.

12.1.1.3 Colocated Care-of Addresses The need for foreign agents makes Mobile IP more difficult to deploy. Already we need a home agent, and need the mobile node to be able to play the role of mobile IP client; with the addition of foreign agents, we would also need every potential visited network to have a foreign agent! Thus, we have a motivation to avoid the use of foreign agents, and seek an alternative.

Such an alternative is found in the concept of *colocated care-of addresses* (Figure 12.3). In this scheme, the MN becomes its own FA! It can no longer rely on there being an FA in the visited network to broadcast an agent advertisement from which it can obtain a care-of address. However, there is no requirement that the care-of address *has* to be that of an FA. It just needs to be an address that is routable (through normal IP routing) to the current visited network, in which the MN is located. Thus, the MN can simply obtain a local IP address in the visited network for use as its care-of address, with the assistance of some other protocol (e.g., DHCP, PPP). (These are standard ways that devices obtain IP addresses when they connect to a network anyway, so new protocols are not needed specifically for an MN to obtain a care-of address.)

Once the MN has obtained a local IP address in the visited network, it can register it with its home agent by sending a mobile IP registration message directly to its home agent (without the need to go through a foreign agent). Then, packets from the HA would be routed to the MN directly, via its colocated care-of address.

Colocated care-of addresses did not make it into the base mobile IP specifications, but are described only in an IETF draft document. The world had to wait for mobile IPv6 to see colocated care-of addresses become the norm.

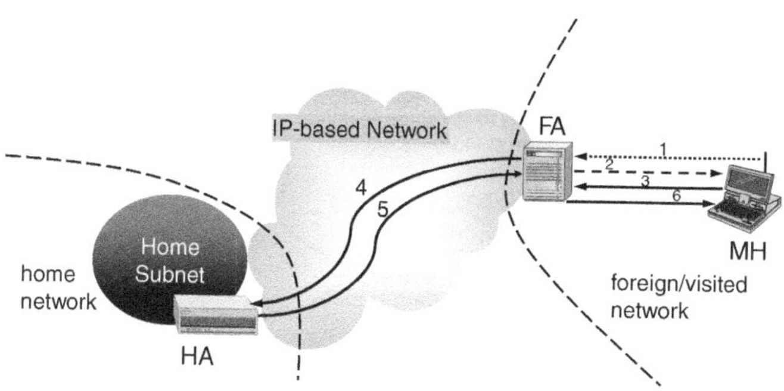

FIGURE 12.3 Mobile IP with colocated care-of address.

12.1.1.4 *Issues with Mobile IP* Mobile IP has the tremendous benefit of not requiring the CN to be aware of mobile IP. It sends packets to the MN as normal. However, this means that packets to the MN always go first to its home network, before being tunneled to the visited network. In some cases, this can result in great inefficiency. For example, suppose that the CN and MN are very close to each other (topographically), perhaps even in the same network. Further suppose that the home network is very far away, perhaps on the other side of the globe. The path from CN to MN would still go around the world to the MN's home network and back again. This inefficiency is often known as *triangular routing*, where the paths from the CN to the HA and from the HA to the MN form two sides of the triangle, and the path from the MN to the CN forms the third.

The encapsulation overhead can be annoying, especially with wireless links where bandwidth efficiency is essential, and even more so with small packets such as VoIP packets.

12.1.1.5 *Mobile IPv6* Mobile IP for IPv6 [4] (mobile IPv6, for short; Figure 12.4) is an enhanced version of mobile IP for IPv6 networks. It takes advantage of new features in IPv6 that provide better support for mobility than IPv4 provides. It addresses various issues with mobile IP in the following ways:

- *The use of FAs*. Colocated care-of addresses are standard in mobile IPv6. There are no more FAs. It is easier for an MN to acquire an IPv6 address in a visited network than it is to acquire an IPv4 address, so there are no impediments for implementing colocated care-of addresses.
- *Triangular routing*. Binding updates sent directly from the MN to the CN inform the CN about the care-of address of the MN, so the CN can send packets directly to the care-of addresses and not have to keep going through the home network and the HA. The major impediment to implementing this in mobile IP is that this requires that all CNs are able to understand the binding updates and act on them accordingly. This would have been unacceptable in mobile IP, as it means that *every* TCP/IP implementation already in existence would need to be upgraded.

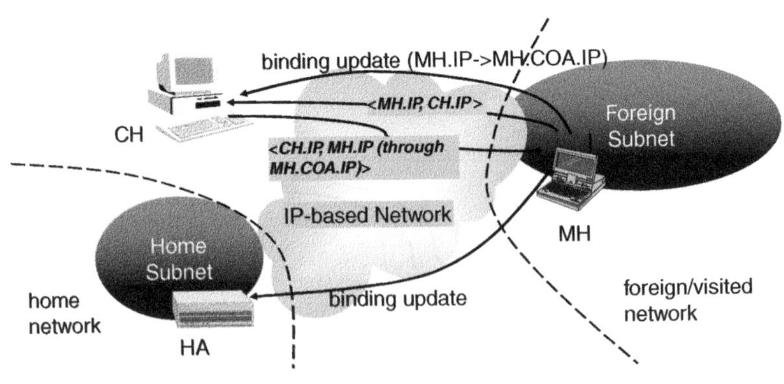

FIGURE 12.4 Mobile IPv6.

With mobile IPv6, though, we have a new protocol (IPv6), so it was reasonable to mandate that every IPv6 node be able to understand and process these binding updates, right from the beginning.

- *Encapsulation overhead*. Instead of tunneling by encapsulation, as in mobile IP, mobile IPv6 takes advantage of the new routing header extension of IPv6, to specify a type of source routing efficiently (Section 10.3.1.2) to the MN through the care-of address.

Mobile IPv6 is not without its own issues, though. As in mobile IP, the MN updates its HA with its latest care-of address when it moves. This can be done securely in both mobile IP and mobile IPv6, as it is reasonable to assume that the HA and MN have a preexisting security assocation. It is not reasonable to make the same assumption of the MN and the CN, since it should work for *any* possible CN. Thus, the binding update from MN to CN is not as secure as the registration with the HA. It uses the *return routability* procedure, which is not foolproof but merely makes it more difficult for an attacker to redirect traffic maliciously by sending fake binding updates.

12.1.2 Header Compression

In a packet-switched network such as an IP-based network, every layer adds overhead, in the form of headers (packet headers, frame headers, etc.) and possibly also trailers at the end of the packets/frames. Headers contain vital information that is needed for the network protocols to operate correctly. However, the presence of these additional bits, which are not user data bits, means that the communication medium will have to transport more bits than the user sends. These additional bits, in headers and trailers, are known as *header overhead*. The higher the ratio of overhead bits to total bits, the more inefficient the transmission of data. It might be said that "the overhead is very high" or "the overhead is too high." We may wish to "reduce the overhead" (i.e., to reduce the ratio of overhead bits to total bits), thus making transmissions more efficient.

One way to reduce the overhead is to compress the header portion of the packets/frames. Indeed, various schemes have been proposed for compressing transport headers, IP headers, and so on, and they do reduce the overhead. But they are not in general widespread use, because there is a price to be paid—header compression/decompression takes work to do. So we only want to use header compression in cases where the gains outweigh the efforts needed to compress and decompress. In most wired networks, for many types of data it is not worth it to use header compression.

However, when transporting voice over IP, over a wireless link, we have a situation where header compression is worth it. Why? Because it combines two motivations to perform header compression, the second of which is especially compelling.

- Voice packets must be small. Otherwise, too much packetization delay is added to the end-to-end delay. Usually, voice codecs produce encoded voice packets that are 10, 20, or 30 ms long. However, header sizes usually are mostly independent of packet size. With a fixed header size, the header will be a much larger percentage of the entire packet, when the packet is small, than when the packet is large. Therefore, small packets will tend to have much higher overhead than large packets.

- Bandwidth is an extremely precious commodity on wireless links. High header overheads are too costly for wireless links.

We elaborate on the need for voice packets to be small (typically, 10 to 30 ms long) below. We then work through a numerical example to get a better feel for the amount of header overheads encountered with VoIP packets. Thus, without header compression, the overheads are excessive and will consume too much of the precious wireless bandwidth if transmitted uncompressed.

12.1.2.1 *Voice Packets Must Be Small* Packetization delay refers to the time it takes to form a packet. Thus, if a voice packet is 100 ms long, the first bit has to wait for 100 ms before the last bit arrives, and the packet is therefore ready for transmission. This 100 ms is in addition to the queuing delays, transmission delays, and other delays that the packet will face in traversing the end-to-end path between sender and receiver.

12.1.2.2 *Worked Example: Packet Sizes, etc.* Suppose that a VoIP system uses G.729. Since G.729 is a very efficient encoder that needs only 8 kbps to encode voice, find the number of bytes of G.729 encoded voice over a 20-ms period. Now add UDP, RTP, and IP header overhead. What percentage of the resulting packet is occupied by the header?

The value 8000 bits per second is equivalent to $(8000 \times 20)/1000 = 160$ bits per 20 ms, which is 20 bytes. The UDP, RTP, and IP headers are at least 12, 8, and 20 bytes each, respectively (the IP header could be longer in some cases, but no shorter than 20 bytes). This is 66.6% of the packet that is occupied by the header, 40 out of 60 bytes.

12.1.2.3 *Robust Header Compression* Various header compression schemes can be used with VoIP packets in wireless systems, but we focus here on *robust header compression* (ROHC [1], Figure 12.5), because of its adoption in UMTS. ROHC is a prominent header compression scheme that was introduced by the IETF in 2001. It tries to squeeze out a lot of redundant information.

In particular, with VoIP packets between a particular source and a particular destination:

- Some header fields are the same in every packet (e.g., the source and destination IP addresses)

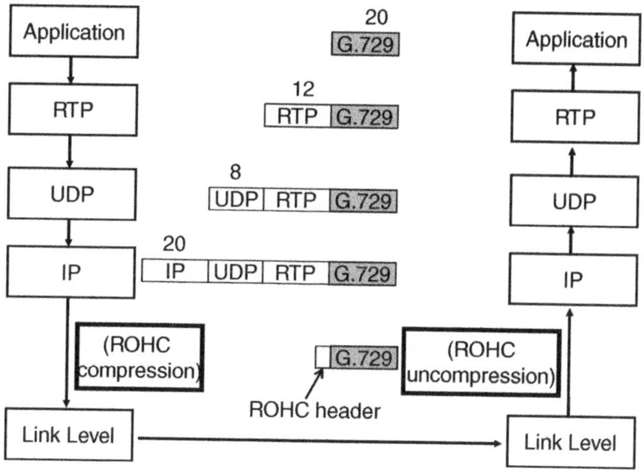

FIGURE 12.5 The benefit of using ROHC.

- Some header fields are incremented in consecutive packets by the same each time (e.g., if the voice codec uses 10-ms packets, the RTP time stamp will be incremented by 10 ms with each successive packet).

Thus, some information can be transmitted just once—at the beginning of a talk spurt—and not have to be repeated (or incremented) in every packet. A more general, but compatible framework is provided in RFC 5795 [6].

12.1.2.4 ROHC in UMTS ROHC made it into UMTS in Release 4. It is a mandatory part of the PDCP layer in the protocol stack. Obviously, the packets with ROHC-compressed headers cannot be routed over the Internet, so ROHC is only used over the radio link.

12.2 GPRS

GSM is optimized for voice traffic rather than for data traffic. It is possible to use GSM for data traffic, albeit with low data rates since the data bits would have to fit within the constraints of the GSM time slots. The GSM network would interface with the external packet data network (such as the Internet) through an *interworking function.* A first step to making GSM more data-friendly is to allow aggregation of time slots (i.e., more than one time slot can be assigned to a user at the same time, thus allowing higher data rates). This is the basis for the *high-speed circuit-switched data* (HSCSD) service in GSM. As its name implies, it is still a circuit-switched solution. Problems with using circuits for data in GSM include:

- Data is bursty. Yet with a circuit-switched solution, time slots (whether aggregated as in HSCSD or not) would be allocated for the data service, and the user would need to be charged for the resources. The meter would be running, so to speak, even while the user was reading a web page for a long time and not doing any data transfer during that time. Thus, the cost to the subscriber would be prohibitively expensive.

- The circuit would need to be set up every time that it was needed, to be used resulting in a delay before connectivity could be established, to the annoyance of users who would be used to "always-on" service from DSL, and so on; alternatively, the circuit could be made "always on," which would be a very poor utilization of radio resources, again too expensive to be practical.

Generalized packet radio service (GPRS) is an enhanced packet service added to GSM systems to address these concerns. The name begins with "generalized" because it was not originally designed just to carry packets of any one packet data network (e.g., TCP/IP). Instead, it was designed to provide a generalized packet radio service that could transport packets from multiple packet data networks, including TCP/IP but not limited to it. IP and X.25 were the original examples of packets that GPRS could transport. Most of the implementations of GPRS have focused on IP traffic, though. GPRS introduces innovations on both the radio access side and the network side.

On the radio access side, GPRS introduces the following features:

- *Division of time slots*. Within the frame, some time slots are assigned to voice and some to GPRS. The assignment is not fixed, and can be changed with time.
- *Higher data rates*. The aggregation of two or more time slots allows the system to be used for traffic to and from one mobile.
- *Dynamic allocation of time slots*. The time slots for GPRS can be shared by multiple devices on an as-needed basis. Thus, when one device is downloading a file, it could be using more of the GPRS time slots, whereas another device that is not doing as much may be using only one GPRS time slot during that time. As we have indicated in Section 8.1.3, GPRS allocation of radio resources is based on units of *radio block*, and each radio block comprises the same time-slot position in four consecutive GSM frames (of eight time slots each).

In addition to allocating the time slots dynamically for different mobiles, another way that GPRS utilizes the radio resources more efficiently is by maintaining states related to the traffic activity of each mobile (the GPRS mobility management states that we discuss in Section 12.2.2). Thus, mobiles in "ready" (active) state have more immediate access to radio resources than do those in "standby" or "idle" states.

On the network side, GPRS introduces a new packet-switched network architecture in parallel with the circuit-switched architecture. Thus, the base stations and base station controllers (BSCs) are shared between voice and GPRS, but from the BSCs, voice traffic goes to/from the MSCs, whereas packet data go to/from the GPRS support nodes (GSNs): namely, the *serving GSNs* (SGSNs) and the *gateway GSNs* (GGSNs). As can be seen in Figure 12.6, we thus have two parallel domains, sometimes called the CS (circuit-switched) domain and the PS (packet-switched) domain. Some network

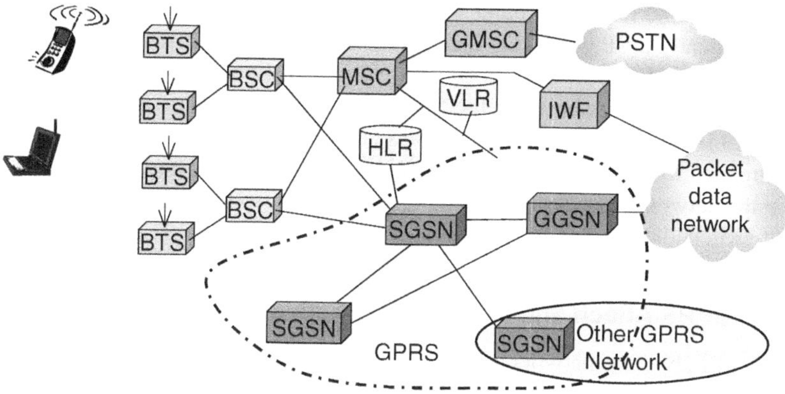

FIGURE 12.6 GSM network architecture with GPRS added.

elements, such as the HLR and AuC, are accessed by both the CS and PS sides, to perform necessary administrative functions. The SGSN is analogous to the MSC/VLR of the CS side. The GGSN is analogous to the GMSC of the CS side. It handles PDP contexts (see Section 12.2.1) and acts as the gateway router for mobiles using GPRS.

The protocol stack for data (a different stack exists for control traffic) is shown in Figure 12.7. Consider the protocol stack in the mobile. The *radio link control* (RLC) and below are based on GSM and modified for GPRS. TCP/IP and above are standard. The *packet data convergence protocol* (PDCP) is a thin layer that allows GPRS to support different packet protocols (besides TCP/IP, there could some other protocols over PDCP) over a common RLC and MAC. ROHC occurs in the PDCP. Another interesting point to notice is that using the PDCP (over the air) and *GPRS tunneling protocol–user plane* (GTP-U) on the network side over the RAN and through the SGSN into the GPRS core network, each IP packet is transported between mobile

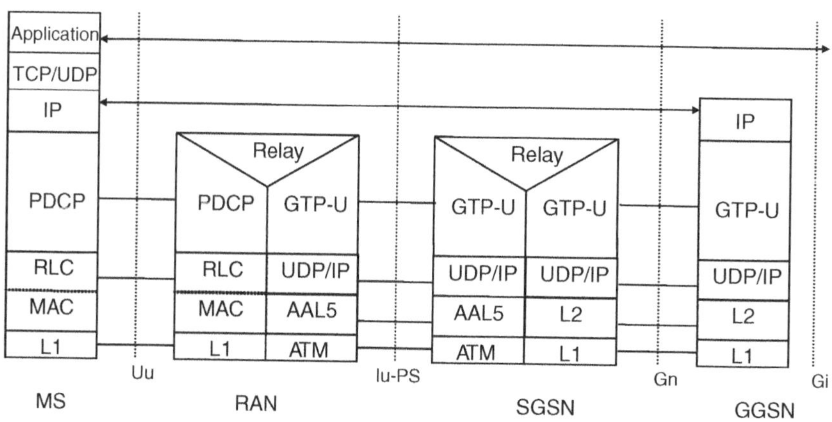

FIGURE 12.7 Protocol stack of GPRS.

and GGSN in one IP hop. In other words, the GGSN appears as the local router for the mobile, and PDCP with GTP-U combine to provide a point-to-point link (i.e., an extended layer 2 service) (Section 10.2.6). Notice also how there is another UDP and IP in the protocol stack between the SGSN and the GGSN. This is because there is an internal IP network between the SGSNs and GGSNs. This is completely separate and unrelated to the IP traffic being transported over GPRS. Thus, both over the air and in the network, GPRS provides more efficient transport of data traffic than with circuit-switched connections.

12.2.1 GPRS Attach and PDP Context Activation

To use GPRS, both the network and the mobile must support it, and the user must have an appropriate subscription or other business arrangement with the operator. Then, when the mobile is powered on, it needs to perform a *GPRS attach* procedure to begin using GPRS (the mobile could be on, but online using voice and text messaging, if it has not yet performed a GPRS attach). The GPRS attach procedure is shown in Figure 12.8 for a case where the mobile has moved to a new SGSN coverage area from a previous SGSN (labeled "old SGSN" in the flow). Notice the similarities with Figure 11.4 and how the SGSN behaves like an MSC/VLR combination, but for PS traffic in GPRS rather than CS traffic in GSM.

As mentioned in our introduction to GPRS, GPRS is a generalized packet radio service, capable of carrying packets from multiple packet data networks. Therefore, it needs a different *context* for each one of them, and the context is called a *packet data*

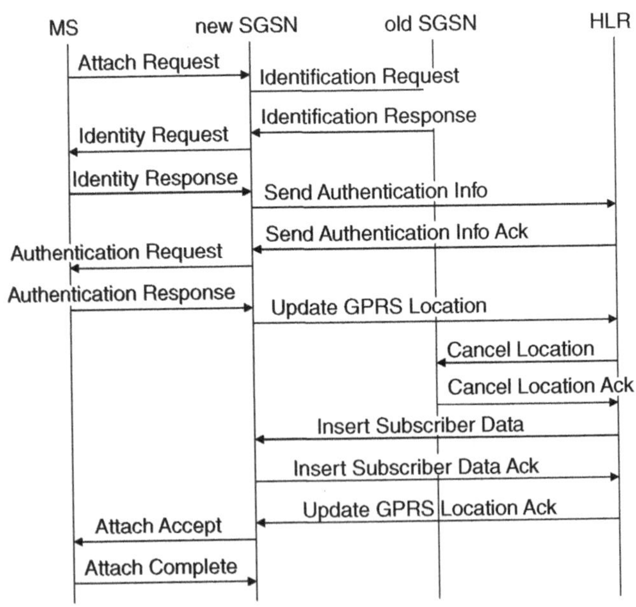

FIGURE 12.8 GPRS attach procedure.

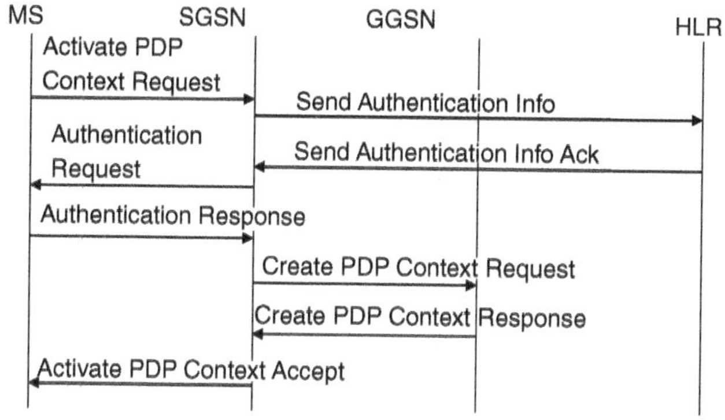

FIGURE 12.9 GPRS PDP context activation procedure.

protocol (PDP) context. In fact, there can be multiple PDP contexts even for IP alone, each having its own IP address, QoS profile, and so on, giving GPRS the flexibility to handle multiple different streams of IP traffic differently (e.g., there can be one PDP context for VoIP data with more stringent delay requirements, another for VoIP signaling, and another for a background file transfer). To create and manage different PDP contexts, GPRS requires a PDP context activation procedure to be performed after GPRS attach. The procedure is shown in Figure 12.9. The procedure can be modified slightly for it to apply to IPv6 as well [7].

12.2.2 GPRS Mobility Management States

The GPRS mobility management states of a mobile are shown in Figure 12.10. On the left of the figure is the state diagram for the mobile, and on the right, the corresponding

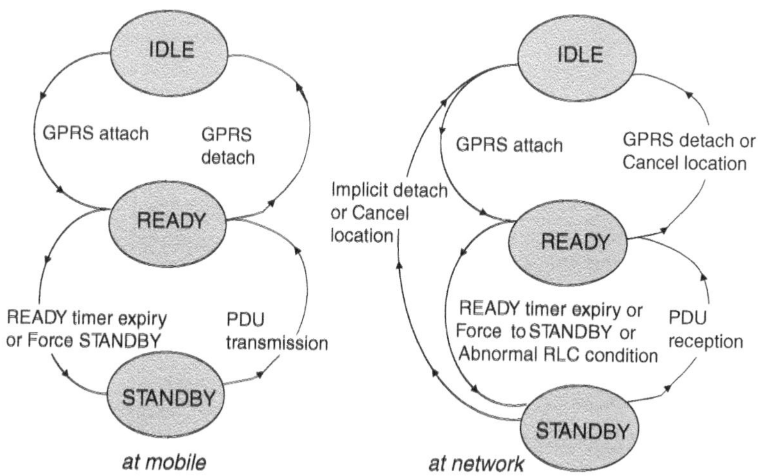

FIGURE 12.10 GPRS mobility management states.

state diagram in the network for the mobile. Unlike GSM on the CS side, where a mobile can be idle or active (roughly corresponding to the "idle" and "ready" states of GPRS), there is an additional in-between state, the "standby" state. This allows shorter access times on the average, because terminals can be in "standby" state ready to quickly switch to the "ready" state when there are packets to communicate. This is faster than going from "idle" to "ready." The mobile moves from "ready" to "idle" state upon expiry of a timer, but when there is a *protocol data unit* (PDU) transmission, it goes back to the "ready" state.

When a mobile is in the "ready" state, the network keeps track of the mobile location on a cell-by-cell basis. When a mobile is in the "standby" state, there is a concept of *routing areas* (RAs) similar to the location areas (LAs) of the CS side. RA updates are performed in a way similar to LA updates when a mobile is idle. For the case of RA updates, though, a mobile not only needs to be powered on, but also needs to be GPRS attached, in the "standby" state.

12.3 EVOLUTION FROM GSM TO UMTS UP TO THE INTRODUCTION OF IMS

Figure 12.11 shows a timeline for evolution from GSM to UMTS. We discuss and summarize some of the highlights of the changes, as GSM evolved to UMTS and more recently to LTE, in the next subsections.

12.3.1 First UMTS: Release '99

The major change in going to UMTS was introduction of the wideband CDMA (WCDMA) air interface. This was a completely new air interface radically different

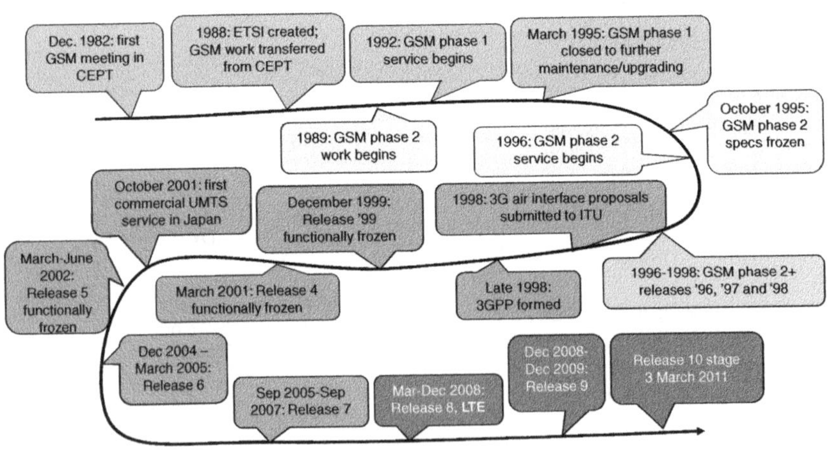

FIGURE 12.11 Timeline for evolution from GSM to UMTS.

TABLE 12.1 Correspondence of New Names in UMTS with GSM Names

GSM Name	UMTS Name	Comment
BTS	Node B	Popularly known as a base station
BSC	Radio network controller (RNC)	
BSS	Radio network subsystem (RNS)	Not a network element but a subsystem
SGSN	3G-SGSN	Or it may just be called SGSN
GGSN	3G-GGSN	Or it may just be called GGSN
MSC	3G-MSC	Or it may just be called MSC
RAN	UMTS terrestrial RAN	The radio access network
MS	UE (user equipment)	
MT	ME (mobile equipment)	
SIM	USIM (UMTS subscriber identity module)	

from the TDMA/FDMA-based GSM air interface. Some parameters were chosen for backward compatibility reasons (e.g., clock rates that were multiples of GSM clock rates to facilitate the construction of dual-mode GSM/UMTS phones). Otherwise, it was a drastic change to a system that supported higher data rates and a greater range of variable data rates.

A number of new network elements were introduced with WCDMA, and these can be mapped to corresponding network elements in GSM, as shown in Table 12.1. The GSM network elements found in this table were introduced in Section 11.1.2. The revised architecture is shown in Figure 12.12. Notice the new interface between RNCs that was not present between BSCs. This is to support soft handoffs, since the UE may be communicating through multiple node B's under different RNCs at the same time.

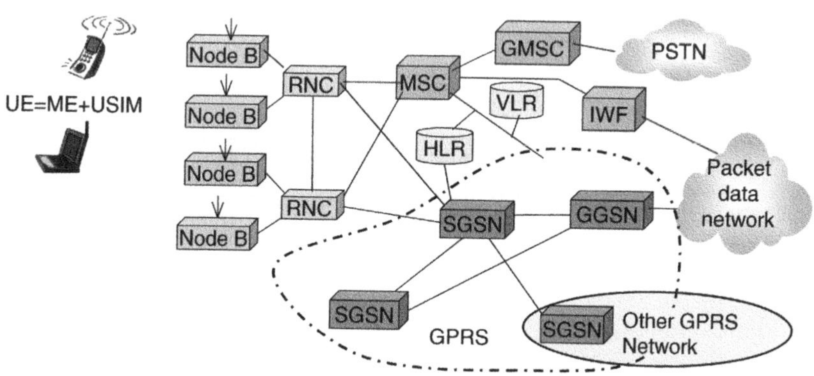

FIGURE 12.12 Network architecture for UMTS (Release '99).

Other features introduced with Release '99 included:

- Terminal-related enhancements
  - *Enhanced messaging service* (EMS), *multimedia messaging service* (MMS), *mobile execution environment* (MExE 1999)
  - AT command enhancements
- Service
  - Multimedia messages
- Core network
  - CAMEL phases 2 and 3
  - GPRS enhancements

Customized applications for mobile enhanced logic (CAMEL) is GSM's version of *intelligent network* (IN) concepts (discussed further in Section 13.2.5).

12.3.2 From Release '99 to Release 4

Even though GPRS had been introduced some years back, until Release 4 came out, voice and data traffic were handled in separate parts of the cellular network infrastructure. Specifically, from the BSC, a circuit-switched voice would go to and from the MSC, whereas data traffic would go to and from the SGSN of GPRS. Having two parallel networks like that (a circuit-switched one for voice and a packet-switched GPRS network for data) is not ideal. So, an important step toward the "all-IP" network was taken in Release 4. From Release 4 onward, the circuit-switched and packet-switched traffic share a common transport network: namely, an internal IP-based network. This change is illustrated in Figure 12.13, which shows a simplified picture

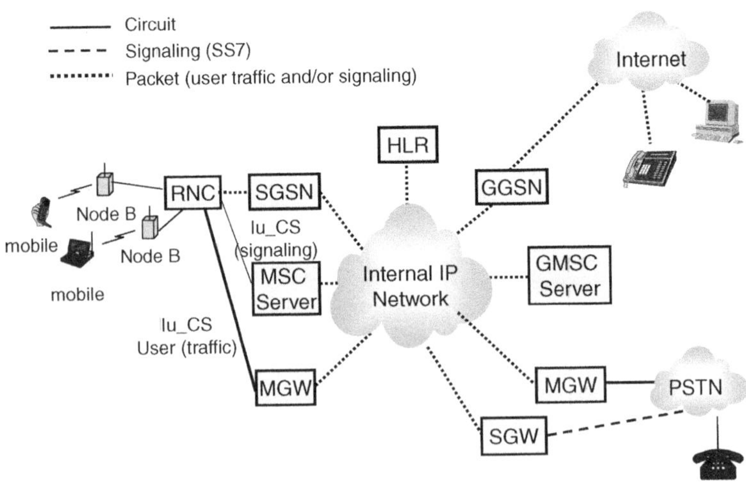

FIGURE 12.13 3GPP Release 4.

of what this change entails to the network architecture. We notice from the figure that circuit-switched elements such as the MSC have been split into the MSC server and *media gateway* (MGW). The interface between RNC and MSC, Iu_CS, is now split into two parts, one going to the MSC server and one to the MGW. The MSC server handles the control signaling aspects, and the MGW handles the format translations of the media from CS to PS, and vice versa. On the other side, interfacing with the PSTN, we similarly have a separation of GMSC server and MGW in place of the old gateway MSC. However, the GMSC server does not interface directly with the PSTN, even though it is sending and receiving SS7 ISUP signaling. This is because the GMSC server is sending and receiving the signals over SCTP (Section 10.3.2.2) over IP. Therefore, signaling between the GMSC server and the PSTN must go through a *signaling gateway* (SGW) to translate between SCTP/IP transport on the IP side and MTP3/MTP2 (SS7) on the PSTN side.

Here is a summary of the changes from Release '99 to Release 4:

- Radio access
  - Changes to UTRAN transport (for IP transport)
  - Other RAN improvements (e.g., robust header compression) (ROHC, RFC 3095, Section 12.1.2.3)
- Terminals
  - Improvements in SMS/EMS/MMS, more AT commands
  - MExE Release 4
- Network and services
  - *Transcoder-free operation* (TFO) introduced
  - *Virtual home environment* (VHE) introduced
  - *Open service access* (OSA) introduced (see Section 13.2.6)

12.3.3 From Release 4 to Release 5

A view of the network architecture of Release 5 (including IMS) is shown in Figure 12.14. Portions related to IMS are discussed in Section 12.4, whereas portions related to the three types of application servers that can be used with IMS (SIP-AS, IM SSF AS, and OSA SCS) are discussed in Section 13.2.8. OSA and CAMEL are also discussed in Chapter 13. Here is a summary of the changes from Release 4 to Release 5:

- Radio access
  - Intradomain connection of RAN nodes to multiple-core network nodes
  - HSDPA (Section 9.3) introduced for more optimized data communications
- Terminals
  - Improvements in MMS
  - MExE Release 5

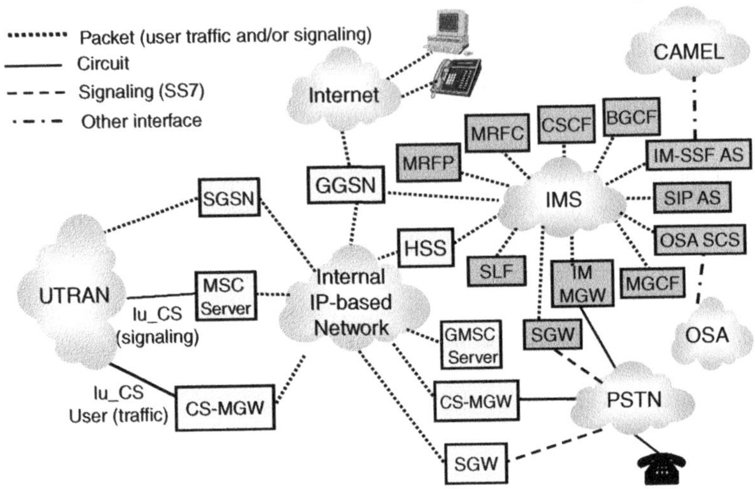

FIGURE 12.14 3GPP Release 5.

- Network and services
  - IP-based multimedia services and the *IP Multimedia Subsystem* (IMS) introduced
  - Enhancements in security, VHE, OSA, and LCS

Why was IMS introduced in Release 5? As explained in Section 12.3.2, a major step toward all-IP wireless networks was the merging of the separate internal transport networks for voice and data into a common packet-switched internal network to transport both voice and data. That step happened in Release 4. However, this was only about transport, not signaling. Release 4 still does not provide a higher-layer framework for supporting services such as voice over the data network. It is focused on the transformation at the network layer. Of course, various third-party application developers could try to fill the gap with their solutions for voice and other services, where they only make use of the network layer services of the wireless network. However, such solutions might vary in quality, especially in terms of QoS. It may be difficult for network operators to bill differently for different types of traffic if the network operator is providing only a transport service and is unaware of what types of packets are being transported. Also, if different applications are created independently by different application developers, it might be difficult to integrate them.

IMS was therefore introduced as a network subsystem that would assist in the creation of such services with the following features:

- It would help to provide a common QoS framework so QoS could be more consistent, expected, and fair in meeting different application requirements.
- It provides a framework that allows network operators more power and flexibility in charging for various services or uses of the network.

- It helps with integration of different services from different sources.

We discuss IMS further in Section 12.4.

12.3.4 From Release 5 to Release 6

Here is a summary of the changes from Release 5 to Release 6:

- Network and services
  - Multimedia broadcast multicast services (MBMS, see Section 13.2.4)
  - UMTS/WLAN interworking
- Radio access
  - HSUPA for enhanced uplink for data traffic
- IMS "phase 2"
  - IMS messaging
  - Core network to circuit-switched interworking in IMS
  - IMS charging

12.3.5 From Release 6 to Release 7

Here is a summary of the changes from Release 6 to Release 7:

- Network and services
  - MBMS enhancements
  - UMTS/WLAN interworking enhancements
- Radio access
  - MIMO
  - 64QAM for HSDPA, 16QAM for HSUPA
- IMS
  - Multimedia conferencing

12.3.6 From Release 7 to Release 8: LTE

Just as the 2G system GSM evolved to the 3G system UMTS as technology progressed and requirements changed, the 3G system UMTS needs to evolve to a 4G system as technology has progressed and requirements keep changing. For the big-picture evolution of UMTS, 3GPP worked on two parallel tracks. One is the evolution of HSPA. Features such as MIMO, dual-cell HSPA, and HSPA+ have continued to enhance HSPA, while maintaining backward compatibility with older devices. The second track is the *long-term evolution* (LTE) track. Unlike HSPA evolution, LTE was given more freedom not to have to be backwardly compatible, but to choose technologies and parameters that are optimized for the latest requirements. For example, unlike

UMTS, LTE does not have to support circuit-switched traffic, so it can be optimized for data traffic.

With regard to LTE, a few acronyms may be seen in the literature, such as *system architecture evolution* (SAE), *evolved packet core* (EPC), and *evolved packet system* (EPS). LTE these days is used to refer to the entire system, including the air interface and network. SAE is the name of the study item for the work done under the second working group of the System Architecture Technical Specification Group (TSG) (see Figure 17.2). The work on SAE was started to complement the work on the air interface of LTE. SAE produced the EPS, which consists of an evolved UTRAN (the e-UTRAN) and an evolved packet core (EPC).

For most of the early years of 3G wireless, there had been two "camps," the UMTS/WCDMA camp and the cdma2000 camp. Work had begun on the next generation of cdma2000, in the *ultra-mobile broadband* (UMB) project. However, with most operators committing to LTE, including large and important operators such as Verizon that had been in the cdma2000 camp, work on UMB was halted. Thus, the hoped for convergence of air interfaces that didn't happen in 3G might happen with LTE. However, WiMAX has meanwhile emerged as an alternative, so it remains to be seen if there will indeed be convergence to one main air interface technology.

12.3.7 Evolved Packet System of LTE

Before going into some details of the evolved packet system (EPS) of LTE, let's look at Figure 12.15. This figure shows a high-level way of looking at the broad sweep of evolution from purely circuit-switched GSM to purely packet-switched EPS. We have seen in Section 12.3.3 that the introduction of IMS did not do away with the circuit-switched part of the network (even though from Release 4 onward, that could be carried over the same internal IP network as the packet-switched traffic). Finally, when we get to EPS, the last remnant of the CS side goes away. However, UMTS and GSM networks will be around for the forseeable future, so EPS still needs to interwork with CS networks. Furthermore, there has been division in the standards

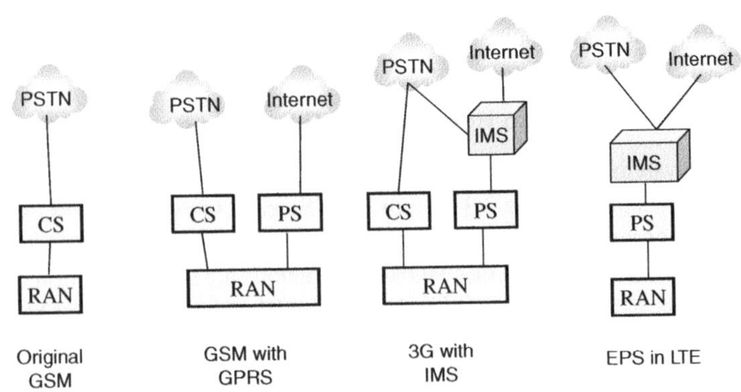

FIGURE 12.15 Evolution to the EPS.

organizations and operators and vendors regarding the right way to evolve to handle voice in the early LTE deployments, because IMS is considered not mature enough to support voice services in a full packet-switched EPS. So if IMS is not ready, what are the alternatives? The options fall mostly into two groups:

- For operators who have an existing UMTS network, voice calls can "fall back" to using the older, circuit-switched networks.
- Circuit-switched traffic over packet options are solutions to carry voice circuit-switched traffic over packets.

A prominent solution for carrying circuit-switched traffic over packets was put forth under the name of *voice over LTE via generic access* (VOLGA), but it did not have as much support as the *OneVoice* solution, which is basically to rush a subset of IMS to maturity so that voice can be handled by that subset of IMS initially, and later evolve to the full set of IMS functionality. OneVoice has been chosen by the GSM Association and 3GPP and is the current choice as of the writing of this book.

Setting aside the handling of voice now, let's examine some of the main features of the EPS. The EPS is famously known for how it "flattened" the radio access network. Unlike in UMTS where there are RNCs and node Bs, the RNCs are removed and some of their functionality goes into the *evolved Node Bs* (eNode Bs), making for a simpler, flatter picture of the evolved UTRAN (e-UTRAN). This flattening of the radio access network can be viewed as the continuing of a trend, where the base stations have been taking on more and more responsibility. In HSPA, for example, one of the changes made was to bring some functions from the RNC to Node B, allowing HARQ and fast scheduling with 2-ms intervals, with multiuser diversity, to be features of HSPA. With EPS, the process continues, and more functionality is moved to the eNode Bs, including the ARQ at the radio link control, compression and ciphering. Handoff measurement processing and decisions are also moved to the eNode Bs. Some of the other functions of RNCs go to the *mobility management entity* (MME) and serving gateway (see Figure 12.16), entities that we discuss shortly. The

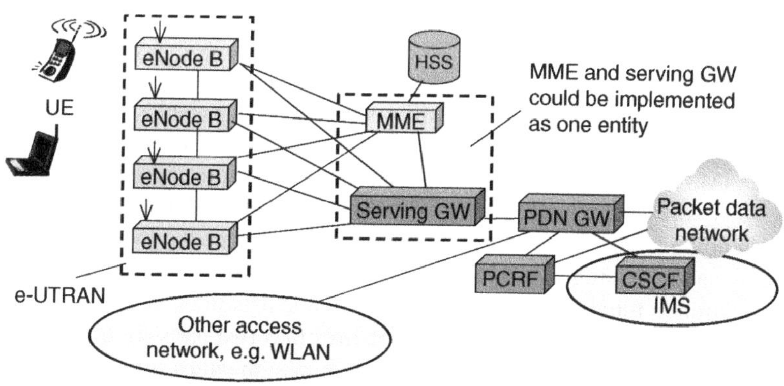

FIGURE 12.16 EPC network architecture.

eNode Bs may be connected to neighboring eNode Bs in a mesh fashion. Unlike the CDMA-based UMTS systems that preceded it, LTE supports hard rather than soft handoffs. So the links between adjacent eNode Bs is not to support soft handoff, but to facilitate smoother hard handoffs.

We move on from the e-UTRAN to the evolved packet core (EPC). The serving gateway can be thought of as an evolved SGSN (of GPRS), and correspondingly, the *PDN gateway* (packet data network gateway) can be thought of as an evolved GGSN of GPRS. The MME handles functions supporting the mobility of mobiles. These include security procedures (as we will see in Section 15.4, wireless and mobility introduce a set of security challenges over and beyond what is found in typical wired networks), and location management of idle mobiles (see Section 11.1.4). As such, the MME needs to communicate with the HSS, as shown in Figure 12.16. EPS is flexible in allowing the MME and serving GW to be implemented as one entity, as indicated in the figure. Another option is for the serving GW and PDN GW to be implemented as one entity (not indicated in the figure).

The PDN gateway is like a GGSN of GPRS, but it also supports access from non-3GPP access networks (e.g., WiFi, WiMAX, or other access networks). The point of connection of these non-3GPP access networks with the EPS would be the PDN gateway, for data transfer, and a 3GPP AAA server (see Section 15.3.2) for AAA procedures. The PDN gateway interfaces with IMS and external packet networks, on the one hand, and also with the *policy and charging rules function* (PCRF), on the other hand. The PCRF is involved in charging (for billing purposes) and QoS policies.

12.4 IP MULTIMEDIA SUBSYSTEM

We have discussed motivations for IMS in Section 12.3.2. IMS was initially introduced in Release 5 of UMTS. However, that version was incomplete, so things like billing support were added in subsequent releases. In the standards documents, the alternative term “IP multimedia core network subsystem” (IM CN SS) is sometimes seen. Although this is arguably a more descriptive name, since it indicates that IMS is a subsystem of the core network, it is not as catchy or popular as the simpler name IMS.

One can argue that IMS belongs to this chapter or the following chapter. Both sides would have valid arguments. With IMS come significant changes in the network architecture, as many new network elements, concepts, and procedures are added. However, IMS also bridges the gap between a network-centric approach toward all-IP wireless networks and a service-centric approach toward all-IP networks. It provides crucial building blocks upon which to build services. In fact, the IMS framework for service management might eventually go beyond just wireless networks and occupy a vital position in future converged wireless/wireline networks.

In this book we divide our discussion of IMS into two parts. In this chapter we discuss the core parts of IMS that allow it (in conjunction with a suitable connectivity access network such as GPRS) to provide a complete VoIP solution roughly comparable to the circuit-switched voice services traditionally provided in cellular systems, but based on IP-centric protocols such as SIP. The SIP servers, known as call state control functions (CSCFs), play a central role for this aspect of IMS. However, IMS is envisioned to be more than just an alternative to the circuit-switched voice solution, and to provide a platform for all types of services in wireless systems. Those aspects of IMS are discussed in Chapter 13, together with service architectures. They include:

- Application servers
- IMS as a service platform

IMS is designed to be used not just from UMTS mobiles, but also with devices accessing the network through other access networks, such as WiFi. The generic term for the access network in IMS is *IP connectivity access network* (CAN). For most of the discussion here, though, we focus on access to IMS through UMTS/GPRS, except where otherwise indicated. In the main subsections in this section we discuss the IMS network functions (Section 12.4.1) and IMS procedures (Section 12.4.2).

12.4.1 Network Functions

In this section we introduce various network functions in IMS. In the process we mention terms like registration a couple of times. Registration and other procedures are discussed in Section 12.4.2. The network functions are shown in Figure 12.14, and the reader may wish to refer back to the figure for reference.

IMS has two databases, the *home subscriber server* (HSS) and the *subscription locator function* (SLF). The HSS contains information about the mobile's subscription. It is analogous to the HLR in the circuit-switched core network. A network could contain multiple HSSs, for example, if there are many subcribers such that a single HSS might not be able to handle efficiently the storage of all the records and respond to queries efficiently. In the case that a network contains more than one HSS, the data for each particular subscriber would not be distributed, but stored in only one HSS. How would the network know which HSS stores a particular subscriber's data? The solution is the *subscription locator function* (SLF), which contains records pointing to the HSS that contains the subscriber data for each IMS subscriber. Hence, it may be queried for the HSS by an I-CSCF during registration and session setup (we explain this shortly). It is needed if the network contains more than one HSS. Otherwise, all the data can be found in one HSS, in which case the SLF is not needed.

The *call state control functions* (CSCFs) are at the heart of the IMS architecture. These are SIP servers that are involved in the SIP-based session control signaling. The CSCFs rely on the assistance of the HSS (and SLF in some cases) for subscription information, and so on, on subscribers. CSCFs can be further subdivided into three roles:

- *S-CSCF (serving CSCF)*. Each mobile, in originating and receiving calls/sessions through the IMS, has a "main" CSCF that serves the mobile, performing various session control functions, acting as SIP registrar for IMS registrations, interacting with the HSS to obtain and use subcriber data, and other functions.

- *P-CSCF (proxy CSCF)*. Whether an MS is at home or roaming, it needs a first contact point within the IMS, and this entry point is always a CSCF. In its role as such an entry point, it is called a P-CSCF.
- *I-CSCF (interrogating CSCF)*. This CSCF is analogous to the gateway MSC in the circuit-switched core network, in the sense that it is the first contact point within an operator's network from outside its network. As such, it plays a crucial role in hiding the internal network topology of the operator's network (since everything goes through this one contact point, routers and devices from the outside see only this one and only contact point).

The S-CSCF is analogous to the serving MSC in the circuit-switched core network, except when the mobile is roaming, in which case the serving MSC role is in a sense roughly split between the S-CSCF and the P-CSCF. The S-CSCF is always a CSCF in the home network of the subcriber, even if it is roaming in another network. As SIP registrar, the S-CSCF maintains the binding between the user's SIP address and its current location (such as an IP address). See Section 12.4.2.2 for more details. As a SIP server, the S-CSCF makes sure that all SIP signaling to/from the IMS terminal goes through it. In this position, the S-CSCF can redirect some of the SIP signaling to one or more application servers (ASs) along the way to the destination, providing a means for flexible service creation that we elaborate on in Section 13.2.8. The S-CSCF also enforces policy by making sure that users can only have capabilities for which they are authorized. Moreover, the S-CSCF provides translation services (e.g., between a phone number and a SIP address) as needed.

As far as SIP is concerned, the P-CSCF acts as an outbound/inbound SIP proxy server. All requests to and from the MS go through the P-CSCF. The P-CSCF is assigned to the MS when IMS registration is performed and remains unchanged during the time the MS remains registered. Between the P-CSCF and the MS is an air interface that could be slow to transmit large SIP messages, so between the MS and P-CSCF (and not elsewhere), SIP messages are compressed. For security purposes, the P-CSCF is the entity that authenticates the user. When the MS is at home, the P-CSCF is also in the home network. When the MS is roaming, the P-CSCF may or may not be in the home network. In the case of IMS access through GPRS, the P-CSCF is located in the same network (home or visited) as the GGSN. Thus, if the MS is roaming and the P-CSCF is in the visited network, P-CSCF, SGSN, and GGSN are all in the visited network. However, if the MS is roaming and the P-CSCF is in the home network, P-CSCF and GGSN are in the home network, but SGSN is still in the network visited. This arrangement has the disadvantage that all media would be routed back to through the GGSN in the home network (which may be on the other side of the world), even if both the MS and the other party are near each other; so there might be unnecessary latency. In the early deployments, the GGSN is usually in the home network, so the P-CSCF is also in the home network. However, with more recent and future deployments, there may be more cases of the GGSN being in the visited network, in which case the P-CSCF would also be found there. The P-CSCF can also perform number analysis and potentially modify a dialed number.

The I-CSCF, as the first contact point of an operator's IMS network, is listed in the DNS records for the operator's network domain. Thus, by normal SIP procedures, the address of the I-CSCF would be obtained by the SIP server in the previous SIP hop. It is responsible for assigning an S-CSCF to a user during registration. It queries the HSS for the S-CSCF address. In the case of multiple HSSs in a single operator network, the I-CSCF needs to query the subscriber locator function (SLF; see Section 12.4.1.2).

12.4.1.1 Interworking with the PSTN For calls to and from the PSTN, there is a *media gateway control function* (MGCF) which controls a *media gateway* (MGW). Besides controlling the MGW, two other main functions in the MGCF are (1) conversion between call control protocols on the packet-switched IMS side (SIP) and the PSTN (SS7 ISUP); and (2) I-CSCF identification, for incoming calls from the PSTN. I-CSCF identification means that when there is an incoming call from the PSTN; it first hits the IMS network at the MGCF, but it is not the MGCF's role to identify the appropriate S-CSCF for the called party. Instead, it will identify the I-CSCF and send it a suitable INVITE message. The I-CSCF then takes over to find the right S-CSCF. The signaling protocol between the MGCF and MGW for media gateway control is H.248 (an ITU standard).

VoIP packets are converted to/from the PSTN format in the MGW. The MGW also performs bearer control on the PSTN side, and additional functions such as echo cancellation and conference bridging. The main difference between the MGCF and the MGW is that the MGCF deals with control signaling, whereas the MGW deals with the user VoIP/data packets.

As for control signaling, there are actually two levels of conversion that need to occur: (1) the application layer and (2) the transport layer. The MGCF handles the application layer conversion [e.g., between SIP and SS7 ISUP (or BICC)]. However, it doesn't handle the transport layer conversion, so the SS7 ISUP messages entering and leaving the MGCF are still carried over IP (with SCTP as the transport protocol). Signaling conversion at the transport layer and below occurs in another gateway, the *signaling gateway* (SGW). The SGW is then the boundary between the IP transport for PSTN signaling (SCTP/IP) and the PSTN network transport layer for signaling (MTP). However, the messages are still ISUP or BICC, and the boundary with SIP is at the MGCF.

The *breakout gateway control function* (BGCF) controls where the "breakout" to the PSTN occurs, in case of calls between an IMS user and a PSTN phone. This is needed because there may be multiple possibilities for where the breakout from IP network to the PSTN occurs. In general, it would be desirable to locate the breakout as close as possible to the location of the phone, so that hopefully, only a local call need be made in the PSTN segment of the call. Just as operators today have roaming agreements with one another to support roaming of their customers between networks, operators with IMS may also make agreements with one another to support breakouts closer to the PSTN phone (even if their own network might be limited in geographical scope). For example, if the destination is a phone in London, and the calling party is an IMS user subscribed to Singtel's service in Singapore, even though Singtel's network may not itself reach London, it may have a roaming agreement with, say, Vodafone, so the breakout occurs from Vodafone's IMS network in London.

Thus, when a call from IMS is to break out to the PSTN, the SIP INVITE message gets passed to the BGCF (in the home IMS network of the calling party), and the BGCF can forward the SIP INVITE message to one of two destinations:

- The BGCF forwards the INVITE message to the MGCF in the same network, and the breakout happens in the same network.
- The BGCF forwards the INVITE message to another BGCF in another IMS network. This destination BGCF then goes through a similar decision-making process (same two options) and either forwards the INVITE message to an MGCF in that same network, or to another BGCF in yet another IMS network.

This goes on until the INVITE gets to an MGCF in one of the IMS networks.

12.4.1.2 Other Functions The application servers (ASs) are used to provide various IM applications and services. We discuss them further in Section 13.2.8. The *media resource function* (MRF) acts as a resource for various multimedia-related functions. These include the following:

- It mixes incoming media streams. This may be necessary in a conference call with multiple parties.
- It acts as a source of some media streams (e.g., certain multimedia announcements).
- It processes media streams using various algorithms to support a variety of applications. For example, it can apply algorithms for voice recognition.

The MRF is divided into a *media resource function controller* (MRFC) and a *media resource function processor* (MRFP). The MRFC is a SIP UA, so an S-CSCF might contact the MRFC using SIP, and the MRFC then controls the resources in the MRFP using H.248 signaling. The MRF is always located in the home network.

Originally, IMS was supposed to support only IPv6, as it was thought that by the time IMS was deployed, IPv6 would be mature enough, and deployed widely enough, that IMS need not support the legacy protocol IPv4 but go straightaway to IPv6. Unfortunately, IPv6 deployment has continued to be slow, so IMS supports both IPv4 and IPv6. Thus, for interworking purposes, there are two network elements, the *IMS application layer gateway* (IMS-ALG) and the *transition gateway* (TrGW). The IMS-ALG translates SIP and SDP messages between IPv4 and IPv6 networks (for those who are very familiar with SIP, the IMS-ALG acts as a B2BUA). For example, it needs to rewrite the SDP portion of SIP messages, changing IP addresses from IPv4 to IPv6, or vice versa. The IMS-ALG interfaces with the S-CSCF for outgoing traffic and I-CSCF for incoming traffic, from the other network (i.e., if the IMS network is

IPv4, the other network is IPv6). The TrGW does the translation between IPv4 and IPv6 for the RTP traffic.

12.4.2 Procedures

Just as there are certain procedures in the circuit-switched domain to enable various network functionality, we also have such procedures in IMS (i.e., procedures such as registration, call delivery, call initiation, etc). We illustrate a few important flows here. Most, but not all, of the messages in the flows are SIP messages. The details of the messages, and how the messages are processed in each node, plus other details, are outside the scope of this book, but can be found in books on IMS [2,5].

12.4.2.1 Getting Connected There are multiple stages of getting connected to IMS (Figure 12.17). When connecting to IMS from UMTS, GPRS is used for the transport connection to IMS. Thus, the first two steps needed are GPRS attach and PDP context activation, as usual for for GPRS. The third step is CSCF discovery, and the fourth is SIP registration (detailed in Section 12.4.2.2).

12.4.2.2 Registration Before registering with IMS, the mobile first needs to discover the IP address of a suitable P-CSCF. There are a number of ways that it may do so. For example, if it connects through GPRS, then when it does the activate PDP context procedure, it not only obtains an IP address that it can use, but it also obtains a P-CSCF IP address. Another way is to obtain the P-CSCF IP address through DHCP.

So the mobile first connects to GPRS or through some other IP-CAN, then locates a P-CSCF and sends a SIP REGISTER message to it. The P-CSCF forwards it to the I-CSCF in the mobile's home network. The I-CSCF locates the appropriate S-CSCF and forwards the REGISTER message to it. To perform appropriate security

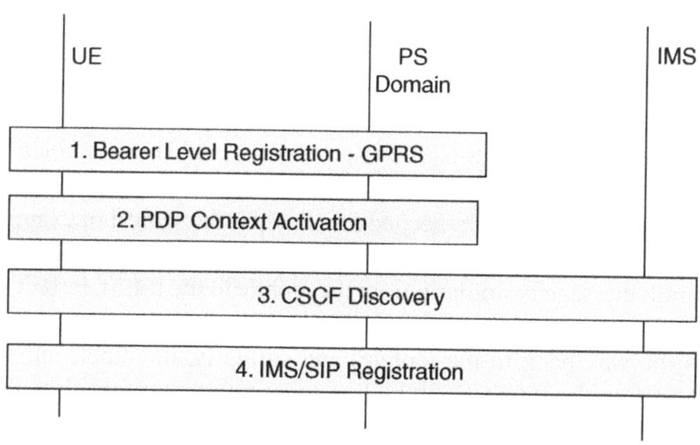

FIGURE 12.17 Multiple stages to get connected to IMS.

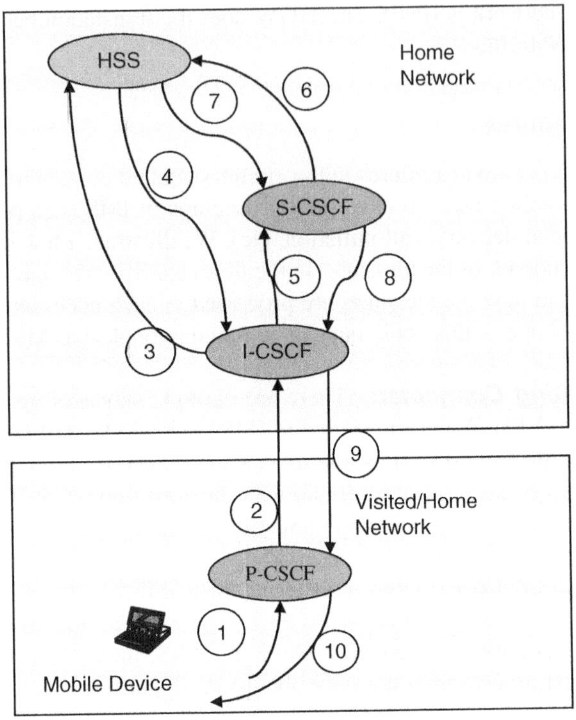

FIGURE 12.18 IMS registration flow.

procedures, the S-CSCF will send a SIP 401 unauthorized response containing a challenge, and the mobile needs to send a second REGISTER message, containing the correct response to the challenge, in order for registration to be successful. A SIP 200 OK message is eventually sent by the S-CSCF to the mobile to indicate successful registration.

The IMS registration signaling is shown in Figure 12.18. The arrows indicate the direction of messages between different elements, and the numbers indicate the sequence of messages. Since there are two round-trips (the SIP REGISTER message is sent twice, and each time the S-CSCF responds), we could have doubled the number of arrows, with a separate number for each arrow. Instead, to reduce the clutter, we just showed the first round-trip, but the second round-trip is the same. For example, the first message (labeled "1") is from the mobile to the P-CSCF (REGISTER the first time), and the eleventh message is similarly from the mobile to the P-CSCF (REGISTER the second time). The ninth message is from the I-CSCF to the P-CSCF (401 unauthorized response on the way back to the mobile) and similarly, the nineteenth message is from the I-CSCF to the P-CSCF (200 OK response on the way back to the mobile). Notice that the HSS is contacted by both the I-CSCF and S-CSCF, because it stores data that they need, and in order to perform security functions. The diagram covers both when the mobile is in its home network and when it is in a visited network.

12.4.2.3 *Call to IMS Device* We consider the case of a call from an IMS phone to another IMS phone. On the calling party side, there is a P-CSCF, followed by the S-CSCF in the home network. When the IMS phone sends the SIP INVITE message, it will always go through these CSCFs. Then the S-CSCF of the calling party will find the I-CSCF of the home network of the called party (the S-CSCF of the calling party has no idea whether the called party is roaming or at home). The I-CSCF will query the HSS to find the correct S-CSCF for the called party. The S-CSCF will know the right P-CSCF to which to forward the INVITE (or subsequent SIP messages). The signaling goes end to end from calling party to called party in several round-trips, as indicated below.

- SIP INVITE goes through all the CSCFs shown in Figure 12.19. The I-CSCF in the home network of the called party queries its HSS using DIAMETER. Each CSCF along the way will send a provisional SIP 100 trying message back to the previous CSCF.
- The called party will return a SIP 183 session progress message. All the CSCFs are involved, but the HSS no longer need be queried.
- The resource reservation phase begins with a provisional acknowledgment (PRACK) send from calling party to called party. The I-CSCF in the home network of the called party can drop out at this time, but all other CSCFs remain in the signaling path, since they would have inserted a `Record-Route` header field pointing to themselves.
- The response to the PRACK is a 200 OK.
- The resource reservation phase completes with another round-trip, with a SIP UPDATE message from calling party to called party, and a 200 OK in response.

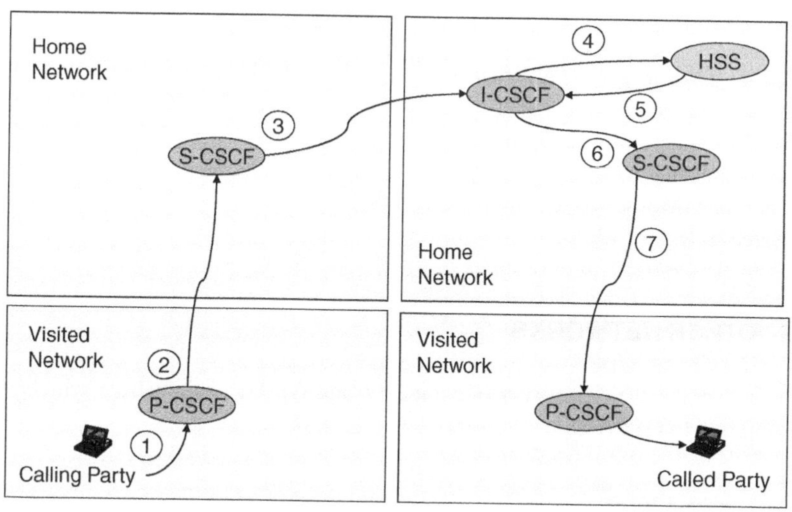

FIGURE 12.19 IMS call flow to roaming mobile.

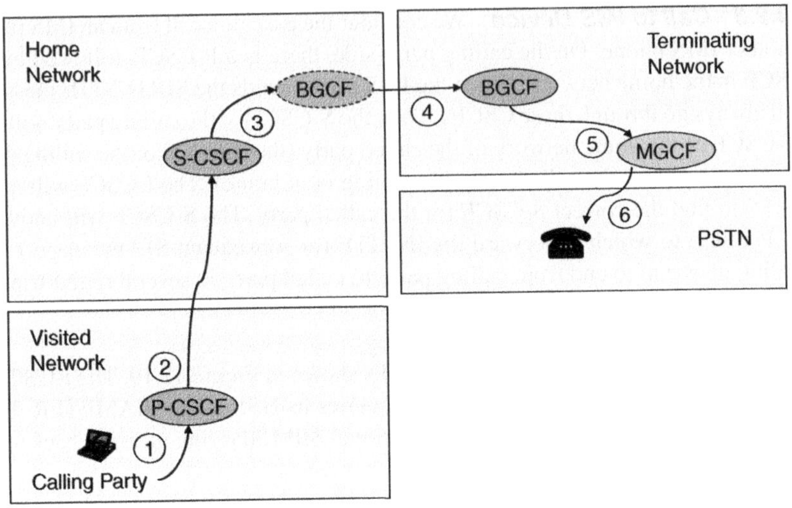

FIGURE 12.20 IMS call flow to PSTN.

Figure 12.19 shows the call flow. To avoid clutter, it just shows the forward direction (from calling party to called party) and does not show the multiple round-trips. In the diagram, the calling party "side" is on the left and the called party "side" is on the right. Both the calling and called parties are roaming, but it is straightforward to see what happens if they are at home (there will just be no visited network, but there still is a P-CSCF). The case of P-CSCF located in the visited network is shown for both sides. All the signaling is SIP signaling except for interactions with the HSS; those interactions use DIAMETER (number 4 and 5 in the sequence). In the diagram, the P-CSCF is shown in the visited network, but it could also be in the home network.

12.4.2.4 Call to a PSTN Phone The call to a PSTN phone is similar to a call to an IMS phone, except that a *breakout* needs to occur, so instead of remaining within IMS from end to end, the signaling needs to break out of IMS to the PSTN. The breakout happens at a local MGCF in the home network of the calling party, or in another IMS network, which we call the *terminating network*. The call flow is illustrated in Figure 12.20.

12.5 OTHER NETWORKS

We introduce the packet network of cdma2000 and WiMAX briefly here, which could be compared to GPRS.

12.5.1 cdma2000

The cdma2000 packet architecture is simpler than the UMTS network architecture, largely because there is no GPRS in it. A simplified picture of the architecture is

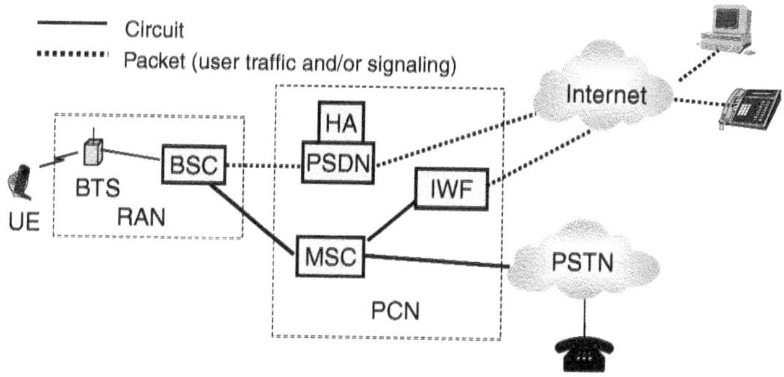

FIGURE 12.21 cdma2000 network.

shown in Figure 12.21. Because it doesn't have a GPRS or GPRS-like portion of the network, it does not have the same capability as UMTS to support general packet radio services. However, this is not a serious impediment in practice, as the main demand is for IP networking support anyway. A few conclusions follow.

- cdma2000's packet architecture does not need to have different types of GSNs (SGSNs and GGSNs); instead, it has a single *packet data service node* (PDSN), which is roughly a combined SGSN and GGSN.
- Without a GPRS network infrastructure, cdma2000 just has the *packet core network* (PCN). The PCN just needs basic functionality such as authentication servers, the PDSN, and mobility management, just to provide access to the Internet. For mobility management, instead of having its own mobility protocols and tunneling protocols as GPRS does, cdma2000 takes advantage of mobile IP, since it is after all an IETF-developed protocol, and it does the job.
- Without a GPRS-based air interface, cdma2000 needs some additional functions to control packet traffic over the air interface, and these functions are located in the *packet control function* (PCF). The PCF can be implemented internal to a BSC, or can be serving multiple BSCs. It maintains a radio resource state associated with the packet data sessions. It also buffers packets for MSs as needed (e.g., if an MS is "dormant," it buffers packets for it until the MS is "active" again).

The functions of the PDSN include:

- It controls a PPP connection between the MS and itself (again, reusing an existing protocol instead of inventing a new one such as GTP in GPRS).
- It acts as a router.
- It acts as a mobile IP FA.

The cdma2000 packet architecture provides an "always-on" connection option. This allows an IP address to be retained by an MS even while it is not actively sending or receiving data. An advantage of this option is that the next time the MS wants to send or receive data, it doesn't have to get a new IP address. A disadvantage of this option is that the IP address retained by the MS cannot be reassigned as long as that MS is "always on."

12.5.2 WiMAX

The IEEE 802.16 standards specify only the physical layer and the MAC layer of WiMAX systems. Aspects of the network layer and above are specified by the WiMAX Forum's *network working group* (NWG). The architecture is shown in Figure 12.22. It reuses IETF protocols extensively and allows for a lot of flexibility in realization of the network. The network is divided into:

- An *access service network* (ASN)
- A *connectivity service network* (CSN)

The ASN is analogous to the base subsystem (BSS) of GSM or radio network subsystem (RNS) of UMTS. The CSN is analogous to the network subsystem (NSS) of GSM. The ASN can be implemented either as an integrated network element or split into a base station (BS) and an *ASN gateway*.

We have seen (Section 9.4.3) that on the access side WiMAX systems support hard handoffs and possibly soft handoffs. For rerouting in the network to support mobility, mobile IP is used, with the home agent in the CSN and the foreign agent in the ASN (ASN gateway, if split into BS and ASN gateways). For movement between base stations under the same ASN gateway, it is considered intra-ASN mobility (or micromobility). Thus, as far as the IP layer is concerned, such mobility is layer 2 mobility, so mobile IP is not involved. For movement between ASNs, however, mobile IP is used with forwarding from the HA to the appropriate FA.

As with traditional cellular systems, idle mode and paging are included in WiMAX to optimize power consumption in mobiles while balancing that with the amount of paging that the network needs to do when it needs to alert the mobile. When the mobile is idle, it performs location updates when crossing between location areas that may

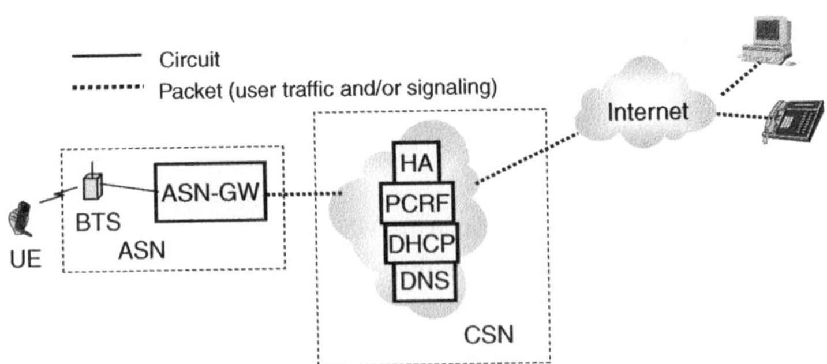

FIGURE 12.22 WiMAX network.

be the size of several base station coverage areas. In the network, a *paging controller* handles paging control, and it may be located in a BS or ASN gateway. Information about idle mobiles is stored in a *location register*.

EXERCISES

12.1 Suppose that mobile B has home address 186.15.25.31, home network prefix 186.15.25.0, and network mask 255.255.255.0. It moves to a foreign network with network prefix 27.0.0.0 and network mask 255.0.0.0. B hears a foreign agent advertisement, through which B obtains the care-of address 27.242.2.9, which is routable to the foreign agent. Fill in the blanks with the correct answers: B needs a home agent in its home network. It has been assigned a home agent, D, with IP address 186.15.25.45. It sends a mobile IP _____ message to its home agent. For security purposes, the mobile-home _____ extension is included in the message. After this, a correspondent node, C (address 179.23.21.11), tries to send packets to the MH. It will send the packets to _____ (enter an IP address). D will intercept the packets. It will add a new IP header to the packets, with source address _____ (enter an IP address) and destination address _____ (enter an IP address). Upon arrival at the foreign agent, the packet will be unencapsulated (meaning, what will happen to the header? _____) and delivered to B.

12.2 What are the three mobility management states of GPRS?

12.3 Which releases of UMTS saw the introduction of (a) IMS; (b) HSDPA; (c) HSUPA; and (d) LTE?

12.4 In what way is the LTE access network (e-UTRAN) said to be flattened as compared to previous access networks such as the RAN of GSM and the UTRAN of UMTS?

12.5 Suppose that a VoIP system uses G.711. G.711 is not as efficient an encoder as G.729 that we saw earlier, so it needs 64 kbps to encode voice. Suppose that the voice is segmented into chunks of 20 ms for transmission. Find the number of bytes of G.711 encoded voice over a 20-ms period. Now add UDP, RTP, and IP header overhead. What percentage of the resulting packet is occupied by the header? How about if the segments are 10 ms each?

REFERENCES

1. C. Bormann, C. Burmeister, M. Degermark, H. Fukushima, H. Hannu, L.-E. Jonsson, R. Hakenberg, T. Koren, K. Le, Z. Liu, A. Martensson, A. Miyazaki, K. Svanbro, T. Wiebke, T. Yoshimura, and H. Zheng. Robust Header Compression (ROHC): framework and four profiles: RTP, UDP, ESP, and uncompressed. RFC 3095, July 2001.
2. G. Camarillo and M.-A. García-Martín. *The 3G IP Multimedia Subsystem (IMS): Merging the Internet and the Cellular Worlds*, 3rd ed. Wiley, Hoboken, NJ, 2008.

3. C. Perkins, editor. IP mobility support for IPv4. RFC 3344, Aug. 2002.
4. D. Johnson, C. Perkins, and J. Arkko. Mobility support in IPv6. RFC 3775, June 2004.
5. M. Poikselkä and G. Mayer. *The IMS: IP Multimedia Concepts and Services*. 3rd ed. Wiley, Hoboken, NJ, 2009.
6. K. Sandlund, G. Pelletier, and L.-E. Jonsson. The Robust header compression (ROHC) framework. RFC 5795, Mar. 2010.
7. K. D. Wong. *Wireless Internet Telecommunications*. Artech House, Norwood, MA, 2005.

13

SERVICE ARCHITECTURES, ALTERNATIVE ARCHITECTURES, AND LOOKING AHEAD

In this chapter we first examine services (Section 13.1) and service architectures (Section 13.2) in mobile wireless networks. Then in Sections 13.3 and 13.4 we consider a number of alternative network architectures, such as mobile ad hoc networks, mesh networks, sensor networks, and vehicular networks. In many application scenarios, these might be connected to, or eventually connected to, a wired network infrastructure; however, communications between nodes in these networks may traverse multiple wireless links in getting from source to destination. Moreover, significant communications between a pair of nodes in such networks may occur without any wired links in the path between the two nodes. In a traditional wireless system, on the other hand, typically only the last hop or last link is wireless, and the wired infrastructure is an essential part of the network architecture that is usually in the path of communications between any two nodes.

13.1 SERVICES

In the early days of second-generation cellular systems, services and service architecture were relatively simple. The main service was voice communications. Other services included fax and text messaging. Even then there was the concept of *bearer services* as a distinct concept from *teleservices*. For example, in GSM, a teleservice is an end-to-end service, which includes the application at the endpoints, including the format and presentation of the information to the users. A bearer service, on the other hand, would be a service internal to GSM, just to bear, or transport, data from one point to another. Here we have used the words *information* and *data* in

Fundamentals of Wireless Communication Engineering Technologies, First Edition. K. Daniel Wong.

a specific way. Data are unformatted bits, whereas in the case of information, the constituent bits are structured, arranged in a meaningful way.

As operators have been evolving their networks to provide more and more support of data traffic (the addition of GPRS, data-optimized access technologies such as HSPA and EV-DO, the evolution toward all-IP networks, and so on, that we have seen in earlier chapters), the space has been growing for all kinds of new services and applications to take advantage of these developments. In this section we introduce various concepts related to services, and discuss selected services. A service can be discussed both from the perspective of a user of the service and from the perspective of how to build/architect the service. In this section we focus more on the former. We *will* talk a little about the architecture of specific services here, but the discussion in this section will be more specific to individual services, whereas we wait until Section 13.2 to discuss frameworks, overall architectures, and related overarching themes.

The explosion of applications (popularly called "apps") in recent years has been causing tremendous changes to the way customers use wireless networks, and it is putting pressure on service providers to upgrade their networks. One challenge is that it is not clear how they will be able to make enough money to pay for these investments. The *average revenue per unit* (ARPU), meanwhile, is going down. Even as access speeds are increasing, it is difficult to get customers to pay more for these improvements. However, services are the most customer-facing aspect of wireless systems, so it is important for customer satisfaction and retention that a wide range of services and applications be available and that they work well. A second challenge is that there has been a major shift in how people use their wireless devices. What used to be primarily mobile *phones* are now increasingly accurately being described as mobile *devices*, or smartphones, and these are being used increasingly as small portable computers. Whereas voice was the major service in the past, voice is increasingly becoming just one of many services and applications for which people use mobile devices. Of course, it also depends on the market and demographics. This transformation of the wireless industry is happening at different speeds in different countries, for example.

Services and service architectures are therefore arguably the most fluid and subject to change of all the topics covered in this book. With the increasing convergence of the wireless devices and computers, various trends in computing and networking, such as social networking, cloud computing, virtualization, and so on, may soon significantly affect applications and services in the wireless world. New developments can be difficult to predict. For example, the rapid emergence of application stores (popularly called "app stores") was not anticipated by many industry observers. Nevertheless, having said how the topic of service architectures is so fluid and rapidly changing, we now attempt to introduce the basics, and then describe the evolution of wireless service architectures in Section 13.2; we resist the temptation to speculate too much on future directions, as it is outside the scope of the book anyway.

Some observations regarding services:

- It is notoriously difficult to predict what services will be successful in the market.

- Services do not have to be flashy to be popular, successful, and/or good revenue generators. Many of the most popular and successful services/applications in the world today are SMS-based services/applications.

What is the difference between a service and an application? In common usage, the terms *service* and *application* are sometimes interchanged. However, some distinctions can be made:

- The concepts denoted by the word *service* are wider and more all-encompassing than the concepts denoted by the word *application*. For example, in the standard view of network layers, each layer provides a *service* to the next-higher layer, using the services of lower layers, in particular as presented by the next-lower layer (see Section 10.1.1).
- One can view an application as a higher-level entity that encompasses one or more underlying services. Thus, an application might make use of a presence service (Section 13.1.1.2) to figure out the reachability of a user's friends through a variety of means, such as push-to-talk over cellular (Section 13.1.1.4), voice call, video call, and voice mail; it may then present the information to the user in a format of its choosing. The user might also be given the option of filtering the results by the location of the friends, where the application makes use of a suitable location-based service (13.1.1.5) to obtain the necessary information.
- The application layer sits on the top of most communications protocol stacks. A service might be a bearer service that is concerned with only the transport layer and below, or a teleservice that is concerned with multiple layers, including the presentation and application layers as well. Applications, in contrast, are more focused on the application layer.
- Usually, a teleservice is communications centric (e.g., voice communications services) whereas we are seeing more and more applications for mobile phones that have little to do with communications, if at all.

13.1.1 Examples of Services

13.1.1.1 Voice Voice services have been around since the first generation of cellular systems. In thinking of services in wireless systems, we may construct an analogy of stores in a shopping mall. The availability and popularity of services are reflected in the size and popularity of the various stores. The first generation of cellular systems is like a shopping mall with one big store, representing voice. By the second generation of cellular systems, voice has become the "anchor tenant" of the shopping mall, but other smaller shops have emerged and some, like text messaging, have become wildly successful. By the third generation and beyond, voice occupies a respected position in the mall, perhaps in one corner of the mall, but there are many other stores that compete for the attention of shoppers.

Traditionally, voice has been handled by circuit-switched networks, but we have seen in the last couple of chapters how the move toward the "all-IP" network concept has led to everything, including voice, moving toward handling by converged packet-switched networks.

13.1.1.2 Presence At a basic level, presence is information about a user's online status (including related statuses such as being busy) and communications capabilities (just voice only, voice and video, etc.). Presence is easily extended to include other status-related items such as location and mood. A *watcher* is another user who is informed about a particular user's presence information. The presence service can be used as a service enabler by other services, such as messaging. In Section 13.2.1 we introduce how presence is implemented in SIP.

13.1.1.3 Messaging By *messaging* we mean not just text messaging but also multimedia messaging. Text messaging is also known as *short message service* (SMS), and it has been a wildly successful service, especially in Asia. *multimedia messaging*, also known as *multimedia message service* (MMS), generalizes the concept of SMS and allows not just text but other forms of media to be sent in the messages.

People have a different expectation of messaging than email, so the messages are usually not stored in servers, as is email. It also has more of a "real-time" expectation, so it usually can be handled as a "session" (like voice), and users will be annoyed if latencies are too high. That is why it is also called *instant messaging*. Messaging goes well with the presence service.

13.1.1.4 Push-to-Talk over Cellular Walkie-talkies have been around since the 1940s. Push-to-talk over cellular (PoC) is a service that mimics the look and feel of the walkie-talkie communications service. In particular:

- A user would push and hold a button to be able to begin talking with one or more parties, unlike in traditional cellular, where a phone number would need to be dialed. Shortly after the user pushes the button, the device indicates (e.g., through beeping) that it is ready, and the user can talk until he or she releases the button.
- It is a half-duplex service. Only one user can talk at a time.

A significant difference between PoC and walkie-talkies is that the walkie-talkies are constrained to be within radio range of each other, whereas PoC is not similarly constrained. PoC is typically enabled through the use of a server (e.g., in the operator's network), and thus the parties in the communications can be in different cities. A second difference between PoC and walkie-talkies is that the group of listeners is "naturally" defined by radio range (those within range of the currently transmitting walkie-talkie), and perhaps the channel (frequency, code, etc.). For PoC, there are various options, such as:

- *One-to-one*: between two users.
- *Ad hoc or prearranged group*: a group of users that is either a transient grouping (that may be selected by a user from a contact list just before the PoC session)

or a more permanent grouping (such as hiking buddies, a list of whom may be stored in their mobile device). Invitations are sent to the group members when the PoC session is initiated.

- *Chat group*: a group that users can join as desired, without the need for an invitation.

13.1.1.5 Location-Based Services Unlike the other services discussed earlier, *location-based services* refers to a category of services that make use of user location information and add value to it. The user location information could be coarse, such as that a user is in a particular location area (which is all the network knows about user location from normal signaling if the mobile device is idle and moving around performing location updates only when it crosses location area boundaries). User location could also be more fine-grained: for example, to within 300 m of a mobile device to meet the FCC's E-911 requirements (the idea being that if a mobile phone user dials the emergency number, 911, the mobile operator's network should be capable of providing such fine-grained location information to the necessary public safety contacts).

An example of a location-based service is a "friend-finder" application/service, where a user can be informed of the location of all of his or her friends. Since privacy issues may arise (many people might not want their location to be available to just anybody, or even to all their friends), such a service is often implemented with an "opt-in" feature, so person A has to allow person B to be informed of person A's location before the system will provide such information to person B. It can be useful for parents who wish to track the location of their children, for example.

Another example of a location-based service is a "personal assistant" application/service which can find the nearest pizza store, the nearest public library, the nearest gas station, and so on, and it can be integrated with a map service or navigation service (which helps the user go to the desired location) or calling service (e.g., it may dial the nearest pizza store for the user).

13.1.1.6 Broadcasting or Multicasting of Multimedia The traditional communication service paradigm in cellular systems is unicast, point-to-point (e.g., as in the classic voice service between two people). However, services such as mobile TV use a different paradigm: broadcast, or at least multicast, of multimedia. As we saw in Section 10.2.4, one of the best ways to use multicasting is if there are multiple recipients for some multimedia data where there is at least partial overlapping of the paths from destination to source.

In Section 13.2.4 we describe briefly how the MBMS architecture in UMTS provides broadcasting or multicasting services for multimedia and that will serve as an example of how such services could be provided by a wireless network.

13.2 SERVICE ARCHITECTURES

What is a service architecture? How are services implemented and delivered to customers? In the past, there were many systems, each with their own network architectures, network protocol stacks, and unique way of providing services. Such an approach resulted in different vertical "silos," which is inefficient in many ways (cost, flexibility, innovation, etc.), so there has been a move toward convergence (Section 10.2.7).

These days, especially in the past decade, the industry thinking on service architecture has moved toward a modular, layered approach. The layering model of network theory is helpful in thinking about and analyzing services. We have already seen how bearer services are built on top of teleservices. The current thinking about services is that they should be constructed from *service enablers*. Higher-level service enablers are built using lower-level service enablers. This allows reuse of different building blocks (service enablers) without the need for a "reinvention of the wheel" whenever a new service is created.

This section is divided into two parts. First, in Sections 13.2.1 to 13.2.4 we give examples of how certain service enablers, such as MBMS, are built. Second, in Sections 13.2.5 to 13.2.8 we look at the larger picture of models of service architecture (e.g., IN, OSA), and see how the industry has advanced from the early days (when the few services that were available were implemented in hardware in switches) to more recent models. In these sections we also elaborate on the benefits of the layered approach to building services using service enablers.

13.2.1 Examples: Presence

In SIP, the model for supporting presence is that presence information is handled by a server called a *presence agent*. Various *presence user agents* (PUAs) will notify the presence agent of the user status using the SIP `PUBLISH` method. On the other side, various watchers use SIP `SUBSCRIBE` to inform the presence agent about what presence information they wish to be notified about. The presence server then uses SIP `NOTIFY` to push the presence information to these watchers.

13.2.2 Examples: Messaging

Instant messaging may be implemented with SIP in two ways. There is a SIP extension, the `MESSAGE` method, that can be used to send a message from one user to another, without starting a session. The other way is session-based (i.e., SIP `INVITE` is used to establish the session). Unlike voice, video, and other such media, instant messaging is not transported with RTP, but with *message session relay protocol* (MSRP, RFC 4975 [2]).

13.2.3 Examples: Location-Based Services

One type of location information that can be obtained about users is the cell they are in (based on latest base station ID), or at least the location area they are in (based on location updates performed when a mobile device is on but idle). To obtain more fine-grained information on user location, additional measurements must be made and additional computations performed. The basic idea is that some form of *triangulation* is done (e.g., measurements are made from three or more surrounding base stations). For example, measurements of *time difference of arrival*, or *angle of arrival*, could be performed. In "fingerprinting" techniques, patterns of measurements (e.g., of signal strength) could be observed beforehand and used to deduce that a mobile device is at or near a particular location [11]. The *global positioning system* (GPS) can also be used, if available. However, GPS cannot always be used: for example, when a device is inside a building or when something is obstructing the paths to the GPS satellites. Thus, hybrid techniques such as *assisted GPS* have been developed that use GPS but do not rely solely on it.

13.2.4 Examples: MBMS

Multimedia broadcast multicast services (MBMS) is the UMTS solution for providing multimedia broadcast and multicast services, which serves as a service enabler for higher-level services such as mobile TV. It was introduced in UMTS Release 6, and relies on a new network element, the *broadcast multicast service center* (BM-SC). The cells are divided into *MBMS service areas*, each of which typically covers multiple cells (but can be as small as one cell). The BM-SC will determine which of the MBMS service areas receive any particular broadcast or multicast multimedia based on the following criteria:

- *For broadcast.* If it is in the MBMS broadcast area, the MBMS service area will receive the broadcast and transmit it within each cell in the MBMS service area, using a point-to-multipoint radio resource.
- *For multicast.* The BM-SC tracks the members of the multicast group and thus determines if each MBMS service area will receive the multicast, if one or more members of the multicast group are in the MBMS service area. Unlike with broadcast, where every cell in that MBMS service area then uses a point-to-multipoint radio resource to transmit the broadcast, in the case of multicast, each cell may individually be using either point-to-point (if there is only one member, or just a few of that multicast group within the cell) or point-to-multipoint (if there are more members of the multicast group within the cell).

13.2.5 The Rise of the Intelligent Network

Originally, switches in the telephone network functioned purely as switches; that is, they would connect an incoming circuit to an outgoing circuit and switch incoming traffic appropriately. As people thought of new services, such as toll-free calling, the most straightforward way to implement these services was to get the switches involved. For example, in the case of toll-free calling, if a switch recognized a certain number (or perhaps just the prefix, such as 800) as a toll-free number, it would then process the call differently. Thus, the switch would need to be replaced so that the new functionality (the *service logic* for the new service) could be added. Gradually, more and more functions were added to switches (Figure 13.1). Some shortcomings of this approach were:

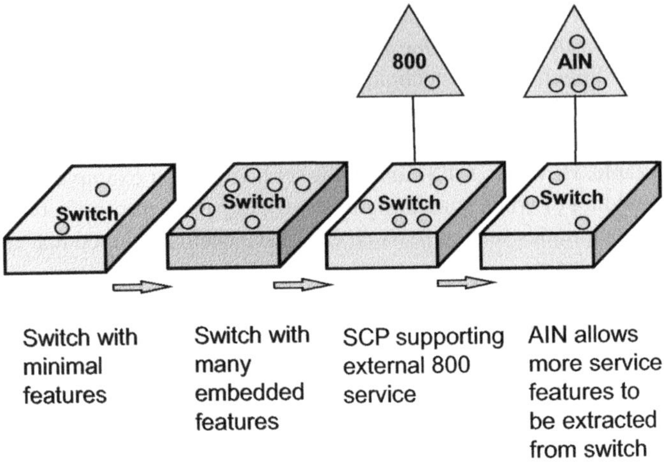

FIGURE 13.1 Service architecture evolution.

- Each new feature meant a change in the switch hardware, so it was inconvenient and impractical to add new features quickly whenever the telco wanted to do so because of long release cycles.
- Since features were implemented in switch hardware, all implementation had to be done by switch vendors.
- In addition to basic switching, the switches now implemented numerous features, with the danger of becoming cumbersome and slow.

A trend was begun to move service logic outside the switches so that the switches could concentrate on switching and on doing it efficiently, whereas the service logic would be off-loaded to *service control points* (SCPs) [or *service control functions* (SCFs)], where various services could be implemented (e.g., toll-free calling). Now, telcos could move more quickly to implement new features without the need to rely on switch vendors to do so, or to upgrade their switches. This idea of separating out the features from the switches is known as *intelligent network* (IN). A variation implemented in the United States by Bellcore is known as the *advanced intelligent network* (AIN). IN can be described as a service architecture where the *service layer* (with the "intelligence") is separated out from the *switching layer*, which remains in the switches. Certain *detection points* are defined in the call processing flow in the switches where processing can be directed to an appropriate SCP if certain conditions are met. Besides SCPs, there might also be *specialized resource functions* (SRFs) or *intelligent peripherals* that can play back a voice message, for example.

One of the early services implemented in IN was to translate a toll-free phone number into a regular phone number for completion of toll-free calls, but later, more complex services, such as the handling of prepaid calls, were also implemented using IN. GSM's implementation of IN concepts is known as *customized applications for mobile enhanced logic* (CAMEL [6]).

13.2.6 Open Service Access

In the early days of cellular systems, the cellular operators controlled the entire system, including all the services. There are a number of advantages to opening the operator's network to third parties to create and provide services.

- Complex services are almost like regular applications on desktop PCs. They are requiring increasing amounts of software expertise to write, and software companies may have more experience in this than cellular operators have.
- It is difficult for the cellular operator to predict what services will become popular with customers. Rather than try out a limited number of services and applications by themselves (a limited number because of limitations in developer resources), it makes sense to open up to third parties so that:
  - The third-party software providers can collectively try out a much larger number of different services.
  - The operator can get a percentage of the profits for whichever services become successful. They do this by providing the platform that the third parties use to create and provide services.
- When there are many third-party providers, competition and innovation are stimulated.

This model works best for the operator (and users) if a large number of third parties are willing and eager to create and provide services for the particular platform. Thus, another fundamental concept of OSA is that the network capabilities are abstracted and presented to the third-party providers through relatively simple *application programming interfaces* (APIs). Consider one of the alternatives, the intelligent network (IN) model. The IN model is tied closely with the telecommunication network, so third parties, for example, would need to get familiar with the various states of call processing, how SS7 works, and so on, to create services with IN. With OSA, on the other hand, the network capabilities are abstracted at a higher level, so many third parties without deep knowledge of telecommunications networks can develop services and applications. Thus, OSA takes advantage of abstraction. This idea is sometimes described as the concept of a *service mediation gateway*. Examples of network capabilities offered through the OSA API are:

1. Call control (including multimedia calls and conference calls)
2. User interaction
3. Mobility
4. Terminal capabilities
5. Data session control

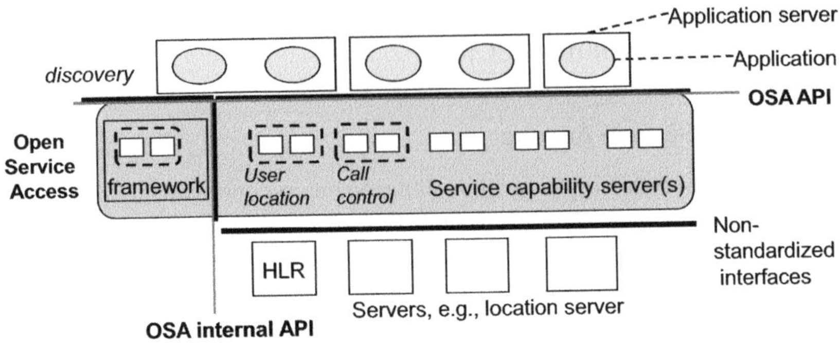

FIGURE 13.2 OSA architecture.

6. Messaging (generic and multimedia)
7. Connectivity manager
8. Account management
9. Charging
10. Presence and availability management

Figure 13.2 shows the OSA architecture. At the top of the diagram are applications residing in application servers. Since these are third-party applications outside the network, OSA needs to provide a way to authenticate these applications and authorize them for access to the network. At the same time, OSA needs to provide a way for these third-party applications to discover the service capabilities that it is making available to these applications. Such matters are handled by the *framework* part of OSA, shown on the left side of the diagram. The framework thus manages trust and security between the OSA network and the applications, and provides discovery features.

To the right of the framework in the diagram, we see a number of service capability servers. These are the service mediation gateways between the applications and the actual servers and network functions that provide the capabilities. The applications access the service capability servers through the OSA API. The interfaces between the service capability servers and the actual servers and network functions that provide the capabilities (bottom of the diagram) are not standardized, and don't have to be, since these are internal to the network and may be implemented in different ways.

13.2.7 Open Mobile Alliance

The Open Mobile Alliance (OMA [1]) is a forum of industry players interested in promoting the use of mobile data services. It facilitates mobile data service by specifying mobile service enablers. A balancing act that OMA attempts is to specify the mobile service enablers with enough detail to support interoperability whie not stifling innovation and service differentiation (i.e., it tries to give room to operators, software developers, and others, to differentiate their products from their competitors' while conforming to the specifications).

In order to be relevant, and to reduce possible resistance to its specifications, it purposefully included members from different stakeholder groups. Thus, it includes members from the following categories:

- Mobile operators
- Mobile device vendors and network equipment suppliers
- Mobile software developers and content providers
- IT companies

In comparison, it could be said that OSA is very operator-centric, resulting in resistance from other groups of stakeholders. Furthermore, differences in implementation of OSA in different operators' networks might reduce interoperability so that applications written for one operator's network might need to be rewritten for other operators' networks. OMA, on the other hand, stresses interoperability from the beginning, across devices, geographies, service providers, operators, and networks.

OMA wisely follows some guiding principles for the benefit of all of its diverse membership:

- OMA specifications are technology neutral (i.e., neither favoring nor precluding any particular device, platform, operating system, programming language, etc.).
- OMA specifications are network technology neutral (i.e., neither favoring nor precluding any network, such as GSM, CDMA, WiFi, WiMAX, etc.).
- OMA allows existing standards to be reused where these provide existing solutions to certain problems that OMA might be trying to address (rather than purposely creating a new solution just for the sake of creating a new solution)—this is, provided that the solution meets the requirements for solving the OMA problem.

Without such principles it is quite likely that certain members might be unhappy about certain decisions, and perhaps break away to form competing forums, which is not what OMA wants.

OMA specifications are compatible with IMS. Just as 3GPP and 3GPP2 refer to IETF protocols for IP-related matters, OMA specifications also refer to IETF documents. OMA comes up with requirements, and the requirements are brought to 3GPP and IETF so solutions can be developed (including modifications of existing protocols).

13.2.8 Services and IMS

We have seen the basic IMS network architecture in Section 12.4 and have seen how IMS can be used to provide basic multimedia session control (e.g., for VoIP). However, this is arguably just a small part of the reason why people have been excited about IMS and why there was considerable hype about IMS, especially when it was introduced. The main innovative aspects of IMS can be said to be in how it is designed as a platform for providing all kinds of innovative new services, not just voice. To extend the capabilities of IMS beyond what the CSCFs provide, in the basic SIP session setup, maintenance, and tear-down, we need to look at the *application servers* (ASs).

A typical IMS deployment would have multiple ASs. IMS defines interfaces between the ASs and the CSCFs: in particular, the S-CSCF and I-CSCF. SIP is used between AS and CSCF. An AS might also support other protocols, such as http, so an IMS user might be able to connect to it using a browser, and configure services easily.

There are three types of application servers in IMS:

- *SIP AS*. This is the main, "native" AS for IMS. If the service(s) supported by the SIP AS depend on information in the HSS, the SIP AS can interact with the HSS using DIAMETER. There is also an option for a SIP AS to be located in a third-party network, but in that case, it would be unable to talk to the HSS.
- *OSA SCS*. This allows IMS to be used with OSA. However, like the SIP AS, the OSA SCS communicates with the S-CSCF using SIP, so the difference between types of AS is transparent to the S-CSCF.
- *IP multimedia service switching function AS* (IM-SSF AS). This brings the range of CAMEL-based services, widely deployed in existing GSM networks, to the world of IMS. The IM-SSF AS interfaces with CAMEL through the gsmSCF, using CAMEL-related protocols that are not part of IMS.

We consider a SIP AS. How does it provide services? It can play many roles, including a user agent (so can be a calling party or called party) and various kinds of SIP server roles. As a user agent, an AS can call an IMS phone at a particular time (e.g., to provide a wake-up call service or reminder service). Other services might be provided when, under certain conditions, an S-CSCF forwards an INVITE to an AS. The AS might then act as one of the SIP servers in the path between the two endpoints, possibly inserting itself with the `Record-Route` method. Under what conditions might the S-CSCF forward messages to an AS? The way these are specified is by filters, as we discuss in Section 13.2.8.1.

13.2.8.1 Filters Various filters may be specified in the service profile of an IMS user. Each is assigned a priority, so the filters are processed in order of priority. The ways to decide whether a SIP message should be forwarded to a certain AS is through the use of *trigger points*. Each trigger point is a group of *service point triggers*. For example, if an INVITE message is being sent, and it is coming from a particular user, the trigger point might contain two service point triggers, one for INVITE and the other for the source as a particular user.

13.3 MOBILE AD HOC NETWORKS

Unlike a traditional wireless network where the wired infrastructure is an important part of the network, a *mobile ad hoc network* (MANET) comprises a group of somewhat independent nodes that communicate between themselves without the need for a wired infrastructure (Figure 13.3). The term *ad hoc* in the name emphasizes the fact that the nodes in the MANET do not have the aid of a traditional network infrastructure (typically wired and relatively fixed) to support their communications. Furthermore, the nodes themselves together form the network infrastructure, but without preplanning. Instead, they need to figure out the network topology on the fly, in an ad hoc manner. Nodes are expected to play the role of routers in addition to being hosts, meaning that they are expected to forward packets that they receive which are meant for some other node. It is therefore sometimes said that in MANET, "every node is a router," unlike in a normal IP-based network, where most of the end devices are just hosts that do not do any forwarding. The word *mobile* in the name emphasizes the fact that the nodes are allowed to move around. As a result, the topology of the MANET is expected to change quite often. Application areas for MANET are military networks, disaster recovery scenarios (where a fixed infrastructure might be damaged or unavailable), vehicular networks, sensor networks, and so on (we discuss some of these, including vehicular networks and sensor networks, in Section 13.4). Because of the ad hoc nature of the MANET network, and the changing topology, a primary area of R&D in MANET is in routing protocols. Unlike more traditional networks, in MANET just finding a path from source to destination becomes a major challenge.

Many routing protocols have been proposed for MANET, and most of them specify ways to deal with the changing topology. The challenges arise because of the ad hoc nature of MANET and because of the mobile nature, resulting in often changing topology. Furthermore, there are many constraints. For example, the routing protocols should be energy efficient, especially since the nodes would often be running on batteries. Moreover, the routing protocol overhead should not consume too much bandwidth, as the links as bandwidth-constrained wireless links. A basic

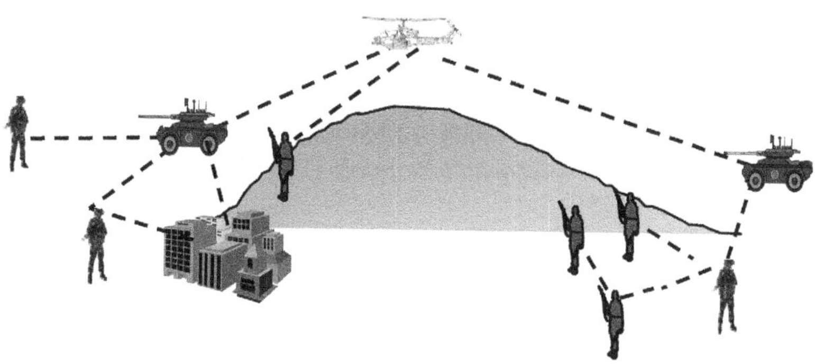

FIGURE 13.3 Mobile ad hoc networks.

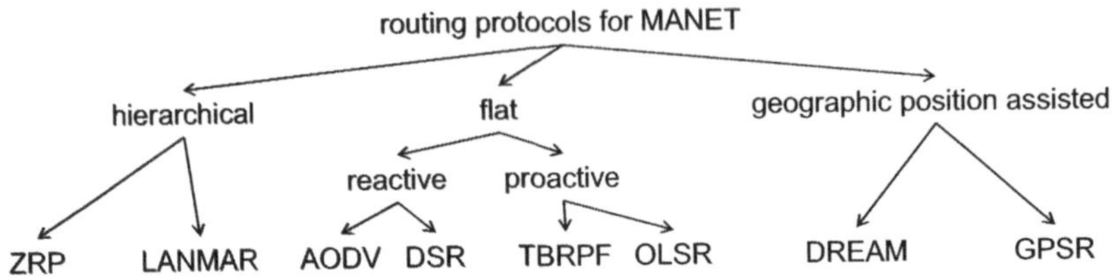

FIGURE 13.4 Classification of mobile ad hoc networks.

classification of ad hoc routing protocols is between proactive and reactive protocols (Figure 13.4). A *proactive routing protocol* is one that tries to maintain an up-to-date routing table at all times (it would therefore want to detect when the network topology changes and update routing tables accordingly). A *reactive routing protocol*, also known as *on-demand*, will find paths to destinations only as needed. Thus, it doesn't "waste" overhead exchanging topology information to maintain an up-to-date routing table (as proactive protocols do), but it takes longer to send packets to destinations because it needs to discover paths to the destinations when needed. Besides proactive and reactive protocols, there are also hybrid protocols that combine some features of both. *Hybrid ad hoc routing protocols* may be proactive for nodes within a short "radius" of a node, and reactive for nodes farther away. *Zone routing protocol* (ZRP [4]) is an example of such a protocol. There are also routing protocols that assume that nodes are GPS-capable, so some kind of position-based routing could be possible.

What if there needs to be communications with a node outside the MANET? Sure, if there needs to be some communications with a node outside the MANET, a node can connect back to the wired infrastructure, but for communications between two nodes in the MANET, the path is typically within the MANET, over one or more wireless links. Besides routing protocols for MANET, are there other active areas of R&D for MANET? Yes, issues such as security can be even more challenging for MANET than for traditional wireless networks. For example, in traditional wireless network, one could often assume that the network infrastructure is somewhat trustworthy, but a MANET cannot rely on that.

Among the more mature mobile ad hoc routing are the four that have been approved as *experimental RFC* by the IETF. In order of RFC number, these include *ad-hoc on-demand vector* (AODV [8]), *optimized link state routing* (OLSR [3]), *topology dissemination based on reverse-path forwarding* (TBRPF [7]), and *dynamic source routing protocol* (DSR [5]).

13.3.1 Example: AODV

AODV is one of the leading ad hoc routing protocols. The creators/developers of AODV are Charles Perkins, Elizabeth Royer, and Samir Das, who were from Nokia, UCSB, and the University of Cincinnati, respectively, at the time of preparation of the RFC for AODV.

13.3.1.1 Protocol Overview AODV is a reactive protocol. It finds routes on demand (assuming the source does not already have a routing table entry for the destination) by a route discovery procedure based on flooding of route requests. It also uses destination sequence numbers, and route lifetimes, to try to avoid using stale routes. The destination sequence numbers also allow AODV to avoid routing loops.

Key features of AODV (but not necessarily unique to AODV) include:

- Destination sequence numbers are used as indicators of how fresh a route is. The higher the sequence number, the more recent the route, and generally the more it is trusted. Hence, the sequence numbers correspond loosely to a time scale. Each node is responsible for maintaining its destination sequence number (i.e., the latest destination sequence number for routes with that node as the destination). Routes in other nodes to a particular node would in general have destination sequence numbers less than or equal to the latest destination sequence maintained by the node itself. If a node obtains multiple routes to the same destination, the one with the highest destination sequence number must be chosen.
- Traditional routing tables are used for routing (not the entire hop-by-hop path to the destination, unlike some other protocols, such as DSR). Hence, less information needs to be stored (per destination) for routing, and packet headers need not be very long when routes are long.
- AODV reacts quickly to network topology changes.

13.3.1.2 Protocol Operation The basic operation of AODV is as follows: When a node wishes to send a packet to another node, it checks if there is a valid route to the destination in its routing table (the concept of valid and invalid routes will be explained in due time). If there is, it uses the route. Otherwise, it performs route discovery. The basic mechanism in route discovery is flooding of the network, in which the source node broadcasts a *route request* (RREQ) message. Only nodes within a limited distance (radio range) can hear the broadcast, but nodes that receive it will rebroadcast, so eventually the RREQ propagates to the destination if the destination is in the set of nodes that can be reached through this type of flooding mechanism.

We show an example in Figure 13.5. The node numbers in the diagram are merely for illustrative purposes (in AODV, the nodes are actually all referred to by their IP addresses). The source node, node 11, is trying to find a route to the destination node, node 46. The first step in the route discovery is to broadcast a route request (RREQ) message. The RREQ contains the source and destination address and a route request ID (RREQ ID). The RREQ ID, in conjunction with the source address, uniquely identifies the particular request (another node could use the RREQ ID in its route discoveries since there is no global synchronization of RREQ IDs, but then the source address would be different). When the destination (node 46) receives one or more RREQs with the same source address and RREQ ID, it will know they are duplicates. Unlike some other protocols (e.g., DSR), where multiple replies are generated, AODV only replies to the first RREQ received.

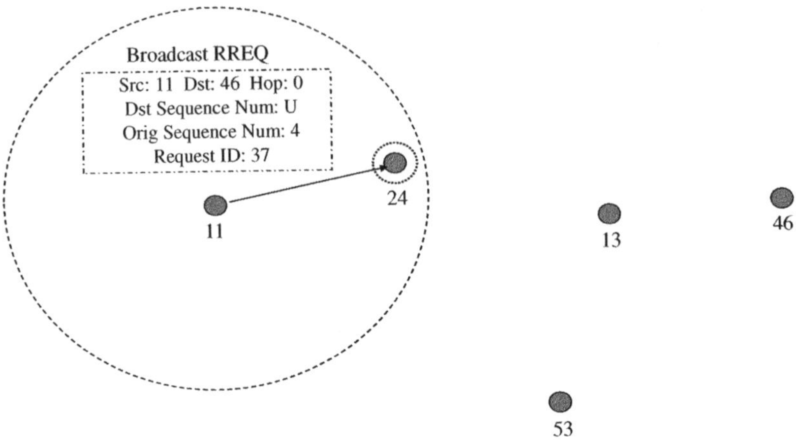

FIGURE 13.5 Route discovery of AODV, propagation of RREQ, part 1.

In addition, the RREQ contains a hop count, which starts at 0 and is incremented as the RREQ propagates through each node. Also carried in the RREQ are the destination sequence number for node 46 that node 11 has, and the originator sequence number for the originator node (node 11). The destination sequence number for node 46 is the largest sequence number that node 11 received for node 46 prior to this route discovery. Sometimes, though, the sequence number is unknown, as when this is the first route discovery for the particular destination, and we indicate this with a "U" in Figure 13.5. As we explain the processing in the intermediate nodes, it will become clearer how the destination sequence number and the originator sequence number are used.

Node 24 receives the RREQ from node 11, and the first thing it does is to create or update a route to node 11. It uses the originator sequence number in the packet as the sequence number for the new (or updated) route in its routing table. Next, node 24 makes sure it has not received a RREQ with the same RREQ ID from node 11 (e.g., through another path; this is quite a likely event, because in addition to being broadcast by the source, the RREQ is also being rebroadcast by other nodes). Otherwise, it would silently discard the packet (this ensures that each node rebroadcasts this packet only once, thus avoiding infinite rebroadcasting loops). Then it increments the hop count and rebroadcasts the packet, as shown in Figure 13.6. Next, node 13 receives the RREQ from node 24, and the processing is the same, except that it creates or updates the route to node 24 as the first step, and then it creates or updates a route to node 11, before it rebroadcasts the packet (see Figure 13.7).

The recipient of a RREQ would generate a route reply (RREP) if it is the destination or if it is an intermediate node that has a valid route to the destination, provided that the destination sequence number (of the route that the intermediate node has) is greater than or equal to that contained in the RREQ. In this example we assume that the intermediate nodes do not have such routing information, so at last the RREQ arrives at node 46 (Figure 13.7).

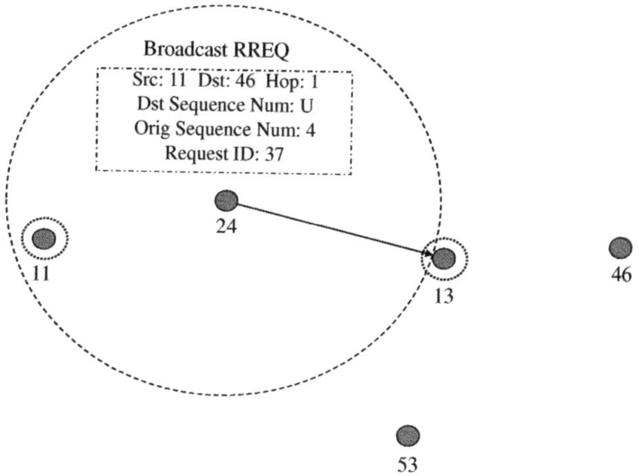

FIGURE 13.6 Route discovery of AODV, propagation of RREQ, part 2.

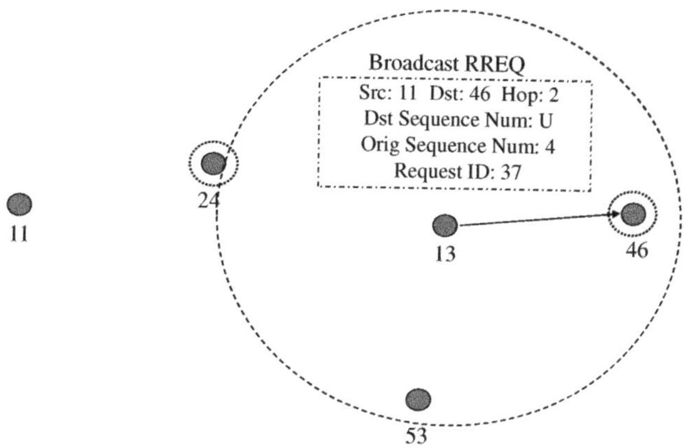

FIGURE 13.7 Route discovery of AODV, propagation of RREQ, part 3.

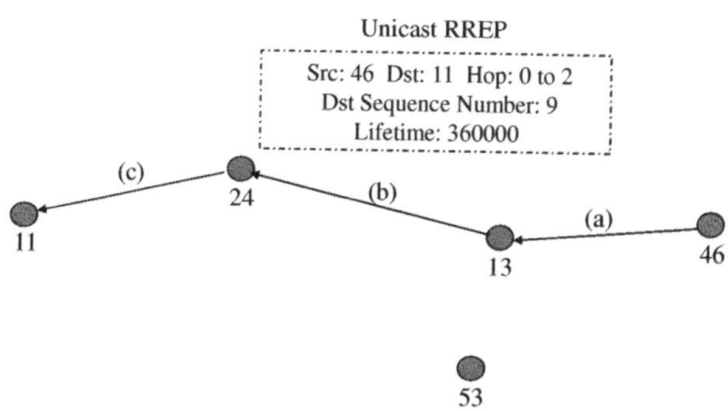

FIGURE 13.8 Route discovery of AODV, propagation of RREP.

Node 46 will notice that it is the destination and will generate a RREP to send back to node 11 (Figure 13.8). Since it has added the reverse route to node 11 as part of the processing when it received the RREQ, it is able to route the RREP back to node 11 by unicasting rather than broadcasting. Suppose that node 46's own sequence number is 9. Since the destination sequence number in the RREQ was "U" (unknown), node 46 must insert its own sequence number, 9, into the RREP. It is the destination node's responsibility to populate the lifetime field of the RREP. The unit of the lifetime is milliseconds, so for our example, we give it a lifetime of 10 hours. It populates the hop count with 0. Each node on the return path will increment the hop count so that it becomes 2 by the time that node 11 receives it.

13.3.1.3 *Other Features of AODV* Another feature of AODV is gratuitous RREP. This is for the case that an intermediate node knows how to route to the destination and it generates a RREP when it receives an RREQ and sends it back to the source. Upon receipt of the RREP, the source knows how to route to the destination. However, since an intermediate node has sent back an RREP instead of rebroadcasting the RREQ, the destination might not discover how to route to the source (in the other case, when the RREQ actually reaches the destination, it can use the route reversal feature to add a route to the source before generating the RREP). The solution in AODV is to perform gratuitous RREP to the destination (from the intermediate node that had sent the RREP to the source), so it learns the route information as well.

There can be a problem when a node reboots, in that in some cases routing loops may occur after it reboots if it loses all its routing information, including its sequence number. Hence, the DELETE_PERIOD feature was introduced so that a rebooted node must wait for a DELETE_PERIOD before responding to routing messages. This is to ensure that all routes that have it as a next hop would expire before it gets back into active involvement in the AODV signaling.

AODV also recommends that each node proactively broadcast "Hello" messages locally to its immediate neighbors if the node is part of an active route. This helps with maintaining local connectivity with active neighbors. Other ways include (1) the use of layer 2 notifications (e.g., nonreceipt of a CTS after an RTS has been sent could be an indication of loss of connectivity); (2) the receipt of any packet from the neighbor (whether "Hello" message or not); and (3) unicasting a RREQ to the neighbor, with the neighbor as the destination.

13.4 MESH, SENSOR, AND VEHICULAR NETWORKS

Each of these alternative wireless network paradigms is not so much a particular well-defined set of requirements and definitions, but more of a cluster of related ideas from which typical characteristics can be drawn. Hence, they might each be describable as an emerging cluster of concepts that continues to evolve as R&D progresses, and various technical and business lessons are learned.

Mesh, sensor, and vehicular networks all have some roots in the earlier work done in MANET, but have evolved and have added lessons learned from other areas of study, to the point where they are each sufficiently distinct from MANET to warrant separate R&D in their own right, and where they have become an independent cluster of concepts with a substantial amount of interest for both industry and academia.

13.4.1 Mesh Networks

A mesh network could be thought of as an extreme case of mobile ad hoc network where there is little or no mobility. The most important common feature of mobile ad hoc networks and mesh networks is the multihop wireless network that is formed by nodes, as shown in Figure 13.9. Typically, beyond the fixed nature of the nodes, mesh networks serve as a fixed wireless extension of the fixed network infrastructure; thus, whereas a single access point or base station provides limited coverage, a mesh network might be able to cover a large area, such as a city area.

In recent years, mesh networks have gained prominence for a specific type of application, and that is providing network connectivity in an area such as a city or town using multihop wireless access. Typically, such mesh networks include most of the following characteristics:

- The topology is in the form of a tree, where the base(s) of the tree are the points of connection to the wired network, typically the Internet. The points of connection to the wired network are sometimes called *Internet gateways* (IGWs).
- The end users are the leaves (sometimes called *mesh clients*), with mesh routers in between.
- There are multiple redundant paths through the mesh network, so if one or more of the mesh routers fails, connectivity is not necessarily lost.
- Usually, the point-to-point wireless links are based on IEEE 802.11.

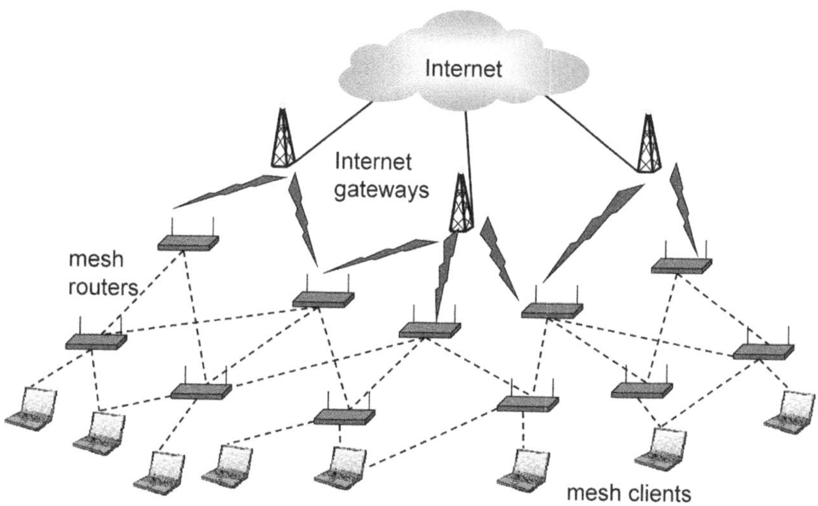

FIGURE 13.9 Mesh networks.

FIGURE 13.10 Mesh router mounted on a lamppost (two views).

Mesh networks can be *community-based* or *fully managed.* A community-based mesh network is one in which the mesh routers and IGWs are managed and operated by different organizations or individuals. They come together and work in the community to provide a community-based mesh network. A fully managed mesh network, on the other hand, is one that is owned and operated by one organization. If the mesh network is somewhere in between community-based and fully managed (e.g., it is operated by a few organizations), it may be described as *semimanaged.* The mesh routers may be relatively small, on the order of the size of a home wireless router, except that they would have to be made more robustly, to withstand outdoor weather conditions, and they might have antennas with higher gain. Figure 13.10 shows a mesh router mounted somewhat inconspicuously on a lamppost.

Whereas in MANET the mobility and constantly changing topology of the nodes is a primary challenge, in mesh networks the topology is relatively static and mesh routers are often stationary. Whereas in MANET power consumption is very critical with nodes running on battery power, this is less of an issue with mesh networks. So other issues are more important in mesh networks. The nodes are more stationary, but robustness and reliability, including rerouting packets when one or more mesh nodes or links have failed, are important. Management of interference between nearby wireless links is a critical challenge. This may involve two phases: (1) deployment and (2) operations. Various choices can have a significant impact on system capacity, robustness, and reliability, including the following choices:

- Placement of IGWs
- Load-balancing algorithms in IGWs
- Placement of mesh routers
- Channel allocation of 802.11 channels between mesh routers
- Channel-switching synchronization

MAC protocols and routing protocols that were designed for MANET might not be optimized for mesh networks, so new MAC protocols and routing protocols have been proposed for mesh networks. Especially in community-based mesh networks, selfish mesh routers can be a serious problem. A selfish mesh routers may intentionally utilize more resources (such as bandwidth) than their fair share, to provide better service for its mesh clients. In some deployment scenarios, ways to detect and correct such behavior is crucial.

13.4.2 Sensor Networks

Wireless sensor networks have been identified as one of the top 10 emerging technologies that will change the world [10]. The field of wireless sensor networking has been developing explosively in recent years. Although much attention has been focused on networking of computers and mobile phones (via the Internet, etc.) over the last 20 years, recent forecasts indicate that the future networks will be dominated by small and embedded devices (Figure 13.11). Networked sensors are an increasingly important segment of the small and embedded devices arena. Some estimates are that in a few years, there might be trillions of networked small embedded devices, compared to billions of mobile phones and computers!

Based on such predictions of where the most growth lies ahead, researchers have been exploring wireless sensor networking (Figure 13.12) in depth over the past decade. Research is active in many areas, including modulation schemes, design of power-efficient hardware, suitable medium access (MAC) protocols that are energy efficient, routing protocols (power efficient, data-centric, perhaps attribute-based

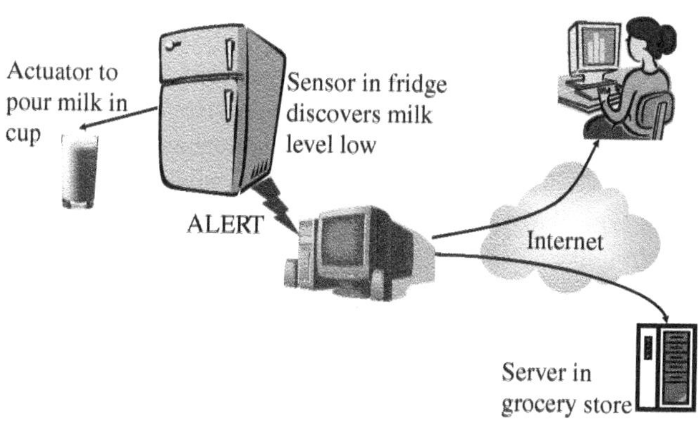

FIGURE 13.11 Application of sensor networks.

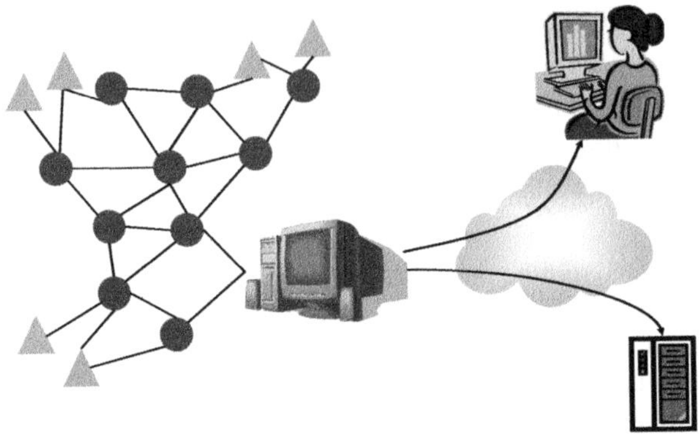

FIGURE 13.12 Sensor network.

addressing and location awareness), data aggregation, novel data dissemination algorithms, and application-layer query and dissemination protocols. Data aggregation has to be with intermediate nodes aggregating data collected from multiple sensors so that the volume of transmissions in the network can be reduced.

Although sensor networks may appear to be just a form of MANET, various differences have been identified that make sensor networking a field in its own right. Some of the characteristics of sensor networks, as compared to a more general MANET, include:

- The number of nodes may be several orders of magnitude higher.
- Sensor networks might be more densely deployed.
- Individual sensors may not have global identification.
- Sensor networks may mainly broadcast, rather than unicast, data.
- Sensors tend to be limited in power, computational capabilities, memory, and so on.
- Individual sensors may have a higher failure rate than a typical node in a MANET.

It is expected that each sensor may be a low-cost specialized device with limited capabilities, and may fail more frequently that a typical MANET node. However, the sensors may be deployed more densely, with more provision for redundancy, so a typical deployment might be able to tolerate loss of a certain percentage of sensors. Since sensor nodes are low-cost specialized devices, they would not be general-purpose machines such as typical MANET nodes might be. For example, they might not have global identification like an IP address, and they may not even implement TCP/IP.

13.4.3 Vehicular Networks

The field of intervehicular communications (IVC [12], also known as vehicle-to-vehicle communications, V2V, or vehicular ad hoc networks, VANET; Figure 13.13) has been gaining momentum in recent years. Numerous emerging communications

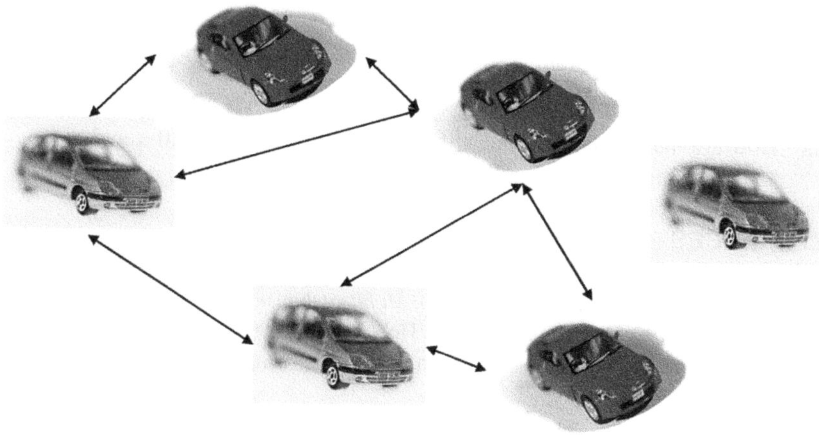

FIGURE 13.13 Vehicular network.

applications are unique to the vehicular setting. These applications include safety applications that will make driving safer and enable mobile commerce and roadside information services that can intelligently inform drivers about congestion, business and services in the vicinity of the vehicle, as well as other types of locally relevant news. Existing forms of entertainment may penetrate the vehicular domain, and new forms of entertainment may emerge, all supported by the intervehicular communications capabilities. These emerging services are not well supported by the limited communication options available on cars today.

Appropriately, the growing importance of intervehicular communications has been recognized by governments, corporations, and the academic community. IVC is recognized as an important component of intelligent transport systems (ITSs) in various national ITS plans. As such, governments have allocated spectrum for IVC and similar applications [e.g., various concepts of DSRC (dedicated short-range communications), such as WAVE (wireless access in vehicle environment)]. Government and industry cooperation has funded large IVC partnerships or projects such as CAMP (crash avoidance metrics partnership), ADASE2 (advanced driver assistance systems) in Europe, Network-on-Wheels, Fleetnet, and Cartalk2000. Academic conferences and workshops on IVC are beginning to grow in popularity (e.g., VANET, Autonet, V2VCOM).

EXERCISES

13.1 What is a service enabler?

13.2 What SIP messages would a presence user agent use to notify a presence agent about user status? How would watchers sign up to receive such information, and how would a presence agent inform the watchers?

13.3 What application servers work with IMS?

13.4 Classify the following mobile ad hoc routing protocols as proactive, reactive, or hierarchical: ZRP, OLSR, AODV.

13.5 How are mesh networks like MANET, and how are they different?

REFERENCES

1. M. Brenner and M. Unmehopa. *The Open Mobile Alliance: Delivering Service Enablers for Next-Generation Applications*. Wiley, Hoboken, NJ, 2008.
2. B. Campbell, R. Mahy, and C. Jennings. The message session relay protocol (MSRP). RFC 4975, Sept. 2007.
3. T. Clausen and P. Jacquet. Optimized link state routing protocol (OLSR). RFC 3626, Oct. 2003.
4. Z. J. Haas, M. R. Pearlman, and P. Samar. The zone routing protocol (ZRP) for ad hoc networks. work-in-progress draft-ietf-manet-zone-zrp-04.txt, July 2002.
5. D. Johnson, Y. Hu, and D. Maltz. The dynamic source routing protocol (DSR) for mobile ad hoc networks for IPv4. RFC 4728, Feb. 2007.
6. R. Noldus. *CAMEL: Intelligent Networks for the GSM, GPRS and UMTS Network*. Wiley, Hoboken, NJ, 2006.
7. R. Ogier, F. Templin, and M. Lewis. Topology dissemination based on reverse-path forwarding (TBRPF). RFC 3684, Feb. 2004.
8. C. Perkins, E. Belding-Royer, and S. Das. Ad hoc on-demand distance vector (AODV) routing. RFC 3561, July 2003.
9. C. E. Perkins. *Ad Hoc Networking*. Addison-Wesley, Reading, MA, 2001.
10. *Technology Review*. 10 emerging technologies that will change the world. http://www.technologyreview.com/Infotech/13060/, Feb. 2003. Retrieved Mar. 2011.
11. K. D. Wong. Geo-location in urban area using signal strength repeatability. *IEEE Communications Letters*, 5(10):411–413, Oct. 2001.
12. K. D. Wong, W. Chen, K. Tepe, and M. Gerla. Inter-vehicular communications. *IEEE Communications*, Special Issue, Oct. 2006.

V

MISCELLANEOUS TOPICS

14

NETWORK MANAGEMENT

In this chapter we introduce network management in Section 14.1. Then in Section 14.2 we describe some of the best known frameworks/models for network management in the industry. We spend the rest of the chapter focusing on a very important protocol for network management, SNMP.

14.1 REQUIREMENTS AND CONCEPTS

What comes to mind when the phrase *network management* is heard? Somebody may say that it is about maintaining the network, keeping it "well oiled" and running smoothly, replacing routers that are failing, upgrading equipment to handle increasing traffic loads as the number of customers increases, and so on. Another person may point out that even aside from the maintenance work, the regular daily operations of the network themselves need management. This person may point out that service providers have *network operation centers* (NOCs) where operators will monitor the network. If an alarm arrives at the NOC, perhaps indicating a link failure somewhere in the network, serious congestion, or something else of that nature, the operators may dispatch service people to take appropriate action. Other operational aspects might include handling of trouble tickets. Yet another person might say that when she thinks of network management, she thinks of handling new subscribers. There must be systematic and orderly processes for adding their information to the subscription and billing databases, and for turning on and activating various features to which they have subscribed. These activities can be described as provisioning. A fourth person might point out that provisioning, operations, and maintenance must be based on a

Fundamentals of Wireless Communication Engineering Technologies, First Edition. K. Daniel Wong.

foundation of good administration. When this person thinks of network management, he thinks of inventory control, customer service reports, and so on. In a way, all four of these views are right, and together, as *operations, administration, maintenance, and provisioning* (OAM&P), one of the popular models of network management (Section 14.2) whose name attempts to capture the scope of its coverage. Before describing some of these models in Section 14.2, we consider a couple of general questions here.

What does network management include? Is it just about managing the network (i.e., the routers, switches, cables, etc.), that make up the network? How about the PCs, servers, and other hosts—are they included? What about billing systems and other such systems that a network operator would need? How about human factors, business decisions, policies, and so on? At one level, "network management" can refer broadly to all these things. However, it can also be used in a narrower sense: for example, referring to management of network layer and network scope issues, as opposed to service layer issues. "Network management" is used in these different contexts, so the reader needs to be aware of context to figure out what is meant when someone talks about network management.

Why is network management important? Can't the operator just buy the equipment from a vendor, get the vendor to help install it, and then just "let it run"? A network is a complex system. Complex systems need to be managed, even if they run smoothly most of the time. Let's consider the analogy of a car, which is another complex system. Most of the time, the car drives as expected. However, a car needs to be operated properly to maximize its utilization. Lots of hard braking wears out the brakes faster than necessary, and abusing the car by driving 100 mph and making very sharp turns can even cause it to overturn or crash. Besides proper and careful operations, proper and careful maintenance, including regular checks on tire pressure, oil levels, and so on, and doing oil changes and other replacements according to schedule, help in the management of the car. The operations and maintenance need to be backed up by careful record keeping and administration. Thus, for a system such as a car, operations, administration, and maintenance are needed, just as in a network. Additionally, provisioning is needed in operator networks that provide services to customers. In the car analogy, this may apply if the car is used as a taxi, for example, and the driver needs to include a meter and other items to provide service to customers.

14.2 NETWORK MANAGEMENT MODELS

Various attempts have been made by various groups to specify and classify the range of things that are encompassed by "network management." The following are common acronyms for these attempts, arranged roughly in chronological order from the earliest to the most recent:

- *OAM&P*: operations, administration, maintainance, and provisioning
- *FCAPS*: fault management, configuration management, accounting management, performance management, and security management

- *TMN*: telecommunications management network
- *TOM*: telecommunications operation map
- *eTOM*: enhanced TOM; includes finance, human resources, etc.

OAM&P came from the traditional telephony world. Sometimes, it is written simply as OAM, or as OAMPT, where "T" stands for "troubleshooting." Operations are about keeping the network running well, administration is about accounting and housekeeping, maintenance is about repairs and upgrades, and provisioning is about the addition of new services. Originally, the *fault management, configuration management, accounting management, performance management, and security management* (FCAPS) model came from OSI in the 1980s. Then, ITU-T incorporated FCAPS in its network management work in the 1990s. Fault management includes networking monitoring, fault diagnosis, and root cause analysis. Trouble ticketing also falls under fault management. Configuration management includes the configuration of equipment and services, auditing, and backup and restoration of network configuration in case of crashes involving loss of configuration. Accounting management includes keeping track of usage and charging accordingly. *Call detail records* (CDRs) are an important part of usage accounting. Fraud detection also falls under accounting management. Performance management involves the monitoring of throughput, delay, and various quality indicators over time. Security management includes making the network more robust against hacking. It also includes intrusion detection and blacklisting of certain addresses.

In its ITU-T incarnation, FCAPS is part of a larger model, the TMN model. TMN breaks down network management into four layers:

- Business management
- Service management
- Network management
- Element management

TMN is not just a framework for classication of functions. It also provides a framework for interoperation between devices from different vendors, through the specification of interface points, and using the *common management information protocol* (CMIP). CMIP is a communication protocol between network managers and agents, roughly analogous to SNMP for IP-based networks. CMIP has many more features than SNMP and is more sophisticated than SNMP. However, as in the case of VoIP session control, where the relatively simple SIP is very popular, being feature-rich is not necessarily the path to success. TMN is generally used for the management of circuit-switched networks (e.g., ISDN or GSM) or virtual circuit-switched networks (e.g., ATM).

More recently, the *telecommunication operation map* (TOM) has emerged [5]. TOM comes from the telecommunications management forum (TMF). TMF is a telecommunication industry forum. TOM starts from the TMN model and defines processes as well as information flows between processes at each layer. Information flows between processes at different layers is also included. Some would say that

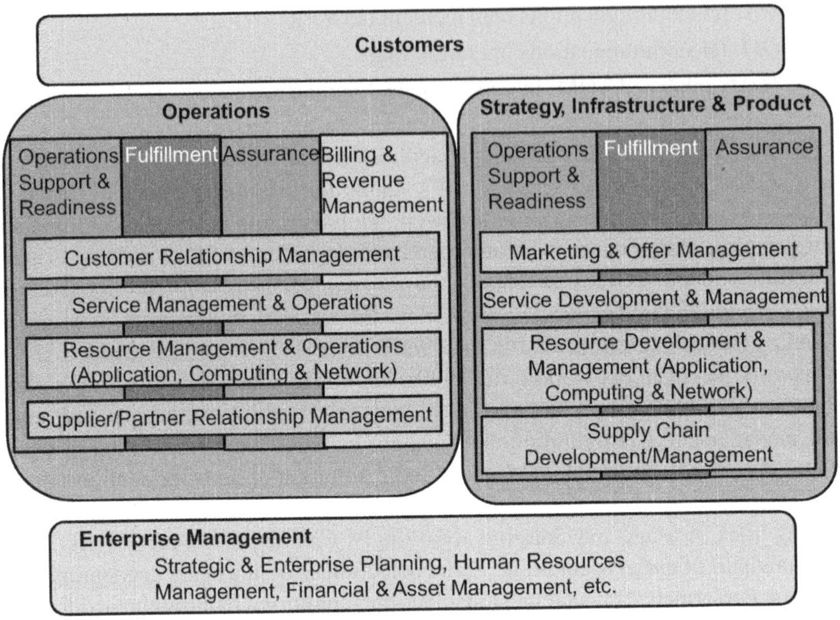

FIGURE 14.1 eTOM.

TOM is a replacement for TMN, but that is not completely correct. TMN is more focused on network management, whereas TOM takes a much broader perspective. As a kind of blueprint for how to run a telecommunications service provider business, TOM also encompasses other functions and processes less often thought of as belonging to network management. TOM helps providers to automate and streamline their processes.

TOM has been enhanced by *enhanced TOM* (eTOM), where eTOM expands the scope of TOM even further. For those who are familiar with ITIL from the IT world, TOM/eTOM can be thought of as an analogous set of industry best practices. Whereas ITIL is for IT practices, TOM/eTOM is for telecommunications network practices. Figure 14.1 shows the eTOM framework. As the figure indicates, good customer service requires good operations, as well as good strategy, infrastructure, and product. Strategy, infrastructure, and product are more long-term focused than operations. For example, resource development falls under strategy, infrastructure, and product, whereas resource operations falls under operations. Similarly, marketing and supply chain development fall under strategy, infrastructure, and product, whereas customer relationship management and supplier/partner relationship management fall under operations. Another dimension in operations is the four-stage life cycle: from operations support and readiness, to fulfillment, assurance, and then billing and revenue management. Each of these have to be executed well for operations to do well. Of the four, fulfillment, assurance, and billing and revenue management are customer facing, whereas operations support and readiness occur less in real time and more behind the scenes. As a group, fulfillment, assurance, and billing and revenue management are sometimes given the acronym FAB. Underlying all these are the enterprise management aspects, including strategic and enterprise planning, human resources management, and so on.

These various frameworks for thinking about network management help us to organize and categorize the scope and functions in network management. In any implementation of a network management system, however, communication protocols would need to be selected. These could be proprietary, or they could be open protocols such as CMIP or SNMP.

14.3 SNMP

SNMP (simple network management protocol) comes from the IETF (see Section 17.2.4). It is meant only for IP-based networks. Unlike CMIP, it is relatively simple and thus is widely implemented in most IP-enabled devices. There are a few versions of SNMP. The oldest, but still the most widely deployed, is SNMPv1 [1]. Enhancements were added with SNMPv2 and SNMPv3, some of which are significant (e.g., the security enhancements with SNMPv3). Since SNMPv1 is still widely deployed, we will base our description here on SNMPv1, with mention of changes introduced with SNMPv2 and SNMPv3 in places as appropriate.

The SNMP network architecture is very simple: There are *managers* and there are *agents*. These are particular roles, not special-purpose devices; for example, it is normal that each IP-enabled device is capable of playing the role of SNMP agent (it may not be turned on by default, though; e.g., the Windows SNMP agent is a Windows component that may not be turned on by default but can easily be added through "`add/remove windows components`"). There might be only a few managers in a network (or even just one), whereas each device that is part of the SNMP-based network management will be an agent.

SNMP uses UDP as its transport protocol. In particular, UDP port 161 is used for sending and receiving requests. UDP port 162 is used for traps. Everything to be managed is broken down into *objects*. Example of objects include the system name, the routing table, the MAC address of a network interface, and so on. The relationship of the objects with the real components of computer systems, and other details about objects, are discussed in Section 14.3.2. Each object is specified in an appropriate management information base (MIB). A MIB can be thought of as a set of variables (the objects), each with their particular data types. An SNMP manager's view of management objects as specified by a MIB is shown in Figure 14.2. Managers query

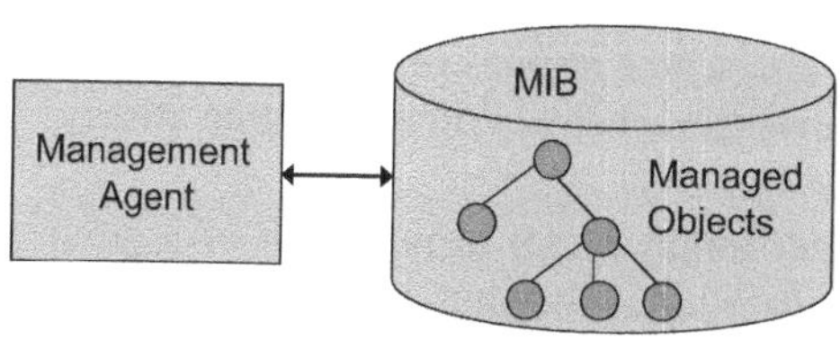

FIGURE 14.2 SNMP management objects.

agents for information about objects, and sometimes set values of objects, using the messages we will now describe.

14.3.1 Messages

There are only five messages in SNMP (version 1; a few more messages are added in versions 2 and 3). Indeed, SNMP is a simple protocol, but the five messages provide a wide scope of capabilities. The messages are:

1. `get-request`: from manager to agent, a query to obtain some specified information (it specifies the object IDs of the information it wishes to obtain; we discuss objects in Section 14.3.2).
2. `get-next-request`: from manager to agent, like `get-request`, but returns the object *after* the specified object, so can be used to step through a MIB, as we will explain shortly.
3. `get-response`: from agent to manager, a response message, sent in response to `get-request` or `get-next-request`.
4. `set-request`: from manager to agent, a command to set (write) a value.
5. `trap`: from agent to manager, sent asynchronously (i.e., not in response to a `get-request` or `get-next-request` message).

The `get-request` specifies one or more variables. When the SNMP agent receives the request, it sends back a `get-response` message. In this message, for each of the variables for which the agent has an exact match of the name, the name and value are returned in the `get-response`.

The `get-next-request` allows traversal of tables (e.g., routing tables). Let's compare it with the `get-request`. The `get-request` is sent to request one or more objects: namely, the objects whose object IDs are specified in the request. The `get-next-request` also specifies an object ID, but the request is for the *next* object in the MIB, the one after the object whose object ID is given in the message (the structure of MIBs is such that the ordering is well defined). Thus, a manager and agent can have an exchange where the manager keeps using the `get-next-request` and each time gets back a `get-response`; it takes the object ID from the `get-response` and puts it into the next `get-next-request`. In this way the manager can step through the entire MIB, or a portion of the MIB, without needing to specify the object IDs of the objects it wants.

The `set-request` allows a manager to set one or more particular variables. The `get-response` is also sent in response to a `set-request`. The `trap` allows the agent to inform the manager when a serious condition arises.

Example. A manager might be set to poll each of a group of routers to obtain information related to who is logging in to each router. Perhaps it does this once an hour and then stores the information at a central location as part of the organization policy of keeping an audit trail of all logins to routers. An SNMP get-request can be sent to do the polls, and each router would respond with an SNMP get-response. Meanwhile, each router could be set to send a trap to the manager as soon as it detects that any of its interfaces or connected links are unexpectedly down. Such a trap could be set to cause the manager software to interrupt a human operator by beeping, flashing, and so on, so that quick action can be taken (how such a received trap is handled by the manager is, of course, not part of SNMP, but is related to the network management policy in effect). If the network management policy allows it (we will see later why this is often not allowed), the manager could be used remotely to change settings such as IP addresses on some router interfaces, by sending appropriate SNMP set-request messages.

When a network manager sends a get-request message to an agent in a network device, it assumes that:

- The network device speaks SNMP (i.e., it is running an SNMP agent).
- The manager and agent speak the same version of SNMP.
- The manager and agent have the same understanding of the *managed objects* (variables, see Section 14.3.2) that the manager can query or set.

Regarding the first issue, most modern computing and network devices speak SNMP (we call them SNMP-capable devices, or SNMP speakers for short). However, some older devices may not be SNMP speakers. An *SNMP proxy* can act as an SNMP agent on behalf of one or more devices that are not SNMP speakers (see Section 14.3.7.1).

Regarding the second issue, the version of SNMP is always specified in the SNMP message header. Regarding the third issue, managers and agents can have the same understanding of the variables through the use of management information bases (MIBs) [7]. We elaborate on the nature of objects in Section 14.3.2 and on object naming in Section 14.3.2.1. More details on MIBs themselves are provided in Section 14.3.3.

14.3.1.1 Additional Messages in SNMPv2 and SNMPv3 SNMPv2 and SNMPv3 add four more messages to the five already introduced with SNMPv1. The new messages are:

1. get-bulk: from manager to agent, an improved way to query for multiple managed objects with one message.
2. inform: from agent to manager, like trap, but more reliable because the manager should respond to inform.
3. notification: agent to manager, like trap, this is sometimes called SNMPv2 Trap.
4. report: from manager to manager, it is asynchronous like a trap, and can be used for manager-to-manager communications.

We say `get-bulk` is an *improved* way to query for multiple managed objects with one message, because even the original `get-request` message can be used to query for multiple managed objects with one message. For example, with the `get-request`, if the message size to return all the MIB objects is too large for the agent to send all at once, it will return an error message instead. With `get-bulk`, however, the agent sends back some of the responses at once, and other parts of the responses later.

Just as the `get-bulk` is an attempt to improve on the functionality of the `get-request`, the `notification` is an attempt to improve on the `trap`. One issue with SNMPv1 is that the `trap` has a different format than `get-request` and `set-request`. `Notification` is a trap that has the same format as `get-request` and `set-request`.

The `inform` adds the capability to acknowledge traps. Previously, with SNMPv1, traps are not acknowledged, so the agent cannot know if a manager has received a trap that it sent.

14.3.2 Managed Objects

The managed objects under SNMP are variables that represent, or are related to, physical objects (such as interfaces) or other properties of the managed entity. Examples of these properties include routing tables, system name, and so on. Whether they are physical objects or properties, we can consider them "real resources." We can conceptually create a divide between a *management plane* and a *real resources plane*, as shown in Figure 14.3, and where the managed objects in a MIB correspond to real resources.

As variables, the managed objects have names (the naming is discussed in Section 14.3.2.1), and they are also typed variables. Some are read-only, others are read-write. Naming and associated data types are specified in RFC 1155 (Structure of

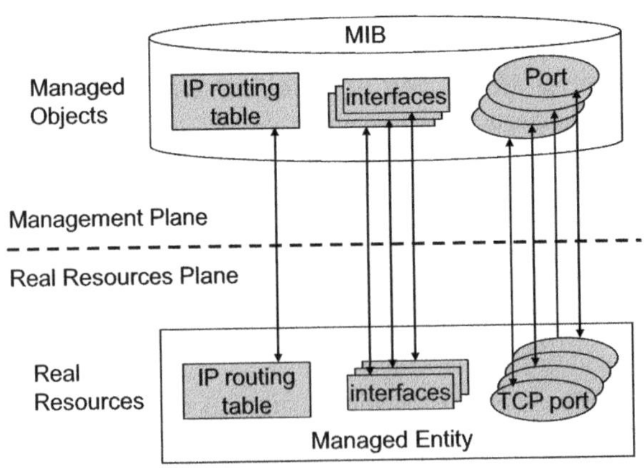

FIGURE 14.3 SNMP management concepts.

Management Information Version 1, SMIv1) [6] and RFC 2578 (Structure of Management Information Version 2, SMIv2) [3]. SMIv2 contains enhancements that came after SNMPv1, but SNMPv2 and SNMPv3 take advantage of these enhancements. It is important to remember that the SMI and SNMP versions are two different things. The version of SMI used for a particular MIB (see Section 14.3.3 for more on MIBs) can be found in the RFC that specifies the MIB.

The data types include INTEGER, OCTET STRING, Counter, OBJECT IDENTIFIER, SEQUENCE, SEQUENCE OF, IpAddress, and so on; Counter is like an INTEGER with specific typical usage (e.g., it should be reset to 0 whenever an agent is rebooted). OBJECT IDENTIFIER refers to the names that we describe next. SEQUENCE is a list of zero or more other data types. SEQUENCE OF is analogous to an array of similar objects.

SMI also specifies the specific encoding for each data type to avoid ambiguities (using syntax and encoding rules from Abstract Syntax Notation 1, ASN.1, which comes from ISO and ITU). ASN.1 is an abstract format for representing, encoding, and decoding data in a machine-independent and precise way that avoids ambiguities; ASN.1 has broader applicability than just SNMP or network management, and SMI uses just a subset of ASN.1. In particular, the "basic encoding rules" of ASN.1 are to be applied for the transmission of objects. This removes ambiguities such as big-endian vs. little-endian encoding of bytes.

14.3.2.1 Naming The names of the managed objects are also called *object identifiers* in SNMP. In MIBs, the object identifiers are arranged in inverted tree structures, called the *structure of management information* (SMI) tree. An example of part of such an SMI tree, showing parts of a MIB called MIB-II, is shown in Figure 14.4. Most of the time in network management, the path from the root node begins with *iso.org.dod.internet*, which numerically is 1.3.6.1; so, the objects encountered with SNMP all have object IDs beginning with 1.3.6.1. Two branches below this are commonly seen in network management: *iso.org.dod.internet.management* (1.3.6.1.2) and *iso.org.dod.internet.private* (1.3.6.1.4). Public objects defined by IETF in MIBs can be found beginning with 1.3.6.1.2, whereas private objects that may be defined by private companies can be found beginning with 1.3.6.1.4. If an equipment vendor specifies private objects, these private vendor-specific objects are presumably supported by the vendor's equipment, but the network manager would also need to understand them to make use of them (in general, the network manager software may come from a provider other than the vendor). Besides 1.3.6.1.2 and 1.3.6.1.4, *iso.org.dod.internet.snmpV2* (1.3.6.1.6) may sometimes be seen for objects that assume supports of the newer data types found in SMIv2, meant to be used with SNMPv2.

Why 1.3.6.1? Why did the Internet Activities Board (IAB) decide to allocate for itself the object identifier 1.3.6.1? It could have decided to create its own tree, thus obviating the need to append the prefix 1.3.6.1 to every object identifier used with SNMP. However, it decided to attach its tree to a higher-level tree that came from ISO and CCITT, perhaps to be better integrated into a unified global naming scheme.

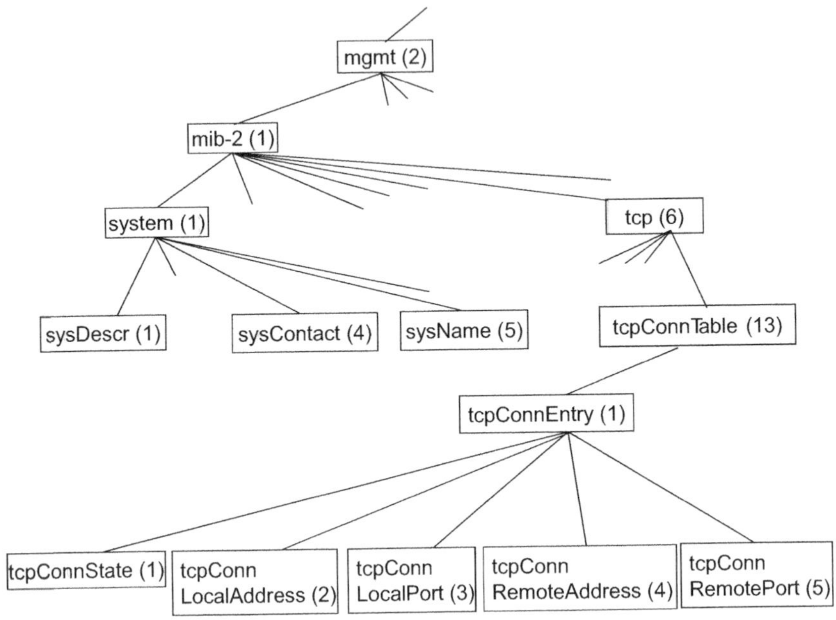

FIGURE 14.4 Portions of the MIB-II MIB.

Perhaps it has something to do with being a good citizen living in peace and harmony with related organizations. Interestingly, RFC 1155 states: "This memo assumes that DoD will allocate a node to the Internet community and a node to the Internet Activities Board (IAB) as follows: . . . that is, the Internet subtree of OBJECT IDENTIFIERS starts with the prefix: 1.3.6.1." So, with this assumed allocation and this statement in RFC 1155, the prefix 1.3.6.1 began its ubiquitous role in network management in the Internet community.

14.3.3 MIBs

How can the SNMP managers and agents have a common understanding of the managed objects? One possible solution is to have one big document that specifies the entire known set of managed objects. This set would be very large, as it would not only include basic information such as system name but would also have to include less common variables. Given the wide range of roles that the devices can take, and the fact that most devices will never be playing many of these roles, we would have a situation where the list is very large, but only a small subset applies to most devices. Moreover, whenever there are new roles, or changes in existing roles, the big document would have to be updated.

An alternative solution, one that is adopted in SNMP, is to have different smaller documents, each of which specifies only subsets of the entire set of managed objects. The most natural grouping is based on roles. Thus, there is one document specifying

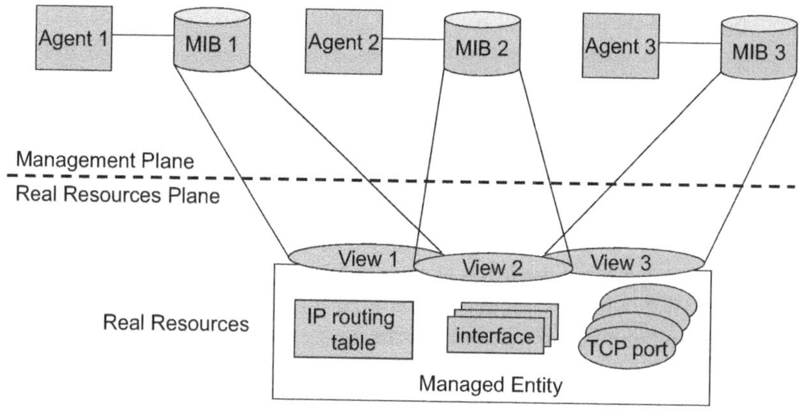

FIGURE 14.5 Multiple views of the managed objects in a device.

the group of objects for DNS servers, one for devices that implement robust header compression, one for devices that support mobile IP, and so on. Furthermore, each device only needs to be concerned with the managed objects related to the roles that it plays, not the many other possible roles. Moreover, when a new role is specified, a new document can be created specifically for that role, and the existing documents for other roles need not be modified.

It is analogous to the description of a woman who may play multiple roles in life, so we may have a collection of attributes related to her role as manager in a company (e.g., attributes such as salary), another collection related to her role as mother (e.g., attributes such as number of children and their names and ages), and another collection of attributes related to her role as Sunday school teacher (e.g., attributes such as dates and titles of lessons). Another woman may have a very different set of roles; for example, she may be a college student (year in college, major, etc.) or a part-time worker at a local fast-food establishment (hours, hourly pay, etc.). With each of them, appropriate attributes can be associated, based on roles and the collections of attributes that come with each role.

Another reason for the SNMP approach is that it makes it easier to divide the managed objects into multiple *views*, where different management agents can be assigned different views, to manage different subsets of the managed objects in the device. This concept is shown in Figure 14.5.

Somebody who is new to SNMP and network management may be surprised to learn that of the first 5000 RFCs, 318 of them contain MIBs (that's more than 6% of all RFCs).

14.3.3.1 Implementing Multiple MIBs A management information base (MIB) is a collection of management information (objects) that represents a subtree of the complete tree of managed objects (some purists may argue that the entire collection of objects is *the* MIB, and all the subtrees that are commonly called MIBs are actually "MIB modules," but we will not make this distinction here). Since each MIB covers only a subset of the complete tree, SNMP speakers typically implement multiple MIBs. By referring to a machine implementing a MIB, we mean that it is prepared to at least be queried on objects specified in that MIB (it may or may not allow a manager to write values).

The objects in the SMI tree are arranged such that objects that are related, and/or are commonly found together, tend to be clustered together in the tree and defined in the same MIB. As just explained, a given machine would typically implement only a small fraction of all the MIBs available. Nevertheless, all SNMP speakers will implement MIB-2, which is a foundational MIB for containing basic objects such as system name and number of network interfaces.

14.3.3.2 Example: A Look at MIB-2 A portion of MIB-2 (RFC 1213 [4]) is reproduced below. In particular, it is the first half of the `Interfaces` group (MIB-2 2). Notice that like all MIBs, it is text based, so even without an understanding of ASN.1 and SMI, we can guess the meaning of much of its contents. For example, comments are prefaced with "–"; each object "ACCESS" could be read-only, read-write, or not accessible; and so on. There is a text description of each object (under "DESCRIPTION"), and each object specification ends with the specification of its parent in the SMI tree. For example, object `ifNumber` is a child of `interfaces`, and it has number 1 (so 1 would be appended to the object ID of interfaces, to get the object ID of `ifNumber`). Also notice how tables are specified. `ifTable` is described as a `SEQUENCE OF IfEntry`. Then each `IfEntry` is described as a sequence including `ifIndex`, `ifDescr`, and so on.

```
-- the Interfaces group

          -- Implementation of the Interfaces group is mandatory for
          -- all systems.

          ifNumber OBJECT-TYPE
              SYNTAX  INTEGER
              ACCESS  read-only
              STATUS  mandatory
              DESCRIPTION
                      "The number of network interfaces (regardless of
                      their current state) present on this system."
              ::= { interfaces 1 }


          -- the Interfaces table

          -- The Interfaces table contains information on the entity's
          -- interfaces.  Each interface is thought of as being
          -- attached to a 'subnetwork'.  Note that this term should
          -- not be confused with 'subnet' which refers to an
```

```
-- addressing partitioning scheme used in the Internet suite
-- of protocols.

ifTable OBJECT-TYPE
    SYNTAX  SEQUENCE OF IfEntry
    ACCESS  not-accessible
    STATUS  mandatory
    DESCRIPTION
            "A list of interface entries.  The number of
            entries is given by the value of ifNumber."
    ::= { interfaces 2 }

ifEntry OBJECT-TYPE
    SYNTAX  IfEntry
    ACCESS  not-accessible
    STATUS  mandatory
    DESCRIPTION
            "An interface entry containing objects at the
            subnetwork layer and below for a particular
            interface."
    INDEX   { ifIndex }
    ::= { ifTable 1 }

IfEntry ::=
    SEQUENCE {
        ifIndex
            INTEGER,
        ifDescr
            DisplayString,
        ifType
            INTEGER,
        ifMtu
            INTEGER,
        ifSpeed
            Gauge,
        ifPhysAddress
            PhysAddress,
        ifAdminStatus
            INTEGER,
        ifOperStatus
            INTEGER,
        ifLastChange
            TimeTicks,
        ifInOctets
            Counter,
        ifInUcastPkts
            Counter,
        ifInNUcastPkts
            Counter,
        ifInDiscards
            Counter,
        ifInErrors
            Counter,
```

```
            ifInUnknownProtos
                Counter,
            ifOutOctets
                Counter,
            ifOutUcastPkts
                Counter,
            ifOutNUcastPkts
                Counter,
            ifOutDiscards
                Counter,
            ifOutErrors
                Counter,
            ifOutQLen
                Gauge,
            ifSpecific
                OBJECT IDENTIFIER
        }

    ifIndex OBJECT-TYPE
        SYNTAX  INTEGER
        ACCESS  read-only
        STATUS  mandatory
        DESCRIPTION
                "A unique value for each interface.  Its value
                ranges between 1 and the value of ifNumber.  The
                value for each interface must remain constant at
                least from one re-initialization of the entity's
                network management system to the next re-
                initialization."
        ::= { ifEntry 1 }

    ifDescr OBJECT-TYPE
        SYNTAX  DisplayString (SIZE (0..255))
        ACCESS  read-only
        STATUS  mandatory
        DESCRIPTION
              "A textual string containing information about the
              interface.  This string should include the name of
              the manufacturer, the product name and the version
              of the hardware interface."
        ::= { ifEntry 2 }

    ifType OBJECT-TYPE
        SYNTAX  INTEGER {
                    other(1),          -- none of the following
                    regular1822(2),
                    hdh1822(3),
                    ddn-x25(4),
                    rfc877-x25(5),
                    ethernet-csmacd(6),
                    iso88023-csmacd(7),
                    iso88024-tokenBus(8),
                    iso88025-tokenRing(9),
```

```
                    iso88026-man(10),
                    starLan(11),
                    proteon-10Mbit(12),
                    proteon-80Mbit(13),
                    hyperchannel(14),
                    fddi(15),
                    lapb(16),
                    sdlc(17),
                    ds1(18),              -- T-1
                    e1(19),               -- european equiv. of T-1
                    basicISDN(20),
                    primaryISDN(21),      -- proprietary serial
                    propPointToPointSerial(22),
                    ppp(23),
                    softwareLoopback(24),
                    eon(25),              -- CLNP over IP [11]
                    ethernet-3Mbit(26),
                    nsip(27),             -- XNS over IP
                    slip(28),             -- generic SLIP
                    ultra(29),            -- ULTRA technologies
                    ds3(30),              -- T-3
                    sip(31),              -- SMDS
                    frame-relay(32)
                }
        ACCESS  read-only
        STATUS  mandatory
        DESCRIPTION
               "The type of interface, distinguished according to
               the physical/link protocol(s) immediately 'below'
               the network layer in the protocol stack."
        ::= { ifEntry 3 }

    ifMtu OBJECT-TYPE
        SYNTAX  INTEGER
        ACCESS  read-only
        STATUS  mandatory
        DESCRIPTION
                "The size of the largest datagram which can be
                sent/received on the interface, specified in
                octets.  For interfaces that are used for
                transmitting network datagrams, this is the size
                of the largest network datagram that can be sent
                on the interface."
        ::= { ifEntry 4 }

    ifSpeed OBJECT-TYPE
        SYNTAX  Gauge
        ACCESS  read-only
        STATUS  mandatory
        DESCRIPTION
               "An estimate of the interface's current bandwidth
               in bits per second.  For interfaces which do not
               vary in bandwidth or for those where no accurate
```

```
          estimation can be made, this object should contain
          the nominal bandwidth."
      ::= { ifEntry 5 }

  ifPhysAddress OBJECT-TYPE
      SYNTAX  PhysAddress
      ACCESS  read-only
      STATUS  mandatory
      DESCRIPTION
          "The interface's address at the protocol layer
          immediately `below' the network layer in the
          protocol stack.  For interfaces which do not have
          such an address (e.g., a serial line), this object
          should contain an octet string of zero length."
      ::= { ifEntry 6 }

  ifAdminStatus OBJECT-TYPE
      SYNTAX  INTEGER {
                  up(1),       -- ready to pass packets
                  down(2),
                  testing(3)   -- in some test mode
              }
      ACCESS  read-write
      STATUS  mandatory
      DESCRIPTION
              "The desired state of the interface.  The
              testing(3) state indicates that no operational
              packets can be passed."
      ::= { ifEntry 7 }

  ifOperStatus OBJECT-TYPE
      SYNTAX  INTEGER {
                  up(1),       -- ready to pass packets
                  down(2),
                  testing(3)   -- in some test mode
              }
      ACCESS  read-only
      STATUS  mandatory
      DESCRIPTION
            "The current operational state of the interface.
            The testing(3) state indicates that no operational
            packets can be passed."
      ::= { ifEntry 8 }

  ifLastChange OBJECT-TYPE
      SYNTAX  TimeTicks
      ACCESS  read-only
      STATUS  mandatory
      DESCRIPTION
              "The value of sysUpTime at the time the interface
              entered its current operational state.  If the
              current state was entered prior to the last re-
              initialization of the local network management
```

```
                subsystem, then this object contains a zero
                value."
        ::= { ifEntry 9 }
```

14.3.4 Security

The security support of SNMPv1 is almost nonexistent. SNMPv1 does not support confidentiality—everything is transmitted in the clear. It only supports authentication, albeit a very weak "community"-based scheme with plaintext passwords.

14.3.5 Traps

In SNMP, managers query agents for information on various objects, and if allowed to, managers may also write values to selected objects in the devices. Whether it is querying or setting values, these are manager-initiated actions. They are inadequate, however, for cases where something unusual occurs and it would be good if the agent could let the manager know. Of course, the manager can simply query more often to reduce the time from the occurrence of such events to when it finds out. This is clearly very inefficient; it would be more reasonable for the agent to be equipped with the capability to initiate contact with the manager in such cases. The SNMP trap allows the agent to do so.

The SNMP trap is sometimes described as asynchronous, meaning that it doesn't have to wait for communications to be initiated by a manager. It allows an agent to send a message to a manager at any time. A manager listens to UDP port 162 for an SNMP trap (unlike SNMP requests and responses, which are sent to port 161 on the agents and managers).

Six generic traps are defined in SNMPv1 (RFC 1157), representing six events that may be expected to occur with some frequency in many networks: (1) coldStart; (2) warmStart; (3) linkDown; (4) linkUp; (5) authenticationFailure; and (6) egpNeighborLoss; these are numbered 0, 1, 2, 3, 4, and 5, respectively. Both coldStart and warmStart indicate that the agent has restarted. The difference is that coldStart implies a reboot where all the SNMP counters and other values are reset, whereas no values are reset in warmStart. linkDown and linkUp indicate when a link goes down or comes up, and are associated with a variable that points to the interfaces table to indicate which link went down or came up. authenticationFailure indicates an SNMP authentication failure (wrong community string was used in a failed attempt to access the device). egpNeighborloss is used when an EGP neighbor goes down. For all other traps, there is one more type, numbered 6, called enterpriseSpecific.

Like other SNMP objects, traps are specified in MIBs. Although not discussed in SMI, traps can be specified in SMIv1 MIBs using the TRAP-TYPE macro (as discussed in RFC 1215) and in SMIv2 MIBs using the NOTIFICATION-TYPE macro.

Example. We have the following trap defined in RFC 1697 (RDBMS MIB):

```
rdbmsOutOfSpace NOTIFICATION-TYPE
    OBJECTS      { rdbmsSrvInfoDiskOutOfSpaces }
```

```
    STATUS      current
    DESCRIPTION
        "An rdbmsOutOfSpace trap signifies that one of the database
         servers managed by this agent has been unable to allocate
         space for one of the databases managed by this agent.  Care
         should be taken to avoid flooding the network with these
         traps."
    ::= { rdbmsTraps 2 }
```

We can scan RFC 1697 to find out more about rdbmsSrvInfoDiskOutOfSpaces. We find the following:

```
rdbmsSrvInfoDiskOutOfSpaces OBJECT-TYPE
    SYNTAX                  Counter32
    MAX-ACCESS              read-only
    STATUS                  current
    DESCRIPTION
       "The total number of times the server has been unable to
        obtain disk space that it wanted, since server startup. This
        would be inspected by an agent on receipt of an
        rdbmsOutOfSpace trap."
    ::= { rdbmsSrvInfoEntry 9 }
```

And by the way, why is the rdbmsOutOfSpace trap specified with the SMIv2 NOTIFICATION-TYPE macro? We can see that in the beginning of the MIB definition, where we read:

```
IMPORTS
     MODULE-IDENTITY, OBJECT-TYPE, NOTIFICATION-TYPE,
      Counter32, Gauge32, Integer32
          FROM SNMPv2-SMI
      DisplayString, DateAndTime, AutonomousType
          FROM SNMPv2-TC
      applIndex, applGroup
          FROM APPLICATION-MIB
      mib-2
          FROM RFC1213-MIB;
```

So we see that NOTIFICATION-TYPE is being imported from SNMPv2-SMI.

14.3.6 Remote Monitoring

Someone who is new to network management and SNMP might wonder if something is missing in the discussions so far. It might appear that we have laid out the foundations for management of devices (computers, routers, switches, servers, and so on) only, but what about management of networks (e.g., the LAN)? How can we obtain LAN statistics (e.g., packet counts)? The answer in SNMP is *remote monitoring* (RMON).

Two versions of RMON are available. RMONv1 is used to obtain packet-level statistics of a LAN or WAN. RMONv2 adds network- and application-level statistics. RMONv1 and RMONv2 are specified in RFCs 2819 [9] and 2021 [8], respectively. Why on earth would RMONv1 have a larger RFC number than RMONv2? It is because RMONv1 was originally specified in RFC 1271 (in 1991), and then RFC 1271 was made obsolete by RFC 1757 (in 1995); RFC 1757 was in turn made obsolete by RFC 2819 in 2000, but in the meantime, RMONv2 (RFC 2021) had come out in 1997. RFC 2021 did not need to be made obsolete by another RFC since it made use of SMIv2, whereas RFC 1757 was made obsolete by RFC 2819, so RMONv1 could be updated to use SMIv2.

Ten groups of objects are specified in RMONv1:

1. `rmon`: the overall RMONv1 group, encompassing the following 9.
2. `statistics`: Ethernet interface statistics.
3. `history`: historical data from the `statistics` group.
4. `alarm`: for specification of polling interval and threshold for any RMONv1 object of interest.
5. `hosts`: host-specific traffic statistics for each host on the network.
6. `hostTopN`.
7. `matrix`.
8. `filter`: for matching/filtering packets; matched packets may be captured; they may also cause an event to be generated; this group specifies the filters.
9. `capture`: to specify that certain packets should be captured if they match a filter in the filter group.
10. `event`: for definition of RMONv1 events.

Unlike normal MIBs, these objects in the RMON MIB do not correspond in a simple fashion to real resources within a machine (the usual case, as shown in Figure 14.3). Instead, the objects have special meanings, so an SNMP manager can specify what the remote monitor should be monitoring by *setting* the appropriate object values (which the remote monitor should take as commands from the SNMP manager, to monitor the appropriate quantities and collect the relevant statistics), and the manager can later read the values.

14.3.7 Other Issues

14.3.7.1 SNMP Proxies Most computing and communications devices, especially SNMP version 1, implement SNMP clients. However, a network administrator may sometimes find a device that does not implement SNMP. In such cases, an *SNMP proxy* can be used to allow the non-SNMP-speaking device(s) to communicate with an SNMP-based network manager. The SNMP proxy may obtain the information from the non-SNMP-speaking device(s) according to whatever means necessary, outside the use of SNMP.

14.3.8 Suggested Activities

Net-snmp is a popular command line snmp manager that runs on multiple platforms, including Windows. If you have time, feel free to work through some of the examples in the online tutorial on net-snmp at http://net-snmp.sourceforge.net/wiki/index.php/Tutorials and refer to the FAQ at http://net-snmp.sourceforge.net/docs/FAQ.html for additional information.

EXERCISES

14.1 The link between two routers in your network suddenly goes down. This results in various alarms being sent to the network management software and displayed in the network operations center. You try to figure out the underlying reason for the link failure. This activity falls under which part of the FCAPS model? How about the OAM&P (or OAMPT) model?

14.2 Suppose that your network uses SNMP for communications of network management–related information, or that at least the two routers in Exercise 14.1 are part of the portion of your network under SNMP-based network management. What SNMP message might have been sent when one or both of the routers noticed there was a problem with the link?

14.3 After completing your analysis of the link failure in Exercise 14.1, you discover that the link hardware is physically OK, but it went down because the configuration on one of the two routers had been corrupted/modified, perhaps maliciously by an outsider. So you now wish to restore the configuration and to install an intrusion detection scheme to detect future modifications of the configuration from outsiders. Which parts of the FCAPS are involved in this? Of OAM&P?

14.4 As part of the intrusion detection scheme in Exercise 14.3, you wish to monitor network traffic statistics. How might this be done using SNMP?

14.5 Refer to the extract from MIB-2 given in Section 14.3.3.2. What is the name and OID of the interfaces table? For a ppp interface, what is the ifType number?

REFERENCES

1. J. Case, M. Fedor, M. Schoffstall, and J. Davin. A simple network management protocol (SNMP). RFC 1157, May 1990.
2. D. Mauro and K. Schmidt. *Essential SNMP*. O'Reilly, Sebastopol, CA, 2005.
3. K. McCloghrie, D. Perkins, and J. Schoenwaelder. Structure of management information version 2 (SMIv2). RFC 2578, Apr. 1999.
4. K. McCloghrie and M. Rose. Management information base for network management of TCP/IP-based internets: MIB-II. RFC 1213, Mar. 1991.

5. K. Misra. *OSS for Telecom Networks*. Springer-Verlag, New York, 2010.
6. M. Rose and K. McCloghrie. Structure and identification of management information for TCP/IP-based internets. RFC 1155, May 1990.
7. M. Rose and K. McCloghrie. Concise MIB definitions. RFC 1212, Mar. 1991.
8. S. Waldbusser. Remote network monitoring management information base version 2 using smiv2. RFC 2021, Jan. 1997.
9. S. Waldbusser. Remote network monitoring management information base. RFC 2819, May 2000.

15

SECURITY

We begin this chapter with a discussion of basic concepts (Section 15.1), including abstract models for network security and typical types of attacks and defenses in the field. Then we briefly introduce cryptography (Section 15.2), since cryptography provides the building blocks for network security protocols, which we discuss in Section 15.3, with an emphasis on IPSec. We round off the chapter with an examination of wireless security (Section 15.4), focusing especially on cellular and WiFi security.

15.1 BASIC CONCEPTS

Security is a large field that includes system, network, and physical security. By *physical security* we mean securing the cables and the network and computing equipment to prevent an attacker from "walking right in" and using consoles in a secure area, tapping cables, and so on. We discuss physical security briefly in Section 16.3.3. When using *system security*, we could include everything, including network and physical security, but often it is used to refer to securing the operations of machines such as routers and servers. Thus, system security is about defending against malicious code that seeks to subvert the regular functioning of the machine (e.g., viruses, trojans, rootkits). System security is also about defending against unauthorized access to services provided by the machine (i.e., against password crackers). By *network security* or *communication security*, we mean securing communications over a network from such attacks as modification of the messages and eavesdropping. In this chapter we focus on network security—wireless communications security, in particular. Thus, we

Fundamentals of Wireless Communication Engineering Technologies, First Edition. K. Daniel Wong.

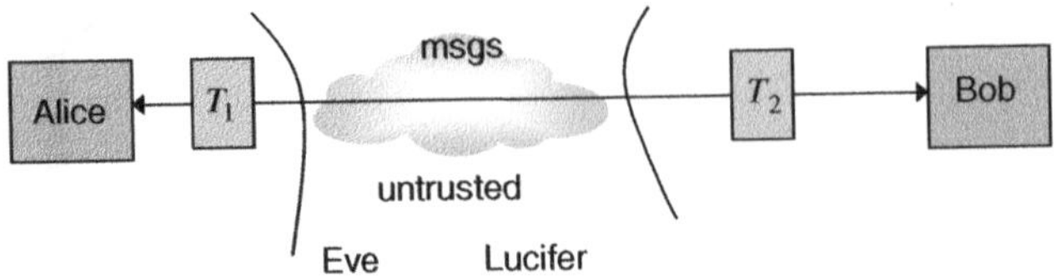

FIGURE 15.1 Abstract model of network security.

do not discuss physical or system security much here. Instead, we focus on attacks such as sniffing for traffic and passwords over the wireless link.

Consider an abstract model such as that shown in Figure 15.1. Two parties, Alice and Bob (call them A and B for short), wish to communicate, but the communications channel is insecure (e.g., something like the public Internet). Typically, to protect against various security attacks, A will apply a transformation, T_1, to the message before it goes over the insecure medium. B will apply a transformation, T_2, to the message received. (NB: We don't say that T_2 is necessarily the inverse of T_1 because it may not be so.) Before being transformed by T_1, the message is called *plain text*, and after being transformed, it is called *cipher text* or *encrypted text*.

If T_1 and T_2 are completely known by a third party, and the third party also has access to the insecure communications medium, the third party can read whatever messages pass from A to B on the channel. In fact, if the third party has complete knowledge of T_2, it can do whatever B does to receive the message. So, for the communications to be secure, do the transformations T_1 and T_2 have to be kept completely secret? In fact, the usual way that T_1 and T_2 are designed is as an algorithm and a key, as shown in Figure 15.2. Let's call the algorithms T_1' and T_2', and let's call the keys used with T_1' and T_2', k_1 and k_2. In the case that T_1' is some kind of encryption scheme meant to prevent eavesdropping by a third party, and T_2' is decryption, we might expect that the keys k_1 and k_2 should be the same. In fact, for hundreds of years of history, various schemes have been used, and always, $k_1 = k_2$. In the 1960s, however, somewhat surprisingly, a new class of encryption and decryption schemes emerged, where k_1 and k_2 are different. Both keys do not need to be different; usually, one of them can be made public, so only one key needs to be secret. These are *asymmetric* schemes, also known as *public key* schemes (discussed further in Section 15.2.2), whereas when $k_1 = k_2$ we have *symmetric* schemes, also known as *shared secret* schemes discussed further in Section 15.2.1).

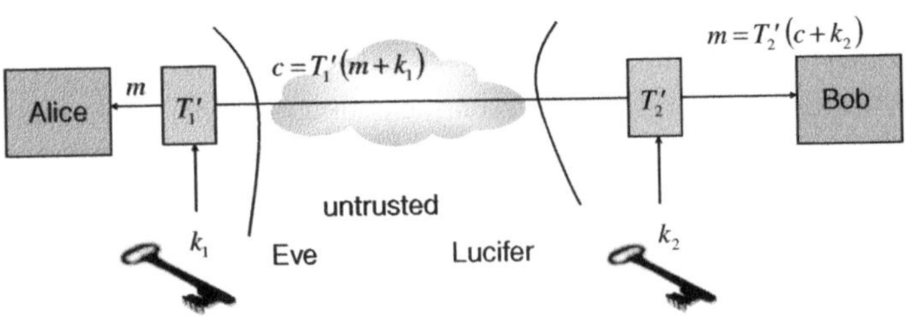

FIGURE 15.2 Abstract model with keys.

So far, we have been focused on creating a system that protects the *privacy* or *confidentiality* of communications between A and B. However, how does A know that B is who A thinks B is, and vice versa? How does each of them know that they are communicating with the intended party? That is where *authentication* comes in. Having the right key is an indirect way to authenticate oneself (a hacker who doesn't have the right key would not be able to apply the appropriate transformation to retrieve the original message). There are also direct ways to authenticate the other party that might involve a special exchange of messages (e.g., *challenge* and *response*). Sometimes, transformation and keys are involved, and our abstract model can cover authentication protocols as well.

We now discuss some of the main types of attacks and defenses. It should be noted that there are sometimes differences in meaning when different people use the same word. For example, some people use *authentication* to refer both to what we (and others) call authentication *as well as* to what we (and others) call *data integrity*.

15.1.1 Attacks

Some attacks could be used to facilitate other attacks. We could thus say that replay attacks, man-in-the-middle attacks, traffic analysis, and so on, are attack tools, or some terminology like that; however, it is sometimes not so clear cut (e.g., a denial-of-service attack might be an end in itself that a hacker is carrying out for "fun"), or it may be to facilitate some other attack (e.g., to make impersonation more successful if the device that the hacker is impersonating is so busy fending off a denial-of-service attack that it is unable to detect the impersonation and warn other devices). So we just call them all attacks, keeping in mind that some of the attacks might sometimes (or often) be used as part of some other attack.

An attacker is also sometimes called a *rogue* network element (e.g., a rogue base station, a rogue router). A human attacker may be called a hacker or a cracker, although the word *hacker* need not necessarily have negative connotations (there is the concept of an *ethical hacker*). Whereas the malicious hacker seeks to do bad things, and possibly break the law, the ethical hacker uses his or her knowledge of security attacks to help organizations defend their networks and systems. This may be done in creative ways; for example, the ethical hacker may try to break into a system for the purpose of discovering and demonstrating vulnerabilities (a process called *penetration testing*). Sometimes, the "good" hackers are called "white hat" hackers, and the "bad" ones are called "black hat" hackers.

A classic attack is eavesdropping. As in our discussion of the abstract model earlier in this chapter, eavesdropping occurs when an attacker tries to breach the confidentiality of the communications between two parties, say, A and B, by listening to the traffic between them in the insecure medium. Since the attacker receives only cipher text, the challenge for the attacker is to somehow be able to recover the corresponding plain text.

Another classic attack is impersonation. Impersonation occurs when an attacker pretends to be who he/she/it is not. For example, an attacker might claim to be A and send a message to B as A. There might be many reasons why an attacker might want to impersonate another entity. For example, it may be part of a larger scheme to eavesdrop on a particular conversation. One class of such schemes is the *man-in-the-middle* class of attacks. In these schemes, the attacker gets between A and B, often impersonating B to A, and impersonating A to B. Impersonation of an IP address is also known as *IP address spoofing*, impersonation of a MAC address is known as *MAC address spoofing*, and so on. However, impersonation is not limited to impersonation of IP, MAC, or other addresses, but can be of any concept of identity. A rogue base station can impersonate a legitimate base station, for example.

Modification of the message is yet another classic attack. Thus, instead of reading "Send help now!", a message might be modified to read "No help needed at the moment."

In denial of service, instead of focusing on individual messages (whether to eavesdrop, to modify the message, or to impersonate), the focus is on disrupting a service. The service could be many things: for example, email server service or access router service.

The attacker might wish to discover or track a person's or device's location, or to know that a particular person or device is currently in a call. This is a type of invasion of privacy and of anonymity.

Traffic analysis is similar or related in some way to tracking of location or use of a person or device, but instead of focusing on the identity of the person or device, it focuses on analyzing the traffic to/from the person or device. Thus, it may be used to detect trends and patterns in the traffic. It may be an end in itself (e.g., to discover that a person often visits certain web sites) or be used to aid in some other attack (e.g., to discover that, the person visits certain online banking web sites, which could be a first step in trying to impersonate the person to access these sites).

Repudiation includes two types of denials. The destination may deny that it has received a message or the source may deny that it has sent a message.

15.1.2 Defenses

We use the terms *protect* and *defend* interchangeably. The defense schemes that we introduce here are also described as security services.

To defend against eavesdropping, the sender needs to apply a transformation (encryption) from plain text to cipher text that only the receiver can untransform (decrypt). Then, even if a hacker is able to retrieve the cipher text that is traveling in the unsecured medium, the hacker is unable to discover the plain text of the original message. Sending messages in plain text, unprotected, over an unsecured medium, is also described as sending it *in the clear* and is considered bad practice, due to the risks of eavesdropping.

To defend against impersonation, the sender should be *authenticated*. We can conceive of different kinds or levels of authentication, depending on what we mean by sender. For example, a mobile device sending a message might not be authenticated to the other side. Sometimes, however, it is not the device but the human user that needs to be authenticated. Thus, the generic term is *authentication*, or *source authentication* to be more specific, but people can, and do, use more precise terms such as *user authentication*, *device authentication*, and so on, as necessary. Authentication services can be built using cryptographic algorithms such as message authentication codes (MACs) and cryptographic hashes (see Section 15.2.4). Authentication services are an integral part of digital signature schemes.

There are numerous variations to authentication. For example, instead of being concerned with the entire mobile device, the network may just want to authenticate that it is a particular subscriber (to mobile communications services) that is accessing the network, because it is the subscription that is associated with the service contract, billing, and payments. The human user can take his or her subscription to another mobile device. The network does not much care which mobile device a subscriber is using, but it does want to identify and authenticate subscribers when they access the network (no matter what mobile device they use to do so).

To defend against modification of the message, schemes can be adopted that provide two outcomes at a receiver that performs certain computations. One of the two outcomes provides a high degree of confidence in *data integrity* (i.e., that the message has not been modified). The other outcome, which is mutually exclusive with the first outcome, indicates that the message was modified. To ensure confidence in the integrity of the data received, such schemes are often called *data integrity schemes*. They are also often lumped together with authentication schemes, since many of the same schemes provide both authentication and data integrity services.

Defense against location and usage tracking is sometimes called *anonymity* (presumably, short for "preserving anonymity"). A good example can be found in the GSM system (Section 15.4.1). Defense against traffic analysis largely overlaps with defense against eavesdropping, and against location and usage tracking. Defense against repudiation, known as *nonrepudiation*, is one of the services provided by digital signatures.

15.2 CRYPTOGRAPHY

We make the distinction between cryptographic protocols and cryptographic algorithms. We use *cryptographic algorithms* to refer to the mathematical algorithms (based on modular exponentiation, etc.) that are the building blocks of cryptographic protocols. We use *cryptographic protocols* to refer to the communication/network protocols that make use of one or more cryptographic algorithms to achieve certain security objectives. We use *network security protocol* synonymously with *cryptographic protocol* and discuss these further in Section 15.3; we introduce cryptographic algorithms briefly in Section 15.2.4. Before we discuss cryptographic algorithms, we discuss the concepts of symmetric vs. asymmetric schemes, and key distribution.

15.2.1 Symmetric Schemes

Symmetric cryptographic schemes are also known as *shared secret* schemes or *private key* schemes, because both the sender and receiver need to possess the same key for encryption and decryption, and the key needs to be kept private (Figure 15.3).

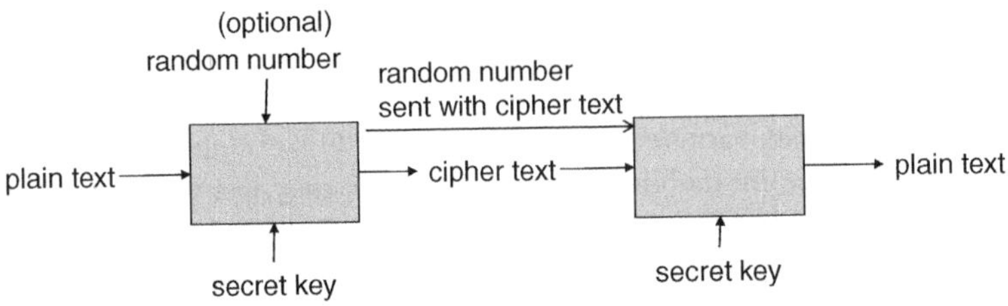

FIGURE 15.3 Basic use of private key cryptography.

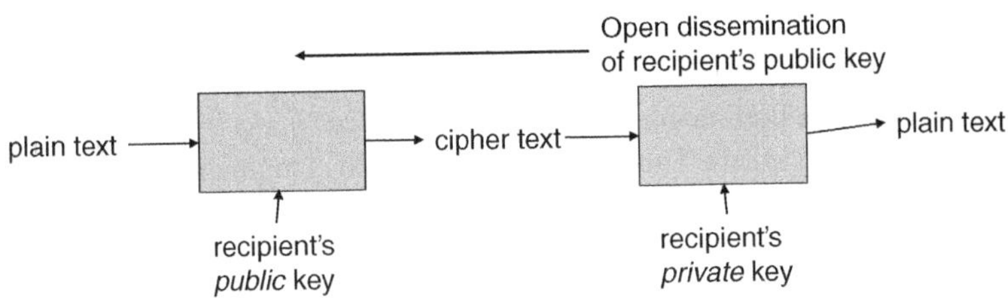

FIGURE 15.4 Basic use of public key cryptography.

15.2.2 Asymmetric Schemes

Asymmetric cryptographic schemes are also known as *public key* schemes because the sender and receiver use different keys. One key is the *public key* and the other is the *private key* (Figure 15.4). The most famous asymmetric scheme is probably the RSA scheme, named after its creators, Rivest, Shamir, and Adleman. Unlike the case with symmetric schemes, it is very tricky to come up with robust asymmetric schemes that can withstand the attempts of the cryptographic research community to break them or find serious weaknesses in them.

A disadvantage of asymmetric schemes is that they are more computationally intensive than symmetric schemes, for any given key size. At least, this is a disadvantage of *currently known* asymmetric schemes. Because of this disadvantage, a common arrangement is to use a symmetric scheme, where the initial few messages for negotiation of the shared secret are protected by asymmetric encryption. Thus, the more computationally intensive asymmetric scheme is used to aid in the *key distribution problem* for the symmetric scheme that is used for subsequent communications.

15.2.3 Key Distribution

Symmetric schemes need a way for the same secret key to be present at both the sender and receiver sides. This is trivial if, for example, the sender and receiver meet somewhere and agree on the key. However, in cases where only one side (whether sender or receiver) has the key, it would somehow need to communicate the key securely to the other side. This may be a challenging problem.

Those who are new to cryptography might be surprised to learn that even public key schemes have a key distribution problem. A helpful way to think about it is that for symmetric schemes, communication of the secret key requires confidentiality, authentication, and data integrity. For asymmetric schemes, communication of the public key does not require confidentiality (since it is a public key), but it *still* requires authentication and data integrity. We want to be sure that it is the correct public key for the other party, not a fake public key from an attacker pretending to be the other party.

15.2.4 Algorithms

Cryptographic algorithms fall into various categories, including:

- *Encryption*, also known as ciphering. These algorithms may be the first thing the newcomer thinks about when hearing the word *cryptography*. Examples of encryption algorithms include shared secret algorithms such as DES, 3DES, AES, IDEA, CAST, and Blowfish, and public key algorithms such as RSA and ElGamal.
- *Computing a message authentication code* (MAC). This is a crytographic checksum of the message.
- *Computing a cryptographic hash*. Whereas a MAC involves keys (shared keys for shared secret schemes or public and private key for private key schemes), a cryptographic hash is a type of hash function with good cryptographic properties. For example, since the hash function would produce output that is generally smaller (and often significantly smaller) than the message, we cannot avoid the existence of other messages that have the same hash function output; however, given a message and the cryptographic hash of that message, it must be very difficult to find another plain text message with the same cryptographic hash (whereas, for an arbitrary hash function, not necessarily a cryptographic hash, it might be relatively easy to find such a message).
- *Key generation*. Many security protocols use shared secret keys. Sometimes it is desirable to generate such shared keys "on-the-fly" as needed. A naive way to do so might be for A to generate a key and send it to B, but this involves sending the key over the air, which is a security risk. The *Diffie Hellman algorithm* is an example of a key generation algorithm that provides a way for both A and B to be able to compute the same secret key *without* sending the key over the insecure network. They only need to pre-agree on some global parameters. Based on the information that is actually sent over the insecure network, it would be computationally infeasible for an eavesdropper to generate the same secret key.

These algorithms are built on other cryptographic algorithms, or on more fundamental building blocks, sometimes called *cryptographic primitives* (e.g., modular exponentiation). Cryptographic primitives, and for the most part cryptographic algorithms (except some knowledge about how they are used in security protocols) are outside the scope of this book, so we do not discuss them further.

15.2.4.1 More on MACs and Cryptographic Hashes A *message authentication code* (MAC) is a cryptographic checksum of a message. It requires that the sender and receiver share a secret key. The MAC is typically shorter than the original message itself, so it can be sent along with the original message as acceptable overhead. Like ordinary checksums, the MAC allows the receiver to discover if the message has been changed. Specifically, if the MAC computed at the receiver does not match the MAC in the message sent, the receiver knows that it has been changed. However, unlike with ordinary checksums, an attacker needs the secret key in order to change the message. Otherwise, the attacker can change the message but cannot create a suitable matching MAC (that would require knowing the secret key). The receiver can compute the MAC of the message received. If it matches the attached MAC, it is highly probable that the message has not been modified (otherwise, we know that the message has been modified). Hence, data integrity is provided.

Only the sender and receiver have the secret key. Therefore, if the receiver computes the same MAC, it can be very confident that the message came from the sender (an attacker would not have been able to attach the correct MAC). Thus, this provides message authentication as well. MACs provide both data integrity and authentication services. An example of a MAC is the data authentication algorithm based on DES.

Cryptographic hash functions are similar to MACs (generally shorter than the original message) except that the sender and receiver do not need to share a secret key. The hash is a function of the message only. The hash function produces a hash of the message, which is also known as a *message digest*. Unlike regular hash functions, cryptographic hash functions are chosen such that it is relatively easy to compute the hash given a message, and it is very difficult to find two messages that result in the same hash. When a message is changed, the hash of the changed message will no longer match the hash attached to the message. Therefore, the change can be detected and data integrity is provided. However, because a secret key is not used in computing the hash, the hash portion of the transmitted message typically needs to be encrypted. Thus, when combined with encryption/decryption, the hash function can provide data integrity and message authentication. Examples of hash functions include MD4, MD5, and SHA-1. Is it possible to modify a cryptographic hash algorithm so that it takes a key and so that it becomes a MAC? Yes, such *keyed hash* functions are also called HMAC [4].

15.3 NETWORK SECURITY PROTOCOLS

Network security protocols are built using various cryptographic algorithms as building blocks. The network security protocols could be implemented at different layers of the protocol stack, as shown in Figure 15.5. For example, for secure web pages or secure email, transport layer protocols such as the *secure sockets layer* (SSL) and *transport layer security* (TLS) could be used to provide security services. In general, implementing security at higher layers has the advantage of making it easier to more selectively protect various specific traffic, whereas implementing security at lower layers (e.g., the IP layer, the link layer) has the advantage of providing broader, more

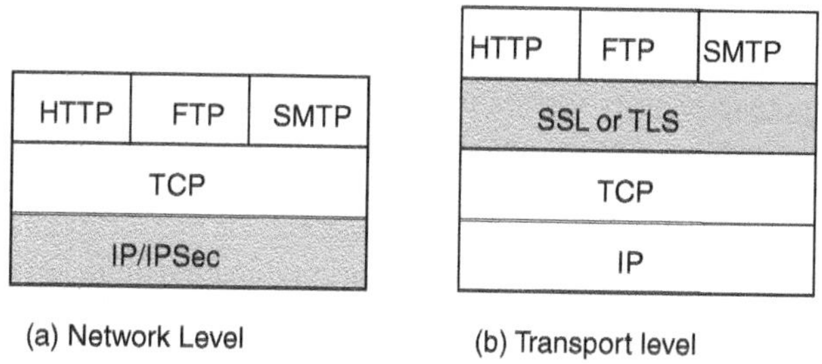

FIGURE 15.5 Security could be provided at different layers.

all-encompassing protection to all traffic passed to that layer from higher layers. In IP networks, a popular suite of network security protocols provides protection at the IP layer. This suite is known as IPSec.

15.3.1 IPSec

IPSec or *IP Security* is an IETF-defined suite of security protocols for IP networks, providing security services at the network layer. Since it is at the network layer, it can be used to protect all traffic between a particular pair of source and destination IP addresses rather than just protecting traffic specific to a particular application as in the case of application layer security, or just protecting traffic specific to a particular session (e.g., a particular SSL or TLS session) as in the case of transport or session layer security. IPsec could also be applied to a more narrow set of packets, and not just in a blanket manner for all traffic between a source and destination address pair (or between a set of source addresses and a set of destination addresses). IPSec was originally designed as part of IPv6, but it has been retrofitted to work with IPv4 as well.

IPSec is designed to be flexible to accommodate different cryptographic protocols and different accompanying parameters for the protected communications between two IPsec peers (we elaborate on the choices in Section 15.3.1.2). Furthermore, the protocols and accompanying parameters from A to B need not be the same as from B to A, and different combinations of protocols and parameters could be used for different (and multiple) IPsec peers. In fact, even for IP packets going from the same source to the same destination, different combinations of protocols and parameters could be applied based on source/destination ports, and so on. To facilitate this type of flexibility, IPsec is designed with the following features:

- There is a protocol to negotiate session parameters.
- There is a concept of *security association* (SA) between two IPsec peers, where an SA is a collection of parameters. Each machine that implements IPsec has a *security association database* (SAD) in which it stores the current SAs.
- Various groups of IP packets get mapped to particular SAs according to the entries in the *security policy database* (SPD).

Just as in the case of establishing a voice-over-IP session, where we need to negotiate session parameters and have SIP to do it for us (Section 11.2.2), in IPSec we need a protocol to negotiate security session parameters. *Internet key exchange* (IKE) plays this role. As the name suggests, among the parameters established by IKE are the keys that will be used for the various cryptographic functions when IPSec protection is active. We discuss IKE further in Section 15.3.1.1.

Every security association (SA) has a unique combination of the following parameters (i.e., you will not find two different SAs with exactly the same values for all three of the parameters):

- Security parameters index (SPI): an identifier with only local significance
- Security protocol identifier: *authentication header* (AH) or *encapsulating security payload* (ESP) (discussed in Section 15.3.1.2)
- IP address of the destination

Besides these three uniquely identifying parameters, each SA would also be associated with other parameters, including:

- Sequence number
- AH information: algorithm, keys, lifetimes of keys, etc.
- ESP information: algorithms, keys, lifetimes of keys, etc.
- Lifetime of the SA
- Mode: tunnel or transport (discussed in Section 15.3.1.3)

15.3.1.1 IKE With IPSec, key management can be manual or automated. If it is manual, the keys would be configured by a system administrator or configured by software not associated with IPSec. If automated, IPSec uses internet key exchange (IKE) to manage the keys. IKE consists of two parts, internet security association and key management protocol (ISAKMP) and Oakley.

Oakley is the part that performs the actual cryptographic algorithms to generate and exchange keys. It is based on the *Diffie Hellman algorithm*, introduced in Section 15.2.4, which lets the two sides generate shared private keys without sending the keys over the insecure network. However, the basic Diffie Hellman algorithm is susceptible to several attacks. Hence, Oakley was designed to improve on Diffie Hellman by authenticating the Diffie Hellman exchange to prevent man-in-the-middle attacks, for example.

Meanwhile, ISAKMP provides a larger framework for key management. With ISAKMP, the two sides can establish, negotiate, modify, and delete security associations.

15.3.1.2 IPSec Options IPSec is very flexible. It allows the use of multiple options for session parameter negotiations (part of ISAKMP). It also allows the use of multiple options for protection of user data traffic. As we have discussed, specific combinations of choices are grouped together into SAs in the security association database and matched with specific IP packets through entries in the security policy database.

Some of the major options are:

- The security protocol (i.e., AH or ESP)
- The mode (i.e., transport mode or tunnel mode)
- The cryptographic protocols to use (i.e., DES, AES, etc.)

We introduce AH vs. ESP here and defer discussion of the transport and tunnel modes to Section 15.3.1.4.

The authentication header (AH) is a set of fields that include control information (such as the security parameter index and a sequence number) and authentication data. The authentication data depend on the particular cryptographic protocols used (which would have been negotiated earlier and are part of the SA, so both sender and destination know which crytographic protocols are being used).

You may have noticed that in discussing the AH, we discussed how it provides authentication and data integrity services but did not say anything about how it might be encrypting the data to protect against eavesdropping. So, is the user data sent in plain text? Yes! We cannot rely on AH to provide confidentiality, but need to turn to ESP. Instead of using AH, the other major choice is encapsulating security payload (ESP), where both confidentiality *and* authentication are provided. Like AH, the ESP also adds a set of fields, but unlike AH, it also encrypts the payload. The ESP authentication data are computed *after* the encryption is performed. In Section 15.3.1.5 we will see which parts of the packet are encrypted and which parts are authenticated.

The location of the AH or ESP fields relative to other headers in the packet, and exactly what parts of the packet are protected, depend on:

- Whether it is for IPv4 or IPv6
- Whether it is in tunnel or transport mode.

Thus, we first need to discuss the tunnel and transport modes (which we do in Section 15.3.1.4), and then in Section 15.3.1.5 we revisit the issue of where the AH or ESP fields are placed and what parts of the packet are authenticated or encrypted.

15.3.1.3 IPsec VPN A *virtual private network* (VPN) is a cost-effective solution to the problem of wanting features of a private network but not wanting to pay the high costs for a truly private network. What do we mean by this? Imagine an organization with multiple locations. Suppose that it has offices throughout the world (Figure 15.6). The organization has secure local networks in all these locations and would like to connect the networks in a secure way. How can it do this? One way is by installing a private network: leased lines interconnecting all these locations. This is expensive because the connections are dedicated to the use of the organization and not shared with anyone else. Furthermore, when an organization has many locations and all need to be connected to one another, the number of dedicated lines between them grows

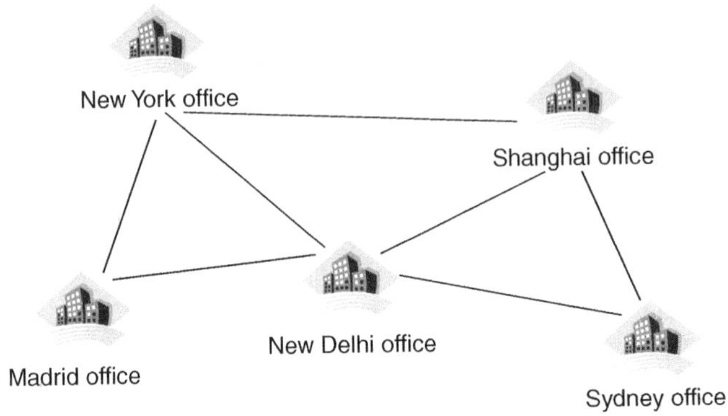

FIGURE 15.6 Dedicated connections between geographically distributed sites.

as the square of the number of locations. Connectivity between the locations can be much cheaper if a public network such as the Internet is used, but such a network is shared with other users and is not secure.

The standard solution is to deploy a VPN over a public network such as the Internet (Figure 15.7). This has the cost benefit of resource sharing with other users that the public network provides. However, instead of transmitting the organization's data across the public network in an unprotected fashion, the data are first protected and then sent across the public network. For example, it may be encrypted at a gateway on the edge of the Sydney network, and then decrypted upon arrival at a gateway on the edge of the New York network. Because of the encryption, some amount of privacy is bestowed on the communications between the two gateways, even though the data are traversing a public network. Since it is not actually a private network, it is called a virtual private network. Another common scenario of VPN use is for remote access

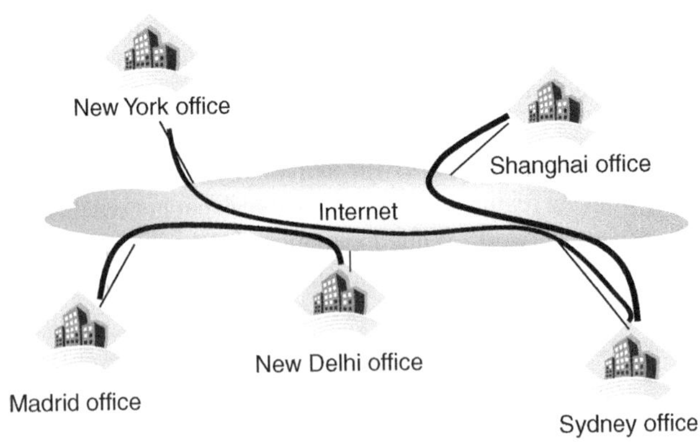

FIGURE 15.7 VPN connections between geographically distributed sites.

(e.g., workers who are telecommuting and want to connect securely to the company network).

There are many ways to create VPNs. Some of the differences have to do with where in the protocol stack the VPN is implemented (e.g., MPLS, SSL). Other differences have to do with the various possibilities for the separation of roles between the service provider and the customer. As far as we are concerned in this book, we focus on just one type of VPN, where IPSec is used to provide the virtual privacy. Such VPNs are called IPSec VPNs.

IPSec VPNs: Four Common Scenarios. In the case of connecting multiple organization locations, there is typically a gateway on the edge of each network, and IPSec is applied at these gateways. Thus, an IP packet might travel unprotected from source to the gateway at the source network, then be encrypted and travel that way across the unsecured public network to the gateway of the destination network, and then travel unprotected from that gateway to the destination. This is a very popular way of using IPSec since organizations that have multiple networks often want to protect their traffic as it traverses an insecure public network such as the Internet. This scenario is illustrated at the top of Figure 15.8. A and B are the end users, and the IPSec protection is applied between gateway routers/firewalls C and D. As for the meaning of "tunnel" in "IPSec tunnel," we will come to that shortly, in Section 15.3.1.4.

A second common scenario is the teleworker scenario, where a person works off-site (e.g., from home) and wants to connect to the corporate network. IPSec is used to protect the traffic over the insecure public network between the teleworker and the gateway router/firewall at the corporate network. Shown in Figure 15.8 at the bottom, this scenario is often called *remote access*. Teleworker A connects to machine B within the corporate network, and IPSec protection is between A and the gateway router/firewall.

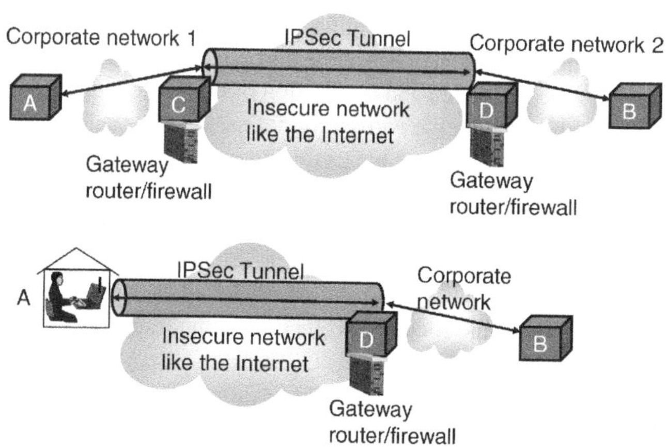

FIGURE 15.8 IPsec VPN scenarios using the tunnel mode.

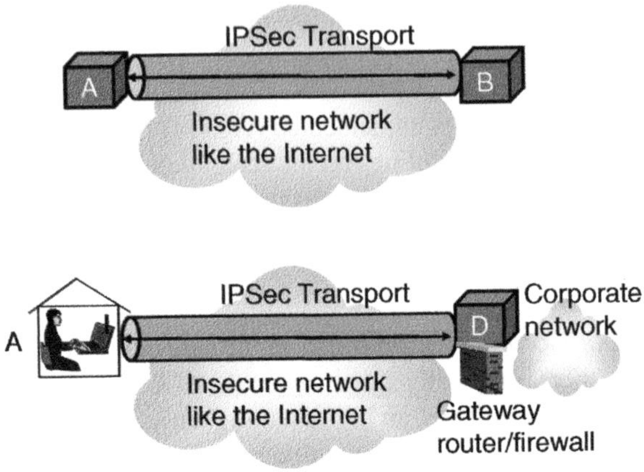

FIGURE 15.9 IPsec VPN scenarios using the transport mode.

A third common scenario is where two hosts (or routers, or any other machines on an IP network) want to protect traffic between themselves, as shown in Figure 15.9 at the top, for communications between A and B. As for the meaning of "transport" in "IPSec transport", we will come to that shortly, in Section 15.3.1.4.

A fourth common scenario is similar to the teleworker/remote access scenario, except that in this case the teleworker might be an IT support staff member whose destination *is* the gateway router/firewall rather than some other machine within the corporate network. This scenario is shown in the lower part of Figure 15.9, for communications between teleworker A and the gateway router/firewall.

15.3.1.4 Transport and Tunnel Modes There are two modes of usage of IPSec:

- *Tunnel mode*. In this mode, the original IP packet is encapsulated in another IP packet with a new IP header. The source and destination addresses of the new IP header are that of the source and destination of the tunnel, respectively. This has the advantage of allowing the *entire* original packet to be encrypted, not just the payload (encrypted headers are not usable by intermediate routers, but when the tunnel mode is used, the intermediate routers only need to work with the new, outer IP header).
- *Transport mode*. In this mode, no tunnel is created. The primary protection is for the higher layers, unlike in the tunnel mode, where the original IP header is also protected. Thus, the focus of protection in the transport mode is on the IP packet *payload*.

The tunnel mode has the advantage of hiding the original IP headers (since it can encrypt the entire original packet, including the original headers), but at the expense of more header overhead. Therefore, it is normally used where the advantage of

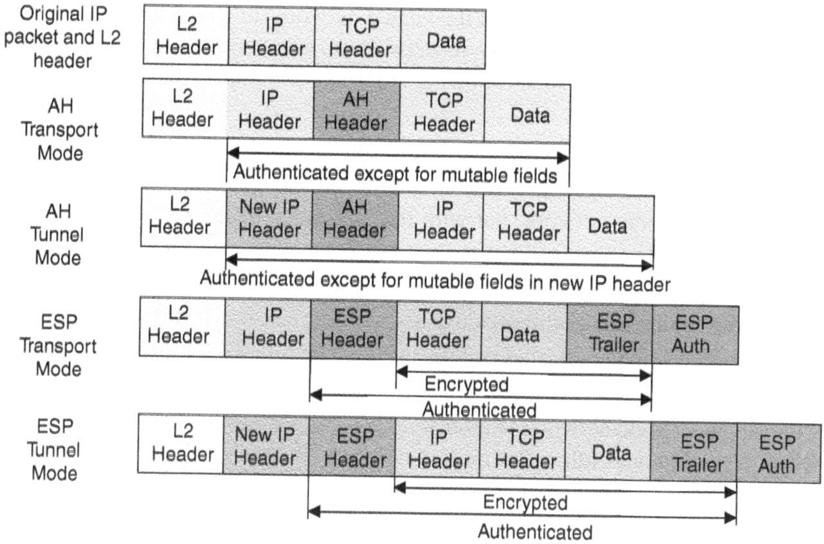

FIGURE 15.10 IPsec modes.

hiding the original IP headers makes sense. For example, the tunnel mode is often used for the first two of the scenarios discussed in Section 15.3.1.3, as shown in Figure 15.8. However, the transport mode is often used for the last two scenarios discussed in Section 15.3.1.3, as shown in Figure 15.8. In these two scenarios, it doesn't make sense to hide the original IP headers, since the source and destination addresses would be visible in the new IP header anyway! In contrast, in scenarios with gateways, only the gateway IP addresses need to be exposed, and the benefits of the tunnel mode become apparent.

15.3.1.5 Location of AH and ESP Headers Having been introduced to AH and ESP and the tunnel and transport modes, we now see four combinations of using AH and ESP with the tunnel and transport modes. Figure 15.10 shows the different scopes of authentication and encryption for different modes of use of IPSec. Mutable fields are such fields as "time-to-live," which will change from hop to hop during normal routing operations, so should not be included in computation of the MAC for AH. These fields are set to zero for purposes of MAC computation. In the case of ESP, the ESP header must not be encrypted because it contains parameters such as the security parameters index (SPI) needed for decrypting the payload. The "ESP Auth" field at the end contains authentication data that are computed after encryption is performed on the rest of the packet, so it must not be encrypted either.

15.3.2 Access Control and AAA

A mobile device roams into a foreign network. We have seen how this roaming is handled in GSM. Part of the procedure for getting service is that the mobile device needs to get authenticated. Note that the visited network (in which the mobile device is roaming) will only allow access to its network, to selected foreign mobile devices. They will be allowed only if the following are true:

- There is a roaming agreement with the home network of the mobile device.
- The mobile device gets properly authenticated. The authentication can be done only by its home network, but with the assistance of the network visited (see Section 15.4.1).
- The service subscription of the mobile allows the roaming service to be used (i.e., the device/user is authorized to use roaming services at the network visited).
- Usage data can be collected and the user can be billed appropriately.

Now, in an IP-style network, is all this necessary? Or can we just assume that any network visited should allow the mobile device to access it? In practice, a mobile device may move into areas that are part of a different network administrative domain than its home network administrative domain. Usually, the mobile node would not be able to connect freely to the network visited, not to mention performing mobile IP registration or doing anything else on the network. One of the creators of mobile IP, Charles Perkins, explains [5] that the designers of mobile IP initially assumed that connectivity would be provided as a courtesy service to visitors, in the same way that free electricity is provided to visitors at any organization to charge their laptops, for example. However, this assumption is increasingly invalid, as the Internet and IP networking matures. Connectivity privileges can be thought of as more like library borrowing privileges than electricity privileges. These are not given casually to visitors. There are good reasons for this. Like library books, network resources are a valuable commodity. It is unlikely that recharging a laptop would be a significant drain on the electricity supply to the organization, whereas a high-data-rate mobile device could consume a significant fraction of the bandwidth available. Therefore, we conclude that even in an IP-style network, something like the list of requirements presented above would also need to satisfied. How? Through ad hoc schemes or through a more unified framework such as AAA.

We discuss AAA first, and then briefly mention some ad hoc alternatives.

15.3.2.1 AAA As discussed earlier (Section 8.3), a cellular system such as GSM is specified more completely than an IP-style system for wireless access to an IP-based network. For example, GSM specifies not just the wireless physical and link layer protocols, but also various network- and higher-layer protocols, including protocols for dealing with mobility, network security, authorization for higher-layer services, and a framework with handling usage details. In an IP-style system, on the other hand, IEEE 802.11 or 802.16 may be used for the physical and link layers, and then other functions are provided by other IP-style protocols. For example, mobile IP may be used to handle mobility, whereas DHCP may be used for autoconfiguration, IPSec for various security services, and so on. Where it comes to functions related to network-wide authentication and authorization, and accounting (proper recording of usage of network resources and services), the IP-style solutions are grouped together under the concept of *authentication, authorization, and accounting* (AAA).

AAA refers to an IETF-defined framework for managing network resources, controlling access to the resources and accounting for their usage, thus helping to prevent and/or detect unauthorized usage (and also enabling proper billing for resource usage). While authentication is about user identity (namely, who is this user?), authorization is about what the user is allowed to do, what resources the user can access. Accounting is about keeping systematic records of resource usage. This allows proper billing, as well as providing an audit trail and data that can be used for fraud detection, among other things. Protocols for AAA include RADIUS [6] and DIAMETER [2], as well as proprietary protocols such as TACACS+.

The access router may allow limited access only to an AAA server and not to other network services. After the exchanges with the AAA server, the access router then "opens up" access to the relevant set of network services as appropriate. AAA servers in the foreign network need to communicate with AAA servers in the home network. This is because the foreign network AAA server probably does not have information about the MH, while the home network server should have such information. A typical arrangement might be that the operators of the two networks have agreed to allow subscribers from each others' networks to access an agreed set of services. This set of services may be small (e.g., just basic IP connectivity with best effort service), or it may include other services such as preferential queuing services in routers. In any case, the AAA server in the foreign network communicates with the AAA server in the home network so the AAA server in the home network can authenticate that the MH is indeed one of the home network's subscribers. Furthermore, since subscribers may not all have the same authorization for services, the server can authorize the MH for the appropriate set of services. Lastly, accounting can be performed so that network usage can be monitored, the subscriber billed appropriately, and so on.

15.3.2.2 Ad Hoc Schemes There are cases where a more complete framework based on AAA might not be necessary. For example, a business such as a cafe or restaurant may choose to provide access to their network (and Internet access from their network) to customers. To make sure that only customers have access to the network, they provide a WEP (see Section 15.4.2) password only to customers, perhaps written on a card that they give out to customers. Noncustomers, even those within range of their AP, would not be able to access their network (of course, we disregard the well-known weakness of WEP security at this time; we are not saying this is a good or foolproof way to authenticate the customers and authorize them to access the network, but it is sufficient for many situations). In this case, the authentication and authorization are simply by means of the WEP password.

Access control protocols can be applied at the link layer and also at the network layer. An example of a link layer access control scheme is the use of WEP, where the user has to know the WEP key to even be able to establish a wireless link. Another example is a WLAN authorization scheme that only allows establishment of link layer connectivity with an AP if the MH has a MAC address that is in a database. With these schemes, link layer access is denied to unauthorized network outsiders, so an

MH would be unable to send any IP packets, not to mention mobile IP registration messages.

An example of a network layer access control scheme, on the other hand, might be to allow wireless links to be established, but to place an access router behind the wireless link, on the network side, that controls further access to network resources.

15.4 WIRELESS SECURITY

Security is a difficult enough problem with wireline communications, where there are already many challenges to handle. When we consider wireless communications, security becomes even more challenging. The wireless environment is a broadcast medium at the physical layer, unlike the wired medium, where the signals are constrained to flow within wires. First, even with directional antennas, the wireless signal can still be received over a relatively large area. Second, except for fixed wireless systems, mobility introduces additional challenges since users are moving around, connecting and reconnecting to the network regularly. Thus, they must be explicitly authenticated regularly. In contrast, for a traditional wired phone system, if a signal is received from a certain physical line, it is assumed to belong to the particular phone line, and corresponding subscription, without requiring explicit authentication.

While this makes authentication more of a challenge, it also introduces the need for additional security services, such as *anonymity*. A subscriber does not usually want it to be known that he or she is using their mobile device. The broadcast nature of the wireless medium makes it difficult to hide the subscriber's identity (IMSI), especially before ciphering is turned on.

15.4.1 Cellular Systems

Here, we use GSM as an example to show how authentication, confidentiality, and anonymity are typically protected in cellular systems. Other cellular systems use similar methods.

15.4.1.1 GSM Authentication At the heart of GSM authentication is a simple challenge/response scheme of the symmetric (shared secret) variety (Figure 15.11). In the authentication procedure, the network (we will get more specific about what elements in the network are involved, in the next paragraph) challenges the mobile device (more specifically, the SIM within the device) with a random number and the device responds with a *signed response* (SRES). The SRES is a cryptographic function of the random number and a secret key. The cryptographic function is called A3, and the secret key is only supposed to be known to the SIM and the authentication center in the subscriber's home network. Thus, if the mobile can compute the same SRES as the network and return it correctly to the network, the mobile gets authenticated.

The challenge/response scheme involving SRES needs certain GSM-specific requirements, thus resulting in it being implemented in the particular way it is implemented (Figure 15.12). The requirements include:

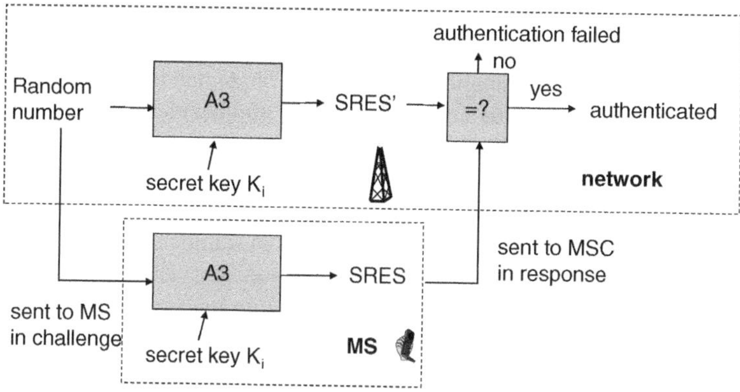

FIGURE 15.11 GSM security, simplified.

- The secret key must not get out of the authentication center (AuC) in the subscriber's home network. Thus, SRESs must be computed in that AuC, even in the case that the subscriber is roaming.
- When the subscriber is roaming, the serving MSC or SGSN is responsible for handling the authentication (sending the challenge, and checking the SRES from the mobile), for circuit-switched and GPRS services, respectively.
- Unnecessary delays, such as for sending the random number and SRES from the AuC in the home network to the network visited when the mobile is roaming, should be avoided.

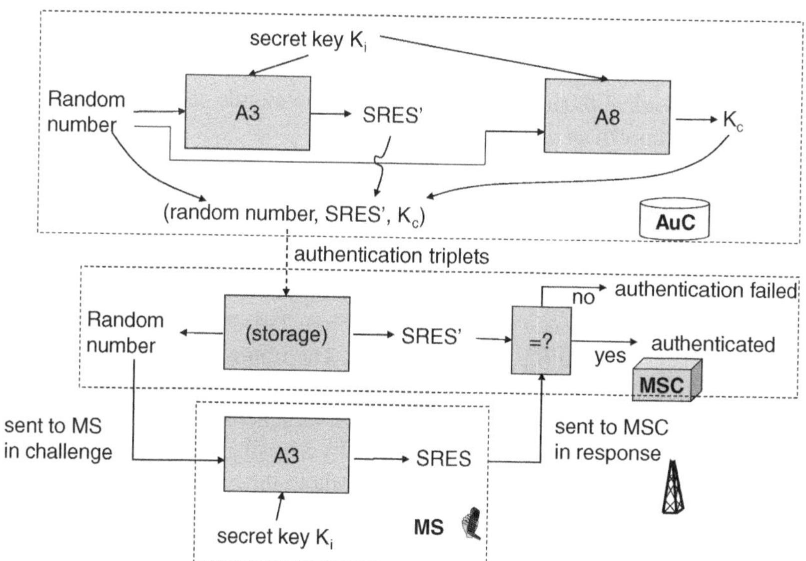

FIGURE 15.12 GSM security.

The solution that meets these requirements that was implemented in GSM is that the computations of SRESs are done in the home network AuC, as required, but to avoid unnecessary delays, the computations are not done on demand. Instead, the multiple SRES computations (each with a different random number) are precomputed. They are then sent, with the corresponding random number, to the serving MSC or SGSN in the network visited, to be used when needed. Precomputation is possible because A3 is a function of only the secret key and the random number, not of any time variable. In fact, multiple SRESs (with corresponding random numbers) are sent in batches, and they do not have to be used in any particular sequence. The only requirement is that each random number used should be matched with its corresponding SRES.

The random number and the SRES are always sent along with another piece of information, K_C. Together, they make up an *authentication triplet*. K_C is the ciphering key used for ciphering of user data over the air. It is computed by another algorithm, A8, which takes as its inputs the same random number and secret key that are input to A3 to compute the SRES. Hence, a given random number corresponds to a particular SRES and particular K_C. Therefore, together they are called an authentication triplet. Often, three to five authentication triplets would be sent at a time [3] from the home network to the VLR, and these can be used one after another, without needing to obtain a new triplet from the home network every time that one is needed. Since the authentication triplets can be used in any order (no particular sequence is required), the VLR can randomly choose from among the triplets it possesses for a mobile station, which triplet to use next.

Notice that the challenge is a random number that is sent over the air to the mobile in plain text, thus raising the question of replay attacks. However, since it is a different random number each time, a rogue base station cannot simply replay an earlier random number. But a rogue base station can simply send *any* random number, and the mobile will compute the SRES and reply! How does the mobile know if the base station sending a challenge is legitimate or rogue? Indeed, this is a serious flaw in the GSM design. The network authenticates the mobile but the mobile cannot authenticate the network. This is known as *one-way authentication*.

15.4.1.2 GSM Confidentiality Like SRES, K_C can only be computed by the two entities that possess the secret key: the AuC and the mobile. However, it can only be used after authentication, when the mobile knows which random number to use to compute K_C. Thus, ciphering is only turned on after authentication is completed.

The ciphering in GSM is a secret key algorithm. Two A5 algorithms are used to generate new keys, S1 and S2, for every frame. The frame number and K_C are used as the inputs to A5. Therefore, S1 and S2 will change from frame to frame, and also, only the mobile and the network should be able to generate S1 and S2. The traffic from the base station to the mobile is encrypted by S1 using exclusive OR (XOR). S2 is used in the same way for traffic from the mobile to the base station.

15.4.1.3 GSM Anonymity Because ciphering is only turned on after authentication has been completed, some important signaling messages are sent unencrypted (before authentication is completed). Information such as the IMSI would be included in such messages. Thus, a subscriber's location can be discovered by others by listening to such messages. What can be done to protect subscriber anonymity? In GSM, the solution is the use of the *temporary mobile subscriber identity* (TMSI). Just like a person might use an alias or pseudonym, or an author might use a pen name to hide their real name, the TMSI is used to hide the IMSI.

Thus, most of the time, only the TMSI is heard over the air. Even when a mobile moves, the signaling procedures are designed to help pass the IMSI internally rather than over the air. For example, in the location area update procedure discussed in Section 11.1.4.1, the MAP_SEND_IDENTIFICATION request from the new VLR to the previous VLR, and the response to the request, allows transfer of the IMSI inside the network between VLRs, without unnecessarily exposing the IMSI over the air.

15.4.1.4 GSM Security Summary We summarize below some of the inherent weaknesses of GSM authentication:

- Authentication is one-way. The network authenticates the mobile, but not the other way around. Rogue base stations are possible.
- Ciphering is only over the air. If the link between BTS and BSC is also wireless, it may be sent in the clear.
- Data integrity is not protected.

Additionally, some other weaknesses in implementations of GSM are:

- Ciphering may not be turned on.
- A weak ciphering algorithm may be used.

15.4.1.5 UMTS UMTS security is also called *authentication and key agreement* (AKA). As with GSM, authentication is based on symmetric cryptography, where the secret key is stored only in the USIM and in the AuC (as part of the HSS). Anonymity is protected through the use of the TMSI as with GSM.

UMTS security improves on GSM security in some ways, especially addressing some of the weaknesses that we have pointed out. As part of the improvement, it replaces the authentication triplets of GSM with authentication quintuplets (five values instead of three). Specifically, it adds:

- An authentication token (AUTN) to allow the mobile to authenticate the base station, thus supporting two-way authentication
- An integrity key to provide integrity protection (mostly, for signaling traffic)

Unlike GSM, where the triplets can be used in any order, the UMTS quintuplets have to be used in sequence. In some books and web pages, the UMTS quintuplets are called quintets, but according to the 3GPP standards, they should be called quintuplets.

Ciphering is stronger in UMTS, with 128-bit keys instead of 64-bits keys as with GSM. Moreover, ciphering is between the mobile and the RNC, thus eliminating the potential unencrypted communications over the air between BTS and BSC that are present in GSM.

15.4.2 802.11 WLAN

When 802.11 was first created, it was recognized that the wireless environment was inherently less secure that the wired LAN environments people were used to. So it was decided to incorporate some security measures into 802.11 so that the new wireless LAN would be brought up to a level comparable to wired LAN, especially with regard to privacy. Thus, wired equivalent privacy (WEP) was born.

WEP uses a 64-bit key, the first 16 bits of which are known as the initialization vector (IV). Meanwhile, an integrity check vector (ICV) is computed from the message and appended to the message. The combination of message and ICV are then operated on by the key through a simple exclusive-OR (XOR) process. The result is the ciphertext, and it is combined with the IV to be sent over the air. This process is illustrated in Figure 15.13.

Much has been published on the weaknesses of WEP, and ways to break it (e.g., through certain brute-force methods). Software is also freely available that can do the job for anyone, within hours or even minutes, depending on factors such as the hardware on which the software is run. Details are outside the scope of this book, but we note some of the problems and weaknesses of WEP:

- The 64-bit key is very short by today's standards.
- The short key was made even weaker by a common practice of using a fixed IV for the IV part of the key, thus effectively leaving only 40 bits for the key. This could be said to be a problem with the 802.11 specification not requiring a way to use dynamic IVs.
- No key management and distribution method is specified, so the key is often entered manually and used for long periods of time without being changed.

Typically, one access point (AP) would be configured with a particular key, and all mobile devices accessing the network through that AP would then all share the same key for that purpose. Clearly, this exposed the key to more risk of falling into the hands of hackers, than with GSM authentication, for example, in which only

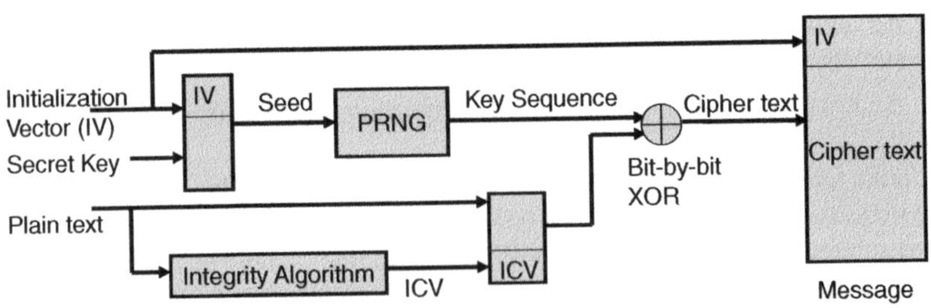

FIGURE 15.13 Wired equivalent privacy.

the SIM and AuC have the key and care is taken never to share it with any other entity. This would be bad enough if the network permits access to only a fixed set of mobile devices. However, in many application scenarios, it is desirable to allow foreign mobile devices (not previously known to the network) to access the network. WEP by itself is not designed to handle this. We will soon describe a solution, IEEE 802.1X, that is included with more recent versions of WiFi security.

IEEE 802.11 authentication can be either *open system authentication* (which means that no authentication is done, and the system/network is open to all users), or *shared key authentication.* The shared key authentication protocol has several weaknesses. Since it uses WEP, it shares the weaknesses of WEP with 802.11 encryption. Like GSM authentication, it is also a one-way authentication protocol, also making it susceptible to rogue base stations.

While the IEEE and others worked on solutions to replace WEP with something more secure, various ad hoc solutions emerged, including the following:

- *Hidden SSID.* APs need not broadcast their SSID, so mobiles need to know the SSID to receive a response from the AP when scanning for an AP. This makes the SSID a form of password, providing weak password protection.
- *MAC address filtering.* APs can be programmed to only allow traffic to/from devices whose MAC address is included in an access control list.
- *Browser hijack.* This class of solution focuses on controlling access by allowing any mobile to access the AP but then redirecting any http request to an access controller. The name *browser hijack* comes from how the web browser is taken to the access controller rather than where the user wanted to browse; thus, the browser is "hijacked."

As shown in Figure 15.14, with browser hijack, the access controller can request user credentials, including a password, to decide whether to offer access. Then there are two possibilities:

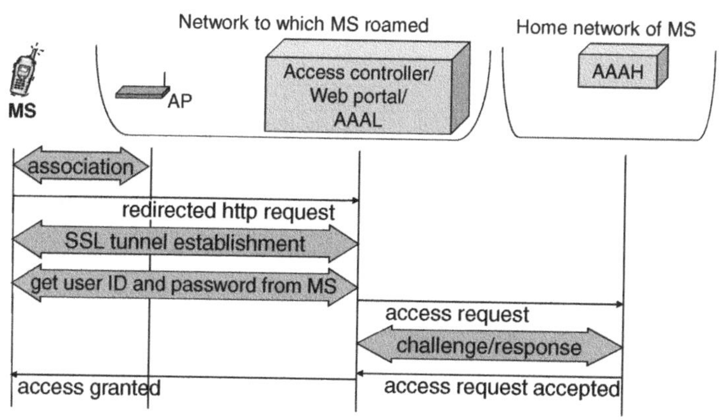

FIGURE 15.14 Browser hijack.

- The access controller might make its own decision as to whether or not to grant access, based on the password received (and without needing the back-end portion shown on the right of the figure). This used to be popular in places such as hotels, where guests could pay for a password to use the network for a certain period of time, and it can still be found in such places.
- The access controller might be colocated with an "AAA local" (AAAL) server, which then defers to a remote "AAA home" (AAAH) server for the authentication decision, using a protocol such as RADIUS to communicate with AAAH. This model might apply in cases where the mobile already has a subscription with a WLAN operator, and his or her operator has a business arrangement with the roaming network to allow its subscribers to use the roaming network.

These are all stop-gap solutions that do not offer strong security. Meanwhile, as IEEE was working on 802.11i to address the security concerns of the original 802.11, the Wi-Fi Alliance came up with *Wi-Fi protected access* (WPA), as an interim solution that included parts of the then work-in-progress draft for 802.11i. Later, WPA2 was created, which included the full suite of enhancements in 802.11i. We discuss WPA in Section 15.4.2.1 and WPA2 and 802.11i in Section 15.4.2.2.

15.4.2.1 WPA WPA saw the introduction of *temporal key integrity protocol* (TKIP). TKIP is not as secure as *advanced encryption standard* (AES) (to be introduced with WPA2 and IEEE 802.11i), but is not as computationally intensive as AES. Hence, it is a more backwardly compatible solution than AES, as it can be run on some older WiFi hardware that cannot handle AES.

With TKIP, the encryption process is as shown in Figure 15.15. As shown in the figure, there is a WEP subsystem, albeit with some add-ons. First, WPA introduces a *pairwise master key* that is 256 bits long (much better than the 40 bits of plain WEP). Furthermore, the master key is not used in the encryption directly, but is used to derive *pairwise transient keys*, each 128 bits long. These transient keys are then mixed (phase 1 and phase 2 key mixing as shown in Figure 15.15), so every packet uses a different

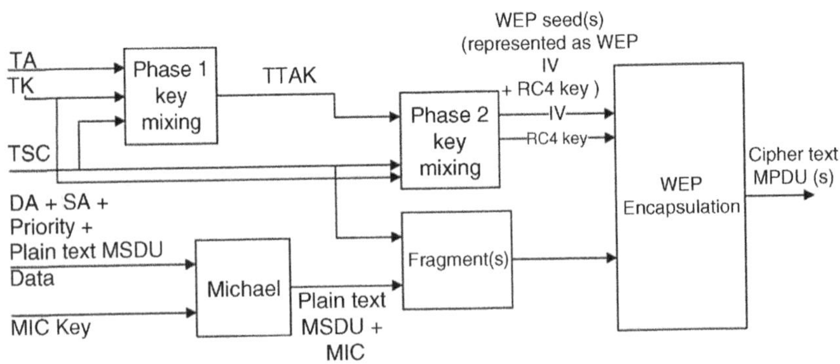

FIGURE 15.15 Encryption with TKIP. (From IEEE 802.11-2007 [7]; copyright © 2007 by IEEE, reprinted with permission.)

104-bit key for RC4 in the WEP subblock. In the figure, from the IEEE 802.11-2007 standard (that incorporates 802.11i), the following acronyms are used, which we now define: SA and DA stand for source address and destination address, TA for transmitter address. TK is a transient key, TSC is a TKIP sequence counter, and TTAK is a TKIP-mixed transmit address and key. The addition of a *message integrity code* (MIC) called *Michael* can also be seen in the diagram. This provides data integrity without a great increase in complexity, since Michael is designed especially for low computational complexity.

Where does the pairwise master key come from? Here, WPA allows for two different possibilities, based on the two main usage scenarios of wireless LANs. For *home users*, the pairwise master key can be a preshared key, just like the WEP key was before. This trades off security for convenience (home users can just enter the same key in their wireless AP and all devices they wish to use on their home wireless network). For *enterprise users*, a more secure authentication and key distribution scheme may be required. WPA uses IEEE 802.1X for this purpose.

802.1X is based on a three-party model borrowed from the IETF's "EAP over LAN" (EAPOL) model.

- The *supplicant* is the entity trying to gain access to the network (i.e., it is the mobile device in the case of 802.1X).
- The *authenticator* is the entity that the supplicant contacts to gain access to the network. The authenticator is in a position to prevent traffic from the supplicant from entering the network until access is authorized.
- The *authentication server* is the other end of the authentication protocol (at least, it is the one that authenticates the mobile; in some cases, there is mutual authentication, so the mobile also authenticates the authentication server).

Extensible Authentication Protocol. The *extensible authentication protocol* (EAP) is a framework for performing authentication in a situation where the three-party model with supplicant, authenticator, and authentication server applies. EAP itself does not specify how to perform authentication, but allows for different *EAP methods* to be used for the actual authentication. This is in accord with the "extensible" in EAP's name—it is extensible to allow for many authentication methods. But why bother with a protocol such as EAP and why not just specify the EAP methods separately? EAP is useful as there are certain common elements that are not specific to authentication methods, but which are needed: for example, negotiation between the supplicant and the authentication server, as to the authentication method to use. For the wireless context, relevant requirements for EAP methods are specified in RFC 4017 [8]. Commonly, RADIUS is used for the authentication server, in which case RFC 3579 [1] applies. In that case, the supplicant and authenticator communicate using EAP messages, the authenticator and authentication server communicate using RADIUS messages, and a logical conversation takes place directly between supplicant and authentication server using the particular EAP method that is agreed upon in negotiations between the authentication server and supplicant.

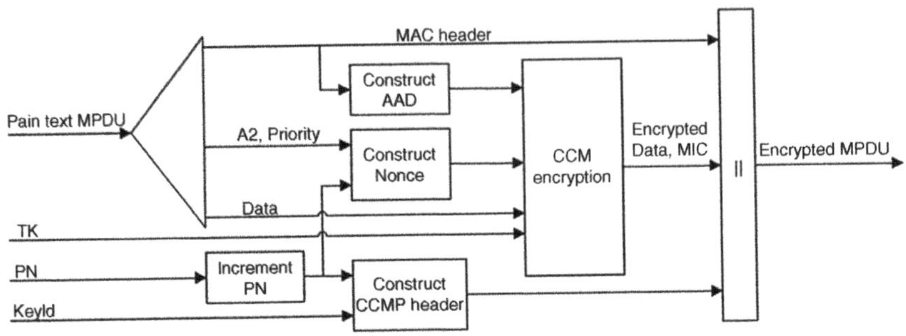

FIGURE 15.16 Encryption with AES/CCM. (From IEEE 802.11-2007 [7]; copyright © 2007 by IEEE, reprinted with permission.)

EAP methods include:

- *EAP-TLS*: TLS stands for transport layer security.
- *EAP-TTLS*: TTLS stands for tunneled TLS.
- *PEAP-TLS*: PEAP stands for protected EAP.
- *EAP-SIM*: EAP using GSM authentication.
- *EAP-AKA*: EAP using UMTS authentication.

The Wi-Fi Alliance specifies a list of EAP methods to be used in WiFi products (see Section 17.2.6.1).

15.4.2.2 WPA2 and IEEE 802.11i The main upgrade in going from WPA to WPA2/802.11i is the change from TKIP to *advanced encryption standard* (AES). AES is used in the *counter with CBC-MAC protocol* (CCMP), where CBC-MAC stands for the cipher-block chaining message authentication code. Use of AES/CCM is shown in Figure 15.16, where AAD is "additional authentication data," TK is "transient key," and PN is "packet number."

15.4.3 Mobile IP Security

Mobile IP supports authentication. A critical question is: How would a HA know if the registration message it receives from a foreign network is really from one of the mobile nodes it serves? It is very important that mobile IP registration messages can be authenticated; otherwise, any third party can send a fake (but valid) registration message to a HA that will result in traffic for one of its mobile nodes being tunneled (by the HA, as an unwitting accomplice to the third party) to any arbitrary care-of address in the world. Conversely, it is also very important that the registration reply messages from the HA be authenticated. Otherwise, a third party (attacker) could intercept and remove the registration message, so the HA does not receive it. Then this attacker could proceed to send a fake (but valid) registration reply purportedly from the HA, claiming that the HA has updated its binding for the MH to the latest COA. Without authentication (of the HA when sending registration replies), the mobile node cannot tell if this is happening. Furthermore, whatever authentication scheme is used must be protected against possible replay attacks.

Therefore, it is mandatory in mobile IP that both registration messages and registration replies be authenticated. A *mobile-home authentication extension* is defined for the two messages (various extensions can be appended to mobile IP messages, some of which are mandatory, like the mobile-home authentication extension, and others of which are optional). The mobile-home authentication extension contains a 4-byte security parameter index (SPI). Together with the home IP address of the mobile node, the SPI uniquely identifies the mobile-home security association. The default authentication algorithm used for this authentication is HMAC-MD5 [4].

Two optional authentication extensions, the mobile-foreign authentication extension and the foreign-home authentication extension, can provide additional security. These could also be attached to mobile IP registration messages and replies. They allow the mobile node and FA to authenticate each other, or the FA and HA to authenticate each other, respectively. They are meaningless in cases of a colocated care-of address where no FA is used. While the mobile-home and mobile-foreign authentication extensions are added by the mobile node, the foreign-home authentication extension is added by the FA (since it is in the path of the registration message from the mobile node to its HA, it can append the extension as it passes through the FA).

As for other services, such as confidentiality and data integrity, these can presumably be handled by other IP protocols and there need not be a separate provision of such services that is Mobile IP specific.

EXERCISES

15.1 In public key cryptography, if A wants to send an encrypted message to B, which key should A use to encrypt, and which key should B use to decrypt?

15.2 Which IPSec mode adds more header overhead, tunnel or transport mode? Why might a mode be useful despite the additional header overhead? What are the source and destination addresses, for packets traversing an IPsec tunnel?

15.3 How is IPsec key management handled?

15.4 In GSM security, why should fresh authentication triplets be used each time the network wants to authenticate the mobile? What if the network tries to reuse an authentication triplet that it has used before?

15.5 Is GSM authentication one- or two-way? Is encryption/ciphering end-to-end? How is anonymity protected?

15.6 What security service does 802.11i provide that WEP does not?

REFERENCES

1. B. Aboba and P. Calhoun. RADIUS (remote authentication dial in user service) support for extensible authentication protocol (EAP). RFC 3579, Sept. 2003.
2. P. Calhoun, J. Loughney, E. Guttman, G. Zorn, and J. Arkko. Diameter base protocol. RFC 3588, Sept. 2003.
3. V. Garg and J. Wilkes. *Principles and Applications of GSM*. Prentice Hall, Upper Saddle River, NJ, 1999.
4. H. Krawczyk, M. Bellare, and R. Canetti. HMAC: keyed-hashing for message authentication. RFC 2104, Feb. 1997.
5. C. Perkins. Mobile IP joins forces with AAA. *IEEE Personal Communications*, pp. 59–61, Aug. 2000.
6. C. Rigney, S. Willens, A. Rubens, and W. Simpson. Remote authentication dial in user service (RADIUS). RFC 2865, June 2000.
7. IEEE Computer Society. IEEE standard for information technology—telecommunications and information exchange between systems—local and metropolitan area networks—specific requirements: 11. Wireless LAN medium access control (MAC) and physical layer (PHY) specifications. IEEE 802.11-2007 (revision of 802.11-1999), June 2007. Sponsored by the LAN/MAN Standards Committee.
8. D. Stanley, J. Walker, and B. Aboba. Extensible authentication protocol (EAP) method requirements for wireless LANs. RFC 4017, Mar. 2005.

16

FACILITIES INFRASTRUCTURE

Facilities infrastructure refers to the supporting facilities and infrastructure that are needed in the real world for a telecommunications network to operate. These include the building, operation, and maintenance of buildings, cabinets, and other structures that house communications equipment, as well as structures, such as communications towers, on which communications equipment can be mounted.

The wired network portion of the facilities infrastructure of wireless networks is very similar to the facilities infrastructure of wired networks. In the case of cellular networks, the mobile switching center is housed in a *mobile telephone switching office* (MTSO), which is analogous to the *central office*. Recall that the MSC is basically a telephone network switch with the addition of mobility support. Thus, a mobile telephone switching office and a central office are similar, except that the *main distribution frame* found in central offices (for terminating local loops or subscriber lines) and related infrastructure would not be found in a mobile telephone switching office.

The wireless network portion of the facilities infrastructure of wireless networks, however, is more interesting in a way, since it brings a different set of challenges. Base stations need to be spread out over the total coverage area of the wireless network, and the antennas need to be mounted high up over the ground [e.g., between 20 to 80 m over the ground (depending on the required coverage area for the particular base station, the terrain conditions around, user density in the area, etc.)]. The antennas are typically mounted on communication towers, which need to be strong enough to support the weight of the mounted equipment and to withstand the elements (wind, rain, sun, etc.), while not posing an undue danger to aircraft. We discuss communications towers in Section 16.1.

Fundamentals of Wireless Communication Engineering Technologies, First Edition. K. Daniel Wong.

Another set of challenges faced by base stations is related to electricity, including adequate provisions for average and peak power consumption and backup power supplies. Also, electrical protection is especially critical for communication towers because of their tendency to attract lightning. Issues related to the electricity supply and electrical protection are discussed in Section 16.2.

Typically, at the base of a communications tower, a sheltered cabin or cabinet (sometimes called a base station cabinet) may be found that houses the communications equipment (in standard 19-inch racks), power supply and backup power supply, temperature control, and monitoring system. The base station cabinet needs to be able to withstand environmental conditions, such as sun, rain, dust, and condensation. We discuss some aspects of these issues briefly in Section 16.3.2 (temperature control, etc.) and Section 16.3.3 (physical security and defense against fire). The communications equipment in the cabinet needs to be connected to the antennas using RF cables, and these we discuss briefly in Section 16.3.1.

16.1 COMMUNICATIONS TOWERS

We use the more general term *communications tower* here rather than *base station*, since "base station" is a more functionally specific term. A communication tower may be where a base station and its antennas are located. It may hold more than one base station. It may also hold various other communications equipment and antennas. These might be at different heights. A popular arrangement is to have base station antennas (such as panel antennas as discussed in Section 4.3.4) as well as a highly directional antenna (such as a parabolic reflector) for the point-to-point microwave link between base station and base station controller (possibly through one or more repeaters). An example of this can be seen in Figure 16.1. In this figure, the base station antennas are mounted toward the top of the tower, and the point-to-point highly directional microwave antennas are farther down.

Higher towers are used for larger coverage areas (cell size), but lower towers with smaller coverage areas could be more useful in densely populated areas with lots of subscribers. In addition to larger coverage areas (when antennas are placed higher up), higher towers also allow more different sets of equipment and antennas to be placed on the same tower. They have more space.

There are a variety of design considerations for towers. They have to support the weight of all the equipment that might be installed on them, and they have to do this under a possibly wide-ranging set of environmental conditions, including:

- Erosion and corrosion from water (rain, etc.), dust, and *salt fog* (i.e., a humid, salt-containing environment), especially when the tower is near a large body of salt water such as the sea.
- Mechanical stresses, such as rain, wind, snow, and ice. For example, ice can add considerable weight to a tower. Wind loading can be a significant factor in some areas.

FIGURE 16.1 Monopole towers.

- Uneven heating from the sun that can cause a tower to flex, possibly changing antenna orientation, and so on.

ANSI/EIA/TIA 222-G, "Structural Standards for Steel Antenna Towers and Antenna Supporting Structures," gives minimum criteria for loading and design of towers. In the past, before the 1950s, towers were often made of timber. However, timber can rot, so by the 1950s, towers started being built of concrete. Concrete is cheaper than steel and can be more rigid than steel towers. However, it is not so easy to mount antennas on concrete towers, and concrete is heavier than steel, leading to the need for heftier and more costly foundations. These days, towers are often built of steel, but sometimes they are built of aluminum or concrete. There are a few types of tower construction in the present day:

- Monopole towers
- Lattice towers
- Guyed tubular masts, guyed lattice masts, guyed monopoles

A *monopole* tower is a simple design, consisting of stacked cyclindrical tube sections whose diameter gets smaller and smaller as we go up the tower. Monopoles are not strong enough to rise too high, with a maximum height of about 200 ft. Higher towers, if built as monopols, would be too heavy and not be able to withstand strong winds as well as other structures, such as lattices and guyed towers. Monopoles are commonly found in urban and suburban areas, where towers don't need to be too high, since the cells are smaller than in rural areas. Besides, land is expensive in those areas, and the compact structure of monopoles does not take up a lot of space.

For towers that need to be sturdier and stronger than a monopole, e.g., to rise over 200 ft, the lattice tower is a popular choice. A lattice tower is built as a three- (or sometimes four-sided) lattice structure (i.e., a structure with widely spaced crossed beams of steel or other materials, often with regular geometrical patterns. The Eiffel Towel in Paris is arguably the most famous lattice tower in the world. Lattice towers for communications, though, are not as elaborate as the Eiffel Tower, but have a more functional, straightforward design. An example of a lattice tower is shown in Figure 16.2.

A *guyed* lattice tower is like a lattice tower with the addition of additional support cables (also known as guy cables or guys, and is often made of a strong material such as steel). The guyed lattice tower is also known as a *lattice mast*, perhaps by analogy with the mast of a ship. Guyed monopole towers are also possible. With the addition of the guy cables, the tower can rise even higher than the lattice tower. Typically, towers above 300 ft would be guyed towers. The disadvantage of guyed towers is that they require a large plot of land to contain the base of all the cables. Since the angle between the ground and each guy might be around 45° to 60°, the radius of the plot of land required is on the order of the height of the tower.

They can be found in rural areas, alongside highways, and so on. In these areas, population density is low, so cells need to be larger. Also, land is cheaper in these areas. An example of such a guyed tower is shown in Figure 16.3. Since the tower is very tall, only the middle section is shown. The faint lines from the tower heading downward at an angle are the guys.

16.1.1 Protecting Planes

There is a danger that low-flying aircraft could fly into a tower. To reduce the chances of such an occurrence, towers above a certain height (e.g., 200 ft, or 61 m, in the United States) must be painted red and white in alternating sections. The towers shown in Figure 16.1 (right side) and Figure 16.2 are examples of where such red and white painting can be found.

A careful look at the monopole on the right side of Figure 16.1 also reveals red lights at the top. A suitable lighting system approved by an authority like the *Federal*

FIGURE 16.2 Lattice tower.

Aviation Administration (FAA) is required, especially for taller towers. One example is the *dual lighting* system, with red lights at night and high- or medium-intensity white flashing lights in the daytime and twilight. Authorities such as the FAA in the United States (and similar authorities in other countries) create regulations for objects such as towers that project into airspace. In the United States, the FCC enforces tower issues. In particular, there is a series of forms, the 7460 forms, that must be filled up for towers higher than 200 ft or towers in the vicinity of airports, and filed with the FAA. The FAA takes its regulations very seriously, so large fines can be imposed for failures to comply. For example:

- Any light failures (for whatever reason, including power failure at the site) have to be reported to the FAA within 30 minutes of occurrence, and the FAA will issue a *notice to airmen* to warn aircraft pilots.
- Even during the construction of a tower, proper lighting must be installed temporarily at each intermediate stage so that the highest part of the partially erected tower is always easily seen by aircraft pilots.

FIGURE 16.3 Part of a guyed tower, also known as (guyed) mast.

16.1.2 Other Considerations

Some considerations as to where to locate towers are related directly to wireless communications: for example, as part of the wireless operator's plan for covering a city or highway or other service region with sufficient capacity, for filling gaps of coverage areas, and so on. Included in those considerations would be things such as the population density, but also terrain factors. For example, one would not want to place a tower in a valley where there are nearby hills or mountains that can obstruct the signal between the tower and devices on the other side of the hill or mountain. Wind loading might be a consideration in places where there may be variations in wind strength. The multipath environment might also be a consideration, so a site with less severe multipath delay spread might be chosen over one with very severe multipath delay spread.

However, there are also other factors not directly related to wireless communications (e.g., how close a proposed site is to airfields, to power supplies, etc.). Soil testing would be important to determine if the soil at the site is strong enough to support the structure and if it is at least adequate for electrical grounding purposes. One might want to avoid swampy, rocky or sandy land, for example. There would be legal and economic considerations, such as the purchase or lease of the land, and whether there are zoning regulations that impinge on the ability to operate communications equipment at the site. Sometimes, zoning regulations, or the desire to foster good relationships with a neighborhood, or for aesthetic reasons, might lead an operator, or even require it, to make towers somewhat hidden, or blended in with the environment.

In cases where towers have to be somewhat hidden or blended with the environment, we show some examples of what can be done, in Section 16.1.2.1. So far up to here, we have been discussing the location of fixed, permanent towers. Sometimes, less permanent towers might be needed, as discussed in Section 16.1.2.2. Finally, we show some creative alternatives based on what is permissible and other factors, in specific localities, in Section 16.1.2.3.

16.1.2.1 Stealth Towers It is quite common to find stealth towers disguised as natural objects in the environment, such as trees. Towers disguised as trees might be painted brown, with fake green leaves near the top to disguise their true nature. We have also seen (Figures 4.21 and 4.22) how antennas can be similarly disguised and made to blend in with their environment. A very creatively designed stealth tower, disguised as a cross next to a church, is shown in Figure 16.4.

16.1.2.2 Portable Towers There are some situations in which it is helpful to have temporary infrastructure to meet the needs of customers. For example, when there is a sporting event in a sports facility, the number of people there may be much, much larger than the average. The existing fixed infrastructure (base stations, etc.) often cannot provide enough capacity to serve the temporarily but significant increase in mobiles within that area. One solution is to bring in temporary infrastructure, such

FIGURE 16.4 Stealth tower disguised as a cross. (Courtesy of Steel in the Air, Inc.)

FIGURE 16.5 Portable tower on a trailer. (Courtesy of Aluma Tower Inc.)

as temporary towers, on which communications equipment can be mounted, to fulfil (at least partially) the temporary increase in service needs in that location. Figures 16.5 and 16.6 show such a portable tower. The tower is integrated with a trailer so it can be driven to where it is needed, and parked there (Figure 16.5). Then the telescoping tower can be extended (typically, it may reach about 100 ft). This particular model has outriggers (i.e., legs that extend outward in all directions) to provide balance and stability, as shown in Figure 16.6.

16.1.2.3 Creative Alternatives Sometimes, instead of finding isolated cell towers, one may find antennas mounted on smaller structures on top of roofs (e.g., roofs of two-story commercial buildings). For example, in Figure 16.7, we see a base station apparently "growing" from the top of a two-story commercial building, where the first story houses an automobile repair shop.

Figure 16.8 show a close-up of the antennas and the short monopole-like structure on which they are mounted on top of a roof. See also Figure 16.9.

16.2 POWER SUPPLIES AND PROTECTION

16.2.1 Power Consumption

The *peak power consumption* is the highest consumption of power from the electricity supplies to the facility. It usually happens around noon or the middle of the day. The peak power consumption should be estimated and provided for, such that the electricity supplies are adequate. For purposes of planning the amount of backup power required to keep the tower equipment running in the event of a failure, the

FIGURE 16.6 Portable tower setup with outriggers. (Courtesy of Aluma Tower Inc.)

average power consumption is more useful. Backup power is often provided by banks of rechargable batteries (see Section 16.2.1.1). However, alternative sources, such as solar panels, wind turbine, hydro generator, and diesel generator, are possible, each with its advantages and disadvantages.

The variation in energy consumption at a tower on a typical day is shown in Figure 16.10. It is from measurements taken at a base station in Alaska [3]. Figure 16.10 shows the power consumption over a week. The peak and average power consumption can be seen from the figure (actually, the plot shows current, but since the voltage is given, power can be derived). Figure 16.11 "zooms in" on one day, showing the variation in power consumption during a single day in more detail. Finally, Figure 16.12 shows how power consumption patterns can change when the technology changes. Whereas TDMA and AMPS were the systems using the tower in the earlier cases, Figure 16.12 plots the power consumption after the addition of a CDMA system to the same tower a few years later.

It is common that the communications equipment in a base station runs on dc (e.g., 24 V dc, as can be seen in the plots, or 48 V dc). Hence, ac power supplied by the electricity supply must be rectified to 24 V dc for this equipment (nevertheless, some other electrical systems at the facility, such as lights, heaters, and air conditioners,

FIGURE 16.7 Rooftop base station.

may run on ac, so there might need to be some circulation of ac as well). Furthermore, dc-to-dc converters are often needed to step down the voltages to appropriate levels for the communications electronics.

16.2.1.1 Batteries and Battery Safety The expected service life and energy storage capacity of batteries is sensitive to temperature conditions.

A safety issue that must be handled carefully is the concentration of hydrogen in the air where the batteries are kept. Lead–acid batteries release hydrogen and oxygen while being charged, especially when there is excessive charging or the room temperature is too high. Since hydrogen is combustible, this can result (and has resulted in numerous cases) in explosions and fires that destroy the tower and its equipment. A rule of thumb is that the hydrogen concentration in the air should not exceed 4%; however, best practice typically puts the limit at 2% or even 1%. For example, an alarm might be raised at 1% hydrogen concentration levels, and corrective action taken immediately at 2%. IEEE Standard 450 provides recommended practices for the maintenance, testing, and replacement of vented lead–acid batteries for stationary applications.

FIGURE 16.8 Rooftop base station, close-up of the antennas.

16.2.2 Electrical Protection

Lightning strikes or other causes of current surges in electrical circuits and devices can result in major damage to communications equipment (e.g., at a base station). In this section we focus mostly on lightning—characterizing it and then discussing protection methods—but we also discuss more general electrical protection methods, [e.g., surge protective devices (SPDs)] that defend against surges whether caused by lightning or something else.

Lightning protection can be divided into two aspects [2]:

- Diversion and shielding (discussed in Sections 16.2.2.3 and 16.2.2.4)
- Surge protection (discussed in Section 16.2.2.5)

16.2.2.1 Characterizing Lightning and Its Effects Lightning is a phenomenon whereby excess charge (usually negative, but sometimes positive) is discharged from clouds to ground. There are four types of lightning [2]:

FIGURE 16.9 Another rooftop base station, on top of a coffee shop and a bookstore.

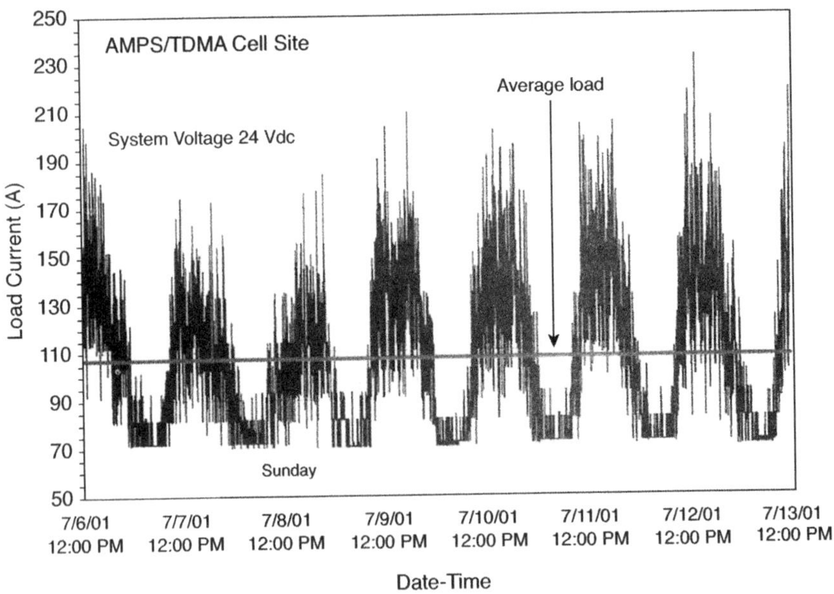

FIGURE 16.10 Power consumption in a BS over a week. (From [3]; reprinted with permission of John Wiley & Sons, Inc.)

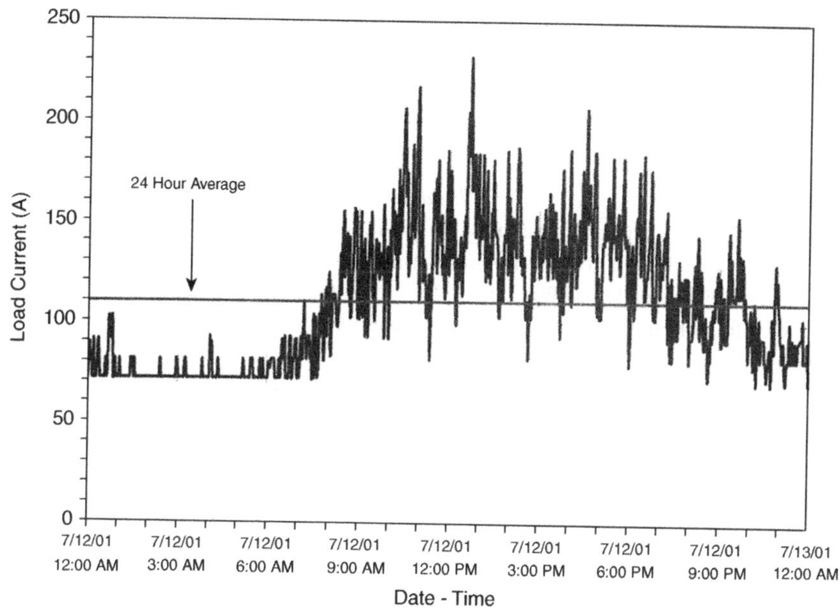

FIGURE 16.11 Power consumption in a BS during one day. (From [3]; reprinted with permission of John Wiley & Sons, Inc.)

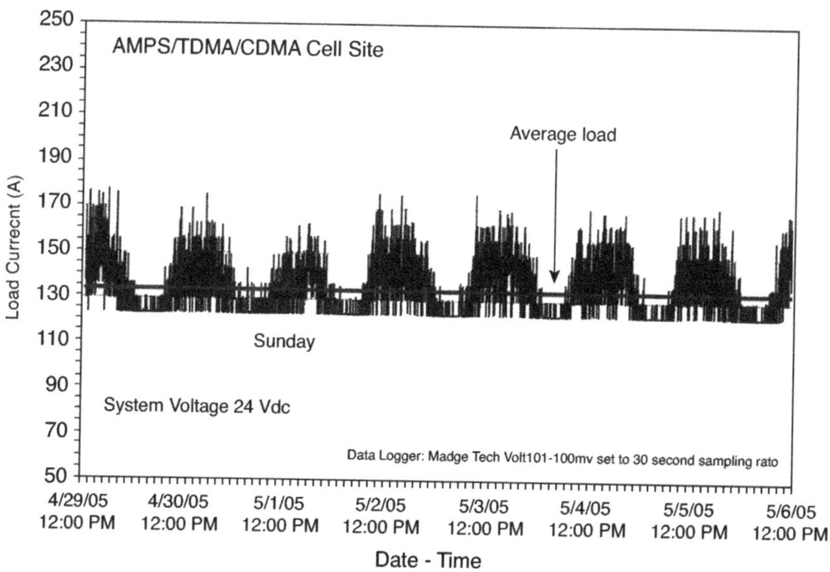

FIGURE 16.12 Power consumption in the same BS a few years later. (From [3]; reprinted with permission of John Wiley & Sons, Inc.)

- Downward negative lightning
- Upward negative lightning
- Downward positive lightning
- Upward positive lightning

The difference between *negative lightning* and *positive lightning* depends on whether the excess charge is negative or positive. Most lightning is of the downward variety, of which a common belief is that 90% is negative and 10% is positive. It may surprise the reader who is new to this topic that upward lightning is possible. However, this may happen in certain cases, usually considered to be where there are tall structures, say 100 m high or higher.

We now consider the most common case of downward negative lightning, in introducing the concepts of leaders and return strokes. A lightning strike typically consists of a *downward leader* and an *upward return stroke*, possibly followed by relatively lower level *continuing current* immediately after. The downward leader creates a conductive path from cloud to ground and it puts negative charges in this path. The return stroke that follows goes on the same path but from ground to cloud. Upward leaders may also be observed in response to the downward leader from it that gets close enough to the ground or other grounded objects. Sometimes, the concept of *striking distance* is brought into the picture. Striking distance is the critical distance of the downward leader from the ground or other grounded objects, where it is close enough that dielectric breakdown occurs and one or more upward connecting leaders are initiated. Figure 16.13 illustrates some of these concepts.

Because of their shape, tall structures such as communications towers unfortunately attract lightning. This can be explained in terms of electric field enhancement near the tip of such structures; that is, there is a tendency for charge to concentrate in sharp points, which results in higher electric field intensity near the tip of the structures (see Section 16.2.2.2).

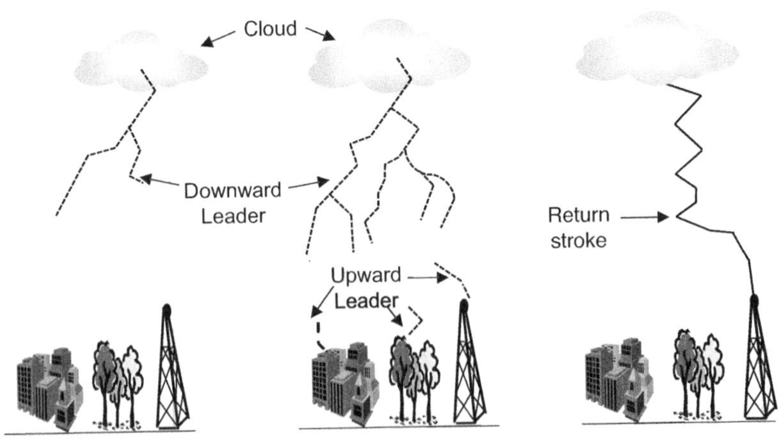

FIGURE 16.13 Aspects of lightning, downward and upward leaders.

The *ground potential rise* (also known as the *earth potential rise*) is a transient phenomenon that is very dangerous for workers or other people in the vicinity of a communication tower. In the event of a lightning strike, a very high current passes into the ground from where the strike hits the structure. Even with a relatively low resistance, the rise in the potential (voltage) between where the strike occurs, and "remote earth" (the ground at any place far away from the lightning strike, so it can be a constant reference point) can be very substantial, due to $V = RI$ and the very large currents involved. The related concept of *step potential rise* refers to how, in such a situation, the voltage between two limbs of a person (e.g., two legs, a "step" apart from each other) that are in contact with parts of the structure *or even with the ground* could be very high, so much so that a dangerous (even fatal) current flows through the person from one limb to the other.

16.2.2.2 Some Intuition on Why Tall Sharp Structures Attract Lightning

We go back to the example in Section 2.2.2.2, and let us assume that $r_1 < r_2$. We focus on comparing the total charge, surface charge, and electric field around the two spheres. From (2.28), we have

$$\frac{Q_1}{Q_2} = \frac{r_1}{r_2} \tag{16.1}$$

So, the ratio of total charge is equal to the ratio of their radii, and the smaller sphere has less total charge. As for the surface charge density, from (2.29) we have

$$\frac{\rho_{s,1}}{\rho_{s,2}} = \frac{r_2}{r_1} \tag{16.2}$$

so

$$\frac{E_{\perp,1}}{E_{\perp,2}} = \frac{r_2}{r_1} \tag{16.3}$$

Therefore, if we view the tip of a tall, thin structure as something like a very small sphere, and a relatively flat, low-curvature surface as like part of a very big sphere, we see how the surface charge density could be much higher at the tip of the tall, thin structure, and the **E** field near it could also be much higher than at a relatively flat surface.

16.2.2.3 Lightning Protection Through Diversion and Shielding: Lightning Rods

The purpose of lightning rods is to intercept the descending lightning leader. As such, they form part of the "diversion and shielding" aspect of lightning protection that we had mentioned earlier. The other components of diversion and shielding are down conductors and ground terminals, as discussed in Section 16.2.2.4. A common alternative to the use of lightning rods is the use of connected horizontal wires to cover the top of a structure. As with lightning rods, the purpose is to intercept the lightning leader. Down conductors and ground terminals are needed in all cases.

Lightning rods are also known as *Franklin rods*, after their inventor, Benjamin Franklin. Sometimes, however, the term *lightning rod* is used more generally to refer to these rods in any orientation, whereas "Franklin rod" might be used specifically for vertically oriented rods. Other terms that are used for lightning rod are *air terminal* and *lightning conductor*.

How do we decide where to place lightning rods and what spacing to use between them? A popular method is the rolling-sphere approach.

Rolling-Sphere Method. The *rolling-sphere method* can be used to find out what parts of a structure (a building, a tower, a combination of objects, etc.) are at high risk of being struck by lightning and what parts are at low risk of being struck by lightning. The idea is that the tip of the downward leader can be imagined to be at the center of an imaginary sphere, the radius of which is the striking distance. We will come to the value of the radius or striking distance shortly, but first, we discuss why the rolling-sphere method contains "rolling-sphere" in its name. Imagine rolling an imaginary sphere over a structure, such that we are always in contact with part of the structure but no part of the structure ever "penetrates" the sphere. Then the surface of the sphere would touch the structure at some points, and these would be high-risk areas (intuitively, because if a downward leader were to get to the center of the sphere, this area would be struck). Conversely, because we don't ever allow any part of the structure to "penetrate" the sphere as we roll it along, there would probably be some areas of the structure that are not touched by the rolling sphere as it rolls along. These areas can be considered relatively protected from lightning strikes. The rolling-sphere method is illustrated in Figure 16.14. For a more "real-world" example of the use of RSM, Figure 16.15 shows the rolling-sphere method being applied to a communications facility, including the various structures in the facility.

What radius should be assigned to the imaginary sphere? The smaller the radius, the less area would be considered to be protected. So it would be desirable to use a larger radius if the larger radius can still account for most lightning strikes. A rule

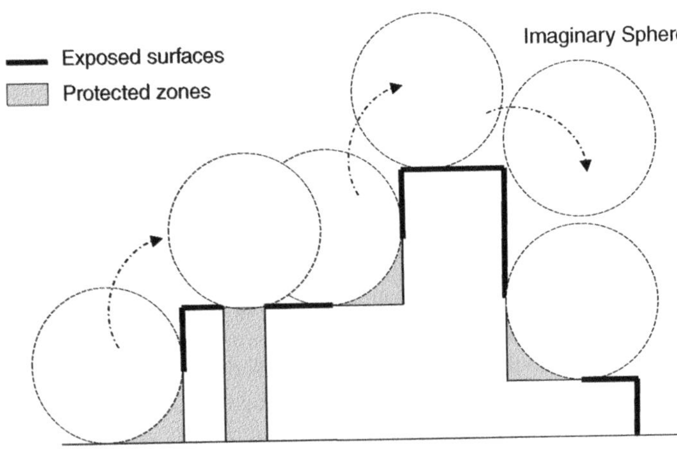

FIGURE 16.14 Rolling-sphere method for an abstract shape.

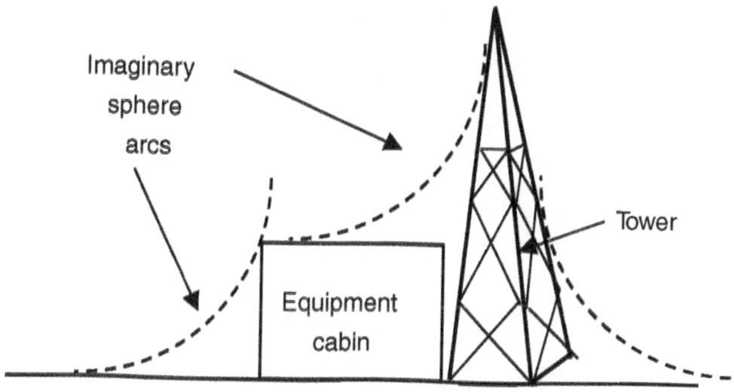

FIGURE 16.15 Rolling-sphere method for communications facilities.

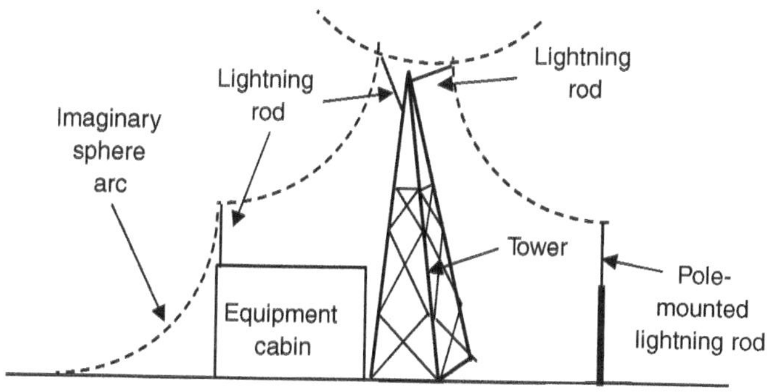

FIGURE 16.16 Same communications facilities protected by lightning rods.

of thumb is that we can account for up to 99% of lightning strikes by using a 20-m radius, and even with a larger, 60-m radius, we can account for up to 84% [2].

The rolling-sphere method can also be used to decide where to put lightning rods, because it can be used to predict how the high-risk exposed areas and the low-risk areas will change. This is illustrated in Figure 16.16. In practice, it is often used to decide where to place lightning rods.

Variations on Lightning Rods. Traditionally, lightning rods have been simple structures; basically, they are long, pointed rods. However, structures of more exotic construction have been proposed as alternatives, based on ideas for dissipating charge before it can accumulate to such an extent that upward leaders form and lightning happens. The intention is that the downward leaders would then find alternative targets elsewhere or disperse before completion of the strike. Figure 16.17 shows two charge dissipation terminals as examples of such alternatives to traditional lightning rods.

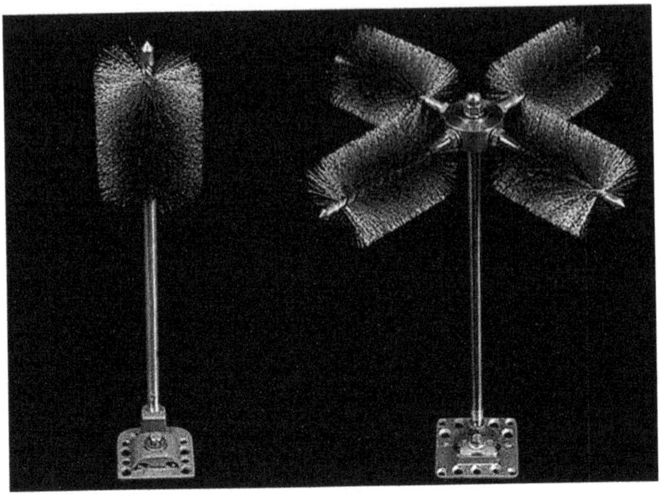

FIGURE 16.17 Variations on a traditional lightning rod. (Courtesy of Alltec Corporation.)

16.2.2.4 Lightning Protection Through Diversion and Shielding: Down Conductors and Grounding Down conductors are the conductors that bring current down from the lightning rods to the ground. While they may easily be overlooked, it should be noted that certain arrangements (typically, more symmetric arrangements) of down conductors are preferable. This is because the down conductors could carry a large time-varying current, and a more asymmetrical arrangement of down conductors may lead to dangerous *induced voltages* in electronic equipment within the structure. Thus, it is better for down conductors to run down all legs of a lattice communications towers rather than just one or two of the legs. Of course, further protection is still needed for the electronic equipment, which can be provided with SPDs (Section 16.2.2.5). Care should be taken that the materials out of which the down conductors (and lightning rods) are made has low enough resistance and can avoid melting from the heat generated by large currents coming from a lightning strike. Loops in down conductors should be avoided; in general, there should be multiple down conductors placed symmetrically and they should take the shortest path between lightning rods and the grounding system.

As a rule of thumb [2], large metallic objects within 5 m of down conductors should be bonded to the conductors, to avoid "side flashes." Additionally, the grounding system is meant to get the lightning current into the earth while minimizing the rise in potential of the part of the structure that is above ground. Commonly, ground rods of length 2 to 3 m are driven into the ground. The grounding system should be well bonded to the down conductors, which should be well bonded to the lightning rods. For such bonding, and other bonding of components of the path from lightning rods down to the ground, various types of specialized welding equipment and systems can be used, an example of which is shown in Figure 16.18. Generally, the lower the grounding resistance, the better. Soil types can make a difference in the grounding

FIGURE 16.18 Welding equipment for bonding conductors. (Courtesy of Alltec Corporation.)

resistance. To minimize differences in potential from one point to another, it is better not to use isolated ground rods but to connect the ground rods together, perhaps through a metal mesh that is buried under the tower.

16.2.2.5 Surge Protective Devices SPDs are also known as *surge arresters*. They are used to protect electrical devices against dangerous surges of current that could severely damage or even destroy them. Thus, they typically have a nonlinear

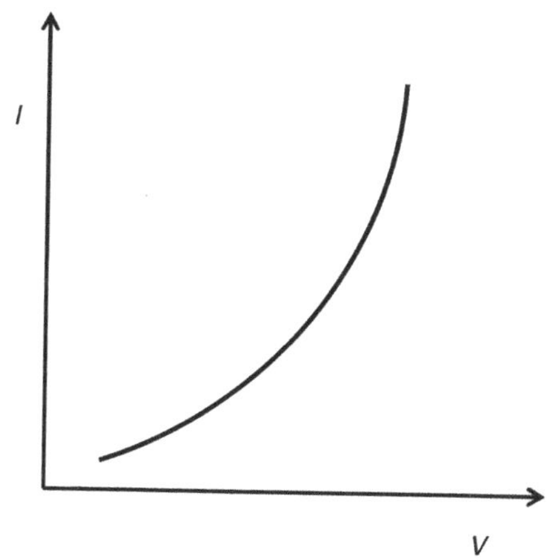

FIGURE 16.19 Nonlinear voltage–current characteristic of an SPD.

voltage–current characteristic, as shown in Figure 16.19, and can be installed in parallel with the device or circuit to be protected. As an increasing voltage is applied across both the SPD and the device/circuit it is protecting, the current drawn by the SPD rises much more quickly than the current flowing into the device/circuit that it is protecting. This has the desired effect: of minimizing the increase in current in the device/circuit being protected. On the other hand, during normal conditions, the SPD will draw very little current, thus wasting very little power. In contrast, if we just had a resistor instead of the SPD, the smaller the resistance, the higher the current drawn during a surge, *but* the more power such a resistor would dissipate unnecessarily during normal conditions.

SPDs are built in accordance with specifications such as the *National Electrical Code* (NEC) in the United States, *Low Voltage Directive* (LVD) in Europe, and *Electrical Appliance and Material Safety Law* (DENAN) in Japan.

16.3 ADDITIONAL TOPICS

In this section we briefly discuss RF cables, such as might be used for feeder and jumper cables to connect the RF equipment to the antennas on a tower, building automation and control systems, and physical security.

16.3.1 RF Cables

RF coaxial cables are sometimes divided into *flexible* and *semirigid* cables [1]. The flexible cables are typically *braided coaxial* cables, but there are a number of variations. By *braided coaxial cable*, we mean a coaxial cable that uses a braided outer conductor. As shown in Figure 16.20, a flexible coaxial cable would consist of a center conductor surrounded by a dielectric. On the outside of the dielectric is the outer conductor. The braided construction is a key reason for its flexibility. The outer covering provides protection against environmental factors. Sometimes, partially for marketing purposes, the term *superflexible* is used for some of the more flexible cables on the market. Traditionally, semirigid cables may be called hard line cables and have a solid outer conductor. Sometimes, the outer conductor is corrugated copper, which makes it easier to bend, but still not as flexible as braided copper.

As mentioned in Section 4.4.2, typically between the antenna(s) and the RF transmitter/receiver equipment, there would be a feeder cable. This main feeder cable would usually be connected on both sides (i.e., cable to/from antenna, and cable

FIGURE 16.20 Flexible coaxial cable with braided outer conductor.

to/from RF equipment) with jumper cables. Jumper cables are typically more flexible than feeder cables, with a smaller bending radius, but also more lossy. Thus, one might expect that feeder cables would generally be of the semirigid variety and jumper cables of the flexible variety. However, there are exceptions; one can find not just flexible jumper cables, but also semirigid jumper cables on the market. As alternative forms of low-loss flexible cables enter the market, these are sometimes being used even as feeder cables.

Cables going out to cell towers also need to be built to withstand the elements, (e.g., to be waterproof, flame retardant, etc.) Cables are sometimes designated by an RG number (e.g., RG6, RG58, RG 213, etc.). RG, which stands for *radio guide*, refers to an old military classification of coaxial cables that is now obsolete but still in popular use. Cables may also be designated by a number that refers to the diameter of the cable.

16.3.2 Building Automation and Control Systems

Building automation and control systems (BACS) consist of systems such as the following:

- Heating, ventilating, and air conditioning (HVAC)
- Energy management
- Fire alarm
- Physical security

The more sophisticated BACS may utilize central control and distributed sensors.

16.3.2.1 HVAC A *heating, ventilating, and air-conditioning* (HVAC) system is typically under the control of a HVAC controller. The HVAC needs to decide whether to circulate hotter air or cooler air, and at what speed. A HVAC controller would take into account such variables as air pressure, rate of airflow, and fan speeds in making decisions. As mentioned in Section 16.2.1.1, good temperature control and ventilation are important in the places where batteries are stored, to avoid deterioration of storage capacity or expected service life, and even to avoid a fire breaking out due to excessive levels of hydrogen. Besides temperature control, a HVAC system would also need to control humidity levels, to avoid condensation that can damage electronic equipment.

16.3.3 Physical Security

In Chapter 15 we distinguished among physical security, system security, and network security.

There are two main categories of physical security threats to a communications facility such as a cell tower. They are:

- Intentional, specific threats from humans seeking to breach the security of the facility, for various malicious purposes (e.g., theft of equipment, cables, etc.); one can also imagine threats from certain humans who have desire to destroy or damage the facility for other purposes (i.e., not necessarily for financial gain), such as arsonists.
- "Unintentional" threats from forest fires, floods, and so on.

Fences, good locks, warning signs, and so on, can help keep the facilities physically secure. Since there will usually not be any human guards present, an electronic alarm system could be installed. Inspections of the fences, locks, and electronic alarm system should be made on a regular basis, depending on operator policy.

The typical components in a fire alarm system are sensors, sprinklers, lights (strobes), and horns. The goals are threefold:

- Detection
- Suppression
- Notification

Detection is about detecting a fire. Sensors are used for detection. The sensors may detect smoke or heat or a combination of the two. Sensors may come with multiple sensitivity levels. A more sensitive sensor may give more false alarms than a less sensitive sensor, but may be quicker than a less sensitive sensor to detect a real fire. *Suppression* is about suppressing the fire. Sprinklers are one of the tools used in suppression. When sufficient smoke/heat is detected, sprinklers will be activated and sprinkle water around. In the case of rooms with electronic equipment such as telecommunications equipment, though, inert gas is often a better alternative than water for combatting fires, since water can damage the equipment. *Notification* is about notifying humans, or a monitoring system, about the fire. Lights and horns are useful for notifying humans.

EXERCISES

16.1 Arrange the following tower types in decreasing order of typical height: lattice, guyed mast, monopole.

16.2 Consider a base station where the average load current is about 135 A, as in Figure 16.12. What is the average power consumption, assuming the same system voltage as indicated in the figure? How much stored energy is needed in batteries for the base station to run normally for 1 day after the regular power supply is cut?

16.3 In using the rolling sphere method, if we have some areas we wish to protect, should those areas be touching or not touching the rolling spheres? Would the use of bigger spheres (larger radius) provide more or less reliability?

16.4 Consider an SPD whose current–voltage relationship is given by $I = 10^{-9}V^4$, installed in parallel with base station equipment running at 24 V dc. What is the current drawn by the SPD under normal conditions? What is the power dissipated under normal conditions? In the event that the voltage across the SPD and base station equipment goes up to 5000 V during a lightning strike, what is the current drawn by the SPD?

16.5 What are the advantages and disadvantages of "flexible" RF cables?

16.6 In telecommunications facilities, why might inert gas be a better substance than water to use in fire extinguishers?

REFERENCES

1. T. S. Laverghetta. *Microwaves and Wireless Simplified*, 2nd ed. Artech House, Norwood, MA, 2005.
2. V. Rakov and M. Uman. *Lightning: Physics and Effects*. Cambridge University Press, New York, 2003.
3. W. Reeve. *DC Power System Design for Telecommunications*. Wiley, Hoboken, NJ, 2006.
4. B. Smith. *Communication Structures*. Thomas Telford, 2006.

17

AGREEMENTS, STANDARDS, POLICIES, AND REGULATIONS

We live in a world where many businesses and organizations provide a wide variety of goods and services. The goods and services provided by a particular business may be affected by, or may depend on, the goods and services provided by another business. Thus, *agreements* are needed between these businesses for mutual benefit.

We also live in a world where there are many ways that technologies can be put together into a system that does something useful. Having a standard way of putting these technologies together can result in various benefits, such as economies of scale, interoperability between equipment from different vendors, ability to roam from system to system and still communicate, and so on. *Standards* are a good way to make this happen. Economies of scale and interoperability notwithstanding, there is still room for choices regarding how technology is deployed. Thus, choices need to be made, and *policies* emerge to guide the decision making on issues such as the level of security to provide in a network and how subscribers will be billed for various services.

We also live in a world with "social contracts" where governments can plan, mandate, and administer certain choices and policies for the common good. *Regulations* are rules that are mandated by a government agency (e.g., on the use of unlicensed bands).

Agreements, standards, policies, and regulations are part of the structure that guides and determines how technology is used. The acronym *ASPR* (agreements, standards, policies, and regulations) has been used in some U.S. government documents. In this chapter we simply follow the acronym ASPR, discussing agreements in Section 17.1, standards in Section 17.2, policies in Section 17.3, and regulations in Section 17.4.

Fundamentals of Wireless Communication Engineering Technologies, First Edition. K. Daniel Wong.

17.1 AGREEMENTS

An agreement is a set of mutually accepted terms that define expectations between two or more parties. Agreements are very similar to contracts, and the two terms are sometimes used interchangeably, especially in American English. However, from a legal perspective, contracts are sometimes viewed more narrowly as a specific class of agreements that meet certain criteria.

Examples of agreements in telecommunications include:

- An agreement between an operator and equipment suppliers (e.g., on quality of the equipment)
- An agreement between operators on how to share a network
- An agreement between operators and subscribers on the service level (also known as a service-level agreement)
- An agreement between network operators on provision of roaming services to each other's subscribers

17.1.1 Service Level Agreements

A *service level agreement* (SLA) between operators, or between operators and subscribers, prescribes a certain level of service, usually in a quantifiable way. For example, it might prescribe:

- The percentage of uptime, and it might even specify precisely what uptime and downtime mean
- Minimum bandwidth, maximum latency, and so on, possibly in more precise terms [e.g., minimum average bandwidth over a specified period of time, maximum average latency (rather than peak latency)]

Along with the agreement might come penalties for failures to meet the agreed-upon performance or availability criteria.

17.1.1.1 Peering and Transit Most ISPs cannot provide global connectivity to their customers without relying on the assistance of other ISPs. Typically, then, larger ISPs may provide *transit* service to smaller ISPs, whereby packets from the smaller ISPs can pass through the larger ISP's network in transit on the way to their destination. The ISP that is obtaining transit service usually pays for this service. In cases where there is a lot of traffic between customers of two similar-size ISPs, the two ISPs would often enter into a *peering* relationship with each other, whereby they install a direct connection between them and use it for forwarding packets between each other (rather than paying for transit services).

17.1.2 Roaming Agreements

No operator has the size and reach to cover every part of the world. Instead, operators cooperate with each other to allow the subscribers of one operator to make use of service from the other operator when the subscriber is in a region that the home operator does not cover. This greatly increases the value of the subscriber's service contract, and people are willing to pay more for roaming services. Roaming involves the technical side, covered in earlier chapters, and the business side, which is specified in *roaming agreements.* Some guidelines regarding policy-related technical issues on roaming may also be found in documents such as the "GPRS Roaming Guidelines" [1] by the GSM Association.

17.2 STANDARDS

Standards enable different vendor devices to work with one another, and different networks to interoperate. Moreover, they provide a kind of assurance of baseline quality and reliability. They also provide a center of attention that multiple vendors can focus on without worrying about infringing on someone else's proprietary technology, while at the same time enjoying the benefits just described. As a result, the standardization of a system in a particular solution space (e.g., wireless LAN) can lead to explosive market growth in that space. Before 802.11, there had been proprietary systems in the wireless LAN space, but these were each by individual companies, whose products did not interoperate. Customers buying any of these products had to deal with the uncertainty of not knowing how the products would perform under various conditions. Once IEEE created a standard for wireless LAN, namely 802.11, the market soon exploded and grew rapidly.

Standards are especially useful in wireless, and global standards even more so. This makes it convenient for subscribers to roam from country to country and still receive service. A situation where each country has its own mobile phone system that is mutually incompatible with that of every other country, is inconvenient for subscribers because they cannot roam from country to country with one phone. Such a situation is not just an imaginary scenario, but has occurred in real life, namely in the first-generation cellular systems in Europe. There were multiple incompatible first-generation cellular systems in Europe, which caused such inconvenience for subscribers that one of the important goals of the creation of GSM was to have a single pan-European standard that would allow easy roaming between all the countries in Europe at least.

It must be noted that participation in standards organization is voluntary and **use of standards is voluntary**, too. However, the benefits of standards (interoperability, etc.) are such that vendors often try to comply, and they may seek certification that their products comply with certain standards where applicable.

Examples of standards organizations relevant to wireless communications include:

- *European Telecommunications Standards Institute* (ETSI). ETSI is perhaps best known for creating GSM.

- *International Telecommunication Union* (ITU). ITU is a United Nations (UN) agency with a global focus.
- *International Standards Organization* (ISO). ISO is perhaps best known for the standard seven-layer protocol stack in networking.
- *Institute of Electrical and Electronics Engineers* (IEEE). IEEE is perhaps best known for creating well-known network standards such as Ethernet (IEEE 802.3), WiFi (IEEE 802.11), and WiMAX (IEEE 802.16).
- *Internet Engineering Task Force* (IETF). IETF creates standards for the Internet.

Some readers might wonder if 3GPP and 3GPP2 should be included in the list. 3GPP and 3GPP2 are, strictly speaking, not standards organization themselves, but umbrella groups for existing standards organization to work together under a collaborative agreement to create global 3G wireless systems.

We now discuss selected specifications groups such as IEEE in more detail. In the course of the discussion, the general specifications process in these groups will also be seen. One exception is amendments, revisions, and other changes, and handling intellectual property, discussed in Sections 17.2.6 and 17.2.7, respectively.

17.2.1 IEEE

The IEEE is a nonprofit organization for electrical engineers and the practice of electrical engineering. IEEE could be said to be the largest professional organization in the world for the advancement of technology. Many of the most respected academic journals in electrical engineering and related fields are published by IEEE.

Standards are developed in the IEEE under the *IEEE Standards Association* (IEEE-SA). IEEE-SA follows ANSI principles of consensus, due process, and openness. It is recognized by ITU-R as a standards body. In IEEE-SA, the drafting of standards is by *sponsor groups*. Each sponsor group is related to one or more of IEEE's technical societies (such as IEEE Communications Society). Perhaps the best known of such sponsor groups is the IEEE 802 LAN/MAN standards committee (LMSC), which has been under the IEEE Computer Society since 1980. There is an 802 Executive Committee that oversees things.

The creation of new standards starts with a *project authorization request* (PAR). A PAR is also required for various changes (see Section 17.2.6). IEEE 802 may establish a study to consider possible standardization, and if the result is positive, a PAR is drafted. The 802 Executive Committee decides whether to approve or not based on criteria such as broad market potential, compatibility with other 802 standards, distinct identity within 802, technical feasibility, and economic feasibility.

The new project may be assigned to an existing working group, or a new working group might be created for it. The decisions in the working group are by voting, with a 75% majority needed. Task groups are typically created to work on specific tasks and come up with drafts. The drafts are voted on. Interestingly, whenever there is a negative vote, comments must be provided on specific changes that will make the document acceptable to the voter. This forces constructive criticism. There may therefore be a number of rounds of voting, followed by changes being made. After voting is passed in IEEE 802, there is a second round of voting at the IEEE-SA level, under its *Review Committee*. Once a standards document is approved, it is professionally edited and typically published within two months.

17.2.2 Example: Standards Development—IEEE 802.16

The IEEE 802.16 working group was created to work on broadband wireless access, after initiation of the project by Roger Marks of NIST. Marks held a first meeting in August 1998 at the IEEE Radio and Wireless Conference with 45 people in attendance. This group drafted a PAR on broadband wireless access for 10 to 66 GHz. It was endorsed by the IEEE 802 Executive Committee in March 1999, and the 802.16 working group was created.

Initially, the group focused on *line-of-sight* (LOS) links between 10 and 66 GHz, and it created the WirelessMAN-SC for this application. Then in March 2000, a PAR for NLOS links under 10 GHz was approved, and this eventually resulted in 802.16a (technically, an amendment of 802.16; see Section 17.2.6 for a discussion on amendments and changes). Although there were some who favored the creation of a new MAC for 802.16a, the direction eventually taken was to use a common MAC, but a sophisticated one, for all cases. Similarly, there were debates on whether to have a separate physical layer for licensed and unlicensed applications, but eventually, these cases were not separated. Meanwhile, the original 802.16 was completed in 2001 and was a single carrier system using TDMA/TDM, with an air interface called WirelessMAN-SC (where the SC stands for "single carrier"). With 802.16-2004 and subsequently with 802.16e-2005, the single carrier option has been retained, in a modified form, as WirelessMAN-SCa. However, multicarrier options have been added, including OFDM/TDMA and OFDMA options. The latest revision of 802.16 is 802.16-2009, where WirelessMAN-SCa has been dropped due to lack of interest.

To help with interoperability and to limit the options to a managable number, the WiMAX forum defines *system profiles*, which are combinations of options from 802.16. Although these system profiles are from a subset of 802.16, it is advantageous for operators and equipment vendors to abide by the profiles from the WiMAX forum, for interoperability purposes and in order to be WiMAX-certified by the WiMAX forum.

17.2.3 ITU

As described on its web page [3], ITU "is the leading United Nations agency for information and communication technology issues, and the global focal point for governments and the private sector in developing networks and services. For 145 years, ITU has coordinated the shared global use of the radio spectrum, promoted international cooperation in assigning satellite orbits, worked to improve telecommunication infrastructure in the developing world, established the worldwide standards that foster seamless interconnection of a vast range of communications systems and addressed the global challenges of our times, such as mitigating climate change and strengthening cybersecurity."

17.2.3.1 Example: The Search for IMT-2000 It was the late 1990s. The 2G networks had been developed by national or regional standards associations, but people were hoping that by the time the world came to the third generation, there could be one global system that would be used everywhere, supporting global roaming. Thus, it was only natural that it was ITU that spearheaded the development of 3G wireless systems under the name IMT-2000. The motivations for IMT-2000 were:

- *Growth of multimedia applications demands more bandwidth.* It was expected that future systems be able to support multimedia applications, including transmission of graphics and sound. Some of these applications would need higher data rates than is necessary for telephone-quality voice transmission.
- *Competitive alternative to wired terminal access is desirable.* Wired terminal access provided higher data rates, higher and more flexible quality of service, at lower costs than did the original second-generation cellular systems. The challenge to wireless terminal access was to provide comparable data rates and quality of service at competitive prices (i.e. without charging too much of a premium for mobility).
- *Projected demand is high and will not be served adequately by current systems.* This was one of the fastest growing markets in the world. New systems would need to provide the supply to meet the growing demand.
- *Smoother interconnecting between different networks, environments, and so on, is desirable.* There were different radio transmission technologies and different networks in use at that time. Separate applications had evolved into different and separate applications such as paging, cordless, and cellular systems that did not interconnect. Integration of these services would be convenient to consumers.
- *Convergence of disparate systems and wireless access technologies is desirable.* More efficient and cost-effective service could be provided if there were fewer competing standards providing similar services. Furthermore, that would ease the provision of the following item, global roaming.
- *Global roaming is desirable.* Global roaming allowed the user to have wireless access (preferably with most of the user's desired features) anywhere in the world, not just at the user's home location. Previously, global roaming was not possible. There were different systems operating at different frequencies in different parts of the world.

These motivations drove the following requirements [5]:

- *High-rate wireless access.* To meet the growing demand for wireless services, which increasingly demanded more bandwidth, high-rate wireless access capabilities were required. An interesting question is what fraction of wireless accesses will consistently require high-rate communications. High-rate wireless access may be especially important in indoor and low-speed environments to enable wireless to be competitive with wired alternatives.

- *Multirate wireless accesses with flexible service requirements (quality, symmetry, and delay).* In order to serve a variety of different multimedia applications with varying demands on bandwidth, quality of service, link symmetry, and delay tolerance, flexibility is essential. This also helps in integration of different services (e.g., voice and data) into a single device. Good solutions to the problems of providing cost-effective high quality of service over radio links are necessary to enable wireless to be competitive with wired alternatives.
- *Small, lightweight, and convenient mobile terminals.* The main motivation for this requirement was to be competitive with wired alternatives. The availability of small, lightweight, and convenient mobile terminals would stimulate demand in addition to the already high projected demand, and it would be one of the factors allowing for economies of scale. Economies of scale would make wireless more competitive.
- *Maximized commonality between the radio interfaces in the different radio environments.* This was another factor that would facilitate economies of scale and drive costs down while providing more integration of services and smoother interconnecting. It would also be an important step toward convergence of disparate systems and wireless access technologies.
- *Global standardization.* This should allow for global roaming.

A major component of any mobile system is the RTT (radio transmission technology). ITU-R requested proposals for candidate RTTs in April 1997, with circular letter 8/LCCE/47 [4]. In the request for proposals, certain required capabilities are specified in the form of an "objectives and requirements template." For terrestrial access, the bearer capability requirements are divided into three basic categories, corresponding to three different radio propagation environments. The environments are *indoor*, *outdoor-to-indoor and pedestrian*, and *vehicular*. These environments are described by Recommendation ITU-R M.1034-1 [6] as follows:

- The indoor environment has a limited transmission range, typically below 100 m. Obstructions of the LOS path result in significant shadow fading losses. Typical rms delay spread is from several tens to several hundreds of nanoseconds. Maximum Doppler shifts are typically less than 10 Hz.
- The outdoor-to-indoor and pedestrian environment has a range larger than the outdoor environment, but limited by building attenuation. Path losses of 10 to 18 dB and 8 to 10 dB for penetration of buildings and cars, respectively, are typical. Delay spread and Doppler shifts are similar to the indoor environment.
- The (terrestrial) outdoor environment has a maximum transmission range from 100 m in urban microcells to 35 km in rural macrocells. Path losses can be modeled by the Okumura–Hata model (see Section 5.2.3) [2]. Delay spread ranges from 1 μs for rural areas to 2 μs for urban areas, but can be higher if reflections from distant hills or distant buildings are considered. Maximum

TABLE 17.1 Minimal Bearer Requirements in Different Radio Propagation Environments for IMT-2000, as Specified by ITU-R

Characteristic	Indoor	Outdoor-to-Indoor and Pedestrian	Vehicular
Maximum range	Below 100 m	Between indoor and vehicular	100 m to 35 km
rms delay spread	0.01–0.5 μs	Less than 1 μs	1–2 μs (more if distant reflections included)
Doppler shift	Less than 10 Hz	Probably less than 10 Hz	10 Hz to 1 kHz

Doppler shifts range from 10 Hz for pedestrian users to about 1 kHz for high-speed vehicular users (about 500 km/h).

The differences in the environments are summarized in Table 17.1.

Because of the differences in the environments, the minimum bearer requirements are different for each of them. The minimum bearer requirements are listed in Table 17.2. Data traffic was subdivided into four categories:

- *Class A*: connection oriented, delay constrained
- *Class B*: connection oriented, delay constrained, variable bit rate
- *Class C*: connection oriented, delay unconstrained
- *Class D*: connectionless, delay unconstrained

TABLE 17.2 Minimal Bearer Requirements in Different Radio Propagation Environments for IMT-2000, as Specified by ITU-R

Test Environment	Indoor	Outdoor-to-Indoor and Pedestrian	Vehicular
Speech	32 kbps, BER $\leq 10^{-3}$, 50% channel activity	32 kbps, BER $\leq 10^{-3}$, 50% channel activity	32 kbps, BER $\leq 10^{-3}$, 50% channel activity
Circuit-switched data	2 mbps, BER $\leq 10^{-6}$, 100% channel activity	384 kbps, BER $\leq 10^{-6}$, 100% channel activity	144 kbps, BER $\leq 10^{-6}$, 100% channel activity
Packet-switched data	2 mbps, BER $\leq 10^{-6}$, exponentially sized packets, Poisson arrivals	384 kpbs, BER $\leq 10^{-6}$, exponentially sized packets, Poisson arrivals	144 kbps, BER $\leq 10^{-6}$, exponentially sized packets, Poisson arrivals

All four classes of data were required to be supported by RTTs. Other bearer requirements included:

- Asymmetric transmission capabilities
- Multimedia capabilities
- Variable-bit-rate capabilities

Results. Things did not turn out the way the ITU had planned. For various political and business reasons, the various groups involved were unable to converge to a single system for 3G. Thus, the groups coalesced around two competing systems: UMTS/WCDMA and cdma2000. The 3G Partnership Project (3GPP, see Section 17.2.5) was formed to create the standards for UMTS/WCDMA, and shortly afterward, the cdma2000 supporters created 3GPP2 to create the cdma2000 standards.

17.2.4 IETF

The Internet Engineering Task Force (IETF) is the organization that creates and maintains the standards documents for the Internet. The standards documents are called *request for comments* (RFCs). A convenient way to determine if a particular RFC has been made obsolete by a more recent one is to go to http://www.faqs.org/rfcs/rfc-obsolete.html and search for the RFC in question.

Worked Example: Finding the Latest Version of the Mobile IP Base Specification. We may have a copy of RFC 2002 on mobile IP, but don't know if it is the latest RFC on mobile IP. We go to http://www.faqs.org/rfcs/rfc-obsolete.html and search for "RFC 2002." We find that it has been made obsolete by RFC 3220. We search for "RFC 3220" and find that it has been made obsolete by RFC 3344. We do not find any RFC that has made RFC 3344 obsolete, so RFC 3344 is the latest.

17.2.5 3GPP

3GPP is not a standards organization, but an umbrella group of standards organizations. Standards created in 3GPP are then brought back to the individual standards organizations making up 3GPP, and adopted as standards by these organizations. The specifications process in 3GPP is shown in Figure 17.1.

The structure of 3GPP is shown in Figure 17.2. A project coordination group coordinates the work that happens in the technical specifications groups (TSGs). Each TSG consists of a number of working groups (WGs).

17.2.6 Revisions, Amendments, Corrections, and Changes

Technology keeps advancing, whereas standards documents are in a sense static entities that are written at a particular time. Changes might need to be made, due to

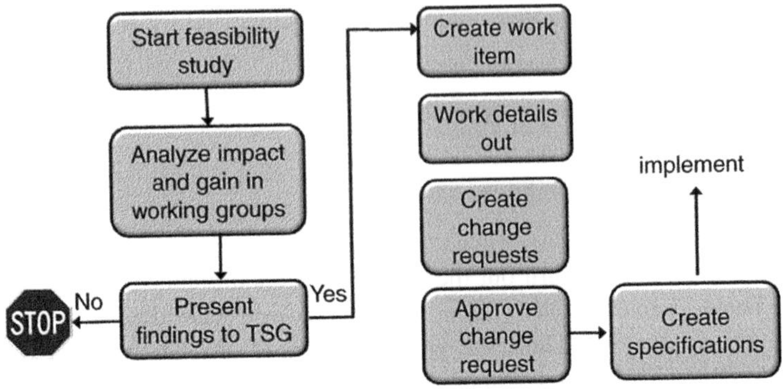

FIGURE 17.1 Specifications process in 3GPP.

errors of an editorial nature (grammar, syntax, etc.) or of a technical nature, or the introduction of additional options or capabilities to the base standard.

There are specific names for different types of changes, to distinguish between them. These include:

- *Revisions*: an update or complete replacement for an existing standard
- *Amendments*: adds new material for an existing standard

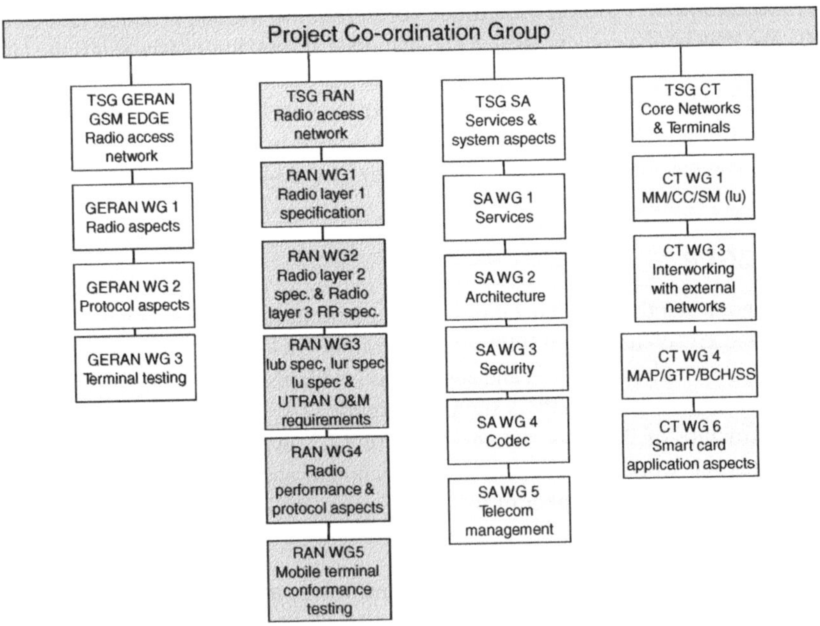

FIGURE 17.2 3GPP structure.

- *Corrections/corrigendum*: technical corrections on existing material
- *Changes*: editorial changes on existing material

The biggest changes are from revisions. Thus, these tend to come out less frequently; for example, IEEE 802.11 first came out in 1997, and it was followed by revision 802.11-1999 and then 802.11-2007. In between, there have been many amendments and corrections. To make changes, typically some type of change control process is needed [e.g., in IEEE, a project authorization request (PAR) must be submitted].

17.2.6.1 Example: IEEE 802.11 IEEE 802.11 was originally published in 1997, six years after work began on it in 1991. A revision appeared in 1999 (with mostly minor alterations from the 1997 standard), and this 802.11-1999 version of the standard had been the baseline version of the standard for many years when various important amendments such as 802.11i appeared. Finally, in 2007, IEEE published a new revision, 802.11-2007 [7], that rolls a number of amendments into the base standard. In particular, the following amendments were rolled into the baseline and are now considered *retired* by IEEE:

- IEEE Std 802.11a-1999 (Amendment 1)
- IEEE Std 802.11b-1999 (Amendment 2)
- IEEE Std 802.11b-1999/Corrigendum 1-2001
- IEEE Std 802.11d-2001 (Amendment 3)
- IEEE Std 802.11g-2003 (Amendment 4)
- IEEE Std 802.11h-2003 (Amendment 5)
- IEEE Std 802.11i-2004 (Amendment 6)
- IEEE Std 802.11j-2004 (Amendment 7)
- IEEE Std 802.11e-2005 (Amendment 8)

Both 802.11a and 802.11b came out in 1999, but 802.11a devices were not compatible with earlier APs since it was using a different frequency band (5 GHz), whereas 802.11b used the same frequency band (2.4 GHz) as the base 802.11. Furthermore, 802.11a devices were initially more expensive than 802.11b devices. Thus, 802.11b came to dominate the WLAN market and would only be replaced by 802.11g some time after that came out in 2003. 802.11g is actually very similar as 802.11a, except that it is in the 2.4-GHz band, and thus is backwardly compatible with 802.11 and 802.11b.

The Wi-Fi Alliance. Supplementary to, and complementary to, the 802.11 standards from IEEE, the Wi-Fi Alliance is an industry association involved in testing and interoperability issues related to 802.11-based products. It is a nonprofit international organization. The Wi-Fi Alliance created WPA and WPA2 (see Sections 15.4.2.1 and 15.4.2.2). Although these are not part of the standard, the Wi-Fi Alliance does have strong leverage over the vendors through its Wi-Fi certification program. For vendor products to be Wi-Fi certified, they have to satisfy the criteria of the Wi-Fi Alliance. Products that pass the Wi-Fi certification testing are allowed to bear the Wi-Fi logo. For example, the inclusion of WPA2 became mandatory to be Wi-Fi certified from 2006 onward. The Wi-Fi Alliance also includes a list of EAP methods in its certification program. Not every EAP method is part of the list.

17.2.7 Intellectual Property

Sometimes, companies or other organizations may own intellectual property (e.g., patents) that relates directly to certain standards contributions that they or someone else may make. Does the existence of a patent, or pending patent, disqualify a technology from being included in a standard? Not necessarily. The typical practice is:

- The intellectual property owner makes a disclosure regarding how some of its intellectual property is related to some standards contribution.
- The intellectual property owner agrees to license its technology to other organizations at reasonable rates.

And that is it! The standards organizations generally are not in a position to, and do not, police such matters, and disputes are handled outside the standards organizations.

17.2.7.1 Example: 802.11 The following is a typical statement on intellectual property, from IEEE 802.11-2007 [7]:

> Attention is called to the possibility that implementation of this standard may require use of subject matter covered by patent rights. By publication of this standard, no position is taken with respect to the existence or validity of any patent rights in connection therewith. The IEEE shall not be responsible for identifying patents for which a license may be required by an IEEE standard or for conducting inquiries into the legal validity or scope of those patents that are brought to its attention. A patent holder has filed a statement of assurance that it will grant licenses under these rights without compensation or under reasonable rates and nondiscriminatory, reasonable terms and conditions to all applications desiring to obtain such licenses. The IEEE makes no representation as to the reasonableness of rates and/or terms and conditions of the license agreements offered by patent holders. Further information may be obtained from the IEEE Standard Department.

17.3 POLICIES

Policies are guiding principles or plans that may influence decisions or actions. Examples include:

- Policies on how to handle requests for information from customers or potential customers

- Policies on how to handle sales pitches from vendors or potential vendors
- Policies on being environmentally friendly
- Policies on the handling of emergencies

"Industry best" practices may be considered to be policies. These may include policies on staffing and running of network operation centers, policies for schedules for maintenance and inspection of various facilities infrastructure, and so on.

Consider the deployment of a WiFi network. There are many decisions that an operator needs to make that would be guided by policies. Examples include:

- Who will be allowed to use the network?
- How will authentication and other security issues be handled? As discussed in Chapter 15, many choices could be made.
- What network architecture will be used for the access network? How many access points will be in each ESS? What technology will be used for the distribution system?
- How will IP addresses be allocated? How often will they need to be renewed? These and other choices relate to policies on DHCP settings.
- Which channels will be used at each access point? How will the channels be chosen to minimize interference between themselves and to/from other users of the same unlicensed band?

To assist operators in making policy decisions for GPRS roaming, for example, the GSM Association has published guidelines [1]. These guidelines are nonbinding. It describes the two main ways of handling GPRS roaming: with both the SGSN and GGSN in the network visited, or with the SGSN in the network visited and the GGSN in the home network. Requirements and recommendations for both scenarios are provided, including how to handle DNS, IP address allocations, and so on. For example, it recommends dynamic IP address configuration over static address configuration, even though both dynamic and static methods are supported by the standard. It also makes recommendations on naming: for example, the "network identifier," which conforms with the 3GPP specification but is more specific. All the recommendations are nonbinding, but can be considered "industry best" practices.

17.4 REGULATIONS

Regulations are rules set by a government agency. They are mandatory. The government agency may impose fines or other penalties, including license suspensions, if regulations are not followed. In many cases, the government agency involved may not check aggressively for violations of regulations, but may be spurred to action upon receipt of complaints from competitors or other entities. We will see an example of how regulations can have a significant impact on the deployment and operation of wireless systems in Section 17.4.1. Examples of regulatory agencies include the

Federal Communications Commission (FCC) in the United States and the *Malaysian Communications and Multimedia Commission* (MCMC) in Malaysia.

Often, the regulatory process is slow, for such reasons as:

- The difficulty in predicting the impact of new technologies
- Lawyers may not understand new technologies, but lawyers need to play a central role because of the legal nature of regulations
- Checks and balances are desired, so many viewpoints are solicited
- Once a regulation is established, it is difficult to change

In Section 17.4.2 we give an example of ultrawideband that illustrates some of these points, as we see how it took years for the FCC to deliberate and make rulings.

Due to the nature of the regulatory process, it is often in the interest of various companies and private-sector organizations to have their voices heard when an issue (e.g., what to do about ultra wideband, or net neutrality, etc.) is being grappled with by a regulatory agency.

17.4.1 Licensed vs. Unlicensed Spectrum

A fundamental choice in the design of any wireless system is the frequency bands in which it is designed to work. Most bands are controlled by *licensing*. Hence, in order to operate a communications system in these bands, the operator is required (by regulation) to have a license. There are some bands that are designated *unlicensed* bands. Anybody can operate a wireless system in these bands. The unlicensed bands, although not licensed, are still regulated, so there are still usage rules, for example, on maximum emission limits. In fact, the rules for unlicensed spectrum are often tighter than for licensed spectrum. For example, a low EIRP value per hertz may be specified, to facilitate sharing the unlicensed band among many users.

The benefits of using licensed spectrum include:

- Other radio systems operated by others will not be allowed to transmit in the band, so the amount of interference from other radio systems is much reduced in licensed bands.
- Since other radio systems are not allowed to use the same band in the same place, the interference environment is more predictable and easier to manage, as the great majority of interference would be intrasystem rather than intersystem. The operator doesn't have to worry, as with unlicensed systems, that the level of interference may rise significantly during times of peak usage, severely affecting communications performance.

The benefits of using unlicensed spectrum include:

- It can be used without paying licensing fees, which are typically very high.
- Deployments can be done faster, since licensing is not needed.

What about other electronic devices that may unintentionally radiate energy at various frequencies? Regulatory agencies sometimes have rules for *unintentional radiators* as well. For example, the FCC in the United States, under Part 15 (more precisely, the Code of Federal Regulations, Title 47, Part 15) specifies emission limits for unintentional radiators. The next example, of UWB, shows a case where some people wanted the FCC to introduce a "third alternative" to licensed and unlicensed systems. It also illustrates how the regulatory process works.

17.4.2 Example: Regulatory Process for Ultrawideband

We begin by introducing briefly the concept of *ultrawideband* (UWB) and its unique features, before concentrating on the regulatory aspects. UWB refers to a family of spread-spectrum technologies characterized by very huge bandwidths. While spread spectrum in general uses large bandwidths, UWB is even more extreme, as illustrated in Figure 17.3. Some would add that another important characteristic of UWB is the large *fractional bandwidths*, and one rule of thumb for this is that

$$\frac{B}{f_c} > 0.25 \tag{17.1}$$

where B is the bandwidth and f_c is the center frequency of the signal, unlike other spread-spectrum systems, where the fractional bandwidth might be on the order of 0.01.

There are many ways that such signals could be generated, but the initial impetus for the FCC coming to examine the question of how to regulate UWB was for the *impulse modulation* variety. In particular, these systems used very narrow pulses (hence sometimes called impulses), on the order of nanoseconds wide (and thus on the order of GHz wide in the frequency domain) to communicate, by pseudo-randomly modulating the position of these pulses. Thus, a pulse train of such narrow pulses, evenly spaced, would convey no information, but modulating that by the PN sequence gave it the random aspect of spread spectrum, plus the additional modulation of the pulse positions by data bits allowed information to be conveyed.

A UWB transmitter for a pulse-position-modulated UWB system might look like what we show in Figure 17.4. Here, PPM stands for pulse position modulation, and the data and PN sequence combine to modulate the position of the pulses. A significant feature of this transmitter is that the pulses are radiated directly through the antenna, **without up-conversion**. Hence, unlike most wireless communications signals that use a sinusoidal carrier, there is no carrier frequency!

FIGURE 17.3 Huge bandwidths in UWB systems.

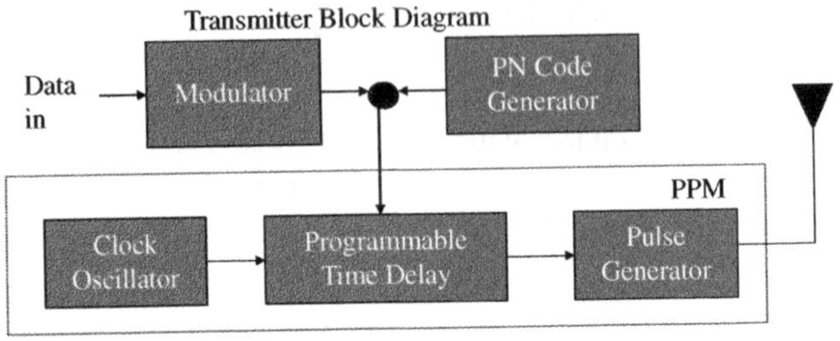

FIGURE 17.4 UWB transmitter.

The corresponding receiver is shown in Figure 17.5. As in the transmitter, the receiver has a radically simplified "RF" portion with no down-conversion. It could also be said that the signals are sent and received practically as baseband signals. One may wonder about the very precise timing and synchronization requirements, but proponents of this type of system claimed that advances in various technologies (precision timing, broadband antennas, etc.) allowed it to be feasible.

Various claims were made of the benefits of such systems, such as:

- Cheap and low power, since there is no need for RF processing in transmitters and receivers
- Small and ubiquitous
- Cause negligible interference to other systems, since the signal power is spread over such a large bandwidth
- Are excellent at rejecting interference, due to high processing gain (on the order of 30 dB larger than for IS-95 CDMA systems, for example)
- Are good at penetrating materials (good for radar, and perhaps for certain communications applications)

However, UWB systems were not covered by existing FCC regulations and would simply not be allowed to operate without changes in the regulations. Thus, proponents

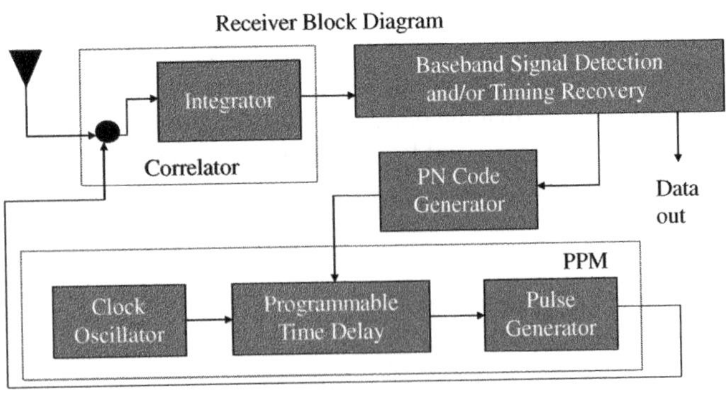

FIGURE 17.5 UWB receiver.

of UWB brought their concerns to the FCC, requesting changes in regulations so that UWB could legally be used.

The regulatory process began with the FCC issuing a *notice of inquiry* (NOI) on a *notice of proposed rule making* (NPRM) in September 1998. The NOI requested comments by December 7, 1998, and replies to those comments by January 4, 1999. All comments and replies to comments were publicly accessible at the FCC web site. The FCC was asking such questions as:

- Is power and interference low enough that UWB radios could operate over all the bands below 5 GHz and not disturb incumbent users of spectrum?
- In certain restricted bands, only "spurious emissions" were permitted. Spurious emissions are unintentional radiation of an incidental nature. Since it would be difficult for UWB systems to avoid these restricted bands, should some FCC rules be changed, so that, for example, not just spurious emissions might be allowed but also UWB radiation that is very small?

The current rules for spurious emissions at that time were in Part 15 of the FCC rules. The NOI elicited many responses on both sides of the issue, from organizations, individual experts, and others. Proponents of UWB were arguing that when the rules had been written, UWB had not been known, so UWB systems had been inadvertently disallowed. They also argued that the rules for spurious emissions were outdated, since it should be not so much the intentional, but the *interference potential*. There were many unlicensed Part 15 devices that were legal because their emissions were spurious, whereas UWB might have much less interference potential, but couldn't be used under Part 15, as they were intentional.

However, opponents of change included powerful organizations, such as the U.S. GPS Industrial Council, the FAA, TV broadcasters, and the Consumer Electronics Manufacturers Association (CEMA). Their arguments included the following:

- UWB systems were still not very well understood at that time and it would be premature to make changes to any rules without further study.
- Certain devices, such GPS or aircraft navigational equipment, were too sensitive to even the slightest interference.
- Even if a single UWB device might not cause a problem (a big "if" to some skeptics nevertheless), what might be the effects of a proliferation of such devices? Who could predict what their collective impact might be?

After a process of deliberation, the FCC proceeded cautiously and in July 1999 granted limited conditional waivers for UWB systems from three companies: Time Domain, U.S. Radar, and Zircon. NB: The rules were not changed, but these companies were granted these waivers so that they could sell UWB systems that didn't obey the rules, provided that the specific conditions specific in the waivers were met. Also, the distribution of these systems was controlled, so records had to be maintained of all sales, and so on. The granting of waivers was widely interpreted as a sign that the

FCC was interested and willing to give it a try under controlled circumstances, but not ready to commit (changing rules would be a commitment that would be difficult to reverse).

In May 2000, the FCC finally issued their NPRM, in which they tentatively proposed some rules, highlights of which are:

- UWB intentional radiators would be subject to the Part 15 emission limits (which previously had been just for unintentional radiators)
- There would be additional restrictions below 2 GHz, such as including a notch for the GPS band, so any transmission that crossed the GPS band would need a sharp filter at that band. Also, below 2 GHz, the emission limit would be 12 dB below the Part 15 limits.

At the same time, the FCC requested more comments.

Then in February 2002, the FCC issued its first report and order. This allowed indoor UWB communications systems to operate between 3.1 and 10.6 GHz, in a peer-to-peer manner, subject to the Part 15 emissions limits. Although the limitations for indoor UWB systems matches the Part 15 emissions limits within 3.1 to 10.6 GHz, it is more severe at other frequencies. There is a sharp notch for GPS between 0.96 and 1.61 GHz: for example, a huge −75 dB/MHz.

Later, in February 2003, the FCC issued a second report and order, with no significant changes.

EXERCISES

17.1 What is the difference between transit and peering arrangements/agreements between Internet service providers?

17.2 Look again at Table 17.1. If 3G systems operate at around 2 GHz, what would be the range of speeds corresponding to Doppler shifts of 10 Hz to 1 kHz? (Compute in m/s and also in km/h.) Is this range reasonable for a vehicular environment?

17.3 Describe some of the pros and cons of licensed and unlicensed spectrum.

17.4 If new material (e.g., a new physical layer option) is to be added to an existing standard, would you expect to see it in a revision, an amendment, or a corrigendum?

17.5 Between regulatory rules and standards, which is mandatory and which is voluntary?

REFERENCES

1. GSM Association. GPRS roaming guidelines. GSMA PRD IR.33, July 2009.
2. M. Hata. Empirical formula for propagation loss in land mobile radio services. *IEEE Transactions on Vehicular Technology*, VT-29(3):317–325, Aug. 1980.

3. ITU. About ITU. http://www.itu.int/net/about/index.aspx, 2011. Retrieved Mar. 11, 2011.
4. ITU-R. Circular letter 8/LCCE/47: Request for submission of candidate radio transmission technologies (RTTs) for IMT-2000/FPLMTS radio interface. Circular letter, Apr. 1997.
5. ITU-R. Principles and approaches on evolution to IMT-2000/FPLMTS. *Handbook on Land Mobile including Wireless Access*, 2, 1997.
6. ITU-R. Recommendation ITU-R M.1034-1: Requirements for the radio interface(s) for future public land mobile telecommunication systems (FPLMTS), 1997.
7. IEEE Computer Society. IEEE standard for information technology—telecommunications and information exchange between systems—local and metropolitan area networks—specific requirements: part 11. Wireless LAN medium access control (MAC) and physical layer (PHY) specifications. IEEE 802.11-2007 (revision of 802.11-1999), June 2007. Sponsored by the LAN/MAN Standards Committee.

EXERCISE SOLUTIONS

Chapter 1

1.1 $\cos 2\pi f_0 nt = \left(e^{j2\pi f_0 nt} + e^{-j2\pi f_0 nt}\right)/2$ and $\sin 2\pi f_0 nt = \left(e^{j2\pi f_0 nt} - e^{-j2\pi f_0 nt}\right)/2j$. Therefore, expanding the cosine and sine functions and collecting terms, we equate the coefficients of the complex exponentials, and we have $c_0 = a_0$ and

$$c_n = \begin{cases} \frac{1}{2}\left(a_n - jb_n\right), & n > 0 \\ \frac{1}{2}\left(a_n + jb_n\right), & n < 0 \end{cases}$$

1.2 It is very similar to the autocorrelation function of the random binary waveform.

$$R_{xx}(\tau) = \sigma^2 \Lambda\left(\tau / T_s\right)$$

1.3 $R_{yy}(\tau) = E\left\{x(t)\cos 2\pi f t \times x(t+\tau)\cos 2\pi f(t+\tau)\right\}$. By the indepedence of $x(t)$ from the sinusoid, we can regroup terms and write

$$E\left\{x(t)x(t+\tau)\right\} \times E\left\{\cos 2\pi f t \cos 2\pi f(t+\tau)\right\}$$

and the result follows.

1.4 We simply take the Fourier transform of $R_{yy}(t)$, given by (1.97), and by the "modulation" property in Table 1.2, $S_y(f)$ is just $S_x(f)$ shifted/translated by the carrier frequency.

1.5 Let $y(t)$ be the output of the matched filter. Then $y(T)$ is the output of the sampling operation following the matched filter. We have

$$y(t) = r(t) * s(T - t) = \int_0^t r(\tau)s\{T - (t - \tau)\}\,d\tau$$

Thus,

$$y(T) = \int_0^T r(\tau)s(\tau)\,d\tau$$

and this is exactly the correlation receiver.

Chapter 2

2.1 Cylindrical: $r = \sqrt{x^2 + y^2}$, $\phi = \arctan y/x$, and $z = z$. Spherical: $R = \sqrt{x^2 + y^2 + z^2}$, $\theta = \arctan(\sqrt{x^2 + y^2}/z)$, and $\phi = \arctan y/x$.

2.2 From cylindrical: $x = r\cos\phi$, $y = r\sin\phi$, $z = z$. From spherical: $x = R\sin\theta\cos\phi$, $y = R\sin\theta\sin\phi$, $z = R\cos\phi$.

2.3 The wave is propagating in the direction of $\mathbf{u_z}$. Since the intrinsic impedance of air is 377 Ω, $H_0 = 1$ mA/m. Therefore, the Poynting vector is $\mathbf{u_z}377\ \mu\text{W/m}^2$. Average power flow per unit area at P is $377/2 = 188.5\ \mu\text{W/m}^2$.

2.4 $1 \le S \le \infty$. So $-1 \le |\Gamma| \le 1$. SWR is $S = 3/1 = 3$. $\Gamma = (3-1)/(3+1) = 1/2$.

Chapter 3

3.1 Applying the Friis formula, we have

$$F = 10^{L/10} + \frac{10^{F_{i+1}/10} - 1}{10^{-L/10}} = 10^{L/10}\left(1 + 10^{F_{i+1}/10} - 1\right) = 10^{L/10}10^{F_{i+1}/10}$$

Convert to decibels and we have the result.

3.2 2.32 = 3.6 dB.

3.3 (b) Noise power contributed by the subsystem, input referenced. Since we multiply by B, it is a noise power and not a noise spectral density. By definition of noise figure, and seeing the derivation, the noise floor comes from the noise contributed by the subsystem, and input referenced. It means the IM3 product, *input referenced*, is equal to the noise floor. This is because the noise floor is input referenced.

3.4 −84 dBm and 59.33 dB; use the cascade formula for the new noise figure, then the sensitivity improves to −88.96 dBm.

3.5 We substitute $f_{\text{LO}} = f_{\text{RF}} + f_{\text{IF}}$ into (354), and we get

$$f_{\text{other},2} = f_{\text{RF}} + \frac{1}{2} f_{\text{IF}}$$

This is the 1/2 IF spur, and $f_{\text{RF}} = 758.1$ MHz, so $f_{1/2-\text{IFspur}} = 793.6$ MHz.

Chapter 4

4.1 For a half-wavelength dipole, $d_{\text{boundary}} = \lambda/2 = L$. For a quarter-wavelength dipole, $d_{\text{boundary}} = \lambda/8 = L/2$.

4.2 47,977.

4.3 $\lambda = 1.5$ m, so $0.95 \times 1.5/2 = 0.7125$ m.

4.4 Factor out $e^{jN\psi/2}$ from the numerator and $e^{j\psi/2}$ from the denominator. If symmetric about the origin, the array elements would be at $z = -(N-1)d/2$ to $z = (N-1)d/2$. Array factor would be $e^{-j(N-1)\psi/2}\left(1 + e^{j\psi} + e^{j2\psi} + \cdots + e^{j(N-1)\psi}\right)$ or $\sin(N\psi/2)/\sin(\psi/2)$.

4.5 It only has one active element, unlike typical antenna arrays.

Chapter 5

5.1 We start from

$$|h| > \sqrt{\frac{\lambda d_1 d_2}{d_1 + d_2}}$$

where λ and d_1, d_2 are in meters. Then $F = c/\lambda \times 10^{-9}$, so $\lambda = c/F \times 10^{-9}$. Multiplying each of d_1, d_2 with 1000 for unit conversions and using $c = 3 \times 10^8$ m/s, and noting that $\sqrt{300} \approx 17.3$, we have the desired result.

5.2

$$\overline{\tau} = \frac{0 + 1 \times (0.1) + 2 \times (0.1) + 4 \times (0.01)}{1 + 0.1 + 0.1 + 0.01} = 0.281\ \mu\text{s}$$

$$\overline{\tau^2} = \frac{0 + 1 \times (0.1) + 2^2 \times (0.1) + 4^2 \times (0.01)}{1 + 0.1 + 0.1 + 0.01} = 0.55\ \mu\text{s}^2$$

so

$$\sigma = \sqrt{0.55 - 0.281^2} = 0.686\ \mu\text{s}$$

and

$$B_c = \frac{1}{2\pi 0.686 \times 10^{-6}} = 232\ \text{kHz}$$

So $B_c < 1/T_s$ for GSM (270.833 kHz signaling rate). It is mildly frequency-selective fading. It is useful to employ the services of an equalizer for a case like this.

5.3 It is positive because the Hata model gives loss rather than received power. When $h_{\text{BS}} = 1$, path loss exponent is 4.49. For the path loss exponent to be exactly equal to 4, we need $44.9 - 6.55\log(h_{\text{BS}}) = 40$, so $h_{\text{BS}} = 5.6$.

5.4 $\lambda = (1/3)$m, so $f_m = 30$ Hz. Average fade duration:

$$\bar{\tau} = \frac{e^{0.5^2} - 1}{(0.5)(30)\sqrt{2\pi}} = 7.55 \text{ ms}$$

Level crossing rate:

$$N_R = \sqrt{2\pi}(30)(0.5)e^{0.5^2/2} = 42.6$$

5.5

$$P(x \leq X) = \int_0^X f_{\text{Rayleigh}}(x)\,dx = 1 - e^{-X^2/2p}$$

We know that this applies to the envelope of the signal. For a given p, γ_j is related to x by $\overline{\gamma_j} = \overline{x^2}/(2N)$, where the factor of half is because x^2 represents the power of the envelope, which is twice the signal power, and N is a scaling factor for SNR. Since x is always nonnegative, and because the function x^2 is monotonic, we can then proceed to write

$$P\left(\frac{x^2}{2N} \leq \frac{X^2}{2N}\right) = 1 - e^{-X^2/2p}$$

Now, $\Gamma = \bar{\gamma} = p/N$ is the average SNR. Furthermore, identifying γ_0 in (5.49) with $X^2/2N$, we have

$$P(\gamma_j \leq \gamma_0) = P\left(\frac{x^2}{2N} \leq \frac{X^2}{2N}\right) = 1 - e^{(-X^2/2N)(N/p)} = 1 - e^{\gamma_0/\Gamma}$$

Finally, we differentiate (5.49) to get (5.48).

5.6 $5 + 7 + 10$ dB $= 22$ dB.

Chapter 6

6.1 We note that $e^{-j2\pi n} = 1$ for any integer n, so

$$X\left(e^{j2\pi(F+1)}\right) = \sum_{n=-\infty}^{\infty} e^{-j2\pi n} x[n] e^{-j2\pi F n} = X\left(e^{j2\pi F}\right)$$

6.2 The distance D can be found by applying the law of cosines to the triangle as shown in Figure S6.2. In this figure the length of the line marked C is the

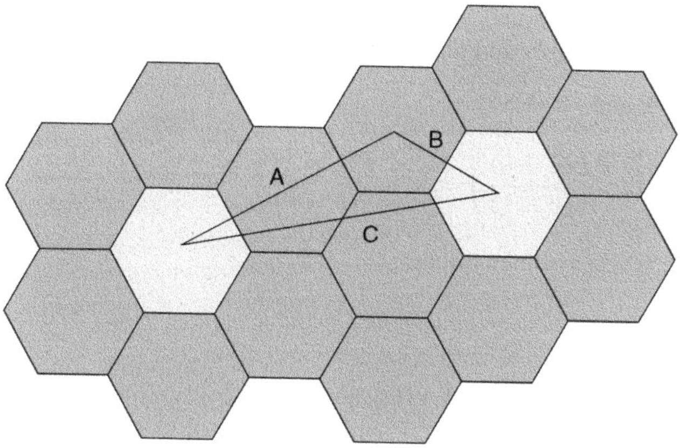

FIGURE S6.2

distance D. We can find the lengths of A and B by first noting that the distance from the center of a hexagon to any edge is $\sqrt{3}/2R$. $2i$ and $2j$ of such center-to-edge segments make up A and B. Thus, the lengths of A and B are $\sqrt{3}\, iR$ and $\sqrt{3}\, jR$. Applying the law of cosines, we have

$$D^2 = 3i^2R^2 + 3j^2R^2 - 3ijR^2 \cos 120^\circ = 3R^2N_s$$

from which the result follows.

6.3 $1/T_s = 20 \times 10^6$, and $N = 64$, so $\Delta f = 1/NT_s = 1/T_s' = 312{,}500 = 312.5$ kHz; the sampling interval is $T_s = 1/(20 \times 10^6) = 50$ ns, so the OFDM symbol is $T_s' = 64 \times T_s = 3.2$ μs; then we have 4.0 μs if the guard interval is included.

6.4 $\Delta f = 110$ Hz, $N\Delta f = 1.76$ kHz$= 1/T_s$; so $T = 568.2$ μs, $NT = 9.091$ ms; $15 \times 110 \times 2$ bps = 3.3 kbps; higher than link 11, with T_s' shorter (9.091 ms), whereas link 11 has T_s' as 13.33, or 22 ms. So for the same subcarrier spacing, OFDM achieves higher data rates with shorter T_s'.

6.5 $\Delta f = 75$ Hz, because it is 2.25 kbps $/(2 \times 15)$; $N\Delta f = 1.2\,\text{kHz} = 1/T_s$; so $T_s = 833$ μs, $T_s' = NT = 13.3$ ms. So OFDM can use the spectrum more compactly.

Chapter 7

7.1 Slotted Aloha.

7.2 Multiplexing is one-to-many, whereas multiple access is many-to-one. Generally, multiple access is more difficult.

7.3 See Section 7.3.1.

7.4 By the singleton bound, $d_{\min} \leq 1 + n - k = 125$, so it could detect at most $\lfloor 125/2 \rfloor = 62$ errors and only correct up to $\lfloor 124/2 \rfloor = 62$ errors.

7.5 0, 1.0986, and -1.0986.

$$\ln\left(\frac{P(k=0)}{P(k=1)}\right) = \ln\left(\frac{P(k=1)}{P(k=0)}\right)^{-1} = -\ln\left(\frac{P(k=1)}{P(k=0)}\right)$$

so there is odd symmetry about $p = 1/2$.

Chapter 8

8.1 Signaling rate $=$ 270.833 kHz, so in a time slot of 0.577 μs, there are $270{,}833 \times 0.000577 = 156.27$ symbols/bits. Indeed, it matches, as $156.27 = 148 + 8.27$. A guard period of 8.25 bits lasts $8.25 \times 1/270{,}833 = 0.0000304$ s (i.e., 30.4 μs), which is larger than the *average* rms delay spread in most places. Nevertheless, the timing advance mechanism should still be used.

8.2 We start with 260 bits and end up with 456, so the average rate is $260/456 \approx 0.57$. The GSM solution is better, because it applies more protection to the more important bits.

8.3 The chip rate is 1.2288 MHz, so the chip period is 0.8138 μs, so 64 chips is about 52 μs. This is larger than the rms delay spread even for an urban environment, which may be around 25 μs at worst.

8.4 1/800 Hz gives us 1.25 ms.

8.5 No, it is stand-alone and independent.

Chapter 9

9.1 (b) 3×3, since the capacity grows roughly linearly with the *minimum* of m and n.

9.2 $(1/2)(0.4) + (1/4)(0.3) + (1/6)(0.2) + (1/8)(0.1) = 0.3208$.

9.3 Because of the asymmetry of power control requirements, certain features of HSDPA cannot be used on HSUPA. See Section 9.3.2.

9.4 Let's list them:

- DL PUSC: 56, with 8 pilot subcarriers and 48 data subcarriers
- UL PUSC: 72, with 24 pilot subcarriers and 48 data subcarriers
- Band AMC: 216, with 24 pilot subcarriers and 192 data subcarriers (and $192 = 48 \times 4$)

9.5 If $N = M$ and the subcarrier mapping is the identity mapping, then, indeed, the DFT and IFFT would cancel each other out. However, usually, $N < M$.

Chapter 10

10.1 Yes, there is overhead and there are inefficiencies, but in many cases, it is worth it, for the order and structure, modularity and simplification, that layering brings. In certain cases, we may wish to optimize the system across layers, which is what is known as *cross-layer optimization* or use special, simplified protocol stacks for specialized applications such as sensor networks. But otherwise, layering is usually worth it.

10.2 See Section 10.2.2.

10.3 If the IP address is 210.78.150.130, this will match the address (210.78.150.128, 255.255.255.128), so it will go out eth0. If the IP address is 210.78.150.133, it will go out eth2; NB: It also matches (210.78.150.128, 255.255.255.128), but (210.78.150.133, 255.255.255.255) is a more specific match.

10.4 fe80:0004:3333:0000:0000:0000:000a:0015.

10.5 $P_b = 0.159$. It drops to 0.008 for $C = 30$ and rises to 0.538 for $C = 10$.

Chapter 11

11.1 When the mobile station is originating, it gets in touch with the network to initiate, so the network has no trouble locating it, whereas for call delivery, the network needs to locate the mobile.

11.2 See Section 11.1.4.

11.3 A SIP proxy forwards SIP messages, whereas a SIP redirect server sends redirection messages back.

11.4 It can insert itself into the `Record-Route` header of the `INVITE` message.

11.5 Buffer starvation is a problem with priority queuing where low-priority queues can get starved of service if the volume of higher-priority traffic is too high. A way to avoid buffer starvation is to use fair queueing.

Chapter 12

12.1 Registration; authentication; 186.15.25.31; 186.15.25.45; 27.242.2.9; the header will be removed.

12.2 Idle, ready, and standby.

12.3 (a) Release 5; (b) release 5; (c) release 6; (d) release 8.

12.4 Beginning with LTE, there is no longer a BSC/RNC (which was present in GSM/UMTS). The base station takes on more functions, as do a couple of other network elements in the core.

12.5 $64,000 \times 20/1000 = 1280$ bits per 20 ms, which is 160 bytes. The UDP, RTP, and IP header are at least 40 bytes together. This is 20% header of the packet that is occupied by the header. If the segments are 10 ms each, that is 80 bytes of G.711 speech, so the header overhead becomes 30%.

Chapter 13

13.1 It is a building block from which other services or other service enablers can be constructed.

13.2 SIP PUBLISH to notify a presence agent about user status, SIP SUBSCRIBE for a watcher to sign up to receive such information, and SIP NOTIFY for the presence agent to push the information to watchers.

13.3 "Native" SIP AS, OSA service capability server (for IMS to be used with OSA), IM-SSF AS (for IMS to be used with CAMEL).

13.4 AODV is reactive, OLSR is proactive, and ZRP is hierarchical.

13.5 Multihop wireless routing is one similarity; small routers is another. MANET nodes tend to be more mobile; mesh networks tend to form a more permanent infrastructure.

Chapter 14

14.1 Fault management in FCAPS; "T" for troubleshooting in OAMPT, otherwise, "O" in OAM&P.

14.2 Trap or notification.

14.3 Configuration management and security management. "O" in OAM&P.

14.4 RMON might be used.

14.5 It is ifTable and the OID is 1.3.6.1.2.1.2.2; ppp – 23.

Chapter 15

15.1 B's public key and B's private key.

15.2 Tunnel mode; useful to hide the original IP packet header with the true source and destination addresses; for packets in an IPsec tunnel, the source and destination addresses are those of the endpoints of the tunnel.

15.3 Using ISAKMP and Oakley.

15.4 To prevent replay attacks. An attacker can eavesdrop on the previous challenge and response, and if the triplet is reused, the attacker can later replay the response and be authenticated.

15.5 One way (BS authenticates MS); no, just between BS and MS; through use of the TMSI.

15.6 Message/data integrity, two-way authentication, secure end-to-end authentication with a remote network (using EAP), and so on.

Chapter 16

16.1 Guyed mast, lattice, monopole.

16.2 The average power consumption is $135 \times 24 = 3240$ W. Stored energy needed $= 24 \times 3240 = 77{,}760$ W-h, or 77.76 kWh.

16.3 They should not be touching. Bigger spheres are less reliable. If an area can be considered protected with a smaller sphere, that estimate of protection is more reliable.

16.4 The current drawn under normal conditions is 0.332 mA; the power dissipated is $IV = 7.96$ mW. When the lightning strike occurs, the current drawn increases to 625 kA. This diverts most of the current away from the base station equipment.

16.5 Flexible RF cables can be bent and twisted more often, and at greater angles, than can semirigid RF cables. Thus, they are good for use as jumper cables. However, they can be more lossy than semirigid cables.

16.6 Water could cause more damage to the equipment.

Chapter 17

17.1 The transit ISP provides a service to the other ISP, so there typically is a financial arrangement in which the transit ISP is paid by the other ISP. Peering is often of mutual benefit, so often is an unpaid arrangement, except when the ratio of traffic starts becomes too imbalanced.

17.2 About 1.5 to 150 m/s (5.5 to 550 km/h). It is reasonable.

17.3 See Section 17.4.1.

17.4 Typically an amendment, and it could later be folded into a revision, but sometimes it might go directly into the latest revision.

17.5 Regulatory rules are mandatory, imposed by a government agency, whereas standards are voluntary.

APPENDIX A

SOME FORMULAS AND IDENTITIES

Here are some useful formulas and identities made use of but not introduced elsewhere in the book.

Euler's identity relating the exponential function to trigonometric functions:

$$e^{j\theta} = \cos\theta + j\sin\theta \tag{A.1}$$

$$\cos\theta = \frac{1}{2}\left(e^{j\theta} + e^{-j\theta}\right) \tag{A.2}$$

$$\sin\theta = \frac{1}{2j}\left(e^{j\theta} - e^{-j\theta}\right) \tag{A.3}$$

Cosine and Sine Multiplication

$$\cos A \cos B = \frac{1}{2}[\cos(A+B) + \cos(A-B)] \tag{A.4}$$

$$\sin A \sin B = \frac{1}{2}[\cos(A-B) - \cos(A+B)] \tag{A.5}$$

$$\sin A \cos B = \frac{1}{2}[\sin(A+B) + \sin(A-B)] \tag{A.6}$$

$$\cos A \sin B = \frac{1}{2}[\sin(A+B) - \sin(A-B)] \tag{A.7}$$

Fundamentals of Wireless Communication Engineering Technologies, First Edition. K. Daniel Wong.

Sum and Difference of Angles

$$\cos(A \pm B) = \cos A \cos B \mp \sin A \sin B \tag{A.8}$$

$$\sin(A \pm B) = \sin A \cos B \pm \cos A \sin B \tag{A.9}$$

Law of Cosines (applicable for any triangle) See Figure A.1.

$$C^2 = A^2 + B^2 - 2AB \cos \theta_{AB} \tag{A.10}$$

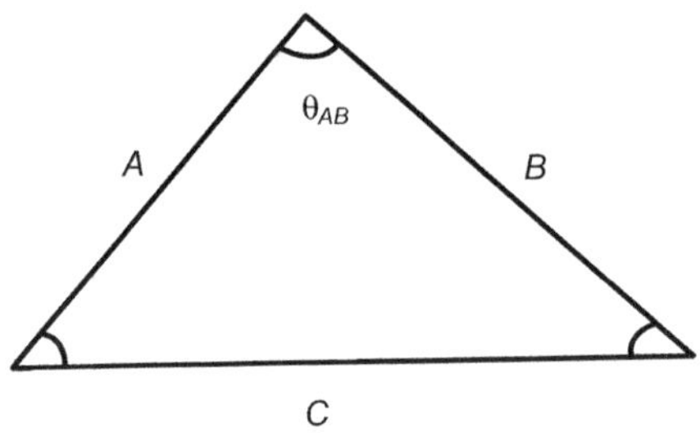

FIGURE A.1 Law of cosines.

APPENDIX B

WCET GLOSSARY EQUATIONS INDEX

We list the equations from the glossary contained in the 2011 WCET candidate's handbook below, and provide references to where the relevant concepts are discussed in this book.

Section 4.1.10, equation (4.23):

$$\frac{P_r}{P_t} = G_t G_r \left(\frac{\lambda}{4\pi d}\right)^2$$

Section 5.1.2.1, equation (5.5):

$$d \approx \sqrt{17h}$$

Section 2.1.1.4, equation (2.14):

$$\lambda = \frac{c}{f}$$

Section 5.3.4, equation (5.41):

$$f_m = \frac{v}{\lambda}$$

Section 5.3.4.1, equation (5.46):

$$N_R = \sqrt{2\pi}\, f_m \rho e^{-\rho^2}$$

Fundamentals of Wireless Communication Engineering Technologies, First Edition. K. Daniel Wong.

Section 5.3.4.1, equation (5.47):

$$\overline{\tau} = \frac{e^{\rho^2} - 1}{\rho f_m \sqrt{2\pi}}$$

Section 10.4.1, equation (10.1):

$$P = \frac{A^C / C!}{\sum_{k=0}^{C} A^k / k!}$$

Section 1.3.3.2, equation (1.56):

$$C = W \log_2 \left(1 + \frac{S}{N}\right)$$

Section 4.1.6, equations (4.9) and (4.12):

$$G = DE_{\text{ant}} = D \frac{R_{\text{rad}}}{R_{\text{rad}} + R_{\text{loss}}}$$

Section 4.2.6, equation (4.24):

$$D = \epsilon_{\text{ap}} \left(\frac{2\pi r}{\lambda}\right)^2$$

Section 4.1.2, equation (4.1):

$$R = \frac{2L^2}{\lambda}$$

Section 3.2.5.2, equation (3.27):

$$F_{\text{sys}} = F_1 + \frac{F_2 - 1}{G_1} + \frac{F_3 - 1}{G_1 G_2} + \cdots$$

APPENDIX C

WCET EXAM TIPS

This appendix is based on information regarding previous exams. We do not know if some or all of the relevant information might change in the future. If so, some or all of these points may not apply.

Some points to note:

- A glossary of formulas is provided. Understand and learn how to use the formulas before the exam.
- The exam is all multiple-choice questions. Prepare accordingly.
- The exam is all text-based, with no diagrams. Prepare accordingly (e.g., without a diagram, the complexity of any system described in words is limited).
- There is no penalty for wrong answers. Therefore, answer all questions.
- The scope of coverage of the exam is very broad. Hence, it is almost certain that there will be questions for which you don't know the answer. Try to use a process of elimination to narrow down the choices and then guess.
- Get plenty of rest before the exam.
- Plan on arriving early at the test center.
- The Windows scientific calculator is the only calculator that is made available during the exam. Learn how to use it!

Fundamentals of Wireless Communication Engineering Technologies, First Edition. K. Daniel Wong.

Figure C.1 shows a screenshot of the Windows scientific calculator. Do you know what all the buttons are for? For example, try the following:

- The calculator has memory storage capabilities. Try using it with the MC, MR, MS, and M+ buttons.
- The Inv checkbox provides inverse function capabilities. Try it.

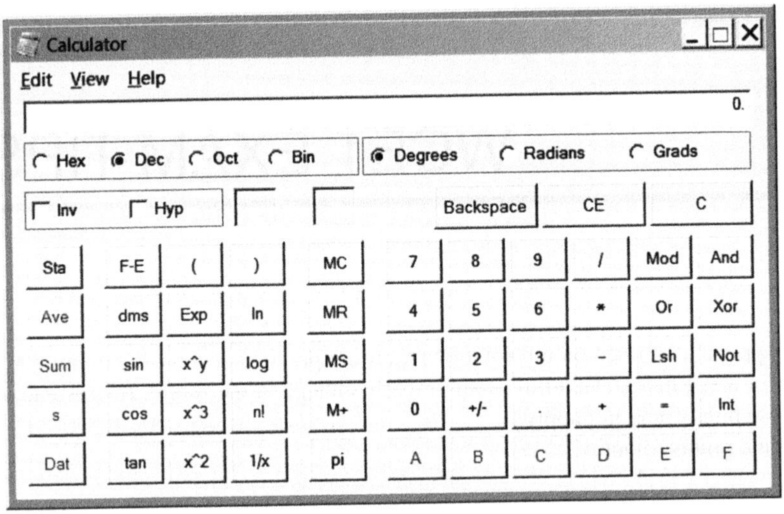

FIGURE C.1 Screenshot of the Windows scientific calculator.

APPENDIX D

SYMBOLS

If a subject area is indicated, it will be in brackets (e.g., [RF]), whereas other comments, such as units for a quantity, will be in parentheses [e.g., (usually, hertz)].

A	amplitude (e.g., of a sinusoid)
A_c	carrier amplitude $A(t)$: time-varying amplitude of modulated sinusoid
$\mathbf{B}$	magnetic flux density (usually, Wb/m^2 or tesla)
B	bandwidth (usually, hertz)
B_b	signal bandwidth at baseband
B_t	transmitted signal bandwidth
B_{channel}	channel bandwidth (usually, hertz)
B_{total}	total bandwidth (usually, hertz; e.g., available for use in a cellular system)
χ_e	electric susceptibility
$\mathcal{C}$	the (infinite) set of complex numbers
C	capacitance (usually, farads)
C	(Shannon) capacity [information theory]
$\{c_n\}$	set of Fourier series coefficients
$\mathbf{D}$	electric displacement field
D	directivity [antennas]
$D(\theta, \phi)$	directive gain [antennas]
d	distance (e.g., between a transmitter and a receiver)
d_{boundary}	boundary between far field and near field [antennas]
Δf	a small difference in frequency
$\delta(t)$	impulse response [signals and systems]

Fundamentals of Wireless Communication Engineering Technologies, First Edition. K. Daniel Wong.

$\mathbf{E}$	electric field
E_{ant}	antenna efficiency [antennas]
E_b/N_0	energy per bit to white noise power spectral density ratio
e	2.71828182846 ... (a constant)
ϵ	permittivity, or absolute permittivity (in contrast to ϵ_r)
ϵ_0	permittivity of free space
ϵ_r	dielectric constant, or relative permittivity (in contrast to ϵ)
η	intrinsic impedance
$\mathbf{F}$	force (usually, newtons)
F	noise figure (linear or dB) [RF]
$\mathcal{F}[\cdots]$	Fourier transform of the function in brackets
f	frequency (usually, hertz)
f_c	carrier frequency
Γ	reflection coefficient, and $\Gamma = \|\Gamma\|e^{j\theta\Gamma}$ (dimensionless)
γ	Euler's constant (also known as the Euler–Mascheroni constant), 0.577215665 (a constant)
γ_c	(continuous-time) PAPR in OFDM systems
γ_d	discrete-time approximation to PAPR in OFDM systems
G	gain [RF]
G	antenna gain [antennas]
$\mathbf{G}$	generating/encoding matrix [FEC]
$\mathbf{H}$	magnetic field
$\mathbf{H}$	parity check matrix [FEC]
$H(f)$	frequency-domain representation of a channel; Fourier transform of $h(t)$
$h(t)$	impulse response of an LTI system
$h(t, \tau)$	impulse response of a time-varying linear system
I	current (usually, amperes); also $i(t)$ [circuits]
I	interference power (as in S/I) [wireless access]
I_N	Norton equivalent current source
$\mathbf{J}$	electric current density (vector; usually, A/m^2)
j	$\sqrt{-1}$ (a constant)
K_c	ciphering key in GSM
k	spatial frequency (waves in space)
k	Boltzmann's constant 3.8×10^{-38} J/K
L	inductance (usually, henries)
L	loss [RF, antennas], path loss [propagation], often in dB
L	maximum dimension of an antenna [antennas]
$\Lambda(t)$	triangle function [signals and systems]
Λ	average SNR (for diversity combining)
λ	wavelength (usually, meters)
λ	arrival rate [teletraffic analysis]
M	number of symbols in a digital communications scheme
$\mathbf{M}$	magnetization

μ	modulation index of an AM signal
μ	departure rate from a single server/channel [teletraffic analysis]
μ_0	permeability of free space, $4\pi \times 10^{-7}$
N	number of points in DFT, FFT, etc., often a power of 2 [signals and systems]
N	noise power (as in S/N)
N_c	number of channels in each channel set
N_{floor}	noise floor
N_{in}	input noise power
$N_{\text{in/Hz}}$	input noise power per hertz
N_{out}	output noise power
N_s	number of channel sets in a cellular system (a.k.a. frequency reuse factor)
n	path loss exponent
$\overline{n}^2(t)$	noise PSD (actually, V^2/Hz; see the text for further explanation)
∇	path loss exponent
Ω	sample space [probability and statistics]
Ω_A	beam area [antennas] ω: angular frequency (usually, rad/s; a.k.a. radial frequency, circular frequency; $\omega = 2\pi f$)
ω	a variable that holds an outcome in Ω [probability and statistics]
ω_i	a specific outcome (ohms)
ϕ	phase of a sinusoid (usually, radians or degrees)
ϕ	in spherical coordinates, the azimuth angle
ϕ_{HP}	half-power beam width (in azimuth plane) [antennas]
π	3.14159 ... (a constant)
$\Pi(t)$	rectangle function [signals and systems]
$\mathbf{P}$	electric polarization vector, or simply, polarization vector
$\mathcal{P}$	poynting vector (usually, Wb/m^2)
P	power (usually, watts)
P_{av}	average power
P_b	blocking probability [teletraffic analysis]
P_d	dropping probability [teletraffic analysis]
$P(\text{event})$	probability function, giving the probability of an event happening [e.g., $P(X = 1)$ or $P(X > 5)$]
P_{in}	input power
$P_{\text{in,min}}$	minimum usable input power
$P_{\text{in,max}}$	maximum usable input power
P_{IIM3}	input-referenced third-order intermodulation point [RF]
P_{IIP3}	input-referenced third-order intercept point [RF]
P_{loss}	ohmic loss in antenna (as opposed to P_{rad}) [antennas]
P_{noise}	noise power (also represented by N, especially in S/N)
P_n	normalized power pattern [antennas]
P_{OIM3}	output-referenced third-order intermodulation product [RF]
P_{OIP3}	output-referenced third-order intercept point [RF]

P_r	received power at a receiver
P_{rad}	radiated power (as opposed to P_{loss}) [antennas]
P_t	transmitted power at a transmitter
$p(t)$	pulse-shaping function
Q	charge (usually, coulombs)
Q	quality factor of an oscillator
ρ	volume charge density (usually, C/m^3)
$\mathcal{R}$	the (infinite) set of real numbers
R	signaling rate
R	resistance (usually, ohms)
R	distance, especially in spherical coordinates
R_b	bit rate (a.k.a. baud rate)
R_N	Norton equivalent resistance
R_{rad}	radiation resistance [antennas]
R_T	Thévenin equivalent resistance
$R_x(\tau)$	autocorrelation of a signal $x(t)$
r	radius of a circle
σ	standard deviation
σ	(electrical) conductivity (usually, A/V·m or S/m)
S	signal power (as in S/N, S/I, etc.)
S	VSWR (dimensionless)
S_{in}	input signal power
S_{mn}	S-parameter
S_{out}	output signal power
$S_x(f)$	power spectral density of signal $x(t)$
SNR	signal-to-noise ratio
SNR_{in}	input SNR [RF]
SNR_{min}	minimum SNR required
SNR_{out}	output SNR [RF]
$\tilde{S}_x(f)$	estimate of $S_x(f)$
sinc(t)	"sinc" function [informally, (sin $x)/x$]
T	period (in time; usually, seconds); $T = 1/f$ [signals, communications]
T	temperature (usually, kelvin) [RF]
T_0	room temperature, usually 290°K or 300°K
T_c	correlation period [statistical signal processing]
T_e	noise temperature (a.k.a. equivalent noise temperature or equivalent temperature) [RF]
T_s	sampling interval or symbol period (except in OFDM and OFDMA, where T_s is sampling interval and T_s' is symbol period)
T_s'	(OFDM) symbol period; sometimes called "useful symbol duration" because it does not include the cyclic prefix τ: time (alternative to t)
t	time (usually, seconds)
θ	an angle; in spherical coordinates, represents the zenith angle
θ_{HP}	half-power beamwidth (an angle, in elevation plane)

$U(\theta, \phi)$	radiation intensity
$u(t)$, $u(f)$	step function (in time, in frequency)
$\mathbf{u}$	unit vector
V	voltage (usually, volts); also $v(t)$
V_n	noise voltage
$V_{n,\mathrm{rms}}$	rms noise voltage
V_T	Thévenin equivalent voltage source
$X(f)$	a signal, frequency-domain representation; often, Fourier transform of $x(t)$
$X_b(f)$	Fourier transform of baseband signal
$X_i(f)$	Fourier transform of in-phase signal
$X_{\mathrm{lp}}(f)$	Fourier transform of low-pass equivalent signal
$X_q(f)$	Fourier transform of quadrature signal
$X_T(f)$	Fourier transform of $x_T(t)$
$x(t)$	a signal (often on the "input" side)
$x_b(t)$	bandpass signal
$x_i(t)$	in-phase signal, in in-phase/quadrature representation
$x_{\mathrm{lp}}(t)$	low-pass equivalent signal
$x_q(t)$	quadrature signal, in in-phase/quadrature representation
$x_T(t)$	truncated version of $x(t)$
$x[n]$	discrete-time signal
$y(t)$	a signal (often on the "output" side)
Z	impedance
∞	infinity

Sometimes, quantities may be qualified by subscripts, such as "rms" for root mean square. Here is a list of such qualifiers used in this book.

0	often a reference value
$1, 2, \ldots$	indices to differentiate objects
av	average
L	load
s	source
rms	root mean square

APPENDIX E

ACRONYMS

3GPP	Third Generation Partnership Project
3GPP2	Third Generation Partnership Project 2
AAA	authentication, authorization and accounting
AAAH	authentication, authorization and accounting home server
AAAL	authentication, authorization and accounting local server
AC	alternating current
ACK	acknowledgment
ADC	analog-to-digital converter
ADSL	asymmetric digital subscriber line
AES	advanced encryption standard
AF	assured forwarding
AGCH	access grant channel
AH	authentication header
AIFS	arbitration interframe space
AIN	advanced intelligent network
AKA	authentication and key agreement
ALR	anti-log-rayleigh
AM	amplitude modulation
AMC	adaptive modulation and coding
AMPS	advanced mobile phone system
AODV	ad hoc on-demand vector routing protocol
AP	access point
API	application programming interface
ARP	address resolution protocol

ARPU	average revenue per unit
ARQ	automatic repeat request
AS	autonomous system
AS	application server
ASK	amplitude shift keying
ASN	access service network
ASN.1	abstract syntax notation 1
ASPR	agreements, standards, policies, and regulations
ATM	asynchronous transfer mode
AuC	authentication center
AWGN	additive white Gaussian noise
BA	behaviour aggregate
BACS	building automation and control systems
BCCH	broadcast control channel
BCH	broadcast channel
BER	bit error rate
BGCF	breakout gateway control function
BGP	border gateway protocol
BICC	bearer independent call control
BM-SC	broadcast multicast service center
BPSK	binary phase shift keying
BS	base station
BSC	base station controller
BSS	basic service set
BTS	base transceiver station *or* base transceiver system
CAMEL	customised applications for mobile enhanced logic
CAN	connectivity access network
CBQ	class-based queuing
CCMP	counter with cipher block chaning message authentication code protocol
CDM	code-division multiplexing
CDMA	code division multiple access
CDR	call detail record
CEMA	consumer electronics manufacturers association
CMIP	common management information protocol
CN	correspondent node
CP	cyclic prefix
CPFSK	continuous-phase frequency shift keying
CQI	channel quality information
CRC	cyclic redundancy check/code
CS	circuit switched
CSMA/CA	carrier sense multiple access with collision avoidance
CSMA/CD	carrier sense multiple access with collision detection
CSN	connectivity service network
CTC	convolutional turbo code

CTI	computer telephony integration
CTS	clear to send
DAB	digital audio broadcast
DC	direct current
DCF	distributed coordination function
DCH	dedicated channel
DDoS	distributed denial-of-service
DES	data encryption standard
DFT	discrete Fourier transform
DFTS-OFDM	DFT-spread OFDM
DHCP	dynamic host configuration protocol
DIFS	distributed coordination function interframe spacing
DL	downlink
DL-SCH	downlink shared channel
DNS	domain name system
DoS	denial of service
DPCCH	dedicated physical control channel
DPSK	differential phase shift keying
DBPSK	differential binary phase shift keying
DQPSK	differential quadrature phase shift keying
DS	distribution system
DS-CDMA	direct sequence code-division multiple access
DSB	double sideband
DSCP	differentiated services code point
DSR	dynamic source routing
DSRC	dedicated short-range communications
DSSS	direct sequence spread spectrum
DTFT	discrete time Fourier transform
DUT	device under test
EAP	extensible authentication protocol
EAPOL	extensible authentication protocol over LAN
EDCF	enhanced distributed coordination function
EDGE	enhanced data rates for GSM evolution
EF	expedited forwarding
EGC	equal gain combining
EIR	equipment identity register
EIRP	effective isotropic radiated power
EM	electromagnetic
EMS	enhanced messaging service
EPC	evolved packet core
EPS	evolved packet system
ERP	effective radiated power
ESP	encapsulating security payload
ESS	extended service set
eTOM	enhanced telecommunications operation map

ETSI	European Telecommunications Standards Institute
EV-DO	evolution data optimized (formerly, evolution data only)
EV-DV	evolution data and voice
FA	foreign agent
FAA	Federal Aviation Administration
FAB	fulfillment, assurance and billing
FBR	front-to-back ratio
FBSS	fast base station switching
FCAPS	fault management, configuration management, accounting management, performance management, and security management
FCC	Federal Communications Commission
FCCH	frequency correction channel
FDD	frequency-division duplexing
FDMA	frequency-division multiple access
FER	frame error rate
FFT	fast Fourier transform
FHSS	frequency hopping spread spectrum
FIFO	first in, first out
FIR	finite impulse response
FM	frequency modulation
FSK	frequency shift keying
FTP	file transfer protocol
FUSC	full usage of subchannels
GEO	geostationary earth orbit
GGSN	gateway GPRS support node
GMSC	gateway mobile switching center
GMSK	Gaussian minimum shift keying
GPRS	general packet radio service
GPS	global positioning system
GSM	global system for mobile communications
GTP	GPRS tunneling protocol
HA	home agent
HARQ	hybrid automatic repeat request
HCF	hybrid coordination function
HDLC	high-level data link control
HLR	home location register
HRPD	high-rate packet data
HSCSD	high-speed circuit-switched data
HSDPA	high-speed downlink packet access
HSPA	high-speed packet access
HSS	home subscriber server
HSUPA	high-speed uplink packet access
HTTP	hypertext transfer protocol
HVAC	heating, ventilating, and air conditioning

IAB	Internet Activities Board
IBSS	independent basic service set
ICMP	Internet control message protocol
ICV	integrity check vector
IDFT	inverse discrete Fourier transform
IEEE	Institute of Electrical and Electronics Engineers
IEEE-SA	Institute of Electrical and Electronics Engineers Standards Association
IETF	Internet Engineering Task Force
IF	intermediate frequency
IFFT	inverse fast Fourier transform
IFS	interframe spacing
IGMP	Internet group management protocol
IIR	infinite impulse response
IKE	Internet key exchange
IMEI	international mobile equipment identity
IMS	IP multimedia subsystem
IMS-ALG	IMS application layer gateway
IMSI	international mobile subscriber identity
IM-SSF AS	IP multimedia service switching function application server
IN	intelligent network
INAP	intelligent network application part
IP	Internet protocol
IP3	third-order intercept point
IPSec	Internet protocol security
I/Q	in-phase/quadrature
IR	infrared
ISAKMP	Internet security association and key management protocol
ISDN	integrated services digital network
ISI	Intersymbol interference
IS-95	Interim Standard 95
IS-IS	intermediate system to intermediate system
ISM (band)	industrial, scientific, and medical (band)
ISO	International Standards Organization
ISP	Internet service provider
ISUP	integrated services digital network (ISDN) user part
ITS	intelligent transport system
ITU	International Telecommunication Union
ITU-R	ITU–Radiocommunication
ITU-T	ITU–Telecommunication
IVC	intervehicular communications
IWF	interworking function
LA	location area
LAN	local area network
LBS	location-based service

LLC	logical link control
LDPC	low-density parity check
LEO	low Earth orbit
LFSR	linear feedback shift register
LLR	log likelihood ratio
LNA	low-noise amplifier
LO	local oscillator
LOS	line of sight
LPD	low probability of detection
LPI	low probability of intercept
LTE	long-term evolution
LTI	linear time invariant
LVD	low-voltage directive
MAC	medium access control
MAC	message authentication code
MAHO	mobile-assisted handoff
MAN	metropolitan area network
MANET	mobile ad hoc network
MAP	mobility application part
MBMS	multimedia broadcast multicast services
MCH	multicast channel
MCHO	mobile controlled handoff
MDHO	macro diversity handoff
MD5	Message Digest 5
MEO	medium Earth orbit
MGCF	media gateway control function
MGW	media gateway
MIB	management information base
MIC	message integrity check *or* message integrity code
MIMO	multiple-input multiple-output
MME	mobility management entity
MMS	multimedia messaging service
MN	mobile node
MR	mesh router
MRC	maximal ratio combining
MRF	media resource function
MRFC	media resource function controller
MRFP	media resource function processor
MS	mobile station
MSC	mobile switching center
MSK	minimum shift keying
MSRN	mobile station roaming number
MSRP	message session relay protocol
MT	mobile terminal
MTBF	mean time between failures

MTP	message transfer part
MTTF	mean time to failure
MTTR	Mean Time To Repair
MTSO	Mobile Telephone Switching Office
NACK	negative acknowledgment
NAV	network allocation vector
NCHO	network-controlled handoff
NCRP	National Council on Radiation Protection
NEC	National Electrical Code
NGN	next-generation network
NLOS	non-line of sight
NMS	network management system
NOC	network operation center
NOI	notice of inquiry
NPRM	notice of proposed rule making
OAM&P	operations, administration, maintenance and provisioning
OFDM	orthogonal frequency-division multiplexing
OFDMA	orthogonal frequency-division multiple access
OLSR	optimized link state routing
OMA	Open Mobile Alliance
OQPSK	offset quadrature phase shift keying
OSA	open service access
OSI	open systems interconnection
OSPF	open shortest path first
OSVF	orthogonal variable spreading factor
PA	power amplifier
PAM	pulse amplitude modulation
PAN	personal area network
PAPR	peak-to-average power ratio
PAR	project authorization request
PCF	point coordination function
PCF	packet control function
PCH	paging channel
PCM	pulse-coded modulation
PCN	packet core network
PCRF	policy and charging rules function
PDCH	packet data channel
PDCP	packet data convergence protocol
PDN GW	packet data network gateway
PDP	packet data protocol
PDSCH	physical downlink shared channel
PDSN	packet data service node
PHB	per hop behavior
PHY	physical (layer)
PIFA	planar inverted-F antenna

PIFS	point coordination function interframe spacing
PN	pseudo-random noise
PoC	push-to-talk over cellular
POP3	Post Office Protocol 3
PPM	pulse position modulation
PPP	point-to-point protocol
PQ	priority queuing
PS	packet switched
PSD	power spectral density
PSK	phase shift keying
PSTN	public-switched telephone network
PSWR	power standing-wave ratio
PUA	presence user agent
PUCCH	physical uplink control channel
PUSC	partial usage of subchannels
PUSCH	physical uplink shared channel
QAM	quadrature amplitude modulation
QoS	quality of service
QPSK	quadrature phase shift keying
RA	routing area
RACH	random access channel
RF	radio frequency
RFC	request for comment
RIP	routing information protocol
RLC	radio link control
RMON	remote network monitoring
RMS	root mean square
RNC	radio network controller
RNS	radio network subsystem
ROHC	robust header compression
RREP	route reply
RREQ	route request
RSA	Rivest, Shamir, and Adleman
RSVP	resource reservation protocol
RTG	Rx/Tx transition gap
RTP	real-time protocol
RTS	request to send
RTT	radio transmission technology
SA	security association
SAD	security association database
SAE	system architecture evolution
SAR	specific absorption rate
SAT	supervisory audio tone
SCCP	signalling connection control part
SCF	service control function

SC-FDMA	single-carrier frequency-division multiple access
SCH	synchronization channel
SCP	service control point
SCTP	stream control transmission protocol
SDP	session description protocol
SFDR	spur-free dynamic range
SGSN	serving GPRS support node
SHA	secure hash algorithm
SIFS	short interframe spacing
SIGTRAN	signal transport
SIM	subscriber identity module
SIP	session initiation protocol
SIR	signal-to-interference ratio
SLA	service-level agreement
SLF	subscription locator function
SMI	structure of management information
SMS	short message service
SMTP	simple mail transfer protocol
SNMP	simple network management protocol
SNR	signal-to-noise ratio
SPD	security policy database
SPD	surge protective devices
SPI	security parameters index
SRES	signed response
SRF	specialized resource function
SS7	signaling system No. 7
SSB	single sideband
SSH	secure shell
SSID	service set ID
SSS	strict-sense stationary
STP	signaling transfer point
SWR	standing-wave ratio
TBRPF	topology dissemination based on reverse-path forwarding
TCP	transmission control protocol
TDD	time-division duplexing
TDM	time-division multiplexing
TDMA	time-division multiple access
TDR	time-domain reflectometer
TFO	transcoder-free operation
TIM	traffic indication map
TKIP	temporal key integrity protocol
TMN	telecommunications management network
TMSI	temporary mobile subscriber identity
TOM	telecommunications operation map
TrGW	transition gateway

TSG	technical specification group
TTG	Tx/Rx transition gap
TTI	transmission time interval
TUSC	tile usage of subcarriers
UDP	user datagram protocol
UE	user equipment
UL	uplink
UL-SCH	uplink shared channel
UMB	ultra mobile broadband
UWB	ultra wideband
UMTS	universal mobile telecommunications system
VANET	vehicular adhoc network
VHE	virtual home environment
VLR	visitor location register
VoIP	voice over internet protocol
VOLGA	voice over LTE via generic access
VPN	virtual privarte network
VSAT	very small aperture terminal
VSWR	voltage standing-wave ratio
WAN	wide area network
WAVE	wireless access in vehicle environment
WCDMA	wideband code-division multiple access
WCET	wireless communication engineering technologies
WEP	wired equivalent privacy
WFQ	weighted fair queuing
Wi-Fi	wireless fidelity
WiMAX	worldwide interoperability for microwave access
WLAN	wireless local area network
WMAN	wireless metropolitan area network
WMN	wireless mesh network
WPA	Wi-Fi protected access
WPAN	wireless personal area network
WRC	world radiocommunication conference
WSS	weak-sense stationary
ZRP	zone routing protocol

INDEX

Fundamentals of Wireless Communication Engineering Technologies, First Edition. K. Daniel Wong.

Printed and bound by CPI Group (UK) Ltd, Croydon, CR0 4YY
16/09/2026
14963412-0007